Lecture Notes in Computer Science　　9691

Commenced Publication in 1973
Founding and Former Series Editors:
Gerhard Goos, Juris Hartmanis, and Jan van Leeuwen

More information about this series at http://www.springer.com/series/7407

Alexander S. Kulikov · Gerhard J. Woeginger (Eds.)

Computer Science – Theory and Applications

11th International Computer Science Symposium
in Russia, CSR 2016
St. Petersburg, Russia, June 9–13, 2016
Proceedings

 Springer

Editors
Alexander S. Kulikov
St. Petersburg Department of Steklov
  Institute of Mathematics of Russian
  Academy of Sciences
St. Petersburg
Russia

Gerhard J. Woeginger
Technische Universiteit Eindhoven
Eindhoven
The Netherlands

ISSN 0302-9743   ISSN 1611-3349 (electronic)
Lecture Notes in Computer Science
ISBN 978-3-319-34170-5   ISBN 978-3-319-34171-2 (eBook)
DOI 10.1007/978-3-319-34171-2

Library of Congress Control Number: 2016938395

LNCS Sublibrary: SL1 – Theoretical Computer Science and General Issues

Printed on acid-free paper

This Springer imprint is published by Springer Nature
The registered company is Springer International Publishing AG Switzerland

Preface

This volume contains the 28 papers presented at CSR 2016, the 11th International Computer Science Symposium in Russia, held during June 9–13, 2016, in St. Petersburg, Russia. The symposium was organized by the Steklov Mathematical Institute at St. Petersburg of the Russian Academy of Sciences (PDMI). The first CSR took place in 2006 in St. Petersburg, and this was then followed by meetings in Ekaterinburg (2007), Moscow (2008), Novosibirsk (2009), Kazan (2010), St. Petersburg (2011), Nizhny Novgorod (2012), Ekaterinburg (2013), Moscow (2014), and Listvyanka (2015). CSR covers a wide range of areas in theoretical computer science and its applications.

The opening lecture at CSR 2016 was given by Christos Papadimitriou (Berkeley). Four other invited plenary lectures were given by Herbert Edelsbrunner (IST Austria), Vladimir Kolmogorov (IST Austria), Orna Kupferman (Hebrew University), and Virginia Vassilevska Williams (Stanford).

We received 71 submissions in total, and out of these the Program Committee selected 28 papers for presentation at the symposium and for publication in the proceedings. Each submission was reviewed by at least three Program Committee members. We expect the full versions of the papers contained in this volume to be submitted for publication in refereed journals. The Program Committee also selected the winners of the two Yandex Best Paper Awards.

Best Paper Award: Meena Mahajan and Nitin Saurabh, "Some Complete and Intermediate Polynomials in Algebraic Complexity Theory"
Best Student Paper Award: Alexander Kozachinskiy, "On Slepian–Wolf Theorem with Interaction"

Many people and organizations contributed to the smooth running and the success of CSR 2016. In particular our thanks go to:

- All authors who submitted their current research to CSR
- Our reviewers and subreferees whose expertise flowed into the decision process
- The members of the Program Committee, who graciously gave their time and energy
- The members of the local Organizing Committee, who made the conference possible
- The EasyChair conference management system for hosting the evaluation process
- Yandex
- The Government of the Russian Federation (Grant 14.Z50.31.0030)
- The Steklov Mathematical Institute at St. Petersburg of the Russian Academy of Sciences
- The European Association for Theoretical Computer Science (EATCS)
- Monomax Congresses and Incentives

June 2016

Alexander S. Kulikov
Gerhard J. Woeginger

Organization

Program Committee

Eric Allender	Rutgers University, USA
Maxim Babenko	Moscow State University, Russia
Marek Chrobak	University of California, Riverside, USA
Volker Diekert	University of Stuttgart, Germany
Leah Epstein	University of Haifa, Israel
Fedor Fomin	University of Bergen, Norway
Lane Hemaspaandra	University of Rochester, USA
Kazuo Iwama	Kyoto University, Japan
Juhani Karhumaki	University of Turku, Finland
Stephan Kreutzer	Technical University of Berlin, Germany
Andrei Krokhin	University of Durham, UK
Piotr Krysta	University of Liverpool, UK
Alexander Kulikov	Steklov Mathematical Institute at St. Petersburg, Russia
Yuri Matiyasevich	Steklov Mathematical Institute at St. Petersburg, Russia
Elvira Mayordomo	Universidad de Zaragoza, Spain
Rolf Niedermeier	Technical University of Berlin, Germany
Vladimir Podolskii	Steklov Mathematical Institute, Russia
Don Sannella	University of Edinburgh, UK
Miklos Santha	CNRS-LRI, UMR 8623, Universitè Paris-Sud, France
Thomas Schwentick	Technical University of Dortmund, Germany
Tatiana Starikovskaya	University of Bristol, UK
Wolfgang Thomas	RWTH Aachen, Germany
Ryan Williams	Stanford University, USA
Gerhard J. Woeginger	Technical University Eindhoven, The Netherlands

Organizing Committee

Asya Gilmanova	Alexander Smal
Alexandra Novikova	Alexander S. Kulikov
Ekaterina Ipatova	

External Reviewers

Akhmedov, Maxim
Alman, Joshua
Anastasiadis, Eleftherios
Artamonov, Stepan
Aumüller, Martin
Averbakh, Igor
Baier, Christel
Bauwens, Bruno
Beliakov, Gleb
Bensch, Suna
Bevern, René Van
Bliznets, Ivan
Brattka, Vasco
Braverman, Mark
Brazdil, Tomas
Buergisser, Peter
Carton, Olivier
Chalopin, Jérémie
Chatzistergiou, Andreas
Chen, Xujin
Chistikov, Dmitry
Chrzaszcz, Jacek
Davydow, Alex
De Paris, Alessandro
de Wolf, Ronald
Duedder, Boris
Durand, Arnaud
Elder, Murray
Escoffier, Bruno
Filiot, Emmanuel
Froese, Vincent
Fukunaga, Takuro
Gairing, Martin
Gastin, Paul
Gawrychowski, Pawel
Geck, Gaetano
Geuvers, Herman
Godin, Thibault
Golovach, Petr
Golovnev, Alexander
Grochow, Joshua
Harju, Tero
Harks, Tobias

Hirvensalo, Mika
Iwamoto, Chuzo
Jancar, Petr
Johnson, Matthew
Kannan, Sampath
Kara, Ahmet
Karpov, Dmitri
Karpov, Nikolay
Kashin, Andrei
Katoen, Joost-Pieter
Kawamura, Akitoshi
Klimann, Ines
Knop, Alexander
Kociumaka, Tomasz
Kolesnichenko, Ignat
Komusiewicz, Christian
Kulkarni, Raghav
Kumar, Mrinal
Kuske, Dietrich
Kutten, Shay
Kuznetsov, Stepan
Laekhanukit, Bundit
Lange, Klaus-Joern
Libkin, Leonid
Löding, Christof
Mahajan, Meena
Maneth, Sebastian
Manquinho, Vasco
Maslennikova, Marina
Mayr, Ernst W.
Mnich, Matthias
Mundhenk, Martin
Nichterlein, André
Nishimura, Harumichi
Nutov, Zeev
Obua, Steven
Okhotin, Alexander
Oliveira, Rafael
Oparin, Vsevolod
Pasechnik, Dmitrii
Pastor, Alexei
Petersen, Holger
Pilipczuk, Michał

Pin, Jean-Eric
Pouzyrevsky, Ivan
Qiao, Youming
Randall, Dana
Reynier, Pierre-Alain
Riveros, Cristian
Romashchenko, Andrei
Rosén, Adi
Saarela, Aleksi
Savchenko, Ruslan
Schnoor, Henning
Shen, Alexander
Sokolov, Dmitry
Souto, André
Talmon, Nimrod
Tamaki, Suguru
Terekhov, Andrey

Thielen, Clemens
van Rooijen, Lorijn
Vassilevska Williams, Virginia
Vegh, Laszlo
Vildhøj, Hjalte Wedel
Volkovich, Ilya
Vortmeier, Nils
Vyalyi, Mikhail
Wahlström, Magnus
Walter, Tobias
Watrous, John
Weinzierl, Tobias
Wächter, Jan Philipp
Zanetti, Luca
Zimand, Marius
Zivny, Standa

Abstracts of Invited Talks

Topological Data Analysis
with Bregman Divergences

Herbert Edelsbrunner
(joint work with Hubert Wagner)

IST Austria (Institute of Science and Technology Austria),
Am Campus 1, 3400 Klosterneuburg, Austria

Given a finite set in a metric space, the topological analysis assesses its multi-scale connectivity quantified in terms of a 1-parameter family of homology groups. Going beyond Euclidean distance and really beyond metrics, we show that the basic tools of topological data analysis also apply when we measure distance with Bregman divergences. While these violate two of the three axioms of a metric, they have been found more effective for high-dimensional data. Examples are the Kullback–Leibler divergence, which is commonly used for text and images, and the Itakura–Saito divergence, which is popular for speech and sound.

Complexity Classifications of Valued Constraint Satisfaction Problems

Vladimir Kolmogorov

IST Austria, Am Campus 1, 3400 Klosterneuburg, Austria
vnk@ist.ac.at

Classifying complexity of different classes of optimization problems is an important research direction in Theoretical Computer Science. One prominent framework is Valued Constraint Satisfaction Problems (VCSPs) in which the class is parameterized by a "language" Γ, i.e. a set of cost functions over a fixed discrete domain D. A instance of VCSP(Γ) is an arbitrary sum of functions from Gamma (possibly with overlapping variables), and the goal is to minimize the sum. The complexity of VCSP (Γ) depends on how "rich" the set Γ is. If, for example, Γ contains only submodular functions then any instance in VCSP(Γ) can be solved in polynomial time. If, on the other hand, Γ contains e.g. the "not-equal" relation then VCSP(Γ) can express the $|D|$-coloring problem and thus is NP-hard when $|D| > 2$.

I will show that establishing complexity classification for plain CSPs (i.e. when functions in Γ only take values in $\{0, \infty\}$) would immediately give the classification for general VCSPs. The key algorithmic tool that we use is a certain LP relaxation of the problem combined with the assumed algorithm for plain CSPs.

In the second part of the talk I will consider a version where we additionally restrict the structure of the instance to be *planar*. More specifically, I will describe a generalization of the Edmonds's blossom-shrinking algorithm from "perfect matching" constraints to arbitrary "even Δ-matroid" constraints. As a consequence of this, we settle the complexity classification of planar Boolean CSPs started by Dvořák and Kupec.

Based on joint work with Alexandr Kazda, Andrei Krokhin, Michal Rolínek, Johann Thapper and Stanislav Živný [1–3].

References

1. Kazda, A., Kolmogorov, V., Rolínek, M.: Even Delta-matroids and the complexity of planar Boolean CSPs (2016). arXiv: 1602.03124v2
2. Kolmogorov, V., Krokhin, A., Rolínek, M.: The complexity of general-valued CSPs. In: IEEE 56th Annual Symposium on Foundations of Computer Science. FOCS 2015, pp. 1246–1258 (2015)
3. Kolmogorov, V., Thapper, J., Živný, S.: The power of linear programming for general-valued CSPs. SIAM J. Comput. **44**(1), 1–36 (2015)

On High-Quality Synthesis

Orna Kupferman[(⊠)]

School of Computer Science and Engineering,
The Hebrew University, Jerusalem, Israel
orna@cs.huji.ac.il

Abstract. In the synthesis problem, we are given a specification ψ over input and output signals, and we synthesize a system that realizes ψ: with every sequence of input signals, the system associates a sequence of output signals so that the generated computation satisfies ψ. The above classical formulation of the problem is Boolean. First, correctness is Boolean: a computation satisfies the specification ψ or does not satisfy it. Then, other important and interesting measures like the size of the synthesized system, its robustness, price, and so on, are ignored. The paper surveys recent efforts to address and formalize different aspects of quality of synthesized systems. We start with multi-valued specification formalisms, which refine the notion of correctness and enable the designer to specify quality, and continue to the quality measure of sensing: the detail in which the inputs should be read in order to generate a correct computation. The first part is based on the articles [1–3]. The second part is based on [4, 5].

The research leading to these results has received funding from the European Research Council under the European Union's Seventh Framework Programme (FP7/2007-2013)/ERC grant agreement no. 278410, and from The Israel Science Foundation (grant no. 1229/10).

Algorithm as a Scientific Weltanschauung

Christos Papadimitriou

UC Berkeley, Berkeley, USA
christos@berkeley.edu

The idea of the algorithm, present in the work of Euclid, Archimedes, and Al Khorizmi, and formalized by Alan Turing only eight decades ago, underlies much of the realm of science — physical, life, or social. Algorithmic processes are present in the great objects of scientific inquiry — the cell, the universe, the market, the brain — as well as in the models developed by scientists over the centuries for studying them. During the past quarter century this algorithmic point of view has helped make important progress in science, for example in statistical physics through the study of phase transitions in terms of the convergence of Markov chain Monte carlo algorithms, and in quantum mechanics through the lens of quantum computing.

In this talk I will recount a few more instances of this mode of research. Algorithmic considerations, as well as ideas from computational complexity, revealed a conceptual flaw in the solution concept of Nash equilibrium ubiquitous in economics. In the study of evolution, a new understanding of century-old questions has been achieved through purely algorithmic ideas. Finally, current work in theoretical neuroscience suggests that the algorithmic point of view may be invaluable in the central scientific question of our era, namely understanding how behavior and cognition emerge from the structure and activity of neurons and synapses.

Fine-Grained Algorithms and Complexity

Virginia Vassilevska Williams

Stanford University, Stanford, USA
virgi@cs.stanford.edu

A central goal of algorithmic research is to determine how fast computational problems can be solved in the worst case. Theorems from complexity theory state that there are problems that, on inputs of size n, can be solved in $t(n)$ time but not in $t(n)^{1-\varepsilon}$ time for $\varepsilon > 0$. The main challenge is to determine where in this hierarchy various natural and important problems lie. Throughout the years, many ingenious algorithmic techniques have been developed and applied to obtain blazingly fast algorithms for many problems. Nevertheless, for many other central problems, the best known running times are essentially those of the classical algorithms devised for them in the 1950s and 1960s.

Unconditional lower bounds seem very difficult to obtain, and so practically all known time lower bounds are conditional. For years, the main tool for proving hardness of computational problems have been NP-hardness reductions, basing hardness on P$\neq$NP. However, when we care about the exact running time (as opposed to merely polynomial vs non-polynomial), NP-hardness is not applicable, especially if the running time is already polynomial. In recent years, a new theory has been developed, based on *"fine-grained reductions"* that focus on exact running times. The goal of these reductions is as follows. Suppose problem A is solvable in $a(n)$ time and problem B in $b(n)$ time, and no $a(n)^{1-\varepsilon}$ and $b(n)^{1-\varepsilon}$ algorithms are known for A and B respectively. The reductions are such that whenever A is fine-grained reducible to B (for $a(n)$ and $b(n)$), then a $b(n)^{1-\varepsilon}$ time algorithm for B (for any $\varepsilon > 0$) implies an $a(n)^{1-\varepsilon'}$ algorithm for A (for some $\varepsilon' > 0$). Now, mimicking NP-hardness, the approach is to (1) select a key problem X that is conjectured to require $t(n)^{1-o(1)}$ time, and (2) reduce X in a fine-grained way to many important problems. This approach has led to the discovery of many meaningful relationships between problems, and even sometimes to equivalence classes.

In this talk I will give an overview of the current progress in this area of study, and will highlight some new exciting developments.

Contents

On High-Quality Synthesis

Orna Kupferman[✉]

School of Computer Science and Engineering, The Hebrew University,
Jerusalem, Israel
orna@cs.huji.ac.il

Abstract. In the synthesis problem, we are given a specification ψ over input and output signals, and we synthesize a system that realizes ψ: with every sequence of input signals, the system associates a sequence of output signals so that the generated computation satisfies ψ. The above classical formulation of the problem is Boolean. First, correctness is Boolean: a computation satisfies the specification ψ or does not satisfy it. Then, other important and interesting measures like the size of the synthesized system, its robustness, price, and so on, are ignored. The paper surveys recent efforts to address and formalize different aspects of quality of synthesized systems. We start with multi-valued specification formalisms, which refine the notion of correctness and enable the designer to specify quality, and continue to the quality measure of sensing: the detail in which the inputs should be read in order to generate a correct computation. The first part is based on the articles [1–3]. The second part is based on [4,5].

1 Introduction

Synthesis is the automated construction of a system from its specification. The basic idea is simple and appealing: instead of developing a system and verifying that it adheres to its specification, we would like to have an automated procedure that, given a specification, constructs a system that is correct by construction. The first formulation of synthesis goes back to Church [11]. The modern approach to synthesis was initiated by Pnueli and Rosner, who introduced LTL (linear temporal logic) synthesis [26]: We are given an LTL formula ψ over sets I and O of input and output signals, and we synthesize a finite-state system that *realizes ψ*. At each moment in time, the system reads a truth assignment, generated by the environment, to the signals in I, and it generates a truth assignment to the signals in O. Thus, with every sequence of inputs, the system associates a sequence of outputs. The system realizes ψ if all the computations that are generated by the interaction satisfy ψ. Synthesis has attracted a lot of research and interest [30].

The research leading to these results has received funding from the European Research Council under the European Union's Seventh Framework Programme (FP7/2007-2013)/ERC grant agreement no. 278410, and from The Israel Science Foundation (grant no. 1229/10).

© Springer International Publishing Switzerland 2016
A.S. Kulikov and G.J. Woeginger (Eds.): CSR 2016, LNCS 9691, pp. 1–15, 2016.
DOI: 10.1007/978-3-319-34171-2_1

One weakness of automated synthesis in practice is that it pays no attention to the quality of the synthesized system. Indeed, the classical setting is Boolean: a computation satisfies a specification or does not satisfy it. Accordingly, while the synthesized system is correct, there is no guarantee about its quality. The formulation of the synthesis problem is Boolean not only because of the Boolean nature of correctness. It also ignores other important aspects of the synthesized systems: their size, robustness, price, and so on. This is a crucial drawback, as designers would be willing to give-up manual design only if automated-synthesis algorithms return systems of comparable quality.

In recent years, researchers have considered several extensions and variants of the classical setting of synthesis. One class of extensions stays in the Boolean setting but releases the requirements that all interactions should satisfy the specification. For example, it allows the user to add assumptions on the behavior of the environment. An assumption may be direct, say given by an LTL formula that restricts the set of possible sequences of inputs [9], or conceptual, say rationality from the side of the environment, which may have its own objectives [17]. Another class of extensions moves to a quantitative setting, where a specification may have different satisfaction values in different systems. The quantitative setting arises directly in systems with quantitative aspects (multi-valued/probabilistic/fuzzy) [14–16,23,25], but is applied also with respect to Boolean systems, where it origins from the semantics of the specification formalism itself [12]. Consider for example the specification $\psi = \mathsf{G}(request \rightarrow \mathsf{F}(response\_grant \vee response\_deny))$; that is, every request is eventually responded, with either a grant or a denial. When we evaluate ψ, there should be a difference between a computation that satisfies it with responses generated soon after requests and one that satisfies it with long waits. Moreover, there should be a difference between grant and deny responses, or cases in which no request is issued.

The issue of generating high-quality hardware and software systems attracts a lot of attention [19,29]. One approach to formalize quality is introduced in [8]. There, the input to the synthesis problem includes also Mealy machines that grade different realizing systems. The first part of the paper surveys another approach, which we introduce in [1–3], and in which quality is part of the specification formalism. Our working assumption is that different ways of satisfying a specification should induce different levels of quality, which should be reflected in the satisfaction value of the specification in the system. Using our approach, a user can specify quality formally, according to the importance she gives to components such as security, maintainability, runtime, delays, and more. Specifically, we introduce and study the temporal logics LTL[$\mathcal{F}$] and LTL[$\mathcal{D}$], as well as their combination. The logic LTL[$\mathcal{F}$] extends LTL with propositional quality operators, which prioritize and weight different ways to satisfy the specification. The logic LTL[$\mathcal{D}$] refines the "eventually" operators of the specification formalism with discounting operators, whose semantics takes into an account the delay incurred in their satisfaction. In both logics, the satisfaction value of a specification is a number in $[0, 1]$, which describes the quality of the satisfaction.

We demonstrate the usefulness of both extensions and study the decidability and complexity of the decision and search problems for them as well as for extensions of LTL that combine both types of operators.

The second approach to formalizing quality refers to other measures that are naturally considered when evaluating systems, like their size or price. Here, researchers have studied synthesis with a bound on the size of the environment and/or the generated system [21,27], synthesis that takes into account the robustness of the generated system [7], and synthesis that refers to the cost of components from which the system is composed. Here, cost can refer to parameters defined by the designer as well as to the actual price, which takes into an account economic considerations, like their sharing by other designers [6]. In the second part of this work we focus on a quality measure that was introduced on [4] and that is based on the amount of *sensing* required to the system in order to realize the specification. Intuitively, the sensing cost quantifies the detail in which a random input word has to be read in order to decide its membership in the language. In [5], we studied the application of the notion of sensing as a quality measure in monitoring and synthesis. In the first, we are given a computation in an on-line manner, and we have to decide whether it satisfies the specification. The goal is to do it with a minimal number of sensors. In the second, in which we focus here, our goal is to design a transducer that realizes a given specification for all input sequences and minimizes the expected average number of sensors used for reading the inputs. We show that minimizing sensing has a price: the synthesis problem becomes exponentially more complex, and the synthesized systems may be exponentially bigger.

2 Preliminaries

Linear Temporal Logic. The linear temporal logic LTL enables the specification of on-going behaviors. Let AP be a set of Boolean atomic propositions. An LTL formula is one of the following:

- True, False, or p, for $p \in AP$.
- $\neg \varphi_1$, $\varphi_1 \vee \varphi_2$, $X\varphi_1$, or $\varphi_1 U\varphi_2$, for LTL formulas φ_1 and φ_2.

We define the semantics of LTL formulas with respect to infinite computations over AP. A *computation* is a word $\pi = \pi_0, \pi_1, \ldots \in (2^{AP})^\omega$. We use π^i to denote the suffix $\pi_i, \pi_{i+1}, \ldots$. The semantics maps a computation π and an LTL[$\mathcal{F}$] formula φ to the *satisfaction value* of φ in π, denoted $[\![\pi, \varphi]\!]$. The satisfaction value is a value in $\{\text{True}, \text{False}\}$, defined inductively as follows.

- $[\![\pi, \text{True}]\!] = \text{True}$.
- $[\![\pi, p]\!] = \begin{cases} \text{True} & \text{if } p \in \pi_0, \\ \text{False} & \text{if } p \notin \pi_0. \end{cases}$
- $[\![\pi, \varphi_1 \vee \psi_2]\!] = [\![\pi, \varphi_1]\!] \vee [\![\pi, \varphi_2]\!]$
- $[\![\pi, \varphi U\psi]\!] = \bigvee_{i \geq 0} ([\![\pi^i, \varphi_2]\!] \wedge \bigwedge_{0 \leq j < i} [\![\pi^j, \varphi_1]\!])$.

- $[\![\pi, \text{False}]\!] = \text{False}$.
- $[\![\pi, \neg \varphi_1]\!] = \neg [\![\pi, \varphi_1]\!]$.
- $[\![\pi, X\varphi]\!] = [\![\pi^1, \varphi]\!]$.

Automata. A *nondeterministic automaton on infinite words* is $\mathcal{A} = \langle \Sigma, Q, Q_0, \delta, \alpha \rangle$, where Q is a finite set of states, $q_0 \subseteq Q$ is a set of initial states, $\delta : Q \times \Sigma \nrightarrow 2^Q$ is a transition function, and α is an acceptance condition. A run of $\mathcal{A}$ on a word $w = \sigma_1 \cdot \sigma_2 \cdots \in \Sigma^\omega$ is a sequence of states $q_0, q_1, \ldots$ such that $q_{i+1} \in \delta(q_i, \sigma_{i+1})$ for all $i \geq 0$. When $|Q_0| = 1$ and $|\delta(q, \sigma)| \leq 1$ for all $q \in Q$ and $\sigma \in \Sigma$, we say that $\mathcal{A}$ is *deterministic*. Note that a deterministic automaton has at most one run on a word. A run is accepting if it satisfies the acceptance condition. A word $w \in \Sigma^\omega$ is accepted by $\mathcal{A}$ if $\mathcal{A}$ has an accepting run on w. The language of $\mathcal{A}$, denoted $L(\mathcal{A})$, is the set of words that $\mathcal{A}$ accepts.

We consider three acceptance conditions. In a *Büchi* automaton, $\alpha \subseteq Q$ and a run is accepting if it visits α infinitely often. In a *looping* automaton, every run is accepting. Thus, a looping automaton is a special case of a Büchi automaton with $\alpha = Q$. Since every run is accepting, we omit the acceptance condition in looping automata and write $\mathcal{A} = \langle \Sigma, Q, q_0, \delta \rangle$. In a *parity* automaton, we have that $\alpha : Q \rightarrow \{1, \ldots, k\}$ label each state by a color. A run is accepting if the minimal color that is visited infinitely often is even. We use three letter acronyms to denote the different types of automata, with the first letter (N or D) refers to the branching mode of the automaton, the second (B, L, or P) to the acceptance condition, and the third – W, indicates we consider word automaton. So, for example, NBWs are nondeterministic Büchi word automata and DLW are deterministic looping automata.

A language $L \subseteq \Sigma^\omega$ is a *safety* language if every word $w \notin L$ has a prefix x such that for all suffixes $y \in \Sigma^\omega$, we have that $x \cdot y \notin L$. It is well known that DLWs can recognize exactly all safety languages [22, 28].

Transducers. For finite sets I and O of input and output signals, respectively, an *I/O transducer* is $\mathcal{T} = \langle I, O, Q, q_0, \delta, \rho \rangle$, where Q is a finite set of states, $q_0 \in Q$ is an initial state, $\delta : Q \times 2^I \rightarrow Q$ is a total transition function, and $\rho : Q \rightarrow 2^O$ is a labeling function on the states. The run of $\mathcal{T}$ on a word $w = i_0 \cdot i_1 \cdots \in (2^I)^\omega$ is the sequence of states $q_0, q_1, \ldots$ such that $q_{k+1} = \delta(q_k, i_k)$ for all $k \geq 0$. The *output* of $\mathcal{T}$ on w is then $o_1, o_2, \ldots \in (2^O)^\omega$ where $o_k = \rho(q_k)$ for all $k \geq 1$. Note that the first output assignment is that of q_1, and we do not consider $\rho(q_0)$. This reflects the fact that the environment initiates the interaction. The *computation* of $\mathcal{T}$ on w is then $\mathcal{T}(w) = i_0 \cup o_1, i_1 \cup o_2, \ldots \in (2^{I \cup O})^\omega$.

Note that each I/O-transducer $\mathcal{T}$ has an underlying DLW $\mathcal{A}_\mathcal{T}$ over the alphabet 2^I with a total transition relation. The language of the DLW is $(2^I)^\omega$, reflecting the receptiveness of $\mathcal{T}$.

Markov Chains and Decision Processes. A Markov chain $\mathcal{M} = \langle S, P \rangle$ consists of a finite state space S and a stochastic transition matrix $P : S \times S \rightarrow [0, 1]$. That is, for all $s \in S$, we have $\sum_{s' \in S} P(s, s') = 1$. Given an initial state s_0, consider the vector v^0 in which $v^0(s_0) = 1$ and $v^0(s) = 0$ for every $s \neq s_0$. The *limiting distribution* of $\mathcal{M}$ is $\lim_{n \rightarrow \infty} \frac{1}{n} \sum_{m=0}^{n} v^0 P^m$. The limiting distribution satisfies $\pi P = \pi$, and can be computed in polynomial time [18].

A Markov decision process (MDP) is $\mathcal{M} = \langle S, s_0, (A_s)_{s \in S}, \mathrm{P}, cost \rangle$ where S is a finite set of states, $s_0 \in S$ is an initial state, A_s is a finite set of actions that are available in state $s \in S$. Let $A = \bigcup_{s \in S} A_s$. Then, $\mathrm{P} : S \times A \times S \rightarrow [0, 1]$ is a partial transition probability function, defining for every two states $s, s' \in S$ and action $a \in A_s$, the probability of moving from s to s' when action a is taken. Accordingly, $\sum_{s' \in S} \mathrm{P}(s, a, s') = 1$. Finally, $cost : S \times A \rightarrow \mathbb{N}$ is a partial cost function, assigning each state s and action $a \in A_s$, the cost of taking action a in state s.

An MDP can be thought of as a game between a player who chooses the actions and nature, which acts stochastically according to the transition probabilities.

A *policy* for an MDP $\mathcal{M}$ is a function $f : S^* \times S \rightarrow A$ that outputs an action given the history of the states, such that for $s_0, \ldots, s_n$ we have $f(s_0, \ldots, s_n) \in A_{s_n}$. Policies correspond to the strategies of the player. The *cost* of a policy f is the expected average cost of a random walk in $\mathcal{M}$ in which the player proceeds according to f. Formally, for $m \in \mathbb{N}$ and for a sequence of states $\tau = s_0, \ldots, s_{m-1}$, we define $\mathrm{P}_f(\tau) = \prod_{i=1}^{m-1} \mathrm{P}(s_{i-1}, f(s_0 \cdots s_{i-1}), s_i)$. Then, $cost_m(f, \tau) = \frac{1}{m} \sum_{i=1}^{m} cost(s_i, f(s_1 \cdots s_i))$ and we define the cost of f as $cost(f) = \liminf_{m \rightarrow \infty} \frac{1}{m} \sum_{\tau:|\tau|=m} cost_m(f, \tau) \cdot \mathrm{P}_f(\tau)$.

A policy is *memoryless* if it depends only on the current state. We can describe a memoryless policy by $f : S \rightarrow A$. A memoryless policy f induces a Markov chain $\mathcal{M}^f = \langle S, P_f \rangle$ with $P_f(s, s') = \mathrm{P}(s, f(s), s')$. Let π be the limiting distribution of $\mathcal{M}^f$. It is not hard to prove that $cost(f) = \sum_{s \in S} \pi_s cost(s, f(s))$. Let $cost(\mathcal{M}) = \inf\{cost(f) : f \text{ is a policy for } \mathcal{M}\}$. That is, $cost(M)$ is the expected cost of a game played on $\mathcal{M}$ in which the player uses an optimal policy.

3   Formalizing Quality: Multi-valued Specification Formalisms

As opposed to traditional verification, where one considers the question of whether a system satisfies, or not, a given specification, reasoning about quality addresses the question of *how well* the system satisfies the specification. We distinguish between two approaches to specifying quality. The first, *propositional quality*, extends the specification formalism with propositional quality operators, which prioritize and weight different satisfaction possibilities. The second, *temporal quality*, refines the "eventually" operators of the specification formalism with discounting operators, whose semantics takes into an account the delay incurred in their satisfaction.

In this section, based on [3], we introduce two quantitative extensions of LTL, one by propositional quality operators and one by discounting operators. In both logics, the satisfaction value of a specification is a number in $[0, 1]$, which describes the quality of the satisfaction. We demonstrate the usefulness of both extensions and study the decidability and complexity of the decision and search problems for them as well as for extensions of LTL that combine both types of operators.

3.1   Propositional Quality: The Temporal Logic LTL[$\mathcal{F}$]

The linear temporal logic LTL[$\mathcal{F}$] generalizes LTL by replacing the Boolean operators of LTL by arbitrary functions over $[0, 1]$. The logic is actually a family of logics, each parameterized by a set $\mathcal{F}$ of functions.

Let AP be a set of Boolean atomic propositions, and let $\mathcal{F} \subseteq \{f : [0, 1]^k \to [0, 1] \mid k \in \mathbb{N}\}$ be a set of functions over $[0, 1]$. Note that the functions in $\mathcal{F}$ may have different arities. An LTL[$\mathcal{F}$] formula is one of the following:

– True, False, or p, for $p \in AP$.
– $f(\varphi_1, ..., \varphi_k)$, $\mathsf{X}\varphi_1$, or $\varphi_1 \mathsf{U} \varphi_2$, for LTL[$\mathcal{F}$] formulas $\varphi_1, ..., \varphi_k$ and $f \in \mathcal{F}$.

We define the semantics of LTL[$\mathcal{F}$] formulas with respect to infinite computations over AP. The semantics maps a computation π and an LTL[$\mathcal{F}$] formula φ to the *satisfaction value* of φ in π, denoted $[\![\pi, \varphi]\!]$, which is a value in $[0, 1]$. The satisfaction value is defined inductively as follows.[1]

– $[\![\pi, \text{True}]\!] = 1$.   – $[\![\pi, \text{False}]\!] = 0$.

– $[\![\pi, p]\!] = \begin{cases} 1 & \text{if } p \in \pi_0, \\ 0 & \text{if } p \notin \pi_0. \end{cases}$ – $[\![\pi, f(\varphi_1, ..., \varphi_k)]\!] = f([\![\pi, \varphi_1]\!], ..., [\![\pi, \varphi_k]\!])$.

– $[\![\pi, \mathsf{X}\varphi]\!] = [\![\pi^1, \varphi]\!]$.   – $[\![\pi, \varphi \mathsf{U} \psi]\!] = \max_{i \geq 0}\{\min\{[\![\pi^i, \varphi_2]\!], \min_{0 \leq j < i}[\![\pi^j, \varphi_1]\!]\}\}$.

Note that the satisfaction value of $\varphi_1 \mathsf{U} \varphi_2$ in π is obtained by going over all suffixes of π, searching for a position $i \geq 0$ that maximizes the minimum between the satisfaction value of φ_2 in π^i (that is, the satisfaction value of the eventuality) and all the satisfaction values of φ_1 in π^j for $0 \leq j < i$ (that is, the satisfaction value of φ_1 until the eventuality is taken into account).

It is not hard to prove, by induction on the structure of the formula, that for every computation π and formula φ, it holds that $[\![\pi, \varphi]\!] \in [0, 1]$, and that each formula has only finitely many possible satisfaction values.

The logic LTL coincides with the logic LTL[$\mathcal{F}$] for $\mathcal{F}$ that corresponds to the usual Boolean operators. For simplicity, we use these operators as abbreviation, as described below. In addition, we introduce notations for some useful functions. Let $x, y \in [0, 1]$ be satisfaction values and $\lambda \in [0, 1]$ be a parameter. Then,

- $\neg x = 1 - x$
- $x \wedge y = \min\{x, y\}$
- $x \vee y = \max\{x, y\}$
- $x \to y = \max\{1 - x, y\}$
- $\nabla_\lambda x = \lambda \cdot x$
- $x \oplus_\lambda y = \lambda \cdot x + (1 - \lambda) \cdot y$

To see that LTL indeed coincides with LTL[$\mathcal{F}$] for $\mathcal{F} = \{\neg, \vee, \wedge\}$, note that for this $\mathcal{F}$, all formulas are mapped to $\{0, 1\}$ in a way that agrees with the semantics of LTL. In particular, observe that under these notations, we can write the semantics of $[\![\pi, \varphi_1 \mathsf{U} \varphi_2]\!]$ as $\bigvee_{i \geq 0}([\![\pi^i, \varphi_2]\!] \wedge \bigwedge_{0 \leq j < i}[\![\pi^j, \varphi_1]\!])$, which coincides with the semantics of LTL.

---

[1] The observant reader may be concerned by our use of max and min where sup and inf are in order. In [3] we prove that there are only finitely many satisfaction values for a formula φ, thus the semantics is well defined.

Example 1. Consider a scheduler that receives requests and generates grants and consider the LTL[$\mathcal{F}$] formula $\varphi = \varphi_1 \wedge \varphi_2$, with $\varphi_1 = \mathsf{G}(req \rightarrow \mathsf{X}(grant \oplus_{\frac{2}{3}} \mathsf{X}grant))$ and $\varphi_2 = \neg(\nabla_{\frac{3}{4}} \mathsf{G}\neg req)$. The satisfaction value of the formula φ_1 is 1 if every request is granted in the next cycle and the grant lasts for two consecutive cycles. If the grant lasts for only one cycle, then the satisfaction value is reduced to $\frac{2}{3}$ if it is the cycle right after the request, and to $\frac{1}{3}$ if it is the next one. In addition, the conjunction with φ_2 implies that if there are no requests, then the satisfaction value is at most $\frac{1}{4}$. The example demonstrates how LTL[$\mathcal{F}$] can conveniently prioritize different scenarios, as well as embody vacuity considerations in the formula. □

3.2 Temporal Quality: The Logic LTL[$\mathcal{D}$]

The linear temporal logic LTL[$\mathcal{D}$] generalizes LTL by adding discounting temporal operators. The logic is actually a family of logics, each parameterized by a set $\mathcal{D}$ of discounting functions.

Let $\mathbb{N} = \{0, 1, ...\}$. A function $\eta : \mathbb{N} \rightarrow [0, 1]$ is a *discounting function* if $\lim_{i \rightarrow \infty} \eta(i) = 0$, and η is strictly monotonic-decreasing. Examples for natural discounting functions are $\eta(i) = \lambda^i$, for some $\lambda \in (0, 1)$, and $\eta(i) = \frac{1}{i+1}$. Note that the strict monotonicity implies that $\eta(i) > 0$ for all $i \in \mathbb{N}$.

Given a set of discounting functions $\mathcal{D}$, we define the logic LTL[$\mathcal{D}$] as follows. The syntax of LTL[$\mathcal{D}$] adds to LTL a *discounting-Until* operator $\varphi \mathsf{U}_\eta \psi$ for every function $\eta \in \mathcal{D}$. Thus, a LTL[$\mathcal{D}$] formula is one of the following:

- **True**, or p, for $p \in AP$.
- $\neg \varphi_1$, $\varphi_1 \vee \varphi_2$, $\mathsf{X}\varphi_1$, $\varphi_1 \mathsf{U}\varphi_2$, or $\varphi_1 \mathsf{U}_\eta \varphi_2$, for LTL[$\mathcal{D}$] formulas φ_1 and φ_2, and a function $\eta \in D$.

Recall that a logic in the family LTL[$\mathcal{F}$] need not have functions that correspond to the usual Boolean operators, in particular $\mathcal{F}$ need not contains negation. On the other hand, the logic LTL[$\mathcal{D}$] does include the Boolean operators $\neg$ and $\vee$.

The semantics of LTL[$\mathcal{D}$] is defined with respect to a *computation* $\pi \in (2^{AP})^\omega$ and agrees with that of LTL[$\mathcal{F}$] on all shared operators. In particular $\neg$, $\vee$, and X. For U and U_η, the semantics is as follows.

- $[\![\pi, \varphi \mathsf{U}\psi]\!] = \sup_{i \geq 0}\{\min\{[\![\pi^i, \psi]\!], \min_{0 \leq j < i}\{[\![\pi^j, \varphi]\!]\}\}\}$.
- $[\![\pi, \varphi \mathsf{U}_\eta \psi]\!] = \sup_{i \geq 0}\{\min\{\eta(i)[\![\pi^i, \psi]\!], \min_{0 \leq j < i}\{\eta(j)[\![\pi^j, \varphi]\!]\}\}\}$.

Note that the semantics for U actually coincides with that of LTL[$\mathcal{F}$], except that here there may be infinitely many different satisfaction values in the different suffixes, which requires the use of sup and inf instead of max and min. The intuition of the discounted-until operator is that events that happen in the future have a lower influence, and the rate by which this influence decreases depends on the function η.[2] For example, the satisfaction value of a formula $\varphi \mathsf{U}_\eta \psi$ in

---

[2] Observe that in our semantics the satisfaction value of future events tends to 0. One may think of scenarios where future events are discounted towards another value in $[0, 1]$ (e.g., discounting towards $\frac{1}{2}$ as ambivalence regarding the future).

a computation π depends on the best (supremum) value that ψ can get along the entire computation, while considering the discounted satisfaction of ψ at a position i, as a result of multiplying it by $\eta(i)$, and the same for the value of φ in the prefix leading to the i-th position.

We add the standard abbreviations $\mathsf{F}\varphi \equiv \mathsf{True}\mathsf{U}\varphi$ and $\mathsf{G}\varphi = \neg\mathsf{F}\neg\varphi$, as well as their quantitative counterparts: $\mathsf{F}_\eta\varphi \equiv \mathsf{True}\mathsf{U}_\eta\varphi$, and $\mathsf{G}_\eta\varphi = \neg\mathsf{F}_\eta\neg\varphi$. Note that $[\![\pi, \mathsf{F}_\eta\varphi]\!] = \sup_{i \geq 0}\{\min\{\eta(i)[\![\pi^i, \varphi]\!], \min_{0 \leq j < i}\{\eta(j) \cdot 1\}\}\}$. Since η is decreasing and $i > j$, the latter becomes $\sup_{i \geq 0}\{\eta(i)[\![\pi^i, \varphi]\!]\}$. From this we also get $[\![\pi, \mathsf{G}_\eta\varphi]\!] = \inf_{i \geq 0}\{1 - \eta(i)(1 - [\![\pi^i, \varphi]\!])\}$.

Example 2. Consider a lossy-disk: every moment in time there is a chance that some bit would flip its value. Fixing flips is done by a global error-correcting procedure. This procedure manipulates the entire content of the disk, such that initially it causes more errors in the disk, but the longer it runs, the more bits it fixes.

Let *init* and *terminate* be atomic propositions indicating when the error-correcting procedure is initiated and terminated, respectively. The quality of the disk (that is, a measure of the amount of correct bits) can be specified by the formula $\varphi = \mathsf{GF}_\eta(init \wedge \neg\mathsf{F}_\mu terminate)$ for some appropriate discounting functions η and μ. Intuitively, φ gets a higher satisfaction value the shorter the waiting time is between initiations of the error-correcting procedure, and the longer the procedure runs (that is, not terminated) in between these initiations. Note that the "worst case" nature of LTL[$\mathcal{D}$] fits here. For instance, running the procedure for a very short time, even once, will cause many errors.

3.3 Combining Propositional and Temporal Quality

The logic LTL[$\mathcal{F}, \mathcal{D}$] is parameterized by both propositional quality operators and discounting functions and enables the specification of both propositional and temporal quality. As studied in [3], some combinations lead to undecidability of search and decision problems for the logic. We shall get back to this point in Sect. 3.5.

3.4 The Search and Decision Questions

In the Boolean setting, an LTL formula maps computations to $\{\mathtt{True}, \mathtt{False}\}$. In the quantitative setting, an LTL[$\mathcal{F}, \mathcal{D}$] formula maps computations to $[0, 1]$. Classical decision problems, such as model checking, satisfiability, synthesis, and equivalence, are accordingly generalized to their quantitative analogues, which are search or optimization problems. Below we specify these questions with respect to LTL[$\mathcal{F}, \mathcal{D}$] and its fragments.

Satisfiability and Validity. In the Boolean setting, the satisfiability problem asks, given an LTL formula φ, whether φ is satisfiable. In the quantitative setting, it asks what the optimal way to satisfy φ is. Thus, the *satisfiability* problem gets as input an LTL[$\mathcal{F}$] formula φ and returns $\sup\{[\![\pi, \varphi]\!] : \pi$ is a computation$\}$.

Dually, the *validity* problem returns, given an LTL[$\mathcal{F}$] formula φ, the value $\inf\{[\![\pi, \varphi]\!] : \pi$ is a computation$\}$, describing the least favorable way to satisfy the specification.

Model Checking. In the Boolean setting of LTL, a system satisfies a formula φ if all its computations satisfy the formula. Adopting this universal approach, the satisfaction value of an LTL[$\mathcal{F}, \mathcal{D}$] formula φ in a system $\mathcal{K}$, denoted $[\![\mathcal{K}, \varphi]\!]$, is induced by the "worst" computation of $\mathcal{K}$, namely the one in which φ has the minimal satisfaction value. Formally, $[\![\mathcal{K}, \varphi]\!] = \inf\{[\![\pi, \varphi]\!] : \pi$ is a computation of $\mathcal{K}\}$. Accordingly, in the model-checking problem, the goal is to find, given a system $\mathcal{K}$ and an LTL[$\mathcal{F}, \mathcal{D}$] formula φ, the satisfaction value $[\![\mathcal{K}, \varphi]\!] =$. In the Boolean setting, good model-checking algorithms return a counterexample to the satisfaction of the specification when it does not hold in the system. The quantitative counterpart is to return a computation π of K that satisfies φ in the least favorable way.

Realizability and Synthesis. In the Boolean setting, the realizability problem gets as input an LTL formula over $I \cup O$, for sets I and O of input and output signals, and asks for the existence of an (I, O)-transducer all of whose computations satisfy the formula. In the quantitative analogue we seek the generation of high-quality systems. Accordingly, given an LTL[$\mathcal{F}, \mathcal{D}$] formula φ over $I \cup O$, the realizability problem is to find $\max\{[\![\mathcal{T}, \varphi]\!] : \mathcal{T}$ is an (I, O)-transducer$\}$. The synthesis problem is then to find a transducer that attains this value.

Decision Problems. The above questions are search and optimization problems. It is sometimes interesting to consider the decision problems they induce, when referring to a specific threshold. For example, the model-checking decision-problem is to decide, given a system $\mathcal{K}$, a specification φ, and a threshold t, whether $[\![\mathcal{K}, \varphi]\!] \geq t$.

Expected Value. In the above definitions, we refer to the different computations of the system universally, thus the satisfaction value is the infimum with respect to all computations. This is similar to the universal approach taken in model checking and synthesis, where all computations have to satisfy the specification. Alternatively, one could follow a stochastic approach and refer to the expected satisfaction value, assuming some distribution on the different computations (in particular, in the setting of synthesis, assuming some distribution on the sequences of input signals). See [10] for studies in this direction.

3.5 Solving the Questions

In the Boolean setting, the automata-theoretic approach has proven to be very useful in reasoning about LTL specifications. The approach is based on translating LTL formulas to nondeterministic Büchi automata on infinite words [31]. In the quantitative approach, it seems natural to translate formulas to *weighted automata* [13,24]. However, these extensively-studied models are complicated and many problems become undecidable for them (e.g., the universality problem – [20]).

In [3], we show that we can construct Boolean automata parameterized by satisfaction values and that decision problems about formulas can be reduced to questions about such automata. In the case of LTL[$\mathcal{F}$], we can use the approach taken in [16], bound the number of possible satisfaction values of LTL[$\mathcal{F}$] formulas, and use this bound in order to translate LTL[$\mathcal{F}$] formulas to Boolean automata. From a technical point of view, the challenge in LTL[$\mathcal{F}$] is to maintain the simplicity and the complexity of the algorithms for LTL, even though the number of possible values is exponential. Things are much more complicated with LTL[$\mathcal{D}$], where one cannot bound the number of possible values. Then, the solution goes via examining lasso-shaped words, in which it is possible to find such a bound, and carefully proving that restricting attention to such words is sound. In both cases, we assume that the calculation of the functions in $\mathcal{F}$ and $\mathcal{D}$ is dominated by the other tasks of the algorithms.

Theorem 1. *Let φ be an* LTL[$\mathcal{F}$] *formula and $P \subseteq [0,1]$ be a predicate. There exists an NBW $\mathcal{A}_{\varphi,P}$ such that for every computation $\pi \in (2^{AP})^\omega$, it holds that $[\![\pi, \varphi]\!] \in P$ iff $\mathcal{A}_{\varphi,P}$ accepts π. Furthermore, $\mathcal{A}_{\varphi,P}$ has at most $2^{(|\varphi|^2)}$ states.*

Theorem 1 implies that the complexities of the questions coincide with these known for LTL. In particular, in model checking, we solve the complement of the problem, namely whether there exists a computation π of $\mathcal{K}$ such that $[\![\pi, \varphi]\!] < v$, which can be solved by taking the product of the NBW $\mathcal{A}_{\varphi,(0,v]}$ from Theorem 1 with the system $\mathcal{K}$ and checking for emptiness on-the-fly. Also, since the various algorithms suggested in the literature for solving the LTL realizability problem [26] are based on a translation of specifications to automata, we can adopt them.

As mentioned above, the case of LTL[$\mathcal{D}$] is more challenging. In particular, the construction of the NBW goes through an intermediate alternating automaton.

Theorem 2. *Given an* LTL[$\mathcal{D}$] *formula φ and a threshold $v \in [0,1]$, there exists an NBW $\mathcal{A}_{\varphi,v}$ such that for every computation π the following hold:*

1. *If $[\![\pi, \varphi]\!] > v$, then $\mathcal{A}_{\varphi,v}$ accepts π.*
2. *If $\mathcal{A}_{\varphi,v}$ accepts π and π is a lasso computation, then $[\![\pi, \varphi]\!] > v$.*

The size of $\mathcal{A}_{\varphi,v}$ depends on the function in $\mathcal{D}$. Essentially, the faster the functions tend to 0, the smaller the state space is. In particular, it is shown in [3] that for exponential-decay functions, the model checking problem can be solved in PSPACE. As for synthesis, [3] provides a partial solution to the decision problems induced from the realizability and synthesis questions, when referring to a specific threshold. Consider an LTL[$\mathcal{D}$] formula φ, and assume a partition of the atomic propositions in φ to input and output signals. Given a threshold $v \geq 0$, we can use the NBW $\mathcal{A}_{\varphi,v}$ in order to address the realizability and synthesis problems, as stated in the following theorem.

Theorem 3. *Consider an* LTL[$\mathcal{D}$] *formula φ over $I \cup O$. If there exists an I/O-transducer all of whose computations π satisfy $[\![\pi, \varphi]\!] > v$, then we can generate a finite-state I/O-transducer all of whose computations τ satisfy $[\![\tau, \varphi]\!] \geq v$.*

As model checking is decidable for LTL[$\mathcal{D}$], one may wish to push the limit and extend the expressive power of the logic. In particular, of great interest is the combination of discounting with propositional quality operators. Interestingly, as it turns out, adding certain propositional quality operators renders the model-checking problem undecidable, while other, simpler operators do not add expressive power, and do not change the decidability status of the logic. Specifically, it is shown in [3] that the fact the weighted average operator $\oplus$ is binary enables its combination with discounting to specify the behavior of a two counter machine, making the validity problem for LTL[$\mathcal{D}$] augmented with $\oplus$ undecidable for every $\mathcal{D} \neq \emptyset$. On the positive side, the construction of the automaton $\mathcal{A}_{\varphi,v}$ from Theorem 2 can be extended to handle the unary ∇_λ operator, making the model checking and synthesis problems decidable for its combination with LTL[$\mathcal{D}$].

4   Sensing as a Quality Measure

In [4], we define regular sensing as a measure for the number of sensors that need to be operated in order to recognize a regular language. Formally, we study languages over an alphabet $\Sigma = 2^P$, for a finite set P of signals. A letter $\sigma \in \Sigma$ corresponds to a truth assignment to the signals, and sensing a signal amounts to knowing its assignment. Describing sets of letters in Σ, it is convenient to use Boolean assertions over P. For example, when $P = \{a, b\}$, the assertion $\neg b$ stands for the set $\{\emptyset, \{a\}\}$ of two letters.

Consider a language L and a deterministic automaton $\mathcal{A} = \langle 2^P, Q, q_0, \delta, \alpha \rangle$ such that $L(\mathcal{A}) = L$. We assume that δ is total. For a state $q \in Q$ and a signal $p \in P$, we say that p is *sensed in* q if there exists a set $S \subseteq P$ such that $\delta(q, S \setminus \{p\}) \neq \delta(q, S \cup \{p\})$. Intuitively, a signal is sensed in q if knowing its value may affect the destination of at least one transition from q. We use $sensed(q)$ to denote the set of signals sensed in q. The *sensing cost* of a state $q \in Q$ is $scost(q) = |sensed(q)|$.[3]

For a finite run $r = q_1, \ldots, q_m$ of $\mathcal{A}$, we define the sensing cost of r, denoted $scost(r)$, as $\frac{1}{m} \sum_{i=0}^{m-1} scost(q_i)$. That is, $scost(r)$ is the average number of sensors that $\mathcal{A}$ uses during r. Now, for a finite word w, we define the sensing cost of w in $\mathcal{A}$, denoted $scost_\mathcal{A}(w)$, as the sensing cost of the run of $\mathcal{A}$ on w. Finally, the sensing cost of $\mathcal{A}$ is the expected sensing cost of words of length that tends to infinity, where we assume that the letters in Σ are uniformly distributed (our results can be adjusted to a setting in which the letters come from a known distribution). Thus, $scost(\mathcal{A}) = \lim_{m \to \infty} |\Sigma|^{-m} \sum_{w \in \Sigma^m} scost_\mathcal{A}(w)$.

In the setting of synthesis, the signals in P are partitioned into sets I and O of input and output signals. An I/O-transducer $\mathcal{T}$ senses only input signals, and we define its sensing cost as the sensing cost of its underlying DLW $\mathcal{A}_\mathcal{T}$. We define the I/O-*sensing cost* of a realizable specification $L \in (2^{I \cup O})^\omega$ as the minimal

---

[3] We note that, alternatively, one could define the *sensing level* of states, with $slevel(q) = \frac{scost(q)}{|P|}$. Then, for all states q, we have that $slevel(q) \in [0, 1]$. All our results hold also for this definition, simply by dividing the sensing cost by $|P|$.

cost of an I/O-transducer that realizes L. Thus, $scost_{I/O}(\mathcal{A}) = \inf\{scost(\mathcal{T}) : \mathcal{T}$ is an I/O-transducer that realizes $L\}$.

The work in [5] focuses on safety specification, given by DLWs. The realizability problem for DLW specifications can be solved in polynomial time. Indeed, given a DLW $\mathcal{A}$, we can view $\mathcal{A}$ as a game between a system, which controls the outputs, and an environment, which controls the inputs. We look for a strategy for the system that never reaches an undefined transition. This amounts to solving a turn-based safety game, which can be done in polynomial time. When sensing is introduced, it is not enough for the system to win this game, as it now has to win while minimizing the sensing cost. Intuitively, not sensing some inputs introduces incomplete information to the game: once the system gives up sensing, it may not know the state in which the game is and knows instead only a set of states in which the game may be. In particular, unlike usual realizability, a strategy that minimizes the sensing need not use the state space of the DLW. We start with an example illustrating this.

Example 3. Consider the DLW $\mathcal{A}$ appearing in Fig. 1. The DLW is over $I = \{p, q\}$ and $O = \{a\}$. A realizing transducer over the structure of $\mathcal{A}$ (see $\mathcal{T}_1$ in Fig. 2) senses p and q, responds with a if $p \wedge q$ was sensed and responds with $\neg a$ if $\neg p \wedge \neg q$ was sensed. In case other inputs are sensed, the response is arbitrary (denoted $*$ in the figure). As $\mathcal{T}_1$ demonstrates, every transducer that is based on the structure of $\mathcal{A}$ senses two input signals (both p and q) every second step, thus its sensing cost is 1. As demonstrated by the transducer $\mathcal{T}_2$ in Fig. 3, it is possible to realize $\mathcal{A}$ with sensing cost of $\frac{1}{2}$ by only sensing p every second step. Indeed, knowing the value of p is enough in order to determine the output. Note that $\mathcal{T}_2$ may output sometimes a and sometimes $\neg a$ after reading assignments that causes $\mathcal{A}$ to reach q_3. Such a behavior cannot be exhibited by a transducer with the state-structure of $\mathcal{A}$.   □

Solving games with incomplete information is typically done by some kind of a subset-construction, which involves an exponential blow up. Unlike usual games with incomplete information, here the strategy of the system should not only take care of the realizability but also decides which input signals should be sensed, where the goal is to obtain a minimally sensing transducer. In order to address these multiple objectives, we first construct an MDP in which the possible policies are all winning for the system, and correspond to different choices of sensing. An optimal policy in this MDP then induces a minimally-sensing transducer.

Theorem 4. *Consider a DLW $\mathcal{A}$ over $2^{I \cup O}$. If $\mathcal{A}$ is realizable, then there exists an MDP $\mathcal{M}$ in which an optimal strategy corresponds to a minimally-sensing I/O-transducer that realizes $\mathcal{A}$. The MDP $\mathcal{M}$ has size exponential in $|\mathcal{A}|$ and can be computed in time exponential in $|\mathcal{A}|$.*

Consider an MDP $\mathcal{M}$. It is well known that $cost(\mathcal{M})$ can be attained by a memoryless policy, which can be computed in polynomial time. Hence, Theorem 4 implies the following.

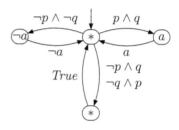

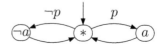

Fig. 2. The transducer $\mathcal{T}_1$ for $\mathcal{A}$.

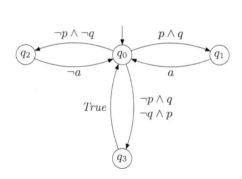

Fig. 1. The DLW $\mathcal{A}$ in Example 3.

Fig. 3. The transducer $\mathcal{T}_2$ for $\mathcal{A}$.

Theorem 5. *Consider a realizable DLW $\mathcal{A}$ over $2^{I \cup O}$. We can calculate $cost_{I,O}(\mathcal{A})$ and return a minimally-sensing I/O-transducer that realizes $\mathcal{A}$ in time exponential in $|\mathcal{A}|$.*

Theorem 4 also implies the upper bound in the following two theorems. In the second, we consider the corresponding decision problem. Given a DLW $\mathcal{A}$ over $2^{I \cup O}$ and a threshold γ, the *synthesis with bounded sensing* problem in the is to decide whether $cost_{I,O}(\mathcal{A}) < \gamma$.

Theorem 6. *A minimally-sensing transducer for a realizable DLW $\mathcal{A}$ has size tightly exponential in $|\mathcal{A}|$.*

Theorem 7. *The synthesis with bounded sensing problem for DLW specifications is EXPTIME-complete.*

As for the lower bounds, they are based on the intuition that small sensing may significantly (that is, exponentially) reduce the state space of the synthesized transducers. To see why, consider for example a sequence $\mathcal{A}_1, \ldots, \mathcal{A}_k$ of DLWs. It is well-known that while a witness to the nonemptiness of each DLW $\mathcal{A}_i$ is linear in its size, a witness to the nonemptiness of the product of the DLWs may be exponential. Assume that we want to synthesize a transducer that gets as input an index $1 \le i \le k$ of a DLW and has to return a word in the language of $\mathcal{A}_i$. When the product of the DLWs is not empty, the transducer can give up reading i and simply return a word in the intersection. A shortest such word, however, may be exponentially longer than words that the transducer can return if it does read i. Thus, giving up sensing i is possible, but required exponentially larger transducers. Moreover, since the transducer apply this idea and give up sensing i only when the intersection of the DLWs is nonempty, checking the emptiness of the intersection of a sequence of DLWs can be reduced to questions about the required sensing. When applied to tree automata, for which the problem of deciding emptiness of the intersection is EXPTIME-hard, we get the lower bound in Theorem 7.

References

1. Almagor, S., Boker, U., Kupferman, O.: Formalizing and reasoning about quality. In: Fomin, F.V., Freivalds, R., Kwiatkowska, M., Peleg, D. (eds.) ICALP 2013, Part II. LNCS, vol. 7966, pp. 15–27. Springer, Heidelberg (2013)
2. Almagor, S., Boker, U., Kupferman, O.: Discounting in LTL. In: Ábrahám, E., Havelund, K. (eds.) TACAS 2014 (ETAPS). LNCS, vol. 8413, pp. 424–439. Springer, Heidelberg (2014)
3. Almagor, S., Boker, U., Kupferman, O.: Formalizing and reasoning about quality. J. ACM (2016, to appear)
4. Almagor, S., Kuperberg, D., Kupferman, O.: Regular sensing. In: Proceedings of the 34th Conference on Foundations of Software Technology and Theoretical Computer Science, LIPIcs, vol. 29, pp. 161–173. Schloss Dagstuhl - Leibniz-Zentrum fuer Informatik, Germany (2014)
5. Almagor, S., Kuperberg, D., Kupferman, O.: The sensing cost of monitoring and synthesis. In: Proceedings of the 35th Conference on Foundations of Software Technology and Theoretical Computer Science, LIPIcs, vol. 35, pp. 380–393. Schloss Dagstuhl - Leibniz-Zentrum fuer Informatik, Germany (2015)
6. Avni, G., Kupferman, O.: Synthesis from component libraries with costs. In: Baldan, P., Gorla, D. (eds.) CONCUR 2014. LNCS, vol. 8704, pp. 156–172. Springer, Heidelberg (2014)
7. Bloem, R., Chatterjee, K., Greimel, K., Henzinger, T.A., Hofferek, G., Jobstmann, B., Könighofer, B., Könighofer, R.: Synthesizing robust systems. Acta Inf. **51**(3–4), 193–220 (2014)
8. Bloem, R., Chatterjee, K., Henzinger, T.A., Jobstmann, B.: Better quality in synthesis through quantitative objectives. In: Bouajjani, A., Maler, O. (eds.) CAV 2009. LNCS, vol. 5643, pp. 140–156. Springer, Heidelberg (2009)
9. Chatterjee, K., Henzinger, T.A., Jobstmann, B.: Environment assumptions for synthesis. In: van Breugel, F., Chechik, M. (eds.) CONCUR 2008. LNCS, vol. 5201, pp. 147–161. Springer, Heidelberg (2008)
10. Chatterjee, K., Henzinger, T.A., Jobstmann, B., Singh, R.: Measuring and synthesizing systems in probabilistic environments. In: Touili, T., Cook, B., Jackson, P. (eds.) CAV 2010. LNCS, vol. 6174, pp. 380–395. Springer, Heidelberg (2010)
11. Church, A.: Logic, arithmetics, and automata. In: Proceedings of the International Congress of Mathematicians, 1962, pp. 23–35. Institut Mittag-Leffler (1963)
12. de Alfaro, L., Faella, M., Henzinger, T.A., Majumdar, R., Stoelinga, M.: Model checking discounted temporal properties. Theor. Comput. Sci. **345**(1), 139–170 (2005)
13. Droste, M., Kuich, W., Vogler, H. (eds.): Handbook of Weighted Automata. Springer, Heidelberg (2009)
14. Droste, M., Rahonis, G.: Weighted automata and weighted logics with discounting. Theor. Comput. Sci. **410**(37), 3481–3494 (2009)
15. Droste, M., Vogler, H.: Weighted automata and multi-valued logics over arbitrary bounded lattices. Theor. Comput. Sci. **418**, 14–36 (2012)

16. Faella, M., Legay, A., Stoelinga, M.: Model checking quantitative linear time logic. Electr. Notes Theor. Comput. Sci. **220**(3), 61–77 (2008)
17. Fisman, D., Kupferman, O., Lustig, Y.: Rational synthesis. In: Esparza, J., Majumdar, R. (eds.) TACAS 2010. LNCS, vol. 6015, pp. 190–204. Springer, Heidelberg (2010)
18. Grinstead, C., Snell, J.L.: 11:Markov chains. In: Introduction to Probability. American Mathematical Society (1997)
19. Kans, S.H.: Metrics and Models in Software Quality Engineering. Addison-Wesley Longman Publishing Co., Boston (2002)
20. Krob, D.: The equality problem for rational series with multiplicities in the tropical semiring is undecidable. Int. J. Algebra Comput. **4**(3), 405–425 (1994)
21. Kupferman, O., Lustig, Y., Vardi, M.Y., Yannakakis, M.: Temporal synthesis for bounded systems and environments. In: Proceedings of the 28th Symposium on Theoretical Aspects of Computer Science, pp. 615–626 (2011)
22. Kupferman, O., Vardi, M.Y.: Model checking of safety properties. Formal Methods Syst. Des. **19**(3), 291–314 (2001)
23. Kwiatkowska, M.Z.: Quantitative verification: models techniques and tools. In: ESEC/SIGSOFT FSE, pp. 449–458 (2007)
24. Mohri, M.: Finite-state transducers in language and speech processing. Comput. Linguist. **23**(2), 269–311 (1997)
25. Moon, S., Lee, K., Lee, D.: Fuzzy branching temporal logic. IEEE Trans. Syst. Man Cybern. Part B **34**(2), 1045–1055 (2004)
26. Pnueli, A., Rosner, R.: On the synthesis of a reactive module. In: Proceedings of the 16th ACM Symposium on Principles of Programming Languages, pp. 179–190 (1989)
27. Schewe, S., Finkbeiner, B.: Bounded synthesis. In: Namjoshi, K.S., Yoneda, T., Higashino, T., Okamura, Y. (eds.) ATVA 2007. LNCS, vol. 4762, pp. 474–488. Springer, Heidelberg (2007)
28. Sistla, A.P.: Safety, liveness and fairness in temporal logic. Formal Aspects Comput. **6**, 495–511 (1994)
29. Spinellis, D.: Code Quality: The Open Source Perspective. Addison-Wesley Professional, Upper Saddle River (2006)
30. Vardi, M.Y.: From verification to synthesis. In: Shankar, N., Woodcock, J. (eds.) VSTTE 2008. LNCS, vol. 5295, p. 2. Springer, Heidelberg (2008)
31. Vardi, M.Y., Wolper, P.: An automata-theoretic approach to automatic program verification. In: Proceedings of the 1st IEEE Symposium on Logic in Computer Science, pp. 332–344 (1986)

Sensitivity Versus Certificate Complexity
of Boolean Functions

Andris Ambainis, Krišjānis Prūsis, and Jevgēnijs Vihrovs[✉]

Faculty of Computing, University of Latvia, Raiņa bulv. 19, Rīga LV-1586, Latvia
jevgenijs.vihrovs@lu.lv

Abstract. Sensitivity, block sensitivity and certificate complexity are basic complexity measures of Boolean functions. The famous sensitivity conjecture claims that sensitivity is polynomially related to block sensitivity. However, it has been notoriously hard to obtain even exponential bounds. Since block sensitivity is known to be polynomially related to certificate complexity, an equivalent of proving this conjecture would be showing that the certificate complexity is polynomially related to sensitivity. Previously, it has been shown that $bs(f) \leq C(f) \leq 2^{s(f)-1}s(f) - (s(f) - 1)$. In this work, we give a better upper bound of $bs(f) \leq C(f) \leq \max\left(2^{s(f)-1}\left(s(f) - \frac{1}{3}\right), s(f)\right)$ using a recent theorem limiting the structure of function graphs. We also examine relations between these measures for functions with 1-sensitivity $s_1(f) = 2$ and arbitrary 0-sensitivity $s_0(f)$.

1 Introduction

Sensitivity and *block sensitivity* are two well-known combinatorial complexity measures of Boolean functions. The sensitivity of a Boolean function, $s(f)$, is just the maximum number of variables x_i in an input assignment $x = (x_1, \ldots, x_n)$ with the property that changing x_i changes the value of f. Block sensitivity, $bs(f)$, is a generalization of sensitivity to the case when we are allowed to change disjoint blocks of variables.

Sensitivity and block sensitivity are related to the complexity of computing f in several different computational models, from parallel random access machines or PRAMs [7] to decision tree complexity, where block sensitivity has been useful for showing the complexities of deterministic, probabilistic and quantum decision trees are all polynomially related [5,6,13].

A very well-known open problem is the *sensitivity vs. block sensitivity conjecture* which claims that the two quantities are polynomially related. This problem is very simple to formulate (so simple that it can be assigned as an undergraduate research project). At the same time, the conjecture appears quite difficult to

---

This work was supported by the European Union Seventh Framework Programme (FP7/2007-2013) under projects RAQUEL (Grant Agreement No. 323970) and ERC Advanced Grant MQC, and the Latvian State Research Programme NeXIT project No. 1.

A.S. Kulikov and G.J. Woeginger (Eds.): CSR 2016, LNCS 9691, pp. 16–28, 2016.
DOI: 10.1007/978-3-319-34171-2_2

solve. It has been known for over 25 years and the best upper and lower bounds are still very far apart. We know that block sensitivity can be quadratically larger than sensitivity [3,14,16] but the best upper bounds on block sensitivity in terms of sensitivity are still exponential [1,11,15].

Block sensitivity is polynomially related to a number of other complexity measures of Boolean functions: *certificate complexity*, *polynomial degree* and the number of queries to compute f either deterministically, probabilistically or quantumly [6]. This gives a number of equivalent formulations for the sensitivity vs. block sensitivity conjecture: it is equivalent to asking whether sensitivity is polynomially related to any one of these complexity measures.

Among the many equivalent forms of the conjecture, relating sensitivity to certificate complexity $C(f)$ might be the combinatorially simplest one. Certificate complexity being at least c simply means that there is an input $x = (x_1, \ldots, x_n)$ that is not contained in an $(n - (c - 1))$-dimensional subcube of the Boolean hypercube on which f is constant. Therefore, in this paper we focus on the "sensitivity vs. certificate complexity" form of the conjecture.

1.1   Related Work

New Approaches to the Sensitivity Conjecture. Recently, there have been multiple developments in various approaches to the sensitivity conjecture. Gilmer et. al. interpret the problem through the cost of a novel communication game [8]. Gopalan et. al. investigate the properties of Boolean functions with low sensitivity [9]. Lin and Zhang give a bound on block sensitivity in terms of sensitivity and the alternating number of the function [12].

Upper Bounds on $bs(f)$ and $C(f)$ in Terms of $s(f)$. There has been a substantial amount of work on reducing the gap between sensitivity and block sensitivity measures. The first non-trivial upper bound is due to Simon [15]:

$$bs(f) \leq 4^{s(f)} s(f). \tag{1}$$

Kenyon and Kutin [11] improved the bound to

$$bs(f) \leq \frac{e}{\sqrt{2\pi}} e^{s(f)} \sqrt{s(f)}. \tag{2}$$

Recently, Ambainis et. al. [1] showed an even better estimate:

$$bs(f) \leq 2^{s(f)-1} s(f) - (s(f) - 1). \tag{3}$$

The essense of this result lies in the following relation between certificate complexity and sensitivity:

$$C_0(f) \leq 2^{s_1(f)-1} s_0(f) - (s_1(f) - 1). \tag{4}$$

Note that any bound for $C_0(f)$ also holds for $C_1(f)$ symmetrically (in this case, $C_1(f) \leq 2^{s_0(f)-1} s_1(f) - (s_0(f) - 1)$).[1]

---

[1] Here, C_0 (C_1) and s_0 (s_1) stand for certificate complexity and sensitivity, restricted to inputs x with $f(x) = 0$ ($f(x) = 1$).

1.2   Our Results

In this work, we give improved upper bounds for the "sensitivity vs. certificate complexity" problem. Our main technical result is

Theorem 1. *Let f be a Boolean function which is not constant. If $s_1(f) = 1$, then $C_0(f) = s_0(f)$. If $s_1(f) > 1$, then*

$$C_0(f) \leq 2^{s_1(f)-1} \left(s_0(f) - \frac{1}{3} \right). \tag{5}$$

A similar bound for $C_1(f)$ follows by symmetry. This implies a new upper bound on block sensitivity and certificate complexity in terms of sensitivity:

Corollary 1. *Let f be a Boolean function. Then*

$$bs(f) \leq C(f) \leq \max \left(2^{s(f)-1} \left(s(f) - \frac{1}{3} \right), s(f) \right). \tag{6}$$

On the other hand, the function of Ambainis and Sun [3] gives the separation of

$$C_0(f) = \left(\frac{2}{3} + o(1) \right) s_0(f) s_1(f) \tag{7}$$

for arbitrary values of $s_0(f)$ and $s_1(f)$. For $s_1(f) = 2$, we show an example of f that achieves

$$C_0(f) = \left\lfloor \frac{3}{2} s_0(f) \right\rfloor = \left\lfloor \frac{3}{4} s_0(f) s_1(f) \right\rfloor. \tag{8}$$

We also study the relation between $C_0(f)$ and $s_0(f)$ for functions with low $s_1(f)$, as we think these cases may provide insights into the more general case.

If $s_1(f) = 1$, then $C_0(f) = s_0(f)$ follows from (4). So, the easiest non-trivial case is $s_1(f) = 2$, for which (4) becomes $C_0(f) \leq 2s_0(f) - 1$.

For $s_1(f) = 2$, we prove a slightly better upper bound of $C_0(f) \leq \frac{9}{5} s_0(f)$. We also show that $C_0(f) \leq \frac{3}{2} s_0(f)$ for $s_1(f) = 2$ and $s_0(f) \leq 6$ and thus our example (8) is optimal in this case. We conjecture that $C_0(f) \leq \frac{3}{2} s_0(f)$ is a tight upper bound for $s_1(f) = 2$.

Our results rely on a recent "gap theorem" by Ambainis and Vihrovs [4] which says that any sensitivity-s induced subgraph G of the Boolean hypercube must be either of size 2^{n-s} or of size at least $\frac{3}{2} 2^{n-s}$ and, in the first case, G can only be a subcube obtained by fixing s variables. Using this theorem allows refining earlier results which used Simon's lemma [15] – any sensitivity-s induced subgraph G must be of size at least 2^{n-s} – but did not use any more detailed information about the structure of such G.

We think that further research in this direction may uncover more interesting facts about the structure of low-sensitivity subsets of the Boolean hypercube, with implications for the "sensitivity vs. certificate complexity" conjecture.

2 Preliminaries

Let $f : \{0,1\}^n \to \{0,1\}$ be a Boolean function on n variables. The i-th variable of an input x is denoted by x_i. For an index set $P \subseteq [n]$, let x^P be the input obtained from an input x by flipping every bit x_i, $i \in P$.

We briefly define the notions of sensitivity, block sensitivity and certificate complexity. For more information on them and their relations to other complexity measures (such as deterministic, probabilistic and quantum decision tree complexities), we refer the reader to the surveys by Buhrman and de Wolf [6] and Hatami et al. [10].

Definition 1. *The* sensitivity complexity $s(f, x)$ *of f on an input x is defined as*

$$s(f,x) = \left| \left\{ i \,\middle|\, f(x) \neq f\left(x^{\{i\}}\right) \right\} \right|. \tag{9}$$

The b-sensitivity $s_b(f)$ of f, where $b \in \{0,1\}$, is defined as $\max(s(f,x) \mid x \in \{0,1\}^n, f(x) = b)$. *The* sensitivity $s(f)$ *of f is defined as* $\max(s_0(f), s_1(f))$.

We say that a vertex x has *full sensitivity* if $s(f, x) = s_{f(x)}(f)$.

Definition 2. *The* block sensitivity $bs(f, x)$ *of f on an input x is defined as the maximum number t such that there are t pairwise disjoint subsets $B_1, \ldots, B_t$ of $[n]$ for which $f(x) \neq f\left(x^{B_i}\right)$. We call each B_i a* block. *The b-block sensitivity $bs_b(f)$ of f, where $b \in \{0,1\}$, is defined as* $\max(bs(f,x) \mid x \in \{0,1\}^n, f(x) = b)$. *The* block sensitivity $bs(f)$ *of f is defined as* $\max(bs_0(f), bs_1(f))$.

Definition 3. *A* certificate c *of f on an input x is defined as a partial assignment $c : P \to \{0,1\}, P \subseteq [n]$ of x such that f is constant on this restriction. We call $|P|$ the length of c. If f is always 0 on this restriction, the certificate is a 0-certificate. If f is always 1, the certificate is a 1-certificate.*

Definition 4. *The* certificate complexity $C(f, x)$ *of f on an input x is defined as the minimum length of a certificate that x satisfies. The b-certificate complexity $C_b(f)$ of f, where $b \in \{0,1\}$, is defined as* $\max(C(f,x) \mid x \in \{0,1\}^n, f(x) = b)$. *The* certificate complexity $C(f)$ *of f is defined as* $\max(C_0(f), C_1(f))$.

In this work we look at $\{0,1\}^n$ as a set of vertices for a graph Q_n (called the *n-dimensional Boolean cube* or *hypercube*) in which we have an edge (x, y) whenever $x = (x_1, \ldots, x_n)$ and $y = (y_1, \ldots, y_n)$ differ in exactly one position. We look at subsets $S \subseteq \{0,1\}^n$ as subgraphs (induced by the subset of vertices S) in this graph.

Definition 5. *Let c be a partial assignment $c : P \to \{0,1\}, P \subseteq [n]$. An $(n - |P|)$-dimensional subcube of Q_n is a subgraph G induced on a vertex set $\{x \mid \forall i \in P\, (x_i = c(i))\}$. It is isomorphic to $Q_{n-|P|}$. We call the value $\dim(G) = n - |P|$ the* dimension *and the value $|P|$ the* co-dimension *of G.*

For example, a subgraph induced on the set $\{x \mid x_1 = 0, x_2 = 1\}$ is a $(n-2)$-dimensional subcube. Note that each certificate of length l corresponds to a subcube of Q_n with co-dimension l.

Definition 6. *Let G be a subcube defined by a partial assignment $c : P \rightarrow \{0,1\}, P \subseteq [n]$. Let $c' : P \rightarrow \{0,1\}$ where $c'(i) \neq c(i)$ for exactly one $i \in P$. Then we call the subcube defined by c' a* neighbour subcube *of G.*

For example, the sets $\{x \mid x_1 = 0, x_2 = 0\}$ and $\{x \mid x_1 = 0, x_2 = 1\}$ induce two neighbouring subcubes, since their union is a subcube induced on the set $\{x \mid x_1 = 0\}$.

We also extend the notion of Hamming distance to the subcubes of Q_n:

Definition 7. *Let G and H be two subcubes of Q_n. Then the* Hamming distance *between G and H is defined as $d(G, H) = \min_{\substack{x \in G \\ y \in H}} d(x, y)$, where $d(x, y)$ is the Hamming distance between x and y.*

Definition 8. *Let G and H be induced subgraphs of Q_n. By $G \cap H$ denote the* intersection *of G and H that is the graph induced on $V(G) \cap V(H)$. By $G \cup H$ denote the* union *of G and H that is the graph induced on $V(G) \cup V(H)$. By $G \setminus H$ denote the* complement *of G in H that is the graph induced by $V(G) \setminus V(H)$.*

Definition 9. *Let G and H be induced subgraphs of Q_n. By $R(G, H)$ denote the* relative size *of G in H:*

$$R(G, H) = \frac{|V(G \cap H)|}{|V(H)|}. \tag{10}$$

We extend the notion of sensitivity to the induced subgraphs of Q_n:

Definition 10. *Let G be a non-empty induced subgraph of Q_n. The* sensitivity *$s(G, Q_n, x)$ of a vertex $x \in Q_n$ is defined as $\left|\left\{i \mid x^{\{i\}} \notin G\right\}\right|$, if $x \in G$, and $\left|\left\{i \mid x^{\{i\}} \in G\right\}\right|$, if $x \notin G$. Then the* sensitivity *of G is defined as $s(G, Q_n) = \max(s(G, Q_n, x) \mid x \in G)$.*

Our results rely on the following generalization of Simon's lemma [15], proved by Ambainis and Vihrovs [4]:

Theorem 2. *Let G be a non-empty induced subgraph of Q_n with sensitivity at most s. Then either $R(G, Q_n) = \frac{1}{2^s}$ and G is an $(n-s)$-dimensional subcube or $R(G, Q_n) \geq \frac{3}{2} \cdot \frac{1}{2^s}$.*

3   Upper Bound on Certificate Complexity in Terms of Sensitivity

In this section we prove Corollary 1. In fact, we prove a slightly more specific result.

Theorem 1. *Let f be a Boolean function which is not constant. If $s_1(f) = 1$, then $C_0(f) = s_0(f)$. If $s_1(f) > 1$, then*

$$C_0(f) \leq 2^{s_1(f)-1} \left(s_0(f) - \frac{1}{3} \right). \tag{11}$$

Note that a similar bound for $C_1(f)$ follows by symmetry. For the proof, we require the following lemma.

Lemma 1. *Let H_1, H_2, ..., H_k be distinct subcubes of Q_n such that the Hamming distance between any two of them is at least 2. Take*

$$T = \bigcup_{i=1}^{k} H_i, \qquad T' = \left\{ x \,\middle|\, \exists i \left(x^{\{i\}} \in T \right) \right\} \setminus T. \tag{12}$$

If $T \neq Q_n$, then $|T'| \geq |T|$.

Proof. If $k = 1$, then the co-dimension of H_1 is at least 1. Hence H_1 has a neighbour cube, so $|T'| \geq |T| = |H_1|$.

Assume $k \geq 2$. Then $n \geq 2$, since there must be at least 2 bit positions for cubes to differ in. We use an induction on n.

Base case. $n = 2$. Then we must have that H_1 and H_2 are two opposite vertices. Then the other two vertices are in T', hence $|T'| = |T| = 2$.

Inductive step. Divide Q_n into two adjacent $(n-1)$-dimensional subcubes Q_n^0 and Q_n^1 by the value of x_1. We will prove that the conditions of the lemma hold for each $T \cap Q_n^b$, $b \in \{0, 1\}$. Let $H_u^b = H_u \cap Q_n^b$. Assume $H_u^b \neq \varnothing$ for some $u \in [k]$. Then either $x_1 = b$ or x_1 is not fixed in H_u. Thus, if there are two non-empty subcubes H_u^b and H_v^b, they differ in the same bit positions as H_u and H_v. Thus the Hamming distance between H_u^b and H_v^b is also at least 2. On the other hand, $Q_n^b \not\subseteq T$, since then k would be at most 1.

Let $T_b = T \cap Q_n^b$ and $T_b' = \left\{ x \,\middle|\, x \in Q_n^b, \exists i \left(x^{\{i\}} \in T_b \right) \right\} \setminus T_b$. Then by induction we have that $|T_b'| \geq |T_b|$. On the other hand, $T_0 \cup T_1 = T$ and $T_0' \cup T_1' \subseteq T'$. Thus

$$|T'| \geq |T_0'| + |T_1'| \geq |T_0| + |T_1| = |T|. \tag{13}$$

$\square$

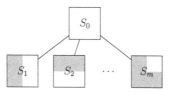

Fig. 1. A schematic representation of the 0-certificate S_0 and its neighbour cubes $S_1, S_2, \ldots, S_m$. The shaded parts represent the vertices in the subcubes for which the value of f is 1.

Proof of Theorem 1. Let z be a vertex such that $f(z) = 0$ and $C(f, z) = C_0(f)$. Pick a 0-certificate S_0 of length $C_0(f)$ and $z \in S_0$. It has $m = C_0(f)$ neighbour subcubes which we denote by $S_1, S_2, \ldots, S_m$ (Fig. 1).

We work with the graph G induced on the vertex set $\{x \mid f(x) = 1\}$. Since S_0 is a minimum certificate for z, $S_i \cap G \neq \varnothing$ for $i \in [m]$.

As S_0 is a 0-certificate, it gives 1 sensitive bit to each vertex in $G \cap S_i$. Then $s(G \cap S_i, S_i) \leq s_1(f) - 1$.

Suppose $s_1(f) = 1$, then for each $i \in [m]$ we must have that $G \cap S_i$ equals to the whole S_i. But then each vertex in S_0 is sensitive to its neighbour in $G \cap S_i$, so $m \leq s_0(f)$. Hence $C_0(f) = s_0(f)$.

Otherwise $s_1(f) \geq 2$. By Theorem 2, either $R(G, S_i) = \frac{1}{2^{s_1(f)-1}}$ or $R(G, S_i) \geq \frac{3}{2^{s_1(f)}}$ for each $i \in [m]$. We call the cube S_i either *light* or *heavy* respectively. We denote the number of light cubes by l, then the number of heavy cubes is $m - l$. We can assume that the light cubes are $S_1, \ldots, S_l$.

Let the average sensitivity of the inputs in S_0 be $as(S_0) = \frac{1}{|S_0|} \sum_{x \in S_0} s_0(x)$. Since each vertex of G in any S_i gives sensitivity 1 to some vertex in S_0, $\sum_{i=1}^{m} R(G, S_i) \leq as(S_0)$. Clearly $as(S_0) \leq s_0(f)$. We have that

$$l\frac{1}{2^{s_1(f)-1}} + (m - l)\frac{3}{2^{s_1(f)}} \leq as(S_0) \leq s_0(f) \tag{14}$$

$$m\frac{3}{2^{s_1(f)}} - l\frac{1}{2^{s_1(f)}} \leq as(S_0) \leq s_0(f). \tag{15}$$

Then we examine two possible cases.

Case 1. $l \leq (s_0(f) - 1)2^{s_1(f)-1}$. Then we have

$$m\frac{3}{2^{s_1(f)}} - (s_0(f) - 1)\frac{2^{s_1(f)-1}}{2^{s_1(f)}} \leq as(S_0) \leq s_0(f) \tag{16}$$

$$m\frac{3}{2^{s_1(f)}} \leq s_0(f) + \frac{1}{2}(s_0(f) - 1) \tag{17}$$

$$m\frac{3}{2^{s_1(f)}} \leq \frac{3}{2}s_0(f) - \frac{1}{2} \tag{18}$$

$$m \leq 2^{s_1(f)-1}\left(s_0(f) - \frac{1}{3}\right). \tag{19}$$

Case 2. $l = (s_0(f) - 1)2^{s_1(f)-1} + \delta$ for some positive integer δ. Since $s_1(f) \geq 2$, the number of light cubes is at least $2(s_0(f) - 1) + \delta$, which in turn is at least $s_0(f)$.

Let $\mathcal{F} = \{F \mid F \subseteq [l], |F| = s_0(f)\}$. Denote its elements by $F_1, F_2, \ldots, F_{|\mathcal{F}|}$. We examine $H_1, H_2, \ldots, H_{|\mathcal{F}|}$ – subgraphs of S_0, where H_i is the set of vertices whose neighbours in S_j are in G for each $j \in F_i$. By Theorem 2, $G \cap S_i$ are subcubes for $i \leq l$. Then so are the intersections of their neighbours in S_0, including each H_i.

Let $N_{i,j}$ be the common neighbour cube of S_i and S_j that is not S_0. Suppose $v \in S_0$. Then by v_i denote the neighbour of v in S_i. Let $v_{i,j}$ be the common neighbour of v_i and v_j that is in $N_{i,j}$.

Next we will show the following:

Proposition 1. *The Hamming distance between any two subcubes H_i and H_j, $i \neq j$ is at least 2.*

Proof. Assume there is an edge (u, v) such that $u \in H_i$ and $v \in H_j$. Then $u_k \in G$ for each $k \in F_i$. Since $i \neq j$, there is an index $t \in F_j$ such that $t \notin F_i$. The vertex u is sensitive to S_k for each $k \in F_i$ and, since $|F_i| = s_0(f)$, has full sensitivity. Thus $u_t \notin G$. On the other hand, since each S_k is light, u_k has full 1-sensitivity, hence $u_{k,t} \in G$ for all $k \in F_i$. This gives full 0-sensitivity to u_t. Hence $v_t \notin G$, a contradiction, since $v \in H_j$ and $t \in F_j$.

Thus there are no such edges and the Hamming distance between H_i and H_j is not equal to 1. That leaves two possibilities: either the Hamming distance between H_i and H_j is at least 2 (in which case we are done), or both H_i and H_j are equal to a single vertex v, which is not possible, as then v would have a 0-sensitivity of at least $s_0(f) + 1$.

Let $T = \bigcup_{i=1}^{|\mathcal{F}|} H_i$. We will prove that $T \neq S_0$. If each of H_i is empty, then $T = \varnothing$ and $T \neq S_0$. Otherwise there is a non-empty H_j. As $s_1(f) \geq 2$, by Theorem 2 it follows that $\dim(G \cap S_k) = \dim(S_k) - s_1(f) + 1 \leq \dim(S_0) - 1$ for each $k \in [l]$. Thus $\dim(H_j) \leq \dim(S_0) - 1$, and $H_j \neq S_0$. Then it has a neighbour subcube H'_j in S_0. But since the Hamming distance between H_j and any other H_i is at least 2, we have that $H'_j \cap H_i = \varnothing$, thus T is not equal to S_0.

Therefore, $H_1, H_2, \ldots, H_{|\mathcal{F}|}$ satisfy all the conditions of Lemma 1. Let T' be the set of vertices in $S_0 \setminus T$ with a neighbour in T. Then, by Lemma 1, $|T'| \geq |T|$ or, equivalently, $R(T', S_0) \geq R(T, S_0)$.

Then note that $R(T', S_0) \geq R(T, S_0) \geq \frac{\delta}{2^{s_1(f)-1}}$, since $R(G, S_i) = \frac{1}{2^{s_1(f)-1}}$ for all $i \in [l]$, there are a total of $(s_0(f) - 1)2^{s_1(f)-1} + \delta$ light cubes and each vertex in S_0 can have at most $s_0(f)$ neighbours in G.

Let S_h be a heavy cube, and $i \in [|\mathcal{F}|]$. The neighbours of H_i in S_h must not be in G, or the corresponding vertex in H_i would have sensitivity $s_0(f) + 1$.

Let $k \in F_i$. As S_k is light, all the vertices in $G \cap S_k$ are fully sensitive, therefore all their neighbours in $N_{k,h}$ are in G. Therefore all the neighbours of H_i in S_h already have full 0-sensitivity. Then all their neighbours must also not be in G.

This means that vertices in T' can only have neighbours in G in light cubes. But they can have at most $s_0(f) - 1$ such neighbours each, otherwise they would be in T, not in T'. As $R(T', S_0) \geq \frac{\delta}{2^{s_1(f)-1}}$, the average sensitivity of vertices in S_0 is at most

$$as(S_0) \leq s_0(f)R(S_0 \setminus T', S_0) + (s_0(f) - 1)R(T', S_0) \tag{20}$$

$$\leq s_0(f)\left(1 - \frac{\delta}{2^{s_1(f)-1}}\right) + (s_0(f) - 1)\frac{\delta}{2^{s_1(f)-1}} \tag{21}$$

$$= s_0(f) - \frac{\delta}{2^{s_1(f)-1}}. \tag{22}$$

Then by inequality (15) we have

$$m\frac{3}{2^{s_1(f)}} - \left((s_0(f) - 1)2^{s_1(f)-1} + \delta\right)\frac{1}{2^{s_1(f)}} \leq s_0(f) - \frac{\delta}{2^{s_1(f)-1}}. \tag{23}$$

Rearranging the terms, we get

$$m \frac{3}{2^{s_1(f)}} \leq \left((s_0(f) - 1)2^{s_1(f)-1} + \delta\right) \frac{1}{2^{s_1(f)}} + s_0(f) - \frac{\delta}{2^{s_1(f)-1}} \tag{24}$$

$$m \frac{3}{2^{s_1(f)}} \leq s_0(f) + \frac{1}{2}(s_0(f) - 1) - \frac{\delta}{2^{s_1(f)}} \tag{25}$$

$$m \frac{3}{2^{s_1(f)}} \leq \frac{3}{2}s_0(f) - \frac{1}{2} - \frac{\delta}{2^{s_1(f)}} \tag{26}$$

$$m \leq 2^{s_1(f)-1} \left(s_0(f) - \frac{1}{3}\right) - \frac{\delta}{3}. \tag{27}$$

$\square$

Theorem 1 immediately implies Corollary 1:

Proof of Corollary 1. If f is constant, then $C(f) = s(f) = 0$ and the statement is true. Otherwise by Theorem 1

$$C(f) = \max(C_0(f), C_1(f)) \tag{28}$$

$$\leq \max_{b \in \{0,1\}} \left(\max\left(2^{s_1 - b(f)-1}\left(s_b(f) - \frac{1}{3}\right), s_b(f)\right)\right) \tag{29}$$

$$\leq \max\left(2^{s(f)-1}\left(s(f) - \frac{1}{3}\right), s(f)\right) \tag{30}$$

On the other hand, $bs(f) \leq C(f)$ is a well-known fact. $\square$

4 Relation Between $C_0(f)$ and $s_0(f)$ for $s_1(f) = 2$

Ambainis and Sun exhibited a class of functions that achieves the best known separation between sensitivity and block sensitivity, which is quadratic in terms of $s(f)$ [3]. This function also produces the best known separation between 0-certificate complexity and 0/1-sensitivity:

Theorem 3. *For arbitrary $s_0(f)$ and $s_1(f)$, there exists a function f such that*

$$C_0(f) = \left(\frac{2}{3} + o(1)\right) s_0(f) s_1(f). \tag{31}$$

Thus it is possible to achieve a quadratic gap between the two measures. As $bs_0(f) \leq C_0(f)$, it would be tempting to conjecture that quadratic separation is the largest possible. Therefore we are interested both in improved upper bounds and in functions that achieve quadratic separation with a larger constant factor.

In this section, we examine how $C_0(f)$ and $s_0(f)$ relate to each other for small $s_1(f)$. If $s_1(f) = 1$, it follows by Theorem 1 that $C_0(f) = s_0(f)$. Therefore we consider the case $s_1(f) = 2$.

Here we are able to construct a separation that is better than (31) by a constant factor.

Theorem 4. *There is a function f with $s_1(f) = 2$ and arbitrary $s_0(f)$ such that*

$$C_0(f) = \left\lfloor \frac{3}{4} s_0(f) s_1(f) \right\rfloor = \left\lfloor \frac{3}{2} s_0(f) \right\rfloor. \tag{32}$$

Proof. Consider the function that takes value 1 iff its 4 input bits are in either ascending or descending sorted order. Formally,

$$\text{SORT}_4(x) = 1 \Leftrightarrow (x_1 \le x_2 \le x_3 \le x_4) \vee (x_1 \ge x_2 \ge x_3 \ge x_4). \tag{33}$$

One easily sees that $C_0(\text{SORT}_4) = 3$, $s_0(\text{SORT}_4) = 2$ and $s_1(\text{SORT}_4) = 2$.

Denote the 2-bit logical AND function by AND_2. We have $C_0(\text{AND}_2) = s_0(\text{AND}_2) = 1$ and $s_1(\text{AND}_2) = 2$.

To construct the examples for larger $s_0(f)$ values, we use the following fact (it is easy to show, and a similar lemma was proved in [3]):

Fact 1. *Let f and g be Boolean functions. By composing them with OR to $f \vee g$ we get*

$$C_0(f \vee g) = C_0(f) + C_0(g), \tag{34}$$
$$s_0(f \vee g) = s_0(f) + s_0(g), \tag{35}$$
$$s_1(f \vee g) = \max(s_1(f), s_1(g)). \tag{36}$$

Suppose we need a function with $k = s_0(f)$. Assume k is even. Then by Fact 1 for $g = \bigvee_{i=1}^{\frac{k}{2}} \text{SORT}_4$ we have $C_0(g) = \frac{3}{2}k$. If k is odd, consider the function $g = \left(\bigvee_{i=1}^{\frac{k-1}{2}} \text{SORT}_4 \right) \vee \text{AND}_2$. Then by Fact 1 we have $C_0(g) = 3 \cdot \frac{k-1}{2} + 1 = \left\lfloor \frac{3}{2}k \right\rfloor$. $\square$

A curious fact is that both examples of (31) and Theorem 4 are obtained by composing some primitives using OR. The same fact holds for the best examples of separation between $bs(f)$ and $s(f)$ that preceded the [3] construction [14,16].

We are also able to prove a slightly better upper bound in case $s_1(f) = 2$.

Theorem 5. *Let f be a Boolean function with $s_1(f) = 2$. Then*

$$C_0(f) \le \frac{9}{5} s_0(f). \tag{37}$$

Proof. Let z be a vertex such that $f(z) = 0$ and $C(f, z) = C_0(f)$. Pick a 0-certificate S_0 of length $m = C_0(f)$ and $z \in S_0$. It has m neighbour subcubes which we denote by $S_1, S_2, \ldots, S_m$. Let $n' = n - m = \dim(S_i)$ for each S_i.

We work with a graph G induced on a vertex set $\{x \mid f(x) = 1\}$. Let $G_i = G \cap S_i$. As S_0 is a minimal certificate for z, we have $G_i \ne \varnothing$ for each $i \in [m]$. Since any $v \in G_i$ is sensitive to S_0, we have $s(G_i, S_i) \le 1$. Thus by Theorem 2 either G_i is an $(n'-1)$-subcube of S_i with $R(G_i : S_i) = \frac{1}{2}$ or $R(G_i : S_i) \ge \frac{3}{4}$. We call S_i *light* or *heavy*, respectively.

Let $N_{i,j}$ be the common neighbour cube of S_i, S_j that is not S_0. Let $G_{i,j} = G \cap N_{i,j}$. Suppose $v \in S_0$. Let v_i be the neighbour of v in S_i. Let $v_{i,j}$ be the neighbour of v_i and v_j in $N_{i,j}$.

Let S_i, S_j be light. By G_i^0, G_j^0 denote the neighbour cubes of G_i, G_j in S_0. We call $\{S_i, S_j\}$ a *pair*, iff $G_i^0 \cup G_j^0 = S_0$. In other words, a pair is defined by a single dimension. Also we have either $z_i \notin G$ or $z_j \notin G$: we call the corresponding cube the *representative* of this pair.

Proposition 2. *Let P be a set of mutually disjoint pairs of the neighbour cubes of S_0. Then there exists a 0-certificate S_0' such that $z \in S_0'$, $\dim(S_0') = \dim(S_0)$ and S_0' has at least $|P|$ heavy neighbour cubes.*

Proof. Let R be a set of mutually disjoint pairs of the neighbour cubes of S_0. W.l.o.g. let $S_1, \ldots, S_{|R|}$ be the representatives of R. Let F_i be the neighbour cube of $S_i \setminus G$ in S_0. Let $B_R = \bigcap_{i=1}^{|R|} F_i$. Suppose $S_0 + x$ is a coset of S_0 and $x_t = 0$ if the t-th dimension is not fixed in S_0: let $B_R(S_0 + x)$ be $B_R + x$.

Pick $R \subseteq P$ with the largest size, such that for each two representatives S_i, S_j of R, $B_R(N_{i,j})$ is a 0-certificate.

Next we prove that the subcube S_0' spanned by $B_R, B_R(S_1), \ldots, B_R(S_{|R|})$ is a 0-certificate. It corresponds to an $|R|$-dimensional hypercube $Q_{|R|}$ where $B_R(S_0 + x)$ corresponds to a single vertex for each coset $S_0 + x$ of S_0.

Let $T \subseteq Q_{|R|}$ be the graph induced on the set $\{v \mid v$ corresponds to $B_R(S_0 + x), B_R(S_0 + x)$ is not a 0-certificate$\}$. Then we have $s(T, Q_{|R|}) \leq 2$. Suppose B_R corresponds to $0^{|R|}$. Let L_d be the set of $Q_{|R|}$ vertices that are at distance d from $0^{|R|}$. We prove by induction that $L_d \cap T = \emptyset$ for each d.

Proof. **Base case.** $d \leq 2$. The required holds since all $B_R, B_R(S_i), B_R(N_{i,j})$ are 0-certificates.

Inductive step. $d \geq 3$. Examine $v \in L_d$. As v has d neighbours in L_{d-1}, $L_{d-1} \cap T = \emptyset$ and $s(T, Q_{|R|}) \leq 2$, we have that $v \notin T$.

Let k be the number of distinct dimensions that define the pairs of R, then $k \leq |R|$. Hence $\dim(S_0') = |R| + \dim(B_R) = |R| + (\dim(S_0) - k) \geq \dim(S_0)$. But S_0 is a minimal 0-certificate for z, therefore $\dim(S_0') = \dim(S_0)$.

Note that a light neighbour S_i of S_0 is separated into a 0-certificate and a 1-certificate by a single dimension, hence we have $s(G, S_i, v) = 1$ for every $v \in S_i$. As S_i neighbours S_0, every vertex in its 1-certificate is fully sensitive. The same holds for any light neighbour S_i' of S_0'.

Now we will prove that each pair in P provides a heavy neighbour for S_0'. Let $\{S_a, S_b\} \in P$, where S_a is the representative. We distinguish two cases:

- $B_R(S_b)$ is a 1-certificate. Since S_b is light, it has full 1-sensitivity. Therefore, $v \in G$ for all $v \in B_R(N_{i,b})$, for each $i \in [|R|]$. Let S_b' be the neighbour of S_0' that contains $B_R(S_b)$ as a subcube. Then for each $v \in B_R(S_b)$ we have $s(G, S_b', v) = 0$. Hence S_b' is heavy.
- Otherwise, $\{S_a, S_b\}$ is defined by a different dimension than any of the pairs in R. Let $R' = R \cup \{S_a, S_b\}$. Examine the subcube $B_{R'}$. By definition of R, there is a representative S_i of R such that $B_{R'}(N_{i,a})$ is not a 0-certificate. Let S_a' be the neighbour of S_0' that contains $B_R(S_a)$ as a subcube. Then there is a vertex $v \in B_{R'}(S_a)$ such that $s(G, S_a', v) \geq 2$. Hence S_a' is heavy. $\qquad \square$

Let $\mathcal{P}$ be the largest set of mutually disjoint pairs of the neighbour cubes of S_0. Let l and $h = m - l$ be the number of light and heavy neighbours of S_0, respectively. Each pair in $\mathcal{P}$ gives one neighbour in G to each vertex in S_0. Now examine the remaining $l - 2|\mathcal{P}|$ light cubes. As they are not in $\mathcal{P}$, no two of them form a pair. Hence there is a vertex $v \in S_0$ that is sensitive to each of them. Then $s_0(f) \geq s_0(f, v) \geq |\mathcal{P}| + (l - 2|\mathcal{P}|) = l - |\mathcal{P}|$. Therefore $|\mathcal{P}| \geq l - s_0(f)$.

Let q be such that $m = q s_0(f)$. Then there are $q s_0(f) - l$ heavy neighbours of S_0. On the other hand, by Proposition 2, there exists a minimal certificate S_0' of z with at least $l - s_0(f)$ heavy neighbours. Then z has a minimal certificate with at least $\frac{(q s_0(f) - l) + (l - s_0(f))}{2} = \frac{q-1}{2} \cdot s_0(f)$ heavy neighbour cubes.

W.l.o.g. let S_0 be this certificate. Then $l = q s_0(f) - h \leq (q - \frac{q-1}{2}) s_0(f) = \frac{q+1}{2} \cdot s_0(f)$. As each $v \in G_i$ for $i \in [m]$ gives sensitivity 1 to its neighbour in S_0,

$$l \frac{1}{2} + h \frac{3}{4} \leq s_0(f). \tag{38}$$

Since the constant factor at l is less than at h, we have

$$\frac{q+1}{2} \cdot s_0(f) \cdot \frac{1}{2} + \frac{q-1}{2} \cdot s_0(f) \cdot \frac{3}{4} \leq s_0(f) \tag{39}$$

By dividing both sides by $s_0(f)$ and simplifying terms, we get $q \leq \frac{9}{5}$. $\square$

This result shows that the bound of Corollary 1 can be improved. However, it is still not tight. For some special cases, through extensive casework we can also prove the following results:

Theorem 6. *Let f be a Boolean function with $s_1(f) = 2$ and $s_0(f) \geq 3$. Then*

$$C_0(f) \leq 2 s_0(f) - 2. \tag{40}$$

Theorem 7. *Let f be a Boolean function with $s_1(f) = 2$ and $s_0(f) \geq 5$. Then*

$$C_0(f) \leq 2 s_0(f) - 3. \tag{41}$$

The proofs of these theorems are available online in the full version of the paper [2].

These theorems imply that for $s_1(f) = 2$, $s_0(f) \leq 6$ we have $C_0(f) \leq \frac{3}{2} s_0(f)$, which is the same separation as achieved by the example of Theorem 4. This leads us to the following conjecture:

Conjecture 1. Let f be a Boolean function with $s_1(f) = 2$. Then

$$C_0(f) \leq \frac{3}{2} s_0(f). \tag{42}$$

We consider $s_1(f) = 2$ to be the simplest case where we don't know the actual tight upper bound on $C_0(f)$ in terms of $s_0(f), s_1(f)$. Proving Conjecture 1 may provide insights into relations between $C(f)$ and $s(f)$ for the general case.

References

1. Ambainis, A., Bavarian, M., Gao, Y., Mao, J., Sun, X., Zuo, S.: Tighter relations between sensitivity and other complexity measures. In: Esparza, J., Fraigniaud, P., Husfeldt, T., Koutsoupias, E. (eds.) ICALP 2014. LNCS, vol. 8572, pp. 101–113. Springer, Heidelberg (2014)
2. Ambainis, A., Prūsis, K., Vihrovs, J.: Sensitivity versus certificate complexity of Boolean functions. CoRR, abs/1503.07691 (2016)
3. Ambainis, A., Sun, X.: New separation between $s(f)$ and $bs(f)$. CoRR, abs/1108.3494 (2011)
4. Ambainis, A., Vihrovs, J.: Size of sets with small sensitivity: a generalization of simon's lemma. In: Jain, R., Jain, S., Stephan, F. (eds.) TAMC 2015. LNCS, vol. 9076, pp. 122–133. Springer, Heidelberg (2015)
5. Beals, R., Buhrman, H., Cleve, R., Mosca, M., de Wolf, R.: Quantum lower bounds by polynomials. J. ACM **48**(4), 778–797 (2001)
6. Buhrman, H., de Wolf, R.: Complexity measures and decision tree complexity: a survey. Theor. Comput. Sci. **288**(1), 21–43 (2002)
7. Cook, S., Dwork, C., Reischuk, R.: Upper and lower time bounds for parallel random access machines without simultaneous writes. SIAM J. Comput. **15**, 87–97 (1986)
8. Gilmer, J., Koucký, M., Saks, M.E.: A communication game related to the sensitivity conjecture. CoRR, abs/1511.07729 (2015)
9. Gopalan, P., Nisan, N., Servedio, R.A., Talwar, K., Wigderson, A.: Smooth Boolean functions are easy: Efficient algorithms forlow-sensitivity functions. In: Proceedings of the 2016 ACM Conference on Innovations inTheoretical Computer Science, ITCS 2016, pp. 59–70. ACM, New York, NY, USA (2016)
10. Hatami, P., Kulkarni, R., Pankratov, D.: Variations on the sensitivity conjecture. In: Number 4 in Graduate Surveys. Theory of Computing Library (2011)
11. Kenyon, C., Kutin, S.: Sensitivity, block sensitivity, and ℓ-block sensitivity of Boolean functions. Inf. Comput. **189**(1), 43–53 (2004)
12. Lin, C., Zhang, S.: Sensitivity conjecture and log-rank conjecture for functions with small alternating numbers. CoRR, abs/1602.06627 (2016)
13. Nisan, N.: CREW PRAMS and decision trees. In: Proceedings of the Twenty-first Annual ACM Symposium on Theory of Computing, STOC 1989, pp. 327–335. ACM, New York, NY, USA (1989)
14. Rubinstein, D.: Sensitivity vs. block sensitivity of Boolean functions. Combinatorica **15**(2), 297–299 (1995)
15. Simon, H.-U.: A tight $\Omega(\log \log N)$-bound on the time for parallel RAM's to compute nondegenerated Boolean functions. In: Karpinski, M. (ed.) FCT 1983. LNCS, vol. 158, pp. 439–444. Springer, London (1983)
16. Virza, M.: Sensitivity versus block sensitivity of Boolean functions. Inf. Process. Lett. **111**(9), 433–435 (2011)

Algorithmic Decidability of Engel's Property for Automaton Groups

Laurent Bartholdi[1,2(✉)]

[1] École Normale Supérieure, Paris, France
laurent.bartholdi@ens.fr, laurent.bartholdi@gmail.com
[2] Georg-August-Universität zu Göttingen, Göttingen, Germany

Abstract. We consider decidability problems associated with Engel's identity ($[\cdots[[x,y],y],\ldots,y] = 1$ for a long enough commutator sequence) in groups generated by an automaton.

We give a partial algorithm that decides, given x, y, whether an Engel identity is satisfied. It succeeds, importantly, in proving that Grigorchuk's 2-group is not Engel.

We consider next the problem of recognizing Engel elements, namely elements y such that the map $x \mapsto [x, y]$ attracts to $\{1\}$. Although this problem seems intractable in general, we prove that it is decidable for Grigorchuk's group: Engel elements are precisely those of order at most 2.

Our computations were implemented using the package FR within the computer algebra system GAP.

1 Introduction

A *law* in a group G is a word $w = w(x_1, x_2, \ldots, x_n)$ such that $w(g_1, \ldots, g_n) = \mathbb{1}$, the identity element, for all $g_1, \ldots, g_n \in G$; for example, commutative groups satisfy the law $[x_1, x_2] = x_1^{-1} x_2^{-1} x_1 x_2$. A *variety* of groups is a maximal class of groups satisfying a given law; e.g. the variety of commutative groups (satisfying $[x_1, x_2]$) or of groups of exponent p (satisfying x_1^p); see [22,23].

Consider now a sequence $\mathscr{W} = (w_0, w_1, \ldots)$ of words in n letters. Say that $(g_1, \ldots, g_n)$ *almost satisfies* $\mathscr{W}$ if $w_i(g_1, \ldots, g_n) = 1$ for all i large enough, and say that G *almost satisfies* $\mathscr{W}$ if all n-tuples from G almost satisfy $\mathscr{W}$. For example, G almost satisfies $(x_1, \ldots, x_1^{i!}, \ldots)$ if and only if G is a torsion group.

The problem of deciding algorithmically whether a group belongs to a given variety has received much attention (see e.g. [17] and references therein); we consider here the harder problems of determining whether a group (respectively a tuple) almost satisfies a given sequence. This has, up to now, been investigated mainly for the torsion sequence above [12].

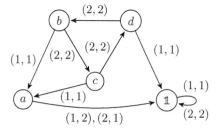

Fig. 1. The Grigorchuk automaton

L. Bartholdi—Partially supported by ANR grant ANR-14-ACHN-0018-01.

A.S. Kulikov and G.J. Woeginger (Eds.): CSR 2016, LNCS 9691, pp. 29–40, 2016.
DOI: 10.1007/978-3-319-34171-2_3

The first Grigorchuk group G_0 is an automaton group which appears prominently in group theory, for example as a finitely generated infinite torsion group [14] and as a group of intermediate word growth [15]; see Sect. 2.1. It is the group of automatic transformations of $\{1,2\}^\infty$ generated by the five states of the automaton from Fig. 1, with input and output written as (in, out).

The Engel law is

$$E_c = E_c(x,y) = [x,y,\ldots,y] = [\cdots[[x,y],y],\ldots,y]$$

with c copies of 'y'; so $E_0(x,y) = x$, $E_1(x,y) = [x,y]$ and $E_c(x,y) = [E_{c-1}(x,y),y]$. See below for a motivation. Let us call a group (respectively a pair of elements) *Engel* if it almost satisfies $\mathscr{E} = (E_0, E_1, \ldots)$. Furthermore, let us call $h \in G$ an *Engel* element if (g,h) is Engel for all $g \in G$.

A concrete consequence of our investigations is:

Theorem 1. *The first Grigorchuk group G_0 is not Engel. Furthermore, an element $h \in G_0$ is Engel if and only if $h^2 = \mathbb{1}$.*

We prove a similar statement for another prominent example of automaton group, the *Gupta-Sidki group*, see Theorem 2.

Theorem 1 follows from a partial algorithm, giving a criterion for an element y to be Engel. This algorithm proves, in fact, that the element ad in the Grigorchuk group is not Engel. We consider the following restricted situation, which is general as far as the Engel property is concerned, see Sect. 2: an *automaton group* is a group G endowed with extra data, in particular with a family of self-maps called *states*, indexed by a set X and written $g \mapsto g@x$ for $x \in X$; it is *contracting* for the word metric $\|\cdot\|$ on G if there are constants $\eta < 1$ and C such that $\|g@x\| \leq \eta\|g\| + C$ holds for all $g \in G$ and all $x \in X$. Our aim is to solve the following **decision problems** in an automaton group G:

Engel(g,h) Given $g,h \in G$, does there exist $c \in \mathbb{N}$ with $E_c(g,h)$?
Engel(h) Given $h \in G$, does Engel(g,h) hold for all $g \in G$?

The algorithm is described in Sect. 3. As a consequence,

Corollary 1. *Let G be an automaton group acting on the set of binary sequences $\{1,2\}^*$, that is contracting with contraction coefficient $\eta < 1$. Then, for torsion elements h of order 2^e with $2^{2^e}\eta < 1$, the property Engel(h) is decidable.*

The Engel property attracted attention for its relation to nilpotency: indeed a nilpotent group of class c satisfies E_c, and conversely among compact [21] and solvable [16] groups, if a group satisfies E_c for some c then it is locally nilpotent. Conjecturally, there are non-locally nilpotent groups satisfying E_c for some c, but this is still unknown. It is also an example of iterated identity, see [3,7]. In particular, the main result of [3] implies easily that the Engel property is decidable in algebraic groups.

It is comparatively easy to prove that the first Grigorchuk group G_0 satisfies no law [1,20]; this result holds for a large class of automaton groups. In fact, if a group satisfies a law, then so does its profinite completion. In the class mentioned above, the profinite completion contains abstract free subgroups, precluding the existence of a law. No such arguments would help for the Engel property: the restricted product of all finite nilpotent groups is Engel, but the unrestricted product again contains free subgroups. This is one of the difficulties in dealing with iterated identities rather than identities.

If $\mathfrak{A}$ is a nil algebra (namely, for every $a \in \mathfrak{A}$ there exists $n \in \mathbb{N}$ with $a^n = 0$) then the set of elements of the form $\{1 + a : a \in \mathfrak{A}\}$ forms a group $1 + \mathfrak{A}$ under the law $(1 + a)(1 + b) = 1 + (a + b + ab)$. If $\mathfrak{A}$ is defined over a field of characteristic p, then $1 + \mathfrak{A}$ is a torsion group since $(1 + a)^{p^n} = 1$ if $a^{p^n} = 0$. Golod constructed in [13] non-nilpotent nil algebras $\mathfrak{A}$ all of whose 2-generated subalgebras are nilpotent (namely, $\mathfrak{A}^n = 0$ for some $n \in \mathbb{N}$); given such an $\mathfrak{A}$, the group $1 + \mathfrak{A}$ is Engel but not locally nilpotent.

Golod introduced these algebras as means of obtaining infinite, finitely generated, residually finite (every non-trivial element in the group has a non-trivial image in some finite quotient), torsion groups. Golod's construction is highly non-explicit, in contrast with Grigorchuk's group for which much can be derived from the automaton's properties.

It is therefore of high interest to find explicit examples of Engel groups that are not locally nilpotent, and the methods and algorithms presented here are a step in this direction.

An important feature of automaton groups is their amenability to computer experiments, and even as in this case of rigorous verification of mathematical assertions; see also [18], and the numerous decidability and undecidability results pertaining to the finiteness property in [2,11,19].

The proof of Theorem 1 relies on a computer calculation. It could be checked by hand, at the cost of quite unrewarding effort. One of the purposes of this article is, precisely, to promote the use of computers in solving general questions in group theory: the calculations performed, and the computer search involved, are easy from the point of view of a computer but intractable from the point of view of a human.

The calculations were performed using the author's group theory package FR, specially written to manipulate automaton groups. This package integrates with the computer algebra system GAP [8], and is freely available from the GAP distribution site

$$\text{http://www.gap-system.org}$$

2   Automaton Groups

An *automaton group* is a finitely generated group associated with an invertible Mealy automaton. We define a *Mealy automaton* $\mathscr{M}$ as a graph such as that in Fig. 1. It has a set of *states* Q and an *alphabet* X, and there are *transitions* between states with *input* and *output* labels in X, with the condition that, at every state, all labels appear exactly once as input and (if $\mathscr{M}$ is to be invertible) once as output on the set of outgoing transitions.

Every state $q \in Q$ of $\mathcal{M}$ defines a transformation, written as exponentiation by q, of the set of words X^* by the following rule: given a word $x_1 \ldots x_n \in X^*$, there exists a unique path in the automaton starting at q and with input labels $x_1, \ldots, x_n$. Let $y_1, \ldots, y_n$ be its corresponding output labels. Then declare $(x_1 \ldots x_n)^q = y_1 \ldots y_n$.

The action may also be defined recursively as follows: if there is a transition from $q \in Q$ to $r \in Q$ with input label $x_1 \in X$ and output label $y_1 \in X$, then $(x_1 x_2 \ldots x_n)^q = y_1 (x_2 \ldots x_n)^r$.

By the *automaton group* associated with the automaton $\mathcal{M}$, we mean the group G of transformations of X^* generated by $\mathcal{M}$'s states. Note that all elements of G admit descriptions by automata; namely, a word of length n in G's generators is the transformation associated with a state in the n-fold product of the automaton of G. See [10] for the abstract theory of automata, and [9] for products more specifically.

The structure of the automaton $\mathcal{M}$ may be encoded in an injective group homomorphism $\psi \colon G \to G^X \rtimes \mathrm{Sym}(X)$ from G to the group of G-*decorated permutations of X*. This last group — the *wreath product* of G with $\mathrm{Sym}(X)$ — is the group of permutations of X, with labels in G on each arrow of the permutation; the labels multiply as the permutations are composed. The construction of ψ is as follows: consider $q \in Q$. For each $x \in X$, let $q@x$ denote the endpoint of the transition starting at q with input label x, and let x^π denote the output label of the same transition; thus every transition in the Mealy automaton gives rise to

$$\underset{q}{\bigcirc} \xrightarrow{\;(x, x^\pi)\;} \underset{q@x}{\bigcirc}$$

The transformation π is a permutation of X, and we set

$$\psi(q) = \langle x \mapsto q@x \rangle \pi,$$

namely the permutation π with decoration $q@x$ on the arrow from x to x^π.

We generalize the notation $q@x$ to arbitrary words and group elements. Consider a word $v \in X^*$ and an element $g \in G$; denote by v^g the image of v under g. There is then a unique element of G, written $g@v$, with the property

$$(v\,w)^g = (v^g)\,(w)^{g@v} \quad \text{for all } w \in X^*.$$

We call by extension this element $g@v$ the *state* of g at v; it is the state, in the Mealy automaton defining g, that is reached from g by following the path v as input; thus in the Grigorchuk automaton $b@1 = a$ and $b@222 = b$ and $(bc)@2 = cd$. There is a reverse construction: by $v * g$ we denote the permutation of X^* (which need not belong to G) defined by

$$(v\,w)^{v*g} = v\,w^g, \qquad w^{v*g} = w \text{ if } w \text{ does not start with } v.$$

Given a word $w = w_1 \ldots w_n \in X^*$ and a Mealy automaton $\mathcal{M}$ of which g is a state, it is easy to construct a Mealy automaton of which $w * g$ is a state: add a path of length n to $\mathcal{M}$, with input and output $(w_1, w_1), \ldots, (w_n, w_n)$ along the path, and ending at g. Complete the automaton with transitions to the identity element. Then the first vertex of the path defines the transformation $w * g$. For example, here is $12 * d$ in the Grigorchuk automaton:

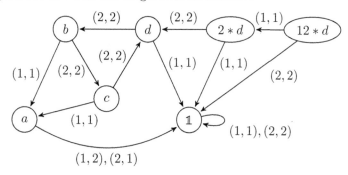

Note the simple identities $(g@v_1)@v_2 = g@(v_1v_2)$, $(v_1v_2) * g = v_1 * (v_2 * g)$, and $(v * g)@v = g$. Recall that we write conjugation in G as $g^h = h^{-1}gh$. For any $h \in G$ we have

$$(v * g)^h = v^h * (g^{h@v}). \tag{1}$$

An automaton group is called *regular weakly branched* if there exists a non-trivial subgroup K of G such that $\psi(K)$ contains K^X. In other words, for every $k \in K$ and every $v \in X^*$, the element $v * k$ also belongs to K, and therefore to G. Abért proved in [1] that regular weakly branched groups satisfy no law.

In this text, we concentrate on the Engel property, which is equivalent to nilpotency for finite groups. In particular, if an automaton group G is to have a chance of being Engel, then its image under the map $G \to G^X \rtimes \mathrm{Sym}(X) \to \mathrm{Sym}(X)$ should be a nilpotent subgroup of $\mathrm{Sym}(X)$. Since finite nilpotent groups are direct products of their p-Sylow subgroups, we may reduce to the case in which the image of G in $\mathrm{Sym}(X)$ is a p-group. A further reduction lets us assume that the image of G is an abelian subgroup of $\mathrm{Sym}(X)$ of prime order. We therefore make the following:

Standing Assumption 1. *The alphabet is* $X = \{1, \ldots, p\}$ *and automaton groups are defined by embeddings* $\psi \colon G \to G^p \rtimes \mathbb{Z}_{/p}$, *with* $\mathbb{Z}_{/p}$ *the cyclic subgroup of* $\mathrm{Sym}(X)$ *generated by the cycle* $(1, 2, \ldots, p)$.

This is the situation considered in the Introduction.

We make a further reduction in that we only consider the Engel property for elements of finite order. This is not a very strong restriction: given h of infinite order, one can usually find an element $g \in G$ such that the conjugates $\{g^{h^n} : n \in \mathbb{Z}\}$ are independent, and it then follows that h is not Engel. This will be part of a later article.

2.1 Grigorchuk's First Group

This section is not an introduction to Grigorchuk's first group, but rather a brief description of it with all information vital for the calculation in Sect. 4. For more details, see e.g. [5].

Fix the alphabet $X = \{1, 2\}$. The first Grigorchuk group G_0 is a permutation group of the set of words X^*, generated by the four non-trivial states a, b, c, d of the automaton given in Fig. 1. Alternatively, the transformations a, b, c, d may be defined recursively as follows:

$$
\begin{aligned}
(1x_2 \ldots x_n)^a &= 2x_2 \ldots x_n, & (2x_2 \ldots x_n)^a &= 1x_2 \ldots x_n, \\
(1x_2 \ldots x_n)^b &= 1(x_2 \ldots x_n)^a, & (2x_2 \ldots x_n)^b &= 2(x_2 \ldots x_n)^c, \\
(1x_2 \ldots x_n)^c &= 1(x_2 \ldots x_n)^a, & (2x_2 \ldots x_n)^c &= 2(x_2 \ldots x_n)^d, \\
(1x_2 \ldots x_n)^d &= 1x_2 \ldots x_n, & (2x_2 \ldots x_n)^d &= 2(x_2 \ldots x_n)^b
\end{aligned}
\tag{2}
$$

which directly follow from $d@1 = \mathbb{1}$, $d@2 = b$, etc.

It is remarkable that most properties of G_0 derive from a careful study of the automaton (or equivalently this action), usually using inductive arguments. For example,

Proposition 1 ([14]). *The group G_0 is infinite, and all its elements have order a power of* 2.

The self-similar nature of G_0 is made apparent in the following manner:

Proposition 2 ([4], Sect. 4). *Define $x = [a, b]$ and $K = \langle x, x^c, x^{ca} \rangle$. Then K is a normal subgroup of G_0 of index* 16, *and $\psi(K)$ contains $K \times K$.*

In other words, for every $g \in K$ and every $v \in X^*$ the element $v * g$ belongs to G_0.

3   A Semi-algorithm for Deciding the Engel Property

We start by describing a semi-algorithm to check the Engel property. It will sometimes not return any answer, but when it returns an answer then that answer is guaranteed correct. It is guaranteed to terminate as long as the contraction property of the automaton group G is strong enough.

Algorithm 1. *Let G be a contracting automaton group with alphabet $X = \{1, \ldots, p\}$ for prime p, with the contraction property $\|g@j\| \leq \eta \|g\| + C$.*

For $n \in p\mathbb{N}$ and $R \in \mathbb{R}$ consider the following finite graph $\Gamma_{n,R}$. Its vertex set is $B(R)^n \cup \{\mathtt{fail}\}$, where $B(R)$ denotes the set of elements of G of length at most R. Its edge set is defined as follows: consider a vertex $(g_1, \ldots, g_n)$ in $\Gamma_{n,R}$, and compute

$$(h_1, \ldots, h_n) = (g_1^{-1} g_2, \ldots, g_n^{-1} g_1).$$

If h_i fixes X for all i, i.e. all h_i have trivial image in $\mathrm{Sym}(X)$, then for all $j \in \{1, \ldots, p\}$ there is an edge from $(g_1, \ldots, g_n)$ to $(h_1@j, \ldots, h_n@j)$, or to $\mathtt{fail}$ if $(h_1@j, \ldots, h_n@j) \notin B(R)^n$. If some h_i does not fix X, then there is an edge from $(g_1, \ldots, g_n)$ to $(h_1, \ldots, h_n)$, or to $\mathtt{fail}$ if $(h_1, \ldots, h_n) \notin B(R)^n$.

Given $g, h \in G$ with $h^n = \mathbb{1}$: Set $t_0 = (g, g^h, g^{h^2}, \ldots, g^{h^{n-1}})$. If there exists $R \in \mathbb{N}$ such that no path in $\Gamma_{n,R}$ starting at t_0 reaches $\mathtt{fail}$, then $Engel(g, h)$ holds if and only if the only cycle in $\Gamma_{n,R}$ reachable from t_0 passes through $(\mathbb{1}, \ldots, \mathbb{1})$.

If the contraction coefficient satisfies $2^n \eta < 1$, then it is sufficient to consider $R = (\|g\| + n\|h\|)2^n C/(1 - 2^n \eta)$.

Given $n \in p\mathbb{N}$: The Engel property holds for all elements of exponent n if and only if, for all $R \in \mathbb{N}$, the only cycle in $\Gamma_{n,R}$ passes through $(\mathbb{1}, \ldots, \mathbb{1})$.

If the contraction coefficient satisfies $2^n \eta < 1$, then it is sufficient to consider $R = 2^n C/(1 - 2^n \eta)$.

Given G weakly branched and $n \in p\mathbb{N}$: If for some $R \in \mathbb{N}$ there exists a cycle in $\Gamma_{n,R}$ that passes through an element of $K^n \setminus \mathbb{1}^n$, then no element of G whose order is a multiple of n is Engel.

If the contraction coefficient satisfies $2^n \eta < 1$, then it is sufficient to consider $R = 2^n C/(1 - 2^n \eta)$.

We consider the graphs $\Gamma_{n,R}$ as subgraphs of a graph $\Gamma_{n,\infty}$ with vertex set G^n and same edge definition as the $\Gamma_{n,R}$.

We note first that, if G satisfies the contraction condition $2^n \eta < 1$, then all cycles of $\Gamma_{n,\infty}$ lie in fact in $\Gamma_{n,2^n C/(1-2^n\eta)}$. Indeed, consider a cycle passing through $(g_1, \ldots, g_n)$ with $\max_i \|g_i\| = R$. Then the cycle continues with $(g_1^{(1)}, \ldots, g_n^{(1)})$, $(g_1^{(2)}, \ldots, g_n^{(2)})$, etc. with $\|g_i^{(k)}\| \leq 2^k R$; and then for some $k \leq n$ we have that all $g_i^{(k)}$ fix X; namely, they have a trivial image in $\mathrm{Sym}(X)$, and the map $g \mapsto g@j$ is an injective homomorphism on them. Indeed, let $\pi_1, \ldots, \pi_n, \pi_1^{(i)}, \ldots, \pi_n^{(i)} \in \mathbb{Z}_{/p} \subset \mathrm{Sym}(X)$ be the images of $g_1, \ldots, g_n, g_1^{(i)}, \ldots, g_n^{(i)}$ respectively, and denote by $S \colon \mathbb{Z}_{/p}^n \to \mathbb{Z}_{/p}^n$ the cyclic permutation operator. Then $(\pi_1^{(n)}, \ldots, \pi_n^{(n)}) = (S-\mathbb{1})^n (\pi_1, \ldots, \pi_n)$, and $(S-\mathbb{1})^n = \sum_j S^j \binom{n}{j} = 0$ since $p|n$ and $S^n = \mathbb{1}$. Thus there is an edge from $(g_1^{(k)}, \ldots, g_n^{(k)})$ to $(g_1^{(k+1)}@j, \ldots, g_n^{(k+1)}@j)$ with $\|g_i^{(k+1)}@j\| \leq \eta\|g_i^{(k)}\| + C \leq \eta 2^n R + C$. Therefore, if $R > 2^n C/(1 - 2^n \eta)$ then $2^n \eta R + C < R$, and no cycle can return to $(g_1, \ldots, g_n)$.

Consider now an element $h \in G$ with $h^n = \mathbb{1}$. For all $g \in G$, there is an edge in $\Gamma_{n,\infty}$ from $(g, g^h, \ldots, g^{h^{n-1}})$ to $([g,h]@v, [g,h]^h@v, [g,h]^{h^{n-1}}@v)$ for some word $v \in \{\varepsilon\} \sqcup X$, and therefore for all $c \in \mathbb{N}$ there exists $d \leq c$ such that, for all $v \in X^d$, there is a length-c path from $(g, g^h, \ldots, g^{h^{n-1}})$ to $(E_c(g, h)@v, \ldots, E_c(g, h)^{h^{n-1}}@v)$ in $\Gamma_{n,\infty}$.

We are ready to prove the first assertion: if $Engel(g, h)$, then $E_c(g, h) = \mathbb{1}$ for some c large enough, so all paths of length c starting at $(g, g^h, \ldots, g^{h^{n-1}})$ end at $(\mathbb{1}, \ldots, \mathbb{1})$. On the other hand, if $Engel(g, h)$ does not hold, then all long enough paths starting at $(g, g^h, \ldots, g^{h^{n-1}})$ end at vertices in the finite graph $\Gamma_{n,2^n C/(1-2^n\eta)}$ so must eventually reach cycles; and one of these cycles is not $\{(\mathbb{1}, \ldots, \mathbb{1})\}$ since $E_c(g, h) \neq \mathbb{1}$ for all c.

The second assertion immediately follows: if there exists $g \in G$ such that $Engel(g, h)$ does not hold, then again a non-trivial cycle is reached starting from $(g, g^h, \ldots, g^{h^{n-1}})$, and independently of g, h this cycle belongs to the graph $\Gamma_{n, 2^n C/(1-2^n \eta)}$.

For the third assertion, let $\bar{k} = (k_1, \ldots, k_n) \in K^n \setminus \mathbb{1}^n$ be a vertex of a cycle in $\Gamma_{n, 2^n C/(1-2^n \eta)}$. Consider an element $h \in G$ of order sn for some $s \in \mathbb{N}$. By the condition that $\#X = p$ is prime and the image of G in $\mathrm{Sym}(X)$ is a cyclic group, sn is a power of p, so there exists an orbit $\{v_1, \ldots, v_{sn}\}$ of h, so labeled that $v_i^h = v_{i-1}$, indices being read modulo sn. For $i = 1, \ldots, sn$ define

$$h_i = (h@v_1)^{-1} \cdots (h@v_i)^{-1},$$

noting $h_i(h@v_i) = h_{i-1}$ for all $i = 1, \ldots, sn$ since $h^{sn} = \mathbb{1}$. Denote by '$i\%n$' the unique element of $\{1, \ldots, n\}$ congruent to i modulo n, and consider the element

$$g = \prod_{i=1}^{sn} \left(v_i * k_{i\%n}^{h_i}\right),$$

which belongs to G since G is weakly branched. Let $(k_1^{(1)}, \ldots, k_n^{(1)})$ be the next vertex on the cycle of $\bar{k}$. We then have, using (1),

$$[g, h] = g^{-1} g^h = \prod_{i=1}^{sn} \left(v_i * k_{i\%n}^{-h_i}\right) \prod_{i=1}^{sn} \left(v_{i-1} * k_{i\%n}^{h_i(h@v_i)}\right) = \prod_{i=1}^{sn} \left(v_i * (k_{i\%n}^{(1)})^{h_i}\right),$$

and more generally $E_c(g, h)$ and some of its states are read off the cycle of $\bar{k}$. Since this cycle goes through non-trivial group elements, $E_c(g, h)$ has a non-trivial state for all c, so is non-trivial for all c, and $\mathrm{Engel}(g, h)$ does not hold.

4   Proof of Theorem 1

The Grigorchuk group G_0 is contracting, with contraction coefficient $\eta = 1/2$. Therefore, the conditions of validity of Algorithm 1 are not satisfied by the Grigorchuk group, so that it is not guaranteed that the algorithm will succeed, on a given element $h \in G_0$, to prove that h is not Engel. However, nothing forbids us from running the algorithm with the hope that it nevertheless terminates. It seems experimentally that the algorithm always succeeds on elements of order 4, and the argument proving the third claim of Algorithm 1 (repeated here for convenience) suffices to complete the proof of Theorem 1.

Below is a self-contained proof of Theorem 1, extracting the relevant properties of the previous section, and describing the computer calculations as they were keyed in.

Consider first $h \in G_0$ with $h^2 = \mathbb{1}$. It follows from Proposition 1 that h is Engel: given $g \in G_0$, we have $E_{1+k}(g, h) = [g, h]^{(-2)^k}$ so $E_{1+k}(g, h) = \mathbb{1}$ for k larger than the order of $[g, h]$.

For the other case, we start by a side calculation. In the Grigorchuk group G_0, define $x = [a, b]$ and $K = \langle x \rangle^{G_0}$ as in Proposition 2, consider the quadruple

$$A_0 = (A_{0,1}, A_{0,2}, A_{0,3}, A_{0,4}) = (x^{-2} x^{2ca}, \; x^{-2ca} x^2 x^{2cab}, \; x^{-2cab} x^{-2}, \; x^2)$$

of elements of K, and for all $n \geq 0$ define

$$A_{n+1} = (A_{n,1}^{-1} A_{n,2},\ A_{n,2}^{-1} A_{n,3},\ A_{n,3}^{-1} A_{n,4},\ A_{n,4}^{-1} A_{n,1}).$$

Lemma 1. *For all $i = 1, \ldots, 4$, the element $A_{9,i}$ fixes 111112, is non-trivial, and satisfies $A_{9,i}@111112 = A_{0,i}$.*

Proof. This is proven purely by a computer calculation. It is performed as follows within GAP:

```
gap> LoadPackage("FR");
true
gap> AssignGeneratorVariables(GrigorchukGroup);;
gap> x2 := Comm(a,b)^2;; x2ca := x2^(c*a);; one := a^0;;
gap> A0 := [x2^-1*x2ca,x2ca^-1*x2*x2ca^b,(x2ca^-1)^b*x2^-1,x2];;
gap> v := [1,1,1,1,1,2];; A := A0;;
gap> for n in [1..9] do A := List([1..4],i->A[i]^-1*A[1+i mod 4]); od;
gap> ForAll([1..4],i->v^A[i]=v and A[i]<>one and State(A[i],v)=A0[i]);
true
```

Consider now $h \in G_0$ with $h^2 \neq 1$. Again by Proposition 1, we have $h^{2^e} = 1$ for some minimal $e \in \mathbb{N}$, which is furthermore at least 2. We keep the notation '$a\%b$' for the unique number in $\{1, \ldots, b\}$ that is congruent to a modulo b.

Let n be large enough so that the action of h on X^n has an orbit $\{v_1, v_2, \ldots, v_{2^e}\}$ of length 2^e, numbered so that $v_{i+1}^h = v_i$ for all i, indices being read modulo 2^e. For $i = 1, \ldots, 2^e$ define

$$h_i = (h@v_1)^{-1} \cdots (h@v_i)^{-1},$$

noting $h_i(h@v_i) = h_{i-1 \% 2^e}$ for all $i = 1, \ldots, 2^e$ since $h^{2^e} = 1$, and consider the element

$$g = \prod_{i=1}^{2^e} \left(v_i * A_{0,i\%4}^{h_i}\right),$$

which is well defined since $4|2^e$ and belongs to G_0 by Proposition 2. We then have, using (1),

$$[g,h] = g^{-1}g^h = \prod_{i=1}^{2^e} \left(v_i * A_{0,i\%4}^{-h_i}\right) \prod_{i=1}^{2^e} \left(v_{i-1\%2^e} * A_{0,i\%4}^{h_i(h@v_i)}\right) = \prod_{i=1}^{2^e} \left(v_i * A_{1,i}^{h_i}\right),$$

and more generally

$$E_c(g,h) = \prod_{i=1}^{2^e} \left(v_i * A_{c,i}^{h_i}\right).$$

Therefore, by Lemma 1, for every $k \geq 0$ we have $E_{9k}(g,h)@v_0(111112)^k = A_{0,1} \neq 1$, so $E_c(g,h) \neq 1$ for all $c \in \mathbb{N}$ and we have proven that h is not an Engel element.

5   Other Examples

Similar calculations apply to the Gupta-Sidki group Γ. This is another example of infinite torsion group, acting on X^* for $X = \{1, 2, 3\}$ and generated by the states of the following automaton:

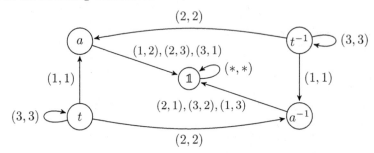

The transformations a, t may also be defined recursively by

$$
\begin{aligned}
(1v)^a &= 2v, & (2v)^a &= 3v, & (3v)^a &= 1v, \\
(1v)^t &= 1v^a, & (2v)^t &= 2v^{a^{-1}}, & (3v)^t &= 3v^t.
\end{aligned}
\tag{3}
$$

The Gupta-Sidki group is contracting, with contraction coefficient $\eta = 1/2$. Again, this is not sufficient to guarantee that Algorithm 1 terminates, but it nevertheless did succeed in proving.

Theorem 2. *The only Engel element in the Gupta-Sidki group Γ is the identity.*

We only sketch the proof, since it follows that of Theorem 1 quite closely. Analogues of Propositions 1 and 2 hold, with $[\Gamma, \Gamma]$ in the rôle of K. An analogue of Lemma 1 holds with $A_0 = ([a^{-1}, t], [a, t]^a, [t^{-1}, a^{-1}])$ and $A_{4,i}@122 = A_{0,i}$.

6   Closing Remarks

It would be dishonest to withhold from the reader how I arrived at the examples given for the Grigorchuk and Gupta-Sidki groups. I started with small words g, h in the generators of G_0, respectively Γ, and computed $E_c(g, h)$ for the first few values of c. These elements are represented, internally to FR, as Mealy automata. A natural measure of the complexity of a group element is the size of the minimized automaton, which serves as a canonical representation of the element.

For some choices of g, h the size of $E_c(g, h)$ increases exponentially with c, limiting the practicality of computer experiments. For others (such as $(g, h) = ((ba)^4c, ad)$ for the Grigorchuk group), the size increases roughly linearly with c, making calculations possible for c in the hundreds. Using these data, I guessed the period p of the recursion (9 in the case of the Grigorchuk group), and searched among the states of $E_c(g, h)$ and $E_{c+p}(g, h)$ for common elements; in the example, I found such common states for $c = 23$. I then took the smallest-size quadruple of states that appeared both in $E_c(g, h)$ and $E_{c+p}(g, h)$ and belonged to

K, and expressed the calculation taking $E_c(g,h)$ to $E_{c+p}(g,h)$ in the form of Lemma 1.

It was already shown by Bludov [6] that the wreath product $G_0^4 \rtimes D_4$ is not Engel. He gave, in this manner, an example of a torsion group in which a product of Engel elements is not Engel. Our proof is a refinement of his argument.

A direct search for elements $A_{0,1}, \ldots, A_{0,4}$ would probably not be successful, and has not yielded simpler elements than those given before Lemma 1, if one restricts them to belong to K; one can only wonder how Bludov found the quadruple $(\mathbb{1}, d, ca, ab)$, presumably without the help of a computer.

Acknowledgments. I am grateful to Anna Erschler for stimulating my interest in this question and for having suggested a computer approach to the problem, and to Ines Klimann and Matthieu Picantin for helpful discussions that have improved the presentation of this note.

References

1. Abért, M.: Group laws and free subgroups in topological groups. Bull. Lond. Math. Soc. **37**(4), 525–534 (2005). arxiv:math.GR/0306364. MR2143732
2. Akhavi, A., Klimann, I., Lombardy, S., Mairesse, J., Picantin, M.: On the finiteness problem for automaton (semi)groups. Internat. J. Algebra Comput. **22**(6), 26 (2012). doi:10.1142/S021819671250052X. 1250052 MR2974106
3. Bandman, T., Grunewald, F., Kunyavskiĭ, B.: Geometry and arithmetic of verbal dynamical systems on simple groups. Groups Geom. Dyn. **4**(4), 607–655 (2010). doi:10.4171/GGD/98. With an appendix by Nathan Jones. MR2727656 (2011k:14020)
4. Bartholdi, L., Grigorchuk, R.I.: On parabolic subgroups and Hecke algebras of some fractal groups. SerdicaMath. J. **28**(1), 47–90 (2002). arxiv:math/9911206. MR1899368 (2003c:20027)
5. Bartholdi, L., Grigorchuk, R.I., Šuni, Z.: Branch groups. In: Handbook of Algebra, vol. 3, pp. 989–1112. North-Holland, Amsterdam (2003). doi:10.1016/S1570-7954(03)80078-5. arxiv:math/0510294. MR2035113 (2005f:20046)
6. Bludov, V.V.: An example of not Engel group generated by Engel elements. In: A Conference in Honor of Adalbert Bovdi's 70th Birthday, 18–23 November 2005, Debrecen, Hungary, pp. 7–8 (2005)
7. Erschler, A.: Iterated identities and iterational depth of groups (2014). arxiv:math/1409.5953
8. The GAP Group, GAP—Groups, Algorithms, and Programming, Version 4.4.10 (2008)
9. Gécseg, F.: Products of Automata EATCS Monographs on Theoretical Computer Science. EATCS Monographs on Theoretical Computer Science, vol. 7. Springer, Berlin (1986). MR88b:68139b
10. Gécseg, F., Csákány, B.: Algebraic Theory of Automata. Akademiami Kiado, Budapest (1971)
11. Gillibert, P.: The finiteness problem for automaton semigroups is undecidable. Int. J. Algebra Comput. **24**(1), 1–9 (2014). doi:10.1142/S0218196714500015. MR3189662

12. Godin, T., Klimann, I., Picantin, M.: On torsion-free semigroups generated by invertible reversible mealy automata. In: Dediu, A.-H., Formenti, E., Martín-Vide, C., Truthe, B. (eds.) LATA 2015. LNCS, vol. 8977, pp. 328–339. Springer, Heidelberg (2015). doi:10.1007/978-3-319-15579-1_25. MR3344813

13. Golod, E.S.: Some problems of Burnside type. In: Proceedings of the International Congress of Mathematicians (Moscow, 1966), Izdat. "Mir", Moscow, 1968, pp. 284–289 (Russian). MR0238880 (39 #240)

14. Grigorchuk, R.I.: On Burnside's problem on periodic groups, Функционал. Анал. и Приложен. **14**(1), 53–54 (1980). English translation: Functional Anal. Appl. **14**, 41–43 (1980). MR81m:20045

15. Grigorchuk, R.I.: On the milnor problem of group growth. Dokl. Akad. Nauk SSSR **271**(1), 30–33 (1983). MR85g:20042

16. Gruenberg, K.W.: The Engel elements of a soluble group. Illinois J. Math. **3**, 151–168 (1959). MR0104730 (21 #3483)

17. Jackson, M.: On locally finite varieties with undecidable equational theory. Algebra Univers. **47**(1), 1–6 (2002). doi:10.1007/s00012-002-8169-0. MR1901727 (2003b:08002)

18. Klimann, I., Mairesse, J., Picantin, M.: Implementing computations in automaton (Semi)groups. In: Moreira, N., Reis, R. (eds.) CIAA 2012. LNCS, vol. 7381, pp. 240–252. Springer, Heidelberg (2012). doi:10.1007/978-3-642-31606-7_21. MR2993189

19. Klimann, I.: The finiteness of a group generated by a 2-letter invertible-reversible Mealy automaton is decidable. In: 30th International Symposium on Theoretical Aspects of Computer Science, LIPIcs. Leibniz International Proceedings in Informatics, vol. 20, Schloss Dagstuhl. Leibniz-Zent. Inform., Wadern, pp. 502–513 (2013). MR3090008

20. Leonov, Y.G.: On identities in groups of automorphisms of trees. Visnyk of Kyiv State University of T.G.Shevchenko **3**, 37–44 (1997)

21. Medvedev, Y.: On compact Engel groups. Israel J. Math. **135**, 147–156 (2003). doi:10.1007/BF02776054. MR1997040 (2004f:20072)

22. Neumann, H.: Varieties of Groups. Springer-Verlag New York, Inc., New York (1967). MR35#6734

23. Straubing, H., Weil, P.: Varieties (2015). arxiv:math/1502.03951

The Next Whisky Bar

Mike Behrisch[1], Miki Hermann[2]([⊠]), Stefan Mengel[3], and Gernot Salzer[1]

[1] Technische Universität Wien, Vienna, Austria
{behrisch,salzer}@logic.at
[2] LIX (UMR CNRS 7161), École Polytechnique, Palaiseau, France
hermann@lix.polytechnique.fr
[3] CRIL (UMR CNRS 8188), Université d'Artois, Lens, France
mengel@cril.fr

Abstract. We determine the complexity of an optimization problem related to information theory. Taking a conjunctive propositional formula over some finite set of Boolean relations as input, we seek a satisfying assignment of the formula having minimal Hamming distance to a given assignment that is not required to be a model (NearestSolution, NSol). We obtain a complete classification with respect to the relations admitted in the formula. For two classes of constraint languages we present polynomial time algorithms; otherwise, we prove hardness or completeness concerning the classes APX, poly-APX, NPO, or equivalence to well-known hard optimization problems.

1 Introduction

We investigate the solution spaces of Boolean constraint satisfaction problems built from atomic constraints by means of conjunction and variable identification. We study the following minimization problems in connection with Hamming distance: Given an instance of a constraint satisfaction problem in the form of a generalized conjunctive formula over a set of atomic constraints, the problem asks to find a satisfying assignment with minimal Hamming distance to a given assignment (NearestSolution, NSol). Note that we do not assume the given assignment to satisfy the formula nor the solution to be different from it as was done in [4], where NearestOtherSolution (NOSol) was studied. This would change the complexity classification (e.g. for bijunctive constraints), and proof techniques would become considerably harder due to inapplicability of clone theory.

The title refers to the *Alabama Song* by Bertolt Brecht (lyrics), Kurt Weill (music), and Elisabeth Hauptmann (English translation). Among the numerous cover versions, the one by Jim Morrison and the Doors became particularly popular in the 1970s.

M. Behrisch, and G. Salzer — Supported by Austrian Science Fund (FWF) grant I836-N23.

M. Hermann — Supported by ANR-11-ISO2-003-01 Blanc International grant ALCOCLAN.

S. Mengel — Research of this author was done during his post-doctoral stay in LIX at École Polytechnique. Supported by a QUALCOMM grant.

A.S. Kulikov and G.J. Woeginger (Eds.): CSR 2016, LNCS 9691, pp. 41–56, 2016.
DOI: 10.1007/978-3-319-34171-2_4

This problem appears in several guises throughout literature. E.g., a common problem in AI is to find solutions of constraints close to an initial configuration; our problem is an abstraction of this setting for the Boolean domain. Bailleux and Marquis [3] describe such applications in detail and introduce the decision problem DistanceSAT: Given a propositional formula φ, a partial interpretation I, and a bound k, is there a satisfying assignment differing from I in no more than k variables? It is straightforward to show that DistanceSAT corresponds to the decision variant of our problem with existential quantification (called $NSol_{pp}^d$ later on). While [3] investigates the complexity of DistanceSAT for a few relevant classes of formulas and empirically evaluates two algorithms, we analyze the decision and the optimization problem for arbitrary semantic restrictions on the formulas.

As is common, these restrictions are given by the set of atomic constraints allowed to appear in the instances of the problem. We give a complete classification of the complexity of approximation with respect to this parameterization, applying methods from clone theory. Despite being classical, for NSol this step requires considerably more non-trivial work than for e.g. satisfiability problems. It turns out that our problem can either be solved in polynomial time, or it is complete for a well-known optimization class, or else it is equivalent to a well-known hard optimization problem.

Our study can be understood as a continuation of the minimization problems investigated by Khanna et al. in [10], especially that of MinOnes. The MinOnes optimization problem asks for a solution of a constraint satisfaction problem with minimal Hamming weight, i.e., minimal Hamming distance to the 0-vector. Our work generalizes this by allowing the given vector to be arbitrary.

Moreover, our work can also be seen as a generalization of questions in coding theory. Our problem NSol restricted to affine relations is the problem NearestCodeword of finding the nearest codeword to a given word, which is the basic operation when decoding messages received through a noisy channel. Thus our work can be seen as a generalization of these well-known problems from affine to general relations.

2   Preliminaries

An n-ary *Boolean relation* R is a subset of $\{0,1\}^n$; its elements $(b_1, \ldots, b_n)$ are also written as $b_1 \cdots b_n$. Let V be a set of variables. An *atomic constraint*, or an *atom*, is an expression $R(\boldsymbol{x})$, where R is an n-ary relation and $\boldsymbol{x}$ is an n-tuple of variables from V. Let Γ be a non-empty finite set of Boolean relations, also called a *constraint language*. A (conjunctive) Γ-*formula* is a finite conjunction of atoms $R_1(\boldsymbol{x_1}) \wedge \cdots \wedge R_k(\boldsymbol{x_k})$, where the R_i are relations from Γ and the $\boldsymbol{x_i}$ are variable tuples of suitable arity.

An *assignment* is a mapping $m \colon V \to \{0,1\}$ assigning a Boolean value $m(x)$ to each variable $x \in V$. If we arrange the variables in some arbitrary but fixed order, say as a vector $(x_1, \ldots, x_n)$, then the assignments can be identified with vectors from $\{0,1\}^n$. The i-th component of a vector m is denoted by $m[i]$ and

Table 1. List of Boolean functions and relations

$x \oplus y = x + y \pmod 2$	$\mathrm{or}^k = \{0,1\}^k \setminus \{0 \cdots 0\}$
$x \equiv y = x + y + 1 \pmod 2$	$\mathrm{nand}^k = \{0,1\}^k \setminus \{1 \cdots 1\}$
$\mathrm{dup}^3 = \{0,1\}^3 \setminus \{010, 101\}$	$\mathrm{even}^4 = \{(a_1, a_2, a_3, a_4) \in \{0,1\}^4 \mid \sum_{i=1}^4 a_i \text{ is even}\}$
$\mathrm{nae}^3 = \{0,1\}^3 \setminus \{000, 111\}$	

corresponds to the value of the i-th variable, i.e., $m[i] = m(x_i)$. The *Hamming weight* $\mathrm{hw}(m) = |\{i \mid m[i] = 1\}|$ of m is the number of 1s in the vector m. The *Hamming distance* $\mathrm{hd}(m, m') = |\{i \mid m[i] \neq m'[i]\}|$ of m and m' is the number of coordinates on which the vectors disagree. The complement $\overline{m}$ of a vector m is its pointwise complement, $\overline{m}[i] = 1 - m[i]$.

An assignment m satisfies a constraint $R(x_1, \ldots, x_n)$ if $(m(x_1), \ldots, m(x_n)) \in R$ holds. It satisfies the formula φ if it satisfies all of its atoms; m is said to be a model or solution of φ in this case. We use $[\varphi]$ to denote the set of models of φ. Note that $[\varphi]$ represents a Boolean relation. In sets of relations represented this way we usually omit the brackets. A *literal* is a variable v, or its negation $\neg v$. Assignments m are extended to literals by defining $m(\neg v) = 1 - m(v)$ (Table 1).

Throughout the text we refer to different types of Boolean constraint relations following Schaefer's terminology [11] (see also the monograph [8] and the survey [6]). A Boolean relation R is (1) *1-valid* if $1 \cdots 1 \in R$ and it is *0-valid* if $0 \cdots 0 \in R$, (2) *Horn* (*dual Horn*) if R can be represented by a formula in conjunctive normal form (CNF) having at most one unnegated (negated) variable in each clause, (3) *monotone* if it is both Horn and dual Horn, (4) *bijunctive* if it can be represented by a CNF having at most two variables in each clause, (5) *affine* if it can be represented by an affine system of equations $Ax = b$ over $\mathbb{Z}_2$, (6) *complementive* if for each $m \in R$ also $\overline{m} \in R$. A set Γ of Boolean relations is called 0-valid (1-valid, Horn, dual Horn, monotone, affine, bijunctive, complementive) if *every* relation in Γ is 0-valid (1-valid, Horn, dual Horn, monotone, affine, bijunctive, complementive). See also Table 3.

A formula constructed from atoms by conjunction, variable identification, and existential quantification is called a *primitive positive formula* (*pp-formula*). We denote by $\langle \Gamma \rangle$ the set of all relations that can be expressed using relations

Table 2. Some Boolean co-clones with bases

$\mathrm{iM}_2\{x \rightarrow y, \neg x, x\}$	$\mathrm{iD}_2\{x \oplus y, x \rightarrow y\}$	$\mathrm{iE}_2\{\neg x \vee \neg y \vee z, \neg x, x\}$
$\mathrm{iS}_0^k\{\mathrm{or}^k\}$	$\mathrm{iL}\{\mathrm{even}^4\}$	$\mathrm{iN}\{\mathrm{dup}^3\}$
$\mathrm{iS}_1^k\{\mathrm{nand}^k\}$	$\mathrm{iL}_2\{\mathrm{even}^4, \neg x, x\}$	$\mathrm{iN}_2\{\mathrm{nae}^3\}$
$\mathrm{iS}_{00}^k\{\mathrm{or}^k, x \rightarrow y, \neg x, x\}$	$\mathrm{iV}\{x \vee y \vee \neg z\}$	$\mathrm{iI}_0\{\mathrm{even}^4, x \rightarrow y, \neg x\}$
$\mathrm{iS}_{10}^k\{\mathrm{nand}^k, \neg x, x, x \rightarrow y\}$	$\mathrm{iV}_2\{x \vee y \vee \neg z, \neg x, x\}$	$\mathrm{iI}_1\{\mathrm{even}^4, x \rightarrow y, x\}$
$\mathrm{iD}_1\{x \oplus y, x\}$	$\mathrm{iE}\{\neg x \vee \neg y \vee z\}$	

from $\Gamma \cup \{=\}$, conjunction, variable identification (and permutation), cylindrification, and existential quantification. The set $\langle \Gamma \rangle$ is called the *co-clone* generated by Γ. A *base* of a co-clone $\mathcal{B}$ is a set of relations Γ, such that $\langle \Gamma \rangle = \mathcal{B}$. All co-clones, ordered by set inclusion, form a lattice. Together with their respective bases, which were studied in [7], some of them are listed in Table 2. In particular the sets of relations being 0-valid, 1-valid, complementive, Horn, dual Horn, affine, bijunctive, 2affine (both bijunctive and affine) and monotone each form a co-clone denoted by iI_0, iI_1, iN_2, iE_2, iV_2, iL_2, iD_2, iD_1, and iM_2, respectively. See also Table 3.

We assume that the reader has a basic knowledge of approximation algorithms and complexity theory, see e.g. [2,8]. For reductions among decision problems we use polynomial-time many-one reduction denoted by $\leq_m$. Many-one equivalence between decision problems is written as $\equiv_m$. For reductions among optimization problems we employ approximation preserving reductions, also called AP-reductions, represented by $\leq_{AP}$. AP-equivalence between optimization problems is stated as $\equiv_{AP}$. Moreover, we shall need the following approximation complexity classes in the hierarchy PO $\subseteq$ APX $\subseteq$ poly-APX $\subseteq$ NPO.

An optimization problem $\mathcal{P}_1$ AP-*reduces* to another optimization problem $\mathcal{P}_2$ if there are two polynomial-time computable functions f, g, and a constant $\alpha \geq 1$ such that for all $r > 1$ on any input x for $\mathcal{P}_1$ the following holds:

- $f(x)$ is an instance of $\mathcal{P}_2$;
- for any solution y of $f(x)$, $g(x,y)$ is a solution of x;
- whenever y is an r-approximate solution for the instance $f(x)$, then $g(x,y)$ provides a $(1 + (r-1)\alpha + o(1))$-approximate solution for x.

If $\mathcal{P}_1$ AP-reduces to $\mathcal{P}_2$ with constant $\alpha \geq 1$ and $\mathcal{P}_2$ has an $f(n)$-approximation algorithm, then there is an $\alpha f(n)$-approximation algorithm for $\mathcal{P}_1$.

We will relate our problems to well-known optimization problems. To this end we make the following convention: For optimization problems $\mathcal{P}$ and $\mathcal{Q}$ we say that $\mathcal{Q}$ is $\mathcal{P}$-*hard* if $\mathcal{P} \leq_{AP} \mathcal{Q}$, i.e. if $\mathcal{P}$ reduces to it. Moreover, $\mathcal{Q}$ is called $\mathcal{P}$-*complete* if $\mathcal{P} \equiv_{AP} \mathcal{Q}$.

To prove our results, we refer to the following optimization problems defined and analyzed in [10]. Like our problems they are parameterized by a constraint language Γ.

Problem MinOnes(Γ). Given a conjunctive formula φ over relations from Γ, any assignment m satisfying φ is a feasible solution. The goal is to minimize the Hamming weight hw(m).

Problem WeightedMinOnes(Γ). Given a conjunctive formula φ over relations in Γ and a weight function $w \colon V \to \mathbb{N}$ on the variables V of φ, solutions are again all assignments m satisfying φ. The objective is to minimize the value $\sum_{x \in V : m(x)=1} w(x)$.

We now define some well-studied problems to which we will relate our problems. Note that these problems do not depend on any parameter.

Problem NearestCodeword. Given a matrix $A \in \mathbb{Z}_2^{k \times l}$ and $m \in \mathbb{Z}_2^l$, any vector $x \in \mathbb{Z}_2^k$ is a solution. The objective is to minimize the Hamming distance hd(xA, m).

Problem MinHornDeletion. For a given conjunctive formula φ over relations from the set $\{[\neg x \vee \neg y \vee z], [x], [\neg x]\}$, an assignment m satisfying φ is sought. The objective is given by the minimum number of unsatisfied conjuncts of φ.

NearestCodeword and MinHornDeletion are known to be NP-hard to approximate within a factor $2^{\Omega(\log^{1-\varepsilon}(n))}$ for every $\varepsilon > 0$ [1,10]. Thus if a problem $\mathcal{P}$ is equivalent to any of these problems, it follows that $\mathcal{P} \notin$ APX unless P $=$ NP.

We also use the classic satisfiability problem SAT(Γ), given a conjunctive formula φ over relations from Γ, asking if φ is satisfiable. Schaefer presented in [11] a complete classification of complexity for SAT. His dichotomy theorem proves that SAT(Γ) is polynomial-time decidable if Γ is 0-valid ($\Gamma \subseteq$ iI$_0$), 1-valid ($\Gamma \subseteq$ iI$_1$), Horn ($\Gamma \subseteq$ iE$_2$), dual Horn ($\Gamma \subseteq$ iV$_2$), bijunctive ($\Gamma \subseteq$ iD$_2$), or affine ($\Gamma \subseteq$ iL$_2$); otherwise it is NP-complete.

3 Results

This section presents the formal definition of the considered problem, parameterized by a constraint language Γ, and our main result; the proofs follow in subsequent sections.

Problem NearestSolution(Γ), NSol(Γ)
Input: A conjunctive formula φ over relations from Γ and an assignment m of the variables occurring in φ, which is not required to satisfy φ.
Solution: An assignment m' satisfying φ (i.e. a codeword of the code described by φ).
Objective: Minimum Hamming distance hd(m, m').

Theorem 1 (illustrated in Fig. 1). *For a given Boolean constraint language Γ the optimization problem* NSol(Γ) *is*

(i) *in* PO *if Γ is*
    (a) *2affine ($\Gamma \subseteq$ iD$_1$) or*
    (b) *monotone ($\Gamma \subseteq$ iM$_2$);*
(ii) APX-*complete if*
    (a) $\langle \Gamma \rangle$ *contains the relation $[x \vee y]$ and $\Gamma \subseteq \langle x_1 \vee \cdots \vee x_k, x \rightarrow y, \neg x, x \rangle$ (iS$_0^2 \subseteq \langle \Gamma \rangle \subseteq$ iS$_{00}^k$) for some $k \in \mathbb{N}$, $k \geq 2$, or*
    (b) Γ *is bijunctive and $\langle \Gamma \rangle$ contains the relation $[x \vee y]$ (iS$_0^2 \subseteq \langle \Gamma \rangle \subseteq$ iD$_2$), or*
    (c) $\langle \Gamma \rangle$ *contains the relation $[\neg x \vee \neg y]$ and $\Gamma \subseteq \langle \neg x_1 \vee \cdots \vee \neg x_k, x \rightarrow y, \neg x, x \rangle$ (iS$_1^2 \subseteq \langle \Gamma \rangle \subseteq$ iS$_{10}^k$) for some $k \in \mathbb{N}$, $k \geq 2$, or*
    (d) Γ *is bijunctive and $\langle \Gamma \rangle$ contains the relation $[\neg x \vee \neg y]$ (iS$_1^2 \subseteq \langle \Gamma \rangle \subseteq$ iD$_2$);*
(iii) NearestCodeword-*complete if Γ is exactly affine (iL $\subseteq \langle \Gamma \rangle \subseteq$ iL$_2$);*
(iv) MinHornDeletion-*complete if Γ is*
    (a) *exactly Horn (iE $\subseteq \langle \Gamma \rangle \subseteq$ iE$_2$) or*
    (b) *exactly dual Horn (iV $\subseteq \langle \Gamma \rangle \subseteq$ iV$_2$);*
(v) *poly-*APX-*complete if Γ does not contain an affine relation and it is*
    (a) *either 0-valid (iN $\subseteq \langle \Gamma \rangle \subseteq$ iI$_0$) or*

(b) 1-valid (iN $\subseteq \langle \Gamma \rangle \subseteq$ iI$_1$); *and*
(vi) NPO-complete otherwise (iN$_2 \subseteq \langle \Gamma \rangle$).

The considered optimization problem can be transformed into a decision problem in the usual way. We add a bound $k \in \mathbb{N}$ to the input and ask if the Hamming distance satisfies the inequality $\text{hd}(m, m') \leq k$. This way we obtain the corresponding decision problem NSol^d. Its complexity follows immediately from the theorem above. All cases in PO become polynomial-time decidable, whereas the other cases, which are APX-hard, become NP-complete. This way we obtain a dichotomy theorem classifying the decision problem as polynomial or NP-complete for all finite sets Γ of Boolean relations.

4   Applicability of Clone Theory and Duality

We show that clone theory is applicable to the problem NSol, as well as a possibility to exploit inner symmetries between co-clones, which shortens several proofs as we continue.

There are two natural versions of $\text{NSol}(\Gamma)$. In one version the formula φ is quantifier free while in the other one we do allow existential quantification. We call the former version $\text{NSol}(\Gamma)$ and the latter $\text{NSol}_{\text{pp}}(\Gamma)$. Fortunately, we will now see that both versions are equivalent.

Let $\text{NSol}^d(\Gamma)$ and $\text{NSol}^d_{\text{pp}}(\Gamma)$ be the decision problems corresponding to $\text{NSol}(\Gamma)$ and $\text{NSol}_{\text{pp}}(\Gamma)$, asking whether there is a satisfying assignment within a given bound.

Lemma 2. *For finite sets Γ we have the equivalences* $\text{NSol}^d(\Gamma) \equiv_m \text{NSol}^d_{\text{pp}}(\Gamma)$ *and* $\text{NSol}(\Gamma) \equiv_{\text{AP}} \text{NSol}_{\text{pp}}(\Gamma)$.

Proof. The reduction from left to right is trivial in both cases. For the other direction, consider first an instance with formula φ, assignment m, and bound k for $\text{NSol}^d_{\text{pp}}(\Gamma)$. Let $x_1, \ldots, x_n$ be the free variables of φ and let $y_1, \ldots, y_\ell$ be the existentially quantified variables, which can be assumed to be disjoint. For each variable z we define a set $B(z)$ as follows:

$$B(z) = \begin{cases} \{x_i^j \mid j \in \{1, \ldots, (n + \ell + 1)^2\}\} & \text{if } z = x_i \text{ for some } i \in \{1, \ldots, n\}, \\ \{y_i\} & \text{if } z = y_i \text{ for some } i \in \{1, \ldots, \ell\}. \end{cases}$$

We construct a quantifier-free formula φ' over the variables $\bigcup_{i=1}^n B(x_i) \cup \bigcup_{i=1}^\ell B(y_i)$ that contains for every atom $R(z_1, \ldots, z_s)$, from φ the atom $R(z'_1, \ldots, z'_s)$ for every combination $(z'_1, \ldots, z'_s)$ from $B(z_1) \times \cdots \times B(z_s)$. Moreover, we construct an assignment $B(m)$ of φ' by assigning to every variable x_i^j the value $m(x_i)$ and to y_i the value 0. Note that because there is an upper bound on the arities of relations from Γ, this is a polynomial time construction.

We claim that φ has a solution m' with $\text{hd}(m, m') \leq k$ if and only if φ' has a solution m'' with $\text{hd}(B(m), m'') \leq k(n+\ell+1)^2+\ell$. First, observe that if m' with the desired properties exists, then there is an extension m'_e of m' to the y_i that satisfies all atoms. Define m'' by setting $m''(x_i^j) := m'(x_i)$ and $m''(y_i) := m_e'(y_i)$ for all i and j. Then m'' is clearly a satisfying assignment of φ'. Moreover, m'' and $B(m)$ differ in at most $k(n+\ell+1)^2$ variables among the x_i^j. Since there exist only ℓ other variables y_i, we get $\text{hd}(m'', B(m)) \leq k(n + \ell + 1)^2 + \ell$ as desired.

Now suppose m'' satisfies φ' with $\text{hd}(B(m), m'') \leq k(n+\ell+1)^2 + \ell$. We may assume for each i that $m''(x_i^1) = \cdots = m''(x_i^{(n+\ell+1)^2})$. Indeed, if this is not the case, then setting all x_i^j to $B(m)(x_i^j) = m(x_i)$ will give us a satisfying assignment closer to $B(m)$. After at most n iterations we get some m'' as desired. Now define an assignment m' to φ by setting $m'(x_i) := m''(x_i^1)$. Then m' satisfies φ, because the variables y_i can be assigned values as in m''. Moreover, whenever $m(x_i)$ differs from $m'(x_i)$, the inequality $B(m)(x_i^j) \neq m''(x_i^j)$ holds for every j. Thus we obtain $(n + \ell + 1)^2 \text{hd}(m, m') \leq \text{hd}(B(m), m'') \leq k(n + \ell + 1)^2 + \ell$. Therefore, we have the inequality $\text{hd}(m, m') \leq k + \ell/(n + \ell + 1)^2$ and hence $\text{hd}(m, m') \leq k$. This completes the many-one reduction.

We claim that the above construction is an AP-reduction, too. To this end, let m'' be an r-approximation for φ' and $B(m)$, i.e., $\text{hd}(B(m), m'') \leq r \cdot \text{OPT}(\varphi', B(m))$. Construct m' as before, so $(n + \ell + 1)^2 \text{hd}(m, m') \leq \text{hd}(B(m), m'') \leq r \cdot \text{OPT}(\varphi', B(m))$. Since $\text{OPT}(\varphi', B(m))$ is at most $(n+\ell+1)^2 \text{OPT}(\varphi, m) + \ell$ as before, we get $(n + \ell + 1)^2 \text{hd}(m, m') \leq r((n + \ell + 1)^2 \text{OPT}(\varphi, m) + \ell)$. This implies the inequality $\text{hd}(m, m') \leq r \cdot \text{OPT}(\varphi, m) + r \cdot \ell/(n + \ell + 1)^2 = (r + o(1)) \cdot \text{OPT}(\varphi, m)$ and shows that the construction is an AP-reduction with $\alpha = 1$. □

Remark 3. Note that in the reduction from $\text{NSol}_{\text{pp}}^{\text{d}}(\Gamma)$ to $\text{NSol}^{\text{d}}(\Gamma)$ we construct the assignment $B(m)$ as an extension of m by setting all new variables to 0. In particular, if m is the constant 0-assignment, then so is $B(m)$. We use this observation as we continue.

We can also show that introducing explicit equality constraints does not change the complexity of our problem.

Lemma 4. *For constraint languages Γ we have $\text{NSol}^{\text{d}}(\Gamma) \equiv_{\text{m}} \text{NSol}^{\text{d}}(\Gamma \cup \{=\})$ and $\text{NSol}(\Gamma) \equiv_{\text{AP}} \text{NSol}(\Gamma \cup \{=\})$.*

Although a proof of this statement can be established by similar methods as those used in Lemma 2, it is a technically rather involved case distinction whose length exceeds the scope of this presentation. The proof is therefore omitted.

Lemmas 2 and 4 are very convenient, because they allow us to freely switch between formulas with quantifiers and equality and those without. This allows us to give all upper bounds in the setting without quantifiers and equality while freely using them in all hardness reductions. In particular it follows that we can use pp-definability when implementing a constraint language Γ by another constraint language Γ'. Hence it suffices to consider Post's lattice of co-clones to characterize the complexity of $\text{NSol}(\Gamma)$ for every finite set of Boolean relations Γ.

Corollary 5. *For constraint languages Γ, Γ' such that $\Gamma' \subseteq \langle \Gamma \rangle$, we have the reductions $\text{NSol}^{\text{d}}(\Gamma') \leq_{\text{m}} \text{NSol}^{\text{d}}(\Gamma)$ and $\text{NSol}(\Gamma') \leq_{\text{AP}} \text{NSol}(\Gamma)$. Thus, if*

$\langle \Gamma' \rangle = \langle \Gamma \rangle$ *is satisfied, then the equivalences* $\mathsf{NSol}^d(\Gamma) \equiv_m \mathsf{NSol}^d(\Gamma')$ *and* $\mathsf{NSol}(\Gamma) \equiv_{\mathrm{AP}} \mathsf{NSol}(\Gamma')$ *hold.*

Next we prove that, in certain cases, unit clauses in the formula do not change the complexity of NSol.

Lemma 6. *We have the equivalence* $\mathsf{NSol}(\Gamma) \equiv_{\mathrm{AP}} \mathsf{NSol}(\Gamma \cup \{[x], [\neg x]\})$ *for any constraint language* Γ *where the problem of finding feasible solutions of* $\mathsf{NSol}(\Gamma)$ *is polynomial-time decidable.*

Proof. The direction from left to right is obvious. For the other direction, we show an AP-reduction from $\mathsf{NSol}(\Gamma \cup \{[x], [\neg x]\})$ to $\mathsf{NSol}(\Gamma \cup \{[x \equiv y]\})$. Since $[x \equiv y]$ is by definition in every co-clone and thus in $\langle \Gamma \rangle$, the result follows from Corollary 5.

The idea of the construction is to introduce two sets of variables $y_1, \ldots, y_{n^2}$ and $z_1, \ldots, z_{n^2}$ such that in any feasible solution all y_i and all z_i take the same value. Then setting $m(y_i) = 1$ and $m(z_i) = 0$ for each i, any feasible solution m' of small Hamming distance to m will have $m'(y_i) = 1$ and $m'(z_i) = 0$ for all i as well, because deviating from this would be prohibitively expensive. Finally, we simulate unary relations x and $\neg x$ by $x \equiv y_1$ and $x \equiv z_1$, respectively. We now describe the reduction formally.

Let the formula φ and the assignment m be a $\Gamma \cup \{[x], [\neg x]\}$-formula over the variables $x_1, \ldots, x_n$ with a feasible solution. We construct a $\Gamma \cup \{[x \equiv y]\}$-formula φ' over the variables $x_1, \ldots x_n, y_1, \ldots, y_{n^2}, z_1, \ldots, z_{n^2}$ and an assignment m'. We get φ' from φ by substituting every occurrence of a constraint $[x_i]$ for some variable x_i by $x_i \equiv y_1$ and substituting every occurrence $[\neg x_i]$ for every variable x_i by $x_i \equiv z_1$. Finally, add $y_i \equiv y_j$ for all $i, j \in \{1, \ldots, n^2\}$ and $z_i \equiv z_j$ for all $i, j \in \{1, \ldots, n^2\}$. Let m' be the assignment of the variables of φ' given by $m'(x_i) = m(x_i)$ for each $i \in \{1, \ldots, n\}$, and $m'(y_i) = 1$ and $m'(z_i) = 0$ for all $i \in \{1, \ldots, n^2\}$. To any feasible solution m'' of φ' we assign $g(\varphi, m, m'')$ as follows.

1. If φ is satisfied by m, we define $g(\varphi, m, m'')$ to be equal to m.
2. Else if $m''(y_i) = 0$ holds for all $i \in \{1, \ldots, n^2\}$ or $m''(z_i) = 1$ for all i in $\{1, \ldots, n^2\}$, we define $g(\varphi, m, m'')$ to be any satisfying assignment of φ.
3. Otherwise, $m''(y_i) = 1$ for all $i \in \{1, \ldots, n^2\}$ and $m''(z_i) = 0$, we define $g(\varphi, m, m'')$ to be the restriction of m'' onto $x_1, \ldots, x_n$.

Observe that all variables y_i and all z_i are forced to take the same value in any feasible solution, respectively, so $g(\varphi, m, m'')$ is always well-defined. The construction is an AP-reduction. Assume that m'' is an r-approximate solution. We will show that $g(\varphi, m, m'')$ is also an r-approximate solution.

Case 1: $g(\varphi, m, m'')$ computes the optimal solution, so there is nothing to show.

Case 2: Observe first that φ has a solution by assumption, so $g(\varphi, m, m'')$ is well-defined and feasible by construction. Observe that m' and m'' disagree on all y_i or on all z_i, so $\mathrm{hd}(m', m'') \geq n^2$ holds. Moreover, since φ has a feasible solution, it follows that $\mathrm{OPT}(\varphi', m') \leq n$. Since m'' is an r-approximate solution, we have that $r \geq \mathrm{hd}(m', m'')/\mathrm{OPT}(\varphi', m') \geq n$. Consequently, the distance $\mathrm{hd}(m, g(\varphi, m, m''))$ is bounded above by $n \leq r \leq r \cdot \mathrm{OPT}(\varphi, m)$, where the last inequality holds because φ is not satisfied by m and thus the distance of the optimal solution from m is at least 1.

Case 3: The variables x_i for which $[x_i]$ is a constraint all have $g(\varphi, m, m'')(x_i) = 1$ by construction. Moreover, we have $g(\varphi, m, m'')(x_i) = 0$ for all x_i for which $[\neg x_i]$ is a constraint of φ. Consequently, $g(\varphi, m, m'')$ is feasible. Again, $\mathrm{OPT}(\varphi', m') \leq n$, so the optimal solution to (φ', m') must set all variables y_i to 1 and all z_i to 0. It follows that $\mathrm{OPT}(\varphi, m) = \mathrm{OPT}(\varphi', m')$. Thus we get

$$\mathrm{hd}(m, g(\varphi, m, m'')) = \mathrm{hd}(m', m'') \leq r \cdot \mathrm{OPT}(\varphi', m') = r \cdot \mathrm{OPT}(\varphi, m),$$

which completes the proof.   □

Given a relation $R \subseteq \{0, 1\}^n$, its *dual* relation is $\mathrm{dual}(R) = \{\overline{m} \mid m \in R\}$, i.e., the relation containing the complements of vectors from R. Duality naturally extends to sets of relations and co-clones. We define $\mathrm{dual}(\Gamma) = \{\mathrm{dual}(R) \mid R \in \Gamma\}$ as the set of dual relations to Γ. Duality is a symmetric relation. If a relation R' (a set of relations Γ') is a dual relation to R (a set of dual relations to Γ), then R (Γ) is also dual to R' (to Γ'). By a simple inspection of the bases of co-clones in Table 2, we can easily see that many co-clones are dual to each other. For instance iE_2 is dual to iV_2. The following lemma shows that it is sufficient to consider only one half of Post's lattice of co-clones.

Lemma 7. *For any set Γ of Boolean relations we have* $\mathsf{NSol}^{\mathrm{d}}(\Gamma) \equiv_{\mathrm{m}} \mathsf{NSol}^{\mathrm{d}}(\mathrm{dual}(\Gamma))$ *and* $\mathsf{NSol}(\Gamma) \equiv_{\mathrm{AP}} \mathsf{NSol}(\mathrm{dual}(\Gamma))$.

Proof. Let φ be a Γ-formula and m an assignment to φ. We construct a $\mathrm{dual}(\Gamma)$-formula φ' by substitution of every atom $R(\boldsymbol{x})$ by $\mathrm{dual}(R)(\boldsymbol{x})$. The assignment m satisfies φ if and only if $\overline{m}$ satisfies φ', where $\overline{m}$ is the complement of m. Moreover, $\mathrm{hd}(m, m') = \mathrm{hd}(\overline{m}, \overline{m'})$.   □

5   Finding the Nearest Solution

This section contains the proof of Theorem 1. We first consider the polynomial-time cases followed by the cases of higher complexity.

5.1   Polynomial-Time Cases

Proposition 8.   *If a constraint language Γ is both bijunctive and affine ($\Gamma \subseteq \mathrm{iD}_1$), then $\mathsf{NSol}(\Gamma)$ can be solved in polynomial time.*

Proof. Since $\Gamma \subseteq \mathrm{iD}_1 = \langle \Gamma' \rangle$ with $\Gamma' := \{[x \oplus y], [x]\}$, we have the reduction $\mathsf{NSol}(\Gamma) \leq_{\mathrm{AP}} \mathsf{NSol}(\Gamma')$ by Corollary 5. Every Γ'-formula φ is equivalent to a linear system of equations over the Boolean ring $\mathbb{Z}_2$ of the type $x \oplus y = 1$ and $x = 1$. Substitute the fixed values $x = 1$ into the equations of the type $x \oplus y = 1$

and propagate. After an exhaustive application of this rule only equations of the form $x \oplus y = 1$ remain. For each of them put an edge $\{x, y\}$ into E, defining an undirected graph $G = (V, E)$, whose vertices V are the unassigned variables. If G is not bipartite, then φ has no solutions, so we can reject the input. Otherwise, compute a bipartition $V = L \dot{\cup} R$. We assume that G is connected; if not perform the following algorithm for each connected component. Assign the value 0 to each variable in L and the value 1 to each variable in R, giving the satisfying assignment m_1. Swapping the roles of 0 and 1 w.r.t. L and R we get a model m_2. Return $\min\{\text{hd}(m, m_1), \text{hd}(m, m_2)\}$.   □

Proposition 9. *If a constraint language Γ is monotone ($\Gamma \subseteq$ iM$_2$), then* NSol(Γ) *can be solved in polynomial time.*

Proof. We have iM$_2 = \langle \Gamma' \rangle$ where $\Gamma' := \{[x \to y], [\neg x], [x]\}$. Thus Corollary 5 and $\Gamma \subseteq \langle \Gamma' \rangle$ imply NSol(Γ) $\leq_{AP}$ NSol(Γ'). The relations $[\neg x]$ and $[x]$ determine the unique value of the considered variable, therefore we can eliminate the unit clauses built from the two latter relations and propagate. We consider formulas φ built only from the relation $[x \to y]$, i.e., formulas containing only binary implicative clauses of the type $x \to y$.

Let V the set of variables of the formula φ. According to the value assigned to the variables by the vector m, we can divide V into two disjoint subsets V_0 and V_1, such that $V_i = \{x \in V \mid m(x) = i\}$. We transform the formula φ to an integer programming problem P. First, for each clause $x \to y$ from φ we add to P the relation $y \geq x$. For each variable $x \in V$ we add the constraints $x \geq 0$ and $x \leq 1$, with $x \in \{0, 1\}$. Finally, we construct the linear function f_φ by defining

$$f_\varphi(m') = \sum_{x_i \in V_0} m'(x_i) + \sum_{x_j \in V_1} (1 - m'(x_j))$$

for assignments m' of φ. Obviously, $f_\varphi(m')$ counts the number of variables changing their parity between m and m', i.e., $f_\varphi(m') = \text{hd}(m, m')$. As P is totally unimodular, the minimum of f_φ can be computed in polynomial time (see e.g. [12]).   □

5.2   Hard Cases

We start off with an easy corollary of Schaefer's dichotomy.

Lemma 10. *Let Γ be a finite set of Boolean relations. If* iN$_2 \subseteq \langle \Gamma \rangle$, *then* NSol($\Gamma$) *is* NPO-complete; *otherwise,* NSol(Γ) $\in$ poly-APX.

Proof. If iN$_2 \subseteq \langle \Gamma \rangle$ holds, finding a solution for NSol(Γ) is NP-hard by Schaefer's theorem [11], hence NSol(Γ) is NPO-complete.

We give an n-approximation algorithm for the other case. Let a formula φ and a model m be an instance of NSol(Γ). If m is a solution of φ, return m. Otherwise, compute an arbitrary solution m' of φ, which is possible by Schaefer's theorem, and return m'.

The approximation ratio of this algorithm is n. Indeed, if m satisfies φ, this is obviously true, because we return the exact solution. Otherwise, we have $\text{OPT}(\varphi, m) \geq 1$ and so, trivially, $\text{hd}(m, m') \leq n$ whence the claim follows. $\qquad\square$

We start with reductions from the optimization version of vertex cover. Since the relation $[x \vee y]$ is a straightforward Boolean encoding of vertex cover, we immediately get the following result.

Proposition 11. $\mathsf{NSol}(\Gamma)$ *is APX-hard for every constraint language* Γ *satisfying the inclusion* $\text{iS}_0^2 \subseteq \langle \Gamma \rangle$ *or* $\text{iS}_1^2 \subseteq \langle \Gamma \rangle$.

Proof. We have $\text{iS}_0^2 = \langle \{[x \vee y]\} \rangle$, whereas $\text{iS}_1^2 = \langle \{[\neg x \vee \neg y]\} \rangle$. So we discuss the former case, the latter one being symmetric and provable from the first one by Corollary 5.

We encode VertexCover into $\mathsf{NSol}(\{[x \vee y]\}) \leq_{AP} \mathsf{NSol}(\Gamma)$ (see Corollary 5). For each edge $\{x, y\} \in E$ of a graph $G = (V, E)$ we add the clause $(x \vee y)$ to the formula φ_G. Every model m' of φ_G yields a vertex cover $\{v \in V \mid m'(v) = 1\}$, and conversely, the characteristic function of any vertex cover satisfies φ_G. Taking $m = \mathbf{0}$, then $\text{hd}(\mathbf{0}, m')$ is minimal if and only if the number of 1s in m' is minimal, i.e., if m' is a minimal model of φ_G, i.e., if m' represents a minimal vertex cover of G. Since VertexCover is APX-complete (see e.g. [2]), the result follows. $\qquad\square$

Proposition 12. *We have* $\mathsf{NSol}(\Gamma) \in \text{APX}$ *for constraint languages* $\Gamma \subseteq \text{iD}_2$.

Proof. As $\{x \oplus y, x \rightarrow z\}$ is a basis of iD_2, it suffices to show that $\mathsf{NSol}(\{x \oplus y, x \rightarrow y\})$ is in APX by Corollary 5. Let (φ, m) be an input of this problem. Feasibility for φ can be written as an integer program as follows: Every constraint $x_i \oplus x_j$ induces a linear equation $x_i + x_j = 1$. Every constraint $x_i \rightarrow x_j$ can be written as $x_i \leq x_j$. If we restrict all variables to $\{0, 1\}$ by the appropriate inequalities, it is clear that any assignment m' satisfies φ if it satisfies the linear system with inequality side conditions. We complete the construction of the linear program by adding the objective function $c(m') := \sum_{i:m(x_i)=0} m'(x_i) + \sum_{i:m(x_i)=1}(1 - m'(x_i))$. Clearly, for every m' we have $c(m') = \text{hd}(m, m')$. The 2-approximation algorithm from [9] for integer linear programs, in which in every inequality at most two variables appear, completes the proof. $\qquad\square$

Proposition 13. *We have* $\mathsf{NSol}(\Gamma) \in \text{APX}$ *for constraint languages* $\Gamma \subseteq \text{iS}_{00}^\ell$ *with* $\ell \geq 2$.

Proof. Due to $\{x_1 \vee \cdots \vee x_\ell, x \rightarrow y, \neg x, x\}$ being a basis of iS_{00}^ℓ and Corollary 5, it suffices to show $\mathsf{NSol}(\{x_1 \vee \cdots \vee x_\ell, x \rightarrow y, \neg x, x\}) \in \text{APX}$. Let formula φ and assignment m be an instance of that problem. We will use an approach similar to that for the corresponding case in [10], again writing φ as an integer program. Every constraint $x_{i_1} \vee \cdots \vee x_{i_\ell}$ is translated to an inequality $x_{i_1} + \cdots + x_{i_\ell} \geq 1$. Every constraint $x_i \rightarrow x_j$ is written as $x_i \leq x_j$. Each $\neg x_i$ is turned into $x_i = 0$, every constraint x_i yields $x_i = 1$. Add $x_i \geq 0$ and

$x_i \leq 1$ for each variable x_i. Again, it is easy to check that feasible Boolean solutions of φ and the linear system coincide. Defining again the objective function $c(m') = \sum_{i:m(x_i)=0} m'(x_i) + \sum_{i:m(x_i)=1}(1 - m'(x_i))$, we have $\mathrm{hd}(m, m') = c(m')$ for every m'. Therefore it suffices to approximate the optimal solution for the linear program.

To this end, let m'' be a (generally non-integer) solution to the relaxation of the linear program which can be computed in polynomial time. We construct m' by setting $m'(x_i) = 0$ if $m''(x_i) < 1/\ell$ and $m'(x_i) = 1$ if $m''(x_i) \geq 1/\ell$. As $\ell \geq 2$, we get $\mathrm{hd}(m, m') = c(m') \leq \ell c(m'') \leq \ell \cdot \mathrm{OPT}(\varphi, m)$. It is easy to check that m' is a feasible solution, which completes the proof. $\qquad\square$

Lemma 14. *We have* $\mathsf{MinOnes}(\Gamma) \leq_{\mathrm{AP}} \mathsf{NSol}(\Gamma)$ *for any constraint language* Γ.

Proof. $\mathsf{MinOnes}(\Gamma)$ is a special case of $\mathsf{NSol}(\Gamma)$ where m is the constant $\mathbf{0}$-assignment. $\qquad\square$

Proposition 15 (Khanna et al. [10, Theorem 2.14]). *The problem* $\mathsf{MinOnes}(\Gamma)$ *is* NearestCodeword-*complete for constraint languages* Γ *satisfying* $\langle \Gamma \rangle = \mathrm{iL}_2$.

Corollary 16. *For a constraint language* Γ *satisfying* $\mathrm{iL} \subseteq \langle \Gamma \rangle$, *the problem* $\mathsf{NSol}(\Gamma)$ *is* NearestCodeword-*hard.*

Proof. Let $\Gamma' := \{\mathrm{even}^4, [x], [\neg x]\}$. Since $\langle \Gamma' \rangle = \mathrm{iL}_2$, NearestCodeword is equivalent to $\mathsf{MinOnes}(\Gamma')$, which reduces to $\mathsf{NSol}(\Gamma')$ by Lemma 14. We have now the AP-equivalence $\mathsf{NSol}(\Gamma') \equiv_{\mathrm{AP}} \mathsf{NSol}(\{\mathrm{even}^4\})$ by appealing to Lemma 6 and the reduction $\mathsf{NSol}(\{\mathrm{even}^4\}) \leq_{\mathrm{AP}} \mathsf{NSol}(\Gamma)$ due to $\mathrm{even}^4 \in \mathrm{iL} \subseteq \langle \Gamma \rangle$ and Corollary 5. $\qquad\square$

Proposition 17. *We have* $\mathsf{NSol}(\Gamma) \leq_{\mathrm{AP}} \mathsf{MinOnes}(\{\mathrm{even}^4, [\neg x], [x]\})$ *for any constraint language* $\Gamma \subseteq \mathrm{iL}_2$.

Proof. The set $\Gamma' := \{\mathrm{even}^4, [\neg x], [x]\}$ is a basis of iL_2, therefore by Corollary 5 it is sufficient to show $\mathsf{NSol}(\Gamma') \leq_{\mathrm{AP}} \mathsf{MinOnes}(\Gamma')$.

We proceed by reducing $\mathsf{NSol}(\Gamma')$ to a subproblem of $\mathsf{NSol}_{\mathrm{pp}}(\Gamma')$, where only instances $(\varphi, \mathbf{0})$ are considered. Then, using Lemma 2 and Remark 3, this reduces to a subproblem of $\mathsf{NSol}(\Gamma')$ with the same restriction on the assignments, which is exactly $\mathsf{MinOnes}(\Gamma')$. Note that $[x \oplus y]$ is equal to $[\exists z \exists z'(\mathrm{even}^4(x, y, z, z') \wedge \neg z \wedge z']$ so we can freely use $[x \oplus y]$ in any Γ'-formula. Let formula φ and assignment m be an instance of $\mathsf{NSol}(\Gamma')$. We copy all clauses of φ to φ'. For each variable x of φ for which $m(x) = 1$, we take a new variable x' and add the constraint $x \oplus x'$ to φ'. Moreover, we existentially quantify x. Clearly, there is a bijection I between the satisfying assignments of φ and those of φ': For every solution s of φ we get a solution $I(s)$ of φ' by setting for each x' introduced in the construction of φ' the value $I(s)(x')$ to the complement of $m(x)$. Moreover, we have that $\mathrm{hd}(m, s) = \mathrm{hd}(\mathbf{0}, I(s))$. This yields a trivial AP-reduction with $\alpha = 1$. $\qquad\square$

Proposition 18 (Khanna et al. [10]). *The problems* $\mathsf{MinOnes}(\{x \vee y \vee \neg z, x, \neg x\})$ *and* $\mathsf{WeightedMinOnes}(\{x \vee y \vee \neg z, x \vee y\})$ *are* $\mathsf{MinHornDeletion}$-*complete.*

Lemma 19. $\mathsf{NSol}(\{x \vee y \vee \neg z\}) \leq_{\mathrm{AP}} \mathsf{WeightedMinOnes}(\{x \vee y \vee \neg z, x \vee y\})$.

Proof. Let formula φ and assignment m be an instance of $\mathsf{NSol}(x \vee y \vee \neg z)$ over the variables $x_1, \ldots, x_n$. If m satisfies φ then the reduction is trivial. We assume in the remainder of the proof that $\mathrm{OPT}(\varphi, m) > 0$. Let $T(m)$ be the set of variables x_i with $m(x_i) = 1$. We construct a $\{x \vee y \vee \neg z, x \vee y\}$-formula from φ by adding for each $x_i \in T(m)$ the constraint $x_i \vee x_i'$ where x_i' is a new variable. We set the weights of the variables of φ' as follows. For $x_i \in T(m)$ we set $w(x_i) = 0$, all other variables get weight 1. To each satisfying assignment m' of φ' we construct the assignment m'' which is the restriction of m' to the variables of φ. This construction is an AP-reduction.

Note that m'' is feasible if m' is. Let m' be an r-approximation of $\mathrm{OPT}(\varphi')$. Note that whenever for $x_i \in T(m)$ we have $m'(x_i) = 0$ then $m'(x_i') = 1$. The other way round, we may assume that whenever $m'(x_i) = 1$ for $x_i \in T(m)$ then $m'(x_i') = 0$. If this is not the case, then we can change m' accordingly, decreasing the weight that way. It follows that $w(m') = n_0 + n_1$ where we have

$$n_0 = |\{i \mid x_i \in T(m), m'(x_i) = 0\}| = |\{i \mid x_i \in T(m), m'(x_i) \neq m(x_i)\}|$$
$$n_1 = |\{i \mid x_i \notin T(m), m'(x_i) = 1\}| = |\{i \mid x_i \notin T(m), m'(x_i) \neq m(x_i)\}|,$$

which means that $w(m')$ equals $\mathrm{hd}(m, m'')$. Analogously, the optima in both problems correspond, that is we have $\mathrm{OPT}(\varphi') = \mathrm{OPT}(\varphi, m)$. From this we deduce the final inequality $\mathrm{hd}(m, m'')/\mathrm{OPT}(\varphi, m) = w(m')/\mathrm{OPT}(\varphi') \leq r$. □

Table 3. Sets of Boolean relations with their names determined by co-clone inclusions

$\Gamma \subseteq \mathrm{iI}_0 \Leftrightarrow \Gamma$ is 0-valid	$\Gamma \subseteq \mathrm{iI}_1 \Leftrightarrow \Gamma$ is 1-valid
$\Gamma \subseteq \mathrm{iE}_2 \Leftrightarrow \Gamma$ is Horn	$\Gamma \subseteq \mathrm{iV}_2 \Leftrightarrow \Gamma$ is dual Horn
$\Gamma \subseteq \mathrm{iM}_2 \Leftrightarrow \Gamma$ is monotone	$\Gamma \subseteq \mathrm{iD}_2 \Leftrightarrow \Gamma$ is bijunctive
$\Gamma \subseteq \mathrm{iL}_2 \Leftrightarrow \Gamma$ is affine	$\Gamma \subseteq \mathrm{iD}_1 \Leftrightarrow \Gamma$ is 2affine
$\Gamma \subseteq \mathrm{iN}_2 \Leftrightarrow \Gamma$ is complementive	$\Gamma \subseteq \mathrm{iI} \Leftrightarrow \Gamma$ is both 0- and 1-valid

Proposition 20. *For every dual Horn constraint language* $\Gamma \subseteq \mathrm{iV}_2$ *we have the reduction* $\mathsf{NSol}(\Gamma) \leq_{\mathrm{AP}} \mathsf{WeightedMinOnes}(\{x \vee y \vee \neg z, x \vee y\})$.

Proof. Since $\{x \vee y \vee \neg z, x, \neg x\}$ is a basis of iV_2, by Corollary 5 it suffices to prove the reduction $\mathsf{NSol}(\{x \vee y \vee \neg z, x, \neg x\}) \leq_{\mathrm{AP}} \mathsf{WeightedMinOnes}(\{x \vee y \vee \neg z, x \vee y\})$. To this end, first reduce $\mathsf{NSol}(\{x \vee y \vee \neg z, x, \neg x\})$ to $\mathsf{NSol}(x \vee y \vee \neg z)$ by Lemma 6 and then use Lemma 19. □

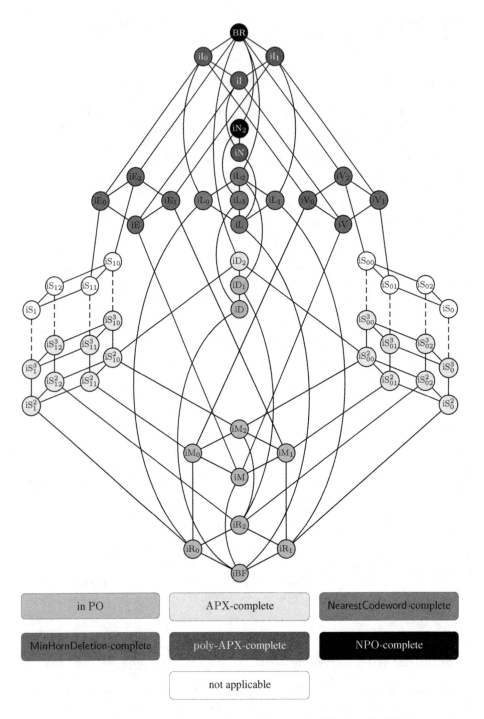

Fig. 1. Lattice of coclones with complexity classification for NSol

Proposition 21. $\mathsf{NSol}(\Gamma)$ *is* $\mathsf{MinHornDeletion}$-*hard for finite* Γ *with* $\mathrm{iV}_2 \subseteq \langle\Gamma\rangle$.

Proof. For $\Gamma' := \{x \vee y \vee \neg z, x, \neg x\}$ we have $\mathsf{MinHornDeletion} \equiv_{\mathrm{AP}} \mathsf{MinOnes}(\Gamma')$ by Proposition 18. Now it follows $\mathsf{MinOnes}(\Gamma') \leq_{\mathrm{AP}} \mathsf{NSol}(\Gamma') \leq_{\mathrm{AP}} \mathsf{NSol}(\Gamma)$ using Lemma 14 and Corollary 5 on the assumption $\Gamma' \subseteq \mathrm{iV}_2 \subseteq \langle\Gamma\rangle$. □

Proposition 22. *The problem* $\mathsf{NSol}(\Gamma)$ *is poly-APX-hard for constraint languages* Γ *verifying* $\mathrm{iN} \subseteq \langle\Gamma\rangle$.

Proof. The constraint language $\Gamma_1 := \{\mathrm{even}^4, x \rightarrow y, x\}$ is a base of iI_1. $\mathsf{MinOnes}(\Gamma_1)$ is poly-APX-hard by Theorem 2.14 of [10] and reduces to $\mathsf{NSol}(\Gamma_1)$ by Lemma 14. Since $(x \rightarrow y) = \mathrm{dup}^3(0, x, y) = \exists z(\mathrm{dup}^3(z, x, y) \wedge \neg z)$, we have the reductions $\mathsf{NSol}(\Gamma_1) \leq_{\mathrm{AP}} \mathsf{NSol}(\Gamma_1 \cup \{\neg x, \mathrm{dup}^3\}) \leq_{\mathrm{AP}} \mathsf{NSol}(\{\mathrm{even}^4, \mathrm{dup}^3, x, \neg x\})$ by Corollary 5. Lemma 6 implies $\mathsf{NSol}(\{\mathrm{even}^4, \mathrm{dup}^3, x, \neg x\}) \equiv_{\mathrm{AP}} \mathsf{NSol}(\{\mathrm{even}^4, \mathrm{dup}^3\})$; the latter problem reduces to $\mathsf{NSol}(\Gamma)$ because of $\{\mathrm{even}^4, \mathrm{dup}^3\} \subseteq \mathrm{iN} \subseteq \langle\Gamma\rangle$ and Corollary 5 □

6   Concluding Remarks

Considering the optimization problem NSol is part of a more general research program (cf. [4,5]) studying the approximation complexity of Boolean constraint satisfaction problems in connection with Hamming distance. The studied problems fundamentally differ in the resulting complexity classification as well as in the methods applicable to them (e.g. stability under pp-definitions and applicability of classical Galois theory for Boolean clones vs. the need for minimal weak bases for weak co-clones).

The problem NSol is in PO for constraints, which are both bijunctive and affine, or both Horn and dual Horn (also called monotone). In the interval of constraint languages starting from those encoding vertex cover up to those encoding hitting set for fixed arity hypergraphs or up to bijunctive constraints, NSol becomes APX-complete. This indicates that the solution structure for these types of constraints is more complex, and it becomes even more complicated for Horn or dual Horn constraints. The next complexity stage of the solution structure is characterized by affine constraints. In fact, these represent the error correcting codes used in real-word applications. Even if we know that the given assignment satisfies the constraint – contrary to the real-word situation in the case of nearest neighbor decoding – the optimization problem NSol is surprisingly equivalent to the one of finding the nearest codeword. The penultimate stage of solution structure complexity is given by 0-valid or 1-valid constraint languages, where one finds poly-APX-completeness. This implies that we cannot get a suitable approximation for these problems. It is implicit in NSol to check for the existence of at least one solution. For the last case, when the constraint language is equivalent to NAESAT, this is hard, where membership in iN_2 implies intractability of the SAT problem. Hence, a polynomial-time approximation is not possible at all.

It can be observed that NSol has a similar complexity classification as the problem MinOnes. However, the relations inhabiting these classification cases are different. For instance, the Horn case is in PO for MinOnes, whereas it is MinHornDeletion-complete for NSol. Another diffence w.r.t. MinOnes is that our complexity classification preserves duality, i.e. that NSol(Γ) and NSol(dual(Γ)) always have the same complexity.

References

1. Arora, S., Babai, L., Stern, J., Sweedyk, Z.: The hardness of approximate optima in lattices, codes, and systems of linear equations. J. Comput. Syst. Sci. **54**(2), 317–331 (1997)
2. Ausiello, G., Crescenzi, P., Gambosi, G., Kann, V., Marchetti-Spaccamela, A., Protasi, M.: Complexity and Approximation: Combinatorial Optimization Problems and Their Approximability Properties. Springer, Heidelberg (1999)
3. Bailleux, O., Marquis, P.: Some computational aspects of Distance-SAT. J. Autom. Reasoning **37**(4), 231–260 (2006)
4. Behrisch, M., Hermann, M., Mengel, S., Salzer, G.: Give me another one!. In: Elbassioni, K., Makino, K. (eds.) ISAAC 2015. LNCS, vol. 9472, pp. 664–676. Springer, Heidelberg (2015). doi:10.1007/978-3-662-48971-0_56
5. Behrisch, M., Hermann, M., Mengel, S., Salzer, G.: As close as it gets. In: Kaykobad, M., Petreschi, R. (eds.) WALCOM 2016. LNCS, vol. 9627, pp. 222–235. Springer, Heidelberg (2016). doi:10.1007/978-3-319-30139-6_18
6. Böhler, E., Creignou, N., Reith, S., Vollmer, H.: Playing with Boolean blocks, part II: constraint satisfaction problems. SIGACT News, Complex. Theor. Column 43 **35**(1), 22–35 (2004)
7. Böhler, E., Reith, S., Schnoor, H., Vollmer, H.: Bases for Boolean co-clones. Inf. Process. Lett. **96**(2), 59–66 (2005)
8. Creignou, N., Khanna, S., Sudan, M.: Complexity Classifications of Boolean Constraint Satisfaction Problems. SIAM Monographs on Discrete Mathematics and Applications. SIAM, Philadelphia (PA) (2001)
9. Hochbaum, D.S., Megiddo, N., Naor, J., Tamir, A.: Tight bounds and 2-approximation algorithms for integer programs with two variables per inequality. Math. Program. **62**(1–3), 69–83 (1993)
10. Khanna, S., Sudan, M., Trevisan, L., Williamson, D.P.: The approximability of constraint satisfaction problems. SIAM J. Comput. **30**(6), 1863–1920 (2000)
11. Schaefer, T.J.: The complexity of satisfiability problems. In: Proceedings of the 10th Symposium on Theory of Computing (STOC 1978), San Diego, pp. 216–226 (1978)
12. Schrijver, A.: Theory of Linear and Integer Programming. John Wiley & Sons, New York (1986)

Parameterizing Edge Modification Problems Above Lower Bounds

René van Bevern[1]([⊠]), Vincent Froese[2], and Christian Komusiewicz[3]

[1] Novosibirsk State University, Novosibirsk, Russian Federation
rvb@nsu.ru
[2] Technische Universität Berlin, Berlin, Germany
vincent.froese@tu-berlin.de
[3] Friedrich-Schiller-Universität Jena, Jena, Germany
christian.komusiewicz@uni-jena.de

Abstract. For a fixed graph F, we study the parameterized complexity of a variant of the F-FREE EDITING problem: Given a graph G and a natural number k, is it possible to modify at most k edges in G so that the resulting graph contains no induced subgraph isomorphic to F? In our variant, the input additionally contains a vertex-disjoint packing $\mathcal{H}$ of induced subgraphs of G, which provides a lower bound $h(\mathcal{H})$ on the number of edge modifications required to transform G into an F-free graph. While earlier works used the number k as parameter or structural parameters of the input graph G, we consider instead the parameter $\ell := k - h(\mathcal{H})$, that is, the number of edge modifications above the lower bound $h(\mathcal{H})$. We show fixed-parameter tractability with respect to ℓ for K_3-FREE EDITING, FEEDBACK ARC SET IN TOURNAMENTS, and CLUSTER EDITING when the packing $\mathcal{H}$ contains subgraphs with bounded solution size. For K_3-FREE EDITING, we also prove NP-hardness in case of edge-disjoint packings of K_3s and $\ell = 0$, while for K_q-FREE EDITING and $q \geq 6$, NP-hardness for $\ell = 0$ even holds for vertex-disjoint packings of K_qs.

Keywords: NP-hard problem · Fixed-parameter algorithm · Subgraph packing · Kernelization · Graph-based clustering · Feedback arc set · Cluster editing

1 Introduction

Graph modification problems are a core topic of algorithmic research [8,21,29]: given a graph G, the aim is to transform G by a minimum number of modifications (like vertex deletions, edge deletions, or edge insertions) into another graph G' fulfilling certain properties. Particularly well-studied are *hereditary*

R. van Bevern—Supported by the Russian Foundation for Basic Research (RFBR) under research project 16-31-60007 mol_a_dk.

C. Komusiewicz—Supported by the DFG, project KO 3669/4-1.

A.S. Kulikov and G.J. Woeginger (Eds.): CSR 2016, LNCS 9691, pp. 57–72, 2016.
DOI: 10.1007/978-3-319-34171-2_5

graph properties, which are closed under vertex deletions and are characterized by *minimal forbidden induced subgraphs*: a graph fulfills such a property if and only if it does not contain a graph F from a property-specific family $\mathcal{F}$ of graphs as induced subgraph. All nontrivial vertex deletion problems and many edge modification and deletion problems for establishing hereditary graph properties are NP-complete [1,2,20,21,29].

One approach to cope with the NP-hardness of these problems are fixed-parameter (tractable) algorithms, which run in $f(k) \cdot n^{O(1)}$ time for a problem-specific parameter k and input size n and can efficiently solve instances in which the parameter k is small, even if the input size n is large. For vertex deletion, edge deletion, and edge modification problems, there is a generic fixed-parameter tractability result: If the desired graph property has a finite forbidden induced subgraph characterization, then the corresponding problems are fixed-parameter tractable with respect to the number of modifications k [8]. When combined with additional data reduction and pruning rules, the corresponding search tree algorithms can yield competitive solvers [17,24]. Nevertheless, the number of modifications is often too large. Thus, smaller parameters should be considered.

A natural approach to obtain smaller parameters is "parameterization above guaranteed values" [11,15,22,23]. The idea is to use a lower bound h on the solution size and to use $\ell := k - h$ as parameter instead of k. This idea has been applied successfully to VERTEX COVER, the problem of finding at most k vertices such that their deletion removes all edges (that is, all K_2s) from G. Since the size of a smallest vertex cover is large in many input graphs, above-guarantee parameterizations have been considered for the lower bounds "size of a maximum matching M in the input graph" and "optimum value L of the LP relaxation of the standard ILP-formulation of VERTEX COVER". After a series of improvements [11,15,22,26], the current best running time is $3^\ell \cdot n^{O(1)}$, where $\ell := k - (2 \cdot L - |M|)$ [15].

We aim to extend this approach to edge modification problems, where the number k of modifications tends to be even larger than for vertex deletion problems. For example, in the case of CLUSTER EDITING, which asks to destroy induced paths on three vertices by edge modifications, the number of modifications is often larger than the number of vertices in the input graph [6]. Hence, above-guarantee parameterization seems even more relevant for edge modification problems. Somewhat surprisingly, this approach has not been considered so far. We thus initiate research on above-guarantee parameterization in this context. As a starting point, we focus on graph properties that are characterized by one small forbidden induced subgraph.

Lower Bounds by Packings of Bounded-Cost Graphs. Following the approach of parameterizing VERTEX COVER above the size of a maximum matching, we can parameterize F-FREE EDITING above a lower bound obtained from packings of induced subgraphs containing F.

Definition 1.1. *A* vertex-disjoint (or edge-disjoint) packing *of induced subgraphs of a graph G is a set $\mathcal{H} = \{H_1, \ldots, H_z\}$ such that each H_i is an induced* *subgraph of G and such that the vertex sets (or edge sets) of the H_i are mutually disjoint.*

While it is natural to consider packings of F-graphs to obtain a lower bound on the solution size, a packing of other graphs that contain F as induced subgraph might yield better lower bounds and thus a smaller above-guarantee parameter. For example, a K_4 contains several triangles and two edge deletions are necessary to make it triangle-free.[1] Thus, if a graph G has a vertex-disjoint packing of h_3 triangles and h_4 K_4s, then at least $h_3 + 2 \cdot h_4$ edge deletions are necessary to make it triangle-free. Moreover, when allowing arbitrary graphs for the packing, the lower bounds provided by vertex-disjoint packings can be better than the lower bounds provided by edge-disjoint packings of F. A disjoint union of h K_4s, for example, has h edge-disjoint triangles but also h vertex-disjoint K_4s. Hence, the lower bound provided by packing vertex-disjoint K_4s is twice as large as the one provided by packing edge-disjoint triangles for this graph.

Motivated by this benefit of vertex-disjoint packings of arbitrary graphs, we mainly consider lower bounds obtained from vertex-disjoint packings, which we assume to receive as input. Thus, we arrive at the following problem, where $\tau(G)$ denotes the minimum size of an F-free editing set for a graph G:

Problem 1.2 (F-FREE EDITING WITH COST-t PACKING).
Input: A graph $G = (V, E)$, a vertex-disjoint packing $\mathcal{H}$ of induced subgraphs
    of G such that $1 \leq \tau(H) \leq t$ for each $H \in \mathcal{H}$, and a natural number k.
Question: Is there an *F-free editing set* $S \subseteq \binom{V}{2}$ of size at most k such
    that $G \Delta S := (V, (E \setminus S) \cup (S \setminus E))$ does not contain F as induced subgraph?

In the context of a concrete variant of F-FREE EDITING, we refer to an F-free editing set as *solution* and call a solution *optimal* if it has minimum size. The special case of F-FREE EDITING WITH COST-t PACKING where only F-graphs are allowed in the packing is called F-FREE EDITING WITH F-PACKING.

From the packing $\mathcal{H}$, we obtain the lower bound $h(\mathcal{H}) := \sum_{H \in \mathcal{H}} \tau(H)$ on the size of an F-free editing set, which allows us to use the excess $\ell := k - h(\mathcal{H})$ over this lower bound as parameter, as illustrated in Fig. 1. Since F is a fixed graph, we can compute the bound $h(\mathcal{H})$ in $f(t) \cdot |G|^{O(1)}$ time using the generic search tree algorithm [8] for each $H \in \mathcal{H}$. In the same time we can also verify whether the cost-t property is fulfilled.

Packings of forbidden induced subgraphs have been used in implementations of fixed-parameter algorithms to prune the corresponding search trees tremendously [17]. By showing fixed-parameter algorithms for the above-guarantee parameters, we hope to explain the fact that these packings help in obtaining fast algorithms.

Our Results. We first state the negative results since they justify the focus on concrete problems and, to a certain extent, also the focus on parameterizing

---

[1] Bounds of this type are exploited, for example, in so-called cutting planes, which are used in speeding up the running time of ILP solvers for concrete problems.

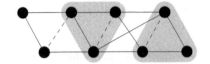 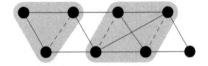

Fig. 1. An instance of TRIANGLE DELETION. The packing graphs have gray background. Left: A vertex-disjoint packing of two triangles giving $\ell = 1$. Right: A vertex-disjoint packing of a triangle and a K_4 giving $\ell = 0$. The solution consists of the three dashed edges.

edge modification problems above vertex-disjoint packings. We show that, if F is a triangle and $\mathcal{H}$ is an *edge-disjoint* packing of h triangles in a graph G, then it is NP-hard to decide whether G has a triangle deletion set of size h (that is, $\ell = 0$). Thus, parameterization by ℓ is hopeless for this packing lower bound. Moreover, we show that K_6-FREE EDITING WITH K_6-PACKING is NP-hard for $\ell = 0$ above *vertex-disjoint* packings. This proves, in particular, that a general fixed-parameter tractability result as it is known for the parameter k [8] cannot be expected. We also show that extending the parameterization "above maximum matching" for VERTEX COVER to d-HITTING SET in a natural way leads to intractable problems. This is achieved by showing that, for all $q \geq 3$, P_q-FREE VERTEX DELETION WITH P_q-PACKING is NP-hard even if $\ell = 0$.

On the positive side, we study three variants of F-FREE EDITING WITH COST-t PACKING in detail, namely the ones in which F is a triangle (that is, a K_3) or a path on three vertices (that is, a P_3). The first case is known as TRIANGLE DELETION, the second one as CLUSTER EDITING. We also consider the case in which the input is a tournament graph and F is a directed cycle on three vertices. This is known as FEEDBACK ARC SET IN TOURNAMENTS. Using a general approach described in Sect. 2, we obtain fixed-parameter algorithms for these variants of F-FREE EDITING WITH COST-t PACKING parameterized by t and ℓ. This implies, in particular, fixed-parameter tractability for F-FREE EDITING WITH F-PACKING parameterized by ℓ. Specifically, we obtain the following positive results:

(i) For TRIANGLE DELETION, we show an $O(t \cdot \ell)$-vertex problem kernel for cost-t packings and a $(2t+3)^{\ell} \cdot 2^{O(t)} \cdot n^{O(1)}$-time algorithm for cost-t packings.

(ii) For CLUSTER EDITING, we obtain a $2^{O(t \cdot \ell)} \cdot n^{O(1)}$ time algorithm and an $O(t \cdot \ell)$-vertex kernel for cost-t packings, and a $4^{\ell} \cdot n^{O(1)}$-time algorithm for P_3-packings.

(iii) For FEEDBACK ARC SET IN TOURNAMENTS, we show an $O(t \cdot \ell)$-vertex problem kernel and a $2^{O(\sqrt{(2t+1)\ell})} \cdot n^{O(1)}$-time algorithm for cost-t packings.

For the kernelization results, we need to assume that $t \in O(\log n)$ to guarantee polynomial running time of the data reduction. Hence, in the context of problem kernelization, we consider t to be part of the problem, whereas for the fixed-parameter algorithms, t can be considered a part of the input.

Due to space constraints, most proofs are deferred to a full version of the paper.

Notation. Unless stated otherwise, we consider only undirected, simple, finite graphs $G = (V, E)$, with a *vertex set* $V(G) := V$ and an *edge set* $E(G) := E \subseteq \binom{V}{2} := \{\{u, v\} \mid u, v \in V \wedge u \neq v\}$. Let $n := |V|$ denote the *order* of the graph and $m := |E|$ its number of edges. A set $S \subseteq \binom{V}{2}$ is an *edge modification set* for G. For an edge modification set S for G, let $G \Delta S := (V, (E \setminus S) \cup (S \setminus E))$ denote the *graph obtained by applying S to G*. If $S \subseteq E$, then S is called an *edge deletion set* and we write $G \setminus S$ instead of $G \Delta S$. The *open neighborhood* of a vertex $v \in V$ is defined as $N_G(v) := \{u \in V \mid \{u, v\} \in E\}$. Also, for $V' \subseteq V$, let $G[V'] := (V', E \cap \binom{V'}{2})$ denote the subgraph of G *induced by V'*. A *directed graph (or digraph)* $G = (V, A)$ consists of a *vertex set* $V(G)$ and an *arc set* $A(G) := A \subseteq \{(u, v) \in V^2 \mid u \neq v\}$. A *tournament* on n vertices is a directed graph (V, A) with $|V| = n$ such that, for each pair of distinct vertices u and v, either $(u, v) \in A$ or $(v, u) \in A$.

For parameterized complexity basics, we refer to [10, 12]. A *kernelization* is a polynomial-time algorithm mapping an instance (x, k) of L to an instance (x', k') of L such that $(x, k) \in L$ if and only if $(x', k') \in L$, $|x'| \leq g(k)$, and $k' \leq h(k)$ for some computable functions g and h. A kernelization often consists of *data reduction rules*, which transform an instance (x, k) in polynomial time into an instance (x', k') and are *correct* if (x, k) is a yes-instance if and only if (x', k') is.

2   General Approach

Recall that $\tau(H)$ is the minimum number of edge modifications required to make a graph H F-free. Our fixed-parameter algorithms for F-FREE EDITING WITH COST-t PACKING parameterized by the combination of t and $\ell := k - h(\mathcal{H})$, where $h(\mathcal{H}) := \sum_{H \in \mathcal{H}} \tau(H)$, are based on the following approach. We show that, for each induced subgraph H of G in a given packing $\mathcal{H}$, we face essentially two situations. Either we find an optimal solution for H that is a subset of an optimal solution for G, or we find a certificate witnessing that (a) H needs to be solved suboptimally or that (b) a vertex pair containing exactly one vertex from H needs to be modified. Thus, we arrive at a classic win-win scenario [13] where we can either apply data reduction or show that the packing size $|\mathcal{H}|$ is bounded.

Lemma 2.1. Let $(G, \mathcal{H}, k)$ be an instance of F-FREE EDITING WITH COST-t PACKING and let S be a size-k solution that contains, for each $H = (W, F) \in \mathcal{H}$,

(a) at least $\tau(H) + 1$ vertex pairs from $\binom{W}{2}$, or
(b) at least one vertex pair $\{v, w\}$ with $v \in V \setminus W$ and $w \in W$.

Then, $|\mathcal{H}| \leq 2\ell$.

Proof. Denote by $\mathcal{H}_a \subseteq \mathcal{H}$ the set of all graphs in $\mathcal{H}$ that fulfill property (a) and let $p_a := |\mathcal{H}_a|$. Let $\mathcal{H}_b := \mathcal{H} \setminus \mathcal{H}_a$ denote the set containing the remaining packing graphs (fulfilling property (b)) and let $p_b := |\mathcal{H}_b|$. Thus, $|\mathcal{H}| = p_a + p_b$.

Furthermore, let $h_a := \sum_{H \in \mathcal{H}_a} \tau(H)$ denote the lower bound obtained from the graphs in $\mathcal{H}_a$ and let $h_b := h(\mathcal{H}) - h_a$ denote the part of the lower bound obtained by the remaining graphs.

The packing graphs in $\mathcal{H}_a$ cause $h_a + p_a$ edge modifications inside of them. Similarly, the packing graphs in $\mathcal{H}_b$ cause at least h_b edge modifications inside of them, and each packing graph in $\mathcal{H}_b$ additionally causes modification of at least one vertex pair that contains exactly one vertex from this graph. Call this vertex pair *crossing* and observe that every such pair can be a crossing pair of at most two different packing graphs (since the packing graphs are pairwise vertex-disjoint). Consequently, at least $h_b + p_b/2$ edge modifications are caused by the graphs in $\mathcal{H}_b$. This implies that

$$k \geq h_a + h_b + p_a + p_b/2$$
$$\Leftrightarrow \qquad k - h(\mathcal{H}) \geq p_a + p_b/2$$
$$\Leftrightarrow \qquad 2\ell \geq 2p_a + p_b \geq |\mathcal{H}|. \qquad \square$$

Lemma 2.1 allows us to upper-bound k in t and ℓ, which we can then exploit in fixed-parameter algorithms.

Lemma 2.2. Let $(G, \mathcal{H}, k)$ be a yes-instance of F-FREE EDITING WITH COST-t PACKING such that $|\mathcal{H}| \leq 2\ell$. Then, $k \leq (2t + 1)\ell$.

Proof. By definition, $k = \ell + h(\mathcal{H}) \leq \ell + t \cdot |\mathcal{H}| \leq \ell + t \cdot 2\ell = (2t + 1)\ell$. $\qquad \square$

3   Triangle Deletion

The first application of our framework is TRIANGLE DELETION, the problem of destroying all induced *triangles* (K_3s) in a graph by at most k edge deletions. TRIANGLE DELETION is NP-complete [29]. It allows for a trivial reduction to 3-HITTING SET since edge deletions do not create new triangles [16]. Combining this approach with the currently fastest known algorithms for 3-HITTING SET [4, 28] gives an algorithm for TRIANGLE DELETION with running time $O(2.076^k + nm)$. Finally, TRIANGLE DELETION admits a problem kernel with at most $6k$ vertices [7].

We show that TRIANGLE DELETION WITH COST-t PACKING is fixed-parameter tractable with respect to the combination of t and $\ell := k - h(\mathcal{H})$. More precisely, we obtain a kernelization and a search tree algorithm. Both make crucial use of the following generic reduction rule for TRIANGLE DELETION WITH COST-t PACKING.

Reduction Rule 3.1. If there is an induced subgraph $H \in \mathcal{H}$ and a set $T \subseteq E(H)$ of $\tau(H)$ edges such that deleting T destroys all triangles of G that contain edges of H, then delete T from G, H from $\mathcal{H}$ and decrease k by $\tau(H)$.

If Reduction Rule 3.1 is not applicable to H, then we define a *certificate* to be a set T of triangles in G, each containing exactly one distinct edge of H, such that $|T| = \tau(H) + 1$ or $\tau(H') > \tau(H) - |T|$, where H' is the subgraph obtained from H by deleting, for each triangle in T, its edge shared with H.

In the following, let $f(G, k)$ denote the running time of the fastest algorithm that returns a minimum triangle-free deletion set of size at most k if it exists. We assume that f is monotonically nondecreasing in both the size of G and in k. Currently $f(G, k) = O(2.076^k + |V(G)| \cdot |E(G)|)$.

Lemma 3.2. Reduction Rule 3.1 is correct. Moreover, in $O(nm + n \cdot f(H^*, t))$ time, we can check and apply it to all $H \in \mathcal{H}$ and output a certificate T if it does not apply to some $H \in \mathcal{H}$. Herein, H^* is the largest graph in $\mathcal{H}$.

Proof. We first show correctness. Let $(G, \mathcal{H}, k)$ be the instance to which Reduction Rule 3.1 is applied and let $(G', \mathcal{H} \setminus \{H\}, k - \tau(H))$ with $G' := G \setminus T$ be the result. We show that $(G, \mathcal{H}, k)$ is a yes-instance if and only if $(G', \mathcal{H} \setminus \{H\}, k - \tau(H))$ is.

First, let S be a solution of size at most k for $(G, \mathcal{H}, k)$. Let $S_H := S \cap E(H)$ denote the set of edges of S that destroy all triangles in H. By definition, $|S_H| \geq \tau(H)$. Since $S_H \subseteq E(H)$, only triangles containing at least one edge of H are destroyed by deleting S_H. It follows that the set of triangles destroyed by S_H is a subset of the triangles destroyed by T. Hence, $(S \setminus S_H) \cup T$ has size at most k and clearly is a solution for $(G, \mathcal{H}, k)$ that contains all edges of T. Thus, deleting T from G, H from $\mathcal{H}$, and decreasing k by $\tau(H)$ yields an equivalent instance.

For the converse direction, let S' be a solution of size at most $k - \tau(H)$ for $(G', \mathcal{H} \setminus \{H\}, k - \tau(H))$. Since $T \subseteq E(H)$, it holds that every triangle contained in G that does not contain any edge of H is also a triangle in G'. Thus, S' is a set of edges whose deletion in G destroys all triangles that do not contain any edge of H. Since T destroys all triangles containing an edge of H, we have that $T \cup S'$ is a solution for G. Its size is k.

We now show the running time. First, in $O(nm)$ time, we compute for all $H \in \mathcal{H}$ all triangles T that contain exactly one edge $e \in E(H)$. These edges are labeled in each $H \in \mathcal{H}$. Then, for each $H \in \mathcal{H}$, in $f(H, t)$ time we determine the size $\tau(H)$ of an optimal triangle-free deletion set for H. Let t^* denote the number of labeled edges of H and let H' denote the graph obtained from H by deleting the labeled edges. If $t^* > \tau(H)$, then we return as certificate $\tau(H) + 1$ triangles of T, each containing a distinct of $\tau(H) + 1$ arbitrary labeled edges. Otherwise, observe that, after deleting the labeled edges, each remaining triangle of G that contains at least one edge of H is contained in H'. Thus, we now determine whether H' can be made triangle-free by $\tau(H) - t^*$ edge deletions in $f(H', \tau(H) - t^*)$ time. If this is the case, then the rule applies and the set T consists of the solution for H' plus the previously deleted edges. Otherwise, destroying all triangles that contain exactly one edge from H leads to a solution which needs more than $\tau(H)$ edge deletions and thus the rule does not apply and we return the certificate T for this $H \in \mathcal{H}$. The overall running time now follows from the monotonicity of f, from the fact that $|\mathcal{H}| \leq n$, and from the fact that one pass over $\mathcal{H}$ is sufficient since deleting edges in each H does not produce new triangles and does not destroy triangles in any $H' \neq H$. □

Observe that Reduction Rule 3.1 never increases the parameter ℓ since we decrease both k as well as the lower bound $h(\mathcal{H})$ by $\tau(H)$. After application of Reduction Rule 3.1, we can upper-bound the solution size k in terms of t and ℓ.

Lemma 3.3. Let $(G, \mathcal{H}, k)$ be a yes-instance of TRIANGLE DELETION WITH COST-t PACKING such that Reduction Rule 3.1 is inapplicable. Then, $k \leq (2t + 1)\ell$.

Proof. Since $(G, \mathcal{H}, k)$ is reduced with respect to Reduction Rule 3.1, for each graph $H = (W, F)$ in $\mathcal{H}$, there is a set of edges between W and $V \setminus W$ witnessing that any optimal solution for H does not destroy all triangles containing at least one edge from H. Consider any optimal solution S. For each graph $H \in \mathcal{H}$, there are two possibilities: Either at least $\tau(H) + 1$ edges inside H are deleted by S, or at least one external edge of H is deleted by S. Therefore, S fulfills the condition of Lemma 2.1 and thus $|\mathcal{H}| \leq 2\ell$. □

This implies a kernelization with respect to ℓ for every fixed value of t. The kernel can be obtained by applying the known kernelization to an instance that is reduced with respect to Reduction Rule 3.1.

Theorem 3.4. *For every fixed $t \geq 1$, TRIANGLE DELETION WITH COST-t PACKING admits a problem kernel with at most $(12t + 6)\ell$ vertices that can be computed in $O(nm + n \cdot f(H, t))$ time, where H is the largest graph in $\mathcal{H}$.*

Proof. Let $(G = (V, E), \mathcal{H}, k)$ be the input instance. First, compute in $O(nm + n \cdot f(H, t))$ time an instance that is reduced with respect to Reduction Rule 3.1. Afterwards, by Lemma 3.3, we can reject if $k > (2t + 1)\ell$.

Otherwise, we apply the known kernelization algorithm for TRIANGLE DELETION to the instance (G, k) (that is, without $\mathcal{H}$). This kernelization produces in $O(m\sqrt{m}) = O(nm)$ time a problem kernel (G', k') with at most $6k \leq (12t + 6)\ell$ vertices and with $k' \leq k$ [7]. Adding an empty packing gives an equivalent instance $(G', \emptyset, k')$ with parameter $\ell' = k' \leq (2t + 1)\ell$ of TRIANGLE DELETION WITH COST-t PACKING. □

We can also devise a search tree algorithm for the combined parameter (t, ℓ).

Theorem 3.5. TRIANGLE DELETION WITH COST-t PACKING *can be solved in $O((2t + 3)^\ell \cdot (nm + n \cdot f(H, t)))$ time, where H is the largest graph in $\mathcal{H}$.*

Proof. First, apply Reduction Rule 3.1 exhaustively in $O(nm + nf(H, t))$ time. Now, consider a reduced instance. If $\ell < 0$, then we can reject the instance. Otherwise, consider the following two cases.

Case 1: $\mathcal{H}$ contains a graph H. Let $t' := \tau(H) \leq t$. Since Reduction Rule 3.1 does not apply to H, there is a certificate $\mathcal{T}$ of $t'' \leq t' + 1$ triangles, each containing exactly one distinct edge of H such that deleting the edges of these triangles contained in H produces a subgraph H' of H that cannot be made triangle-free by $t' - t''$ edge deletions. Thus, branch into the following $(2t'' + 1)$ cases: First, for each triangle $T \in \mathcal{T}$, create two cases, in each deleting a different one of the two edges of T that are not in H. In the remaining case, delete the t'' edges of H and replace H by H' in $\mathcal{H}$.

Case 2: $\mathcal{H} = \emptyset$. Either G is triangle-free, then we are done. Otherwise, pick an arbitrary triangle in G and add it to $\mathcal{H}$.

It remains to show the running time by bounding the search tree size. In Case 2, no branching is performed and the parameter is decreased by at least one. In Case 1, the parameter value is decreased by one in each branch: in the first $2t''$ cases, an edge that is not contained in any packing graph is deleted. Thus, k decreases by one while $h(\mathcal{H})$ remains unchanged. In the final case, the value of k decreases by t'' since this many edge deletions are performed. However, $\tau(H') \geq \tau(H) - t'' + 1$. Hence, the lower bound $h(\mathcal{H})$ decreases by at most $t'' - 1$ and thus the parameter ℓ decreases by at least one. Note that applying Reduction Rule 3.1 never increases the parameter. Hence, the depth of the search tree is at most ℓ. $\square$

For the natural special case $t = 1$, that is, for triangle packings, we immediately obtain the following running time.

Corollary 3.6. TRIANGLE DELETION WITH TRIANGLE PACKING *is solvable in* $O(5^\ell nm)$ *time.*

We complement these positive results by the following hardness result for the case of edge-disjoint triangle packings:

Theorem 3.7. TRIANGLE DELETION WITH EDGE-DISJOINT TRIANGLE PACKING *is NP-hard even if $\ell = 0$.*

We prove Theorem 3.7 using a reduction from 3-SAT.

Problem 3.8 (3-SAT).
Input: A Boolean formula $\phi = C_1 \wedge \ldots \wedge C_m$ in conjunctive normal form over variables $x_1, \ldots, x_n$ with at most three variables per clause.
Question: Does ϕ have a satisfying assignment?

Construction 3.9. Given a Boolean formula ϕ, we create a graph G and an edge-disjoint packing $\mathcal{H}$ of triangles such that G can be made triangle-free by exactly $|\mathcal{H}|$ edge deletions if and only if there is a satisfying assignment for ϕ. We assume that each clause of ϕ contains exactly three pairwise distinct variables. The construction is illustrated in Fig. 2.

For each variable x_i of ϕ, create a triangle X_i on the vertex set $\{x_i^1, x_i^2, x_i^3\}$ with two distinguished edges $x_i^T := \{x_i^1, x_i^2\}$ and $x_i^F := \{x_i^2, x_i^3\}$ and add X_i to $\mathcal{H}$. For each clause $C_j = (l_1, l_2, l_3)$ of ϕ, create a triangle Y_j on the vertex set $\{c_j^1, c_j^2, c_j^3\}$ with three edges $c_j^{l_1}$, $c_j^{l_2}$, and $c_j^{l_3}$. Connect the clause gadget Y_j to the variable gadgets as follows: If $l_t = x_i$, then connect the edge $c_j^{l_t} =: \{u, v\}$ to the edge $x_i^T = \{x_i^1, x_i^2\}$ via two adjacent triangles $A_{ij} := \{u, v, x_i^1\}$ and $B_{ij} := \{v, x_i^1, x_i^2\}$ sharing the edge $\{v, x_i^1\}$. The triangle A_{ij} is added to $\mathcal{H}$.

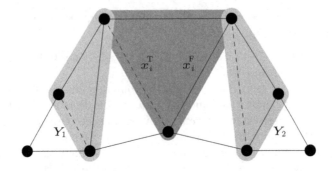

Fig. 2. Construction for the formula $C_1 \wedge C_2$, where $C_1 = (x_i)$ and $C_2 = (\neg x_i)$. The triangles on a gray background are contained in the mutually edge-disjoint triangle packing $\mathcal{H}$. Deleting the dashed edges corresponds to setting x_i to true and, thus, satisfying C_1. Note that it is impossible to destroy triangle Y_2 by $|\mathcal{H}|$ edge deletions if we delete x_i^{T}. This corresponds to the fact that C_2 cannot be satisfied by setting $x_i = 1$.

If $l_t = \neg x_i$, then connect the edge $c_j^{l_t} =: \{u, v\}$ to the edge $x_i^{\mathrm{F}} = \{x_i^2, x_i^3\}$ via two adjacent triangles $A_{ij} := \{u, v, x_i^3\}$ and $B_{ij} := \{v, x_i^2, x_i^3\}$ sharing the edge $\{v, x_i^3\}$. The triangle A_{ij} is added to $\mathcal{H}$.

Proof (of Theorem 3.7). First, observe that Construction 3.9 introduces no edges between distinct clause gadgets or distinct variable gadgets. Thus, under the assumption that each clause contains each variable at most once, the only triangles in the constructed graph are the X_i, the Y_j, the A_{ij} and B_{ij} for all variables x_i and the incident clauses C_j.

Now, assume that ϕ allows for a satisfying assignment. We construct a set of edges S of size $|\mathcal{H}|$ such that $G' := (V, E \setminus S)$ is triangle-free. For each variable x_i that is true, add x_i^{T} to S. For each variable x_i that is false, add x_i^{F} to S. By this choice, the triangle X_i is destroyed in G' for each variable x_i. Additionally, for each clause C_j and its *true* literals $l \in \{x_i, \neg x_i\}$, the triangle B_{ij} is destroyed. To destroy A_{ij}, we add to S the edge of A_{ij} shared with Y_j, which also destroys the triangle Y_j. For each clause C_j containing a *false* literal $l \in \{x_i, \neg x_i\}$, we destroy B_{ij} and, simultaneously, A_{ij}, by adding to S the edge of A_{ij} shared with B_{ij}.

Conversely, assume that there is a set S of size $|\mathcal{H}|$ such that $G' = (V, E \setminus S)$ is triangle-free. We construct a satisfying assignment for ϕ. First, observe that, since the triangles in $\mathcal{H}$ are pairwise edge-disjoint, S contains exactly one edge of each triangle in $\mathcal{H}$. Thus, of each triangle X_i, at most one of the two edges x_i^{F} and x_i^{T} is contained in S. The set S contains at least one edge e of each Y_j. This edge is shared with a triangle A_{ij}. Since $A_{ij} \in \mathcal{H}$ and, with e, S already contains one edge of A_{ij}, S does not contain the edge shared between A_{ij} and B_{ij}. Since $B_{ij} \notin \mathcal{H}$, S has to contain an edge of B_{ij} shared with another triangle in $\mathcal{H}$. If the clause C_j contains x_i, then the only such edge is x_i^{T} and we set x_i to true. If the clause C_j contains $\neg x_i$, then the only such edge is x_i^{F} and we set x_i

to false. In both cases, clause C_j is satisfied. Since at most one of x_i^{T} and x_i^{F} is in S, the value of each variable x_i is well-defined. □

4   Feedback Arc Set in Tournaments

In this section, we study the following problem.

*Problem 4.1 (*FEEDBACK ARC SET IN TOURNAMENTS *(FAST)).*
Input: An n-vertex tournament $G = (V, A)$ and a natural number k.
Question: Does G have a *feedback arc set* $S \subseteq A$, that is, a set S such that
   reversing all arcs in S yields an acyclic tournament, of size at most k?

FAST is NP-complete [1] but fixed-parameter tractable with a currently best known running time of $2^{O(\sqrt{k})} + n^{O(1)}$ [18]. Moreover, a problem kernel with $(2 + \epsilon)k$ vertices for any fixed $\epsilon > 0$ is known [3] as well as a simpler $4k$-vertex kernel [25]. It is well-known that a tournament is acyclic if and only if it does not contain a *directed triangle* (cycle on 3 vertices). Hence, the problem is to find a set of arcs whose reversal leaves no directed triangle in the tournament.

   We show fixed-parameter tractability of FAST WITH COST-t PACKING parameterized by the combination of t and $\ell := k - h(\mathcal{H})$ and a problem kernel with respect to ℓ for fixed t. Recall that $h(\mathcal{H}) := \sum_{H \in \mathcal{H}} \tau(H) \geq |\mathcal{H}|$, where $\tau(G)$ is the size of a minimum feedback arc set for a directed graph G. The approach is the same as for TRIANGLE DELETION in Sect. 3, that is, we upper-bound the solution size k in t and ℓ and apply the fixed-parameter algorithm for k [18].

Reduction Rule 4.2. *If there is a subtournament $H \in \mathcal{H}$ and a feedback arc set $T \subseteq A(H)$ of size $\tau(H)$ such that reversing the arcs in T leaves no directed triangles in G containing arcs of H, then reverse the arcs in T, remove H from $\mathcal{H}$, and decrease k by $\tau(H)$.*

Although Reduction Rule 4.2 is strikingly similar to Reduction Rule 3.1, its correctness proof is significantly more involved.

   Exhaustive application of Reduction Rule 4.2 allows us to show that $k \leq (2t + 1)\ell$ in any yes-instance (analogous to Lemma 3.3). Having established the bound for k, we obtain the following two fixed-parameter tractability results, where $f(G, k)$ denotes the running time of the fastest algorithm that finds a minimum feedback arc set of size at most k for a given tournament G if it exists. We assume that f is monotonically nondecreasing in both the size of G and in k. Currently, $f(G, k) = 2^{O(\sqrt{k})} + |G|^{O(1)}$ [18].

Theorem 4.3. FEEDBACK ARC SET IN TOURNAMENTS WITH COST-t PACKING *is solvable in* $2^{O(\sqrt{(2t+1)\ell})} + n^{O(1)} + nf(G, t)$ *time.*

Theorem 4.4. *For every fixed $t \geq 1$,* FEEDBACK ARC SET IN TOURNAMENTS WITH COST-t PACKING *admits a problem kernel with at most $(8t + 4)\ell$ vertices.*

5   Cluster Editing

We now apply our framework to CLUSTER EDITING, a well-studied edge modification problem in parameterized complexity [5,9,14,19].

*Problem 5.1 (*CLUSTER EDITING*).*
Input: A graph $G = (V, E)$ and a natural number k.
Question: Is there an edge modification set $S \subseteq \binom{V}{2}$ of size at most k such that $G \triangle S$ is a cluster graph, that is, a disjoint union of cliques?

A graph is a cluster graph if and only if it is P_3-free [27]. Thus, CLUSTER EDITING is the problem of destroying all P_3s by few edge modifications. For brevity, we refer to the connected components of a cluster graph (which are cliques) and to their vertex sets as *clusters*. The currently fastest algorithm for CLUSTER EDITING parameterized by the solution size k runs in $O(1.62^k + |G|)$ time [5]. Assuming the exponential-time hypothesis, CLUSTER EDITING cannot be solved in $2^{o(k)} \cdot |G|^{O(1)}$ time [14,19]. CLUSTER EDITING admits a problem kernel with at most $2k$ vertices [9].

Several kernelizations for CLUSTER EDITING are based on the following observation: If G contains a clique such that all vertices in this clique have the same closed neighborhood, then there is an optimal solution that treats these vertices in a similar manner. That is, it puts these vertices into the same cluster. This implies that the edges of this clique are never deleted. The following rule is based on a generalization of this observation.

Reduction Rule 5.2. *If $G = (V, E)$ contains an induced subgraph $H = (W, F)$ having an optimal solution S of size $\tau(H)$ such that for all vertices $u, v \in W$:*

– *$N_G(v) \setminus W = N_G(u) \setminus W$ if u and v are in the same cluster of $H \triangle S$, and*
– *$N_G(v) \setminus W \cap N_G(u) \setminus W = \emptyset$ otherwise,*

then replace G by $G \triangle S$ and decrease k by $\tau(H)$.

Let $f(G, k)$ denote the running time of the fastest algorithm that returns an optimal solution of size at most k if it exists; we assume again that f is monotonically nondecreasing in both the size of G and in k. Currently $f(G, k) = O(1.62^k + |G|)$. The following lemma asserts that the rule is correct.

Lemma 5.3. *Let $(G, \mathcal{H}, k)$ be an instance of* CLUSTER EDITING WITH COST-t PACKING *and let $H = (W, F) \in \mathcal{H}$. Then, in $O(nm + |W|^2 + f(H, t))$ time we can determine whether Reduction Rule 5.2 applies. If this is the case, then $(G, \mathcal{H} \setminus \{H\}, k - \tau(H))$ is an equivalent instance.*

Observe that since k is decreased by $\tau(H)$, the parameter ℓ does not increase when the rule is applied. As for the previous problems, applying the rule to each $H \in \mathcal{H}$ is sufficient for showing fixed-parameter tractability with respect to (t, ℓ).

Theorem 5.4. CLUSTER EDITING WITH COST-t PACKING *is solvable in* $O(2^{O(t \cdot \ell)} \cdot n + nm)$ *time.*

Finally, we observe that, if t is not part of the input, then we obtain a problem kernel with a linear number of vertices.

Theorem 5.5. CLUSTER EDITING WITH COST-t PACKING *admits a problem kernel with* $(4t + 2)\ell$ *vertices which can be computed in* $O(mn + 1.62^t \cdot n)$ *time.*

For CLUSTER EDITING WITH P_3-PACKING, the generic algorithm based on Reduction Rule 5.2 (with $t = 1$) using the currently best CLUSTER EDITING running time leads to a running time of $O(4.26^\ell + nm)$; we can show an improved running time.

Theorem 5.6. CLUSTER EDITING WITH P_3-PACKING *can be solved in* $4^\ell \cdot n^{O(1)}$ *time.*

6 Hardness Results

In this section, we show that there are edge modification problems which are NP-hard even for constant-size forbidden subgraphs and if $\ell = 0$.

Theorem 6.1. *For every fixed* $q \geq 6$, K_q-FREE DELETION WITH K_q-PACKING *is NP-hard for* $\ell = 0$.

We prove Theorem 6.1 by giving a reduction from 3-SAT.

Construction 6.2. Let ϕ be a Boolean formula with variables $x_1, \ldots, x_n$ and clauses $C_1, \ldots, C_m$. We assume that each clause C_j contains exactly three pairwise distinct variables. We create a graph G and a vertex-disjoint K_q-packing $\mathcal{H}$ as follows.

For each variable x_i, add a q-clique X_i to G that has two distinguished disjoint edges x_i^{F} and x_i^{T}. For each clause $C_j = (l_1 \wedge l_2 \wedge l_3)$ with literals l_1, l_2, and l_3, add a q-clique Y_j to G that has three distinguished and pairwise disjoint edges e_{l_1}, e_{l_2}, and e_{l_3} (which exist since $q \geq 6$). Finally, if $l_t = x_i$, then identify the edge e_{l_t} with x_i^{T} and if $l_t = \neg x_i$, then identify the edge e_{l_t} with x_i^{F}. The packing $\mathcal{H}$ consists of all X_i introduced for the variables x_i of ϕ.

Lemma 6.3. Let G be the graph output by Construction 6.2 and let H be an induced K_q in G. Then, H is either one of the X_i or one of the Y_j.

Proof. First, note that the X_i are pairwise vertex-disjoint since Construction 6.2 only identifies edges of Y_js with edges of X_is and no edge in any Y_j is identified with edges in different X_i. For any X_i and Y_j, the vertices in $V(X_i) \setminus V(Y_j)$ are nonadjacent to those in $V(Y_j) \setminus V(X_i)$. Similarly, for Y_i and Y_j, the vertices in $V(Y_i) \setminus V(Y_j)$ are nonadjacent to those in $V(Y_j) \setminus V(Y_i)$ for $i \neq j$. Thus, every clique in G is entirely contained in one of the X_i or Y_j. □

Lemma 6.3 allows us to prove Theorem 6.1.

Proof (of Theorem 6.1). We show that ϕ is satisfiable if and only if G can be made K_q-free by $k = |\mathcal{H}|$ edge deletions (that is, $\ell = 0$).

First, assume that there is an assignment that satisfies ϕ. We construct a K_q-free deletion set S for G as follows: if the variable x_i is set to true, then put x_i^{T} into S. If the variable x_i is set to false, then add x_i^{F} to S. Thus, for each X_i, we add exactly one edge to S. Since $\mathcal{H}$ consists of the X_i, we have $|S| = |\mathcal{H}|$. Moreover, since each clause C_j contains a true literal, at least one edge of each Y_j is contained in S. Thus, $G \setminus S$ is K_q-free, since, by Lemma 6.3, the only K_qs in G are the X_i and Y_j and, for each of them, S contains at least one edge.

Now, assume that G can be made K_q-free by deleting a set S of $|\mathcal{H}|$ edges. Then, S deletes exactly one edge of each X_i and at least one edge of each Y_j. We can assume without loss of generality that S contains either the edge x_i^{T} or x_i^{F} for each X_i since deleting one of these edges instead of another edge in X_i always yields a solution by Construction 6.2. Thus, the deletion set S corresponds to a satisfying assignment for ϕ. □

Finally, one can show NP-hardness of the problem of destroying all induced paths of a fixed length $q \geq 3$ by at most h *vertex* deletions even if a packing of h vertex-disjoint induced P_qs in the input graph G is provided as input.

Theorem 6.4. *For every fixed $q \geq 3$, P_q-FREE VERTEX DELETION WITH P_q-PACKING is NP-hard even if $\ell = 0$.*

7   Conclusion

It is open to extend our framework to further problems. The most natural candidates appear to be COGRAPH EDITING which is the problem of destroying all induced P_4s, K_4-FREE EDITING, and CLAW-FREE EDITING. In the case of vertex-deletion problems, TRIANGLE VERTEX DELETION appears to be the most natural open case. Furthermore, it would be nice to obtain more general theorems separating the tractable from the hard cases for this parameterization. For CLUSTER EDITING and TRIANGLE DELETION improved running times are desirable. Maybe more importantly, it is open to determine the complexity of CLUSTER EDITING and FEEDBACK ARC SET IN TOURNAMENTS parameterized above the size of *edge-disjoint* packings of forbidden induced subgraphs. Finally, our framework offers an interesting tradeoff between running time and power of generic data reduction rules. Exploring such tradeoffs seems to be a rewarding topic for the future. The generic rules presented in this work can be easily implemented, which asks for subsequent experiments to evaluate their effectiveness.

References

1. Alon, N.: Ranking tournaments. SIAM J. Discrete Math. **20**(1), 137–142 (2006)
2. Aravind, N.R., Sandeep, R.B., Sivadasan, N.: Parameterized lower bounds and dichotomy results for the NP-completeness of H-free edge modification problems. In: Kranakis, E., Navarro, G., Chávez, E. (eds.) LATIN 2016: Theoretical Informatics. LNCS, vol. 9644. Springer, Heidelberg (2016)
3. Bessy, S., Fomin, F.V., Gaspers, S., Paul, C., Perez, A., Saurabh, S., Thomassé, S.: Kernels for feedback arc set in tournaments. J. Comput. Syst. Sci. **77**(6), 1071–1078 (2011)
4. van Bevern, R.: Towards optimal and expressive kernelization for d-Hitting Set. Algorithmica **70**(1), 129–147 (2014)
5. Böcker, S.: A golden ratio parameterized algorithm for cluster editing. J. Discrete Algorithms **16**, 79–89 (2012)
6. Böcker, S., Briesemeister, S., Bui, Q.B.A., Truß, A.: Going weighted: parameterized algorithms for cluster editing. Theor. Comput. Sci. **410**(52), 5467–5480 (2009)
7. Brügmann, D., Komusiewicz, C., Moser, H.: On generating triangle-free graphs. In: Proceedings of the DIMAP Workshop on Algorithmic Graph Theory (AGT 2009). pp. 51–58. ENDM, Elsevier (2009)
8. Cai, L.: Fixed-parameter tractability of graph modification problems for hereditary properties. Inf. Process. Lett. **58**(4), 171–176 (1996)
9. Chen, J., Meng, J.: A $2k$ kernel for the cluster editing problem. J. Comput. Syst. Sci. **78**(1), 211–220 (2012)
10. Cygan, M., Fomin, F.V., Kowalik, L., Lokshtanov, D., Marx, D., Pilipczuk, M., Pilipczuk, M., Saurabh, S.: Parameterized Algorithms. Springer, Switzerland (2015)
11. Cygan, M., Pilipczuk, M., Pilipczuk, M., Wojtaszczyk, J.O.: On multiway cut parameterized above lower bounds. ACM T. Comput. Theor. **5**(1), 3 (2013)
12. Downey, R.G., Fellows, M.R.: Fundamentals of Parameterized Complexity. Springer, Heidelberg (2013)
13. Fellows, M.R.: Blow-ups, win/win's, and crown rules: some new directions in FPT. In: Bodlaender, H.L. (ed.) WG 2003. LNCS, vol. 2880, pp. 1–12. Springer, Heidelberg (2003)
14. Fomin, F.V., Kratsch, S., Pilipczuk, M., Pilipczuk, M., Villanger, Y.: Subexponential fixed-parameter tractability of cluster editing, manuscript available on arXiv. arXiv:1112.4419
15. Garg, S., Philip, G.: Raising the bar for vertex cover: fixed-parameter tractability above a higher guarantee. In: Proceedings of the 27th SODA. SIAM (2016)
16. Gramm, J., Guo, J., Hüffner, F., Niedermeier, R.: Automated generation of search tree algorithms for hard graph modification problems. Algorithmica **39**(4), 321–347 (2004)
17. Hartung, S., Hoos, H.H.: Programming by optimisation meets parameterised algorithmics: a case study for cluster editing. In: Jourdan, L., Dhaenens, C., Marmion, M.-E. (eds.) LION 9 2015. LNCS, vol. 8994, pp. 43–58. Springer, Heidelberg (2015)
18. Karpinski, M., Schudy, W.: Faster algorithms for feedback arc set tournament, Kemeny rank aggregation and betweenness tournament. In: Cheong, O., Chwa, K.-Y., Park, K. (eds.) ISAAC 2010, Part I. LNCS, vol. 6506, pp. 3–14. Springer, Heidelberg (2010)
19. Komusiewicz, C., Uhlmann, J.: Cluster editing with locally bounded modifications. Discrete Appl. Math. **160**(15), 2259–2270 (2012)

20. Křivánek, M., Morávek, J.: NP-hard problems in hierarchical-tree clustering. Acta Informatica **23**(3), 311–323 (1986)

21. Lewis, J.M., Yannakakis, M.: The node-deletion problem for hereditary properties is NP-complete. J. Comput. Syst. Sci. **20**(2), 219–230 (1980)

22. Lokshtanov, D., Narayanaswamy, N.S., Raman, V., Ramanujan, M.S., Saurabh, S.: Faster parameterized algorithms using linear programming. ACM Trans. Algorithms **11**(2), 15:1–15:31 (2014)

23. Mahajan, M., Raman, V.: Parameterizing above guaranteed values: MaxSat and MaxCut. J. Algorithms **31**(2), 335–354 (1999)

24. Moser, H., Niedermeier, R., Sorge, M.: Exact combinatorial algorithms and experiments for finding maximum k-plexes. J. Comb. Optim. **24**(3), 347–373 (2012)

25. Paul, C., Perez, A., Thomassé, S.: Conflict packing yields linear vertex-kernels for k-FAST, k-dense RTI and a related problem. In: Murlak, F., Sankowski, P. (eds.) MFCS 2011. LNCS, vol. 6907, pp. 497–507. Springer, Heidelberg (2011)

26. Razgon, I., O'Sullivan, B.: Almost 2-SAT is fixed-parameter tractable. J. Comput. Syst. Sci. **75**(8), 435–450 (2009)

27. Shamir, R., Sharan, R., Tsur, D.: Cluster graph modification problems. Discrete Appl. Math. **144**(1–2), 173–182 (2004)

28. Wahlström, M.: Algorithms, measures and upper bounds for satisfiability and related problems. Ph.D. thesis, Linköpings universitet (2007)

29. Yannakakis, M.: Edge-deletion problems. SIAM J. Comput. **10**(2), 297–309 (1981)

Completing Partial Schedules for Open Shop with Unit Processing Times and Routing

René van Bevern[1(✉)] and Artem V. Pyatkin[1,2]

[1] Novosibirsk State University, Novosibirsk, Russian Federation
rvb@nsu.ru, artem@math.nsc.ru
[2] Sobolev Institute of Mathematics, Novosibirsk, Russian Federation

Abstract. OPEN SHOP is a classical scheduling problem: given a set $\mathcal{J}$ of jobs and a set $\mathcal{M}$ of machines, find a minimum-makespan schedule to process each job $J_i \in \mathcal{J}$ on each machine $M_q \in \mathcal{M}$ for a given amount p_{iq} of time such that each machine processes only one job at a time and each job is processed by only one machine at a time. In ROUTING OPEN SHOP, the jobs are located in the vertices of an edge-weighted graph $\mathcal{G} = (V, E)$ whose edge weights determine the time needed for the machines to travel between jobs. The travel times also have a natural interpretation as sequence-dependent family setup times. ROUTING OPEN SHOP is NP-hard for $|V| = |\mathcal{M}| = 2$. For the special case with unit processing times $p_{iq} = 1$, we exploit Galvin's theorem about list-coloring edges of bipartite graphs to prove a theorem that gives a sufficient condition for the completability of partial schedules. Exploiting this schedule completion theorem and integer linear programming, we show that ROUTING OPEN SHOP with unit processing times is solvable in $2^{O(|V||\mathcal{M}|^2 \log |V||\mathcal{M}|)} \cdot \mathrm{poly}(|\mathcal{J}|)$ time, that is, fixed-parameter tractable parameterized by $|V| + |\mathcal{M}|$. Various upper bounds shown using the schedule completion theorem suggest it to be likewise beneficial for the development of approximation algorithms.

Keywords: NP-hard scheduling problem · Fixed-parameter algorithm · Edge list-coloring · Sequence-dependent family or batch setup times

1 Introduction

One of the most fundamental and classical scheduling problems is OPEN SHOP [18], where the input is a set $\mathcal{J} := \{J_1, \ldots, J_n\}$ of jobs, a set $\mathcal{M} := \{M_1, \ldots, M_m\}$ of machines, and the processing time p_{iq} that job J_i needs on machine M_q; the task is to process all jobs on all machines in a minimum amount of time such that each machine processes at most one job at a time and each job is processed by at most one machine at a time.

R. van Bevern—Supported by the Russian Foundation for Basic Research (RFBR) under research project 16-31-60007 mol_a_dk.
A.V. Pyatkin—Supported by the RFBR under research projects 15-01-00462 and 15-01-00976.

© Springer International Publishing Switzerland 2016
A.S. Kulikov and G.J. Woeginger (Eds.): CSR 2016, LNCS 9691, pp. 73–87, 2016.
DOI: 10.1007/978-3-319-34171-2_6

Averbakh et al. [3] introduced the variant ROUTING OPEN SHOP, where the jobs are located in the vertices of an edge-weighted graph whose edge weights determine the time needed for the machines to travel between jobs. Initially, the machines are located in a depot. The task is to minimize the time needed for processing all jobs by all machines and returning all machines to the depot. ROUTING OPEN SHOP models, for example, tasks where machines have to perform maintenance work on stationary objects in a workshop [3]. ROUTING OPEN SHOP has also been interpreted as a variant of OPEN SHOP with sequence-dependent family or batch setup times [1,36]. Formally, ROUTING OPEN SHOP is defined as follows.

Definition 1.1 (ROUTING OPEN SHOP). An *instance* of ROUTING OPEN SHOP consists of a graph $\mathcal{G} = (V, E)$ with a *depot* $v^* \in V$ and travel times $c \colon E \to \mathbb{N}$, jobs $\mathcal{J} = \{J_1, \ldots, J_n\}$ with locations $\mathcal{L} \colon \mathcal{J} \to V$, machines $\mathcal{M} = \{M_1, \ldots, M_m\}$, and, for each job J_i and machine M_q, a processing time $p_{iq} \in \mathbb{N}$.

A *route with s stays* is a sequence $R := (R_i)_{i=1}^{s}$ of *stays* $R_i = (a_i, v_i, b_i) \in \mathbb{N} \times V \times \mathbb{N}$ from time a_i to time b_i in vertex v_i for $1 \le i \le s$ such that $v_1 = v_s = v^*$, $a_1 = 0$, and $b_i + c(v_i, v_{i+1}) \le a_{i+1} \le b_{i+1}$ for $1 \le i \le s - 1$. The *length* of R is the end b_s of the last stay.

A *schedule* $S \colon \mathcal{J} \times \mathcal{M} \to \mathbb{N}$ is a total function determining the *start time* $S(J_i, M_q)$ of each job J_i on each machine M_q. That is, each job J_i is *processed* by each machine M_q in the half-open time interval $[S(J_i, M_q), S(J_i, M_q) + p_{iq})$. A schedule is *feasible* with respect to routes $(R_{M_q})_{M_q \in \mathcal{M}}$ if

  (i) no machine M_q processes two jobs $J_i \neq J_j$ at the same time, that is, $S(J_i, M_q) + p_{iq} \le S(J_j, M_q)$ or $S(J_j, M_q) + p_{jq} \le S(J_i, M_q)$ for all jobs $J_i \neq J_j$ and machines M_q,
 (ii) no job J_i is processed by two machines M_q, M_r at the same time, that is, $S(J_i, M_q) + p_{iq} \le S(J_i, M_r)$ or $S(J_i, M_r) + p_{ir} \le S(J_i, M_q)$ for all jobs J_i and machines $M_q \neq M_r$,
(iii) machines stay in the location $\mathcal{L}(J_i)$ while executing a job J_i, that is, for each job J_i and machine M_q with route $R_{M_q} = (R_k)_{k=1}^{s}$, there is a $k \in \{1, \ldots, s\}$ such that $R_k = (a_k, \mathcal{L}(J_i), b_k)$ with $a_k \le S(J_i, M_q) \le S(J_i, M_q) + p_{iq} \le b_k$.

A schedule S is *feasible* and has *length* L if there are routes $(R_{M_q})_{M_q \in \mathcal{M}}$ of length L such that S is feasible with respect to $(R_{M_q})_{M_q \in \mathcal{M}}$. An *optimal solution* to a ROUTING OPEN SHOP instance is a feasible schedule of minimum length.

Preemption and Unit Processing Times. OPEN SHOP is NP-hard for $|\mathcal{M}| = 3$ machines [18]. Thus, so is ROUTING OPEN SHOP with $|V| = 1$ vertex and $|\mathcal{M}| = 3$ machines. ROUTING OPEN SHOP remains (weakly) NP-hard even for $|V| = |\mathcal{M}| = 2$ [3]; there are approximation algorithms both for this special and the general case [2,11,27,35]. However, OPEN SHOP is solvable in polynomial time if

(1) job preemption is allowed, or
(2) all jobs J_i have unit processing time $p_{iq} = 1$ on all machines M_q.

It is natural to ask how these results transfer to ROUTING OPEN SHOP.

Regarding (1), Pyatkin and Chernykh [31] have shown that ROUTING OPEN SHOP with allowed preemption is solvable in polynomial time if $|V| = |\mathcal{M}| = 2$, yet NP-hard for $|V| = 2$ and an unbounded number $|\mathcal{M}|$ of machines.

Regarding (2), ROUTING OPEN SHOP with unit processing times models tasks where machines process batches of equal jobs in several locations and where the transportation of machines between the locations takes significantly longer than processing each individual job in a batch. Herein, there are conceivable situations where the number of machines and locations is small.

ROUTING OPEN SHOP with unit processing times clearly is NP-hard even for $|\mathcal{M}| = 1$ machine since it generalizes the metric travelling salesperson problem. It is not obvious whether it is solvable in polynomial time even when both $|V|$ and $|\mathcal{M}|$ are fixed. We show the even stronger result that ROUTING OPEN SHOP with unit processing times is solvable in $2^{O(|V||\mathcal{M}|^2 \log |V||\mathcal{M}|)} \cdot \mathrm{poly}(|\mathcal{J}|)$ time, that is, *fixed-parameter tractable*.

Fixed-Parameter Algorithms. Fixed-parameter algorithms are an approach towards efficiently and optimally solving NP-hard problems: the main idea is to accept the exponential running time for finding optimal solutions to NP-hard problems, yet to confine it to some smaller problem parameter k [12,14,16,30]. A problem with parameter k is called *fixed-parameter tractable (FPT)* if there is an algorithm that solves any instance I in $f(k)\mathrm{poly}(|I|)$ time, where f is an arbitrary computable function. The corresponding algorithm is called *fixed-parameter algorithm*. In contrast to algorithms that merely run in polynomial time for fixed k, fixed-parameter algorithms can potentially solve NP-hard problems optimally and efficiently if the parameter k is small.

Recently, the field of fixed-parameter algorithmics has shown increased interest in scheduling [5,7,9,10,15,23,25,29] and routing [6,8,13,19–22,34], yet fixed-parameter algorithms for routing scheduling problems are unexplored so far.

Our Results. Using Galvin's theorem on list-coloring edges of bipartite graphs [17,33], in Sect. 3 we prove a sufficient condition for the polynomial-time completability of a partial schedule, which does not necessarily assign start times to all jobs on all machines, into a feasible schedule.

We use the schedule completion theorem to prove upper bounds on various parameters of optimal schedules, in particular on their lengths in Sect. 4.

Using these bounds and integer linear programming, in Sect. 5 we show that ROUTING OPEN SHOP with unit processing times is fixed-parameter tractable parameterized by $|V| + |\mathcal{M}|$ (unlike the general case when assuming P $\neq$ NP).

Since the schedule extension theorem is a useful tool for proving upper bounds on various parameters of optimal schedules, we expect the schedule completion theorem to be likewise beneficial for approximation algorithms.

Due to space constraints, some proofs are deferred to a full version of the paper.

Input Encoding. In general, a ROUTING OPEN SHOP instance requires at least $\Omega(|\mathcal{J}| \cdot |\mathcal{M}| + |V| + |E|)$ bits in order to encode the processing time of each job on each machine and the travel time for each edge. We call this the *standard encoding*. In contrast, an instance of ROUTING OPEN SHOP with unit processing times can be encoded using $O(|V|^2 \cdot \log c_{\max} + |V| \cdot \log |\mathcal{J}|)$ bits by simply associating with each vertex in V the number of jobs it contains, where $c_{\max}$ is the maximum travel time. We call this the *compact encoding*.

All running times in this article are stated for computing and outputting a minimum-length schedule, whose encoding requires at least $\Omega(|\mathcal{J}| \cdot |M|)$ bits for the start time of each job on each machine. Thus, outputting the schedule is impossible in time polynomial in the size of the compact encoding. We therefore assume to get the input instance in standard encoding, like for general ROUTING OPEN SHOP.

However, we point out that the *decision version* of ROUTING OPEN SHOP with unit processing times is fixed-parameter tractable parameterized by $|V| + |\mathcal{M}|$ even when assuming the compact encoding: our algorithm is able to decide whether there is a schedule of given length L in $2^{O(|V||\mathcal{M}|^2 \log |V||\mathcal{M}|)} \cdot \text{poly}(|I|)$ time, where I is an instance given in compact encoding. To this end, the algorithm does not apply the schedule completion Theorem 3.4 to explicitly *construct* a schedule but merely to conclude its existence.

2   Preprocessing for Metric Travel Times

In this section, we show how any instance can be transformed into an equivalent instance with travel times satisfying the triangle inequality. This will allow us to assume that, in an optimal schedule, a machine only stays in a vertex if it processes at least one job there: otherwise, it could take a "shortcut" bypassing the vertex.

Lemma 2.1. Let I be a ROUTING OPEN SHOP instance and I' be obtained from I by replacing the graph $\mathcal{G} = (V, E)$ with travel times $c\colon E \to \mathbb{N}$ by a complete graph $\mathcal{G}'$ on the vertex set V with travel times $c' \colon \{v, w\} \mapsto \text{dist}_c(v, w)$, where $\text{dist}_c(v, w)$ is the length of a shortest path between v and w in $\mathcal{G}$ with respect to c.

Then, any schedule for I is a schedule of the same length for I' and vice versa. Moreover, c' satisfies the triangle inequality $c'(\{v, w\}) \leq c'(\{v, u\}) + c'(\{u, w\})$ for all $u, v, w \in V$ and can be computed in $O(|V|^3)$ time.

Lemma 2.2. Let S be a feasible schedule of length L for a ROUTING OPEN SHOP instance satisfying the triangle inequality.

Then, S is feasible with respect to machine routes $(R_{M_q})_{M_q \in \mathcal{M}}$ of length at most L such that, for each route $R = ((a_k, v_k, b_k))_{k=1}^{s}$ and each stay (a_k, v_k, b_k) on R, except, maybe, for $k \in \{1, s\}$, there is a job $J_i \in \mathcal{J}$ with $S(J_i, M_q) \in [a_k, b_k)$.

Clearly, from Lemma 2.2, we get the following:

Observation 2.3. Vertices $v \in V \setminus \{v^*\}$ with $\mathcal{J}_v = \emptyset$ can be deleted from a ROUTING OPEN SHOP instance satisfying the triangle inequality, where v^* is the depot.

From now on, we assume that our input instances of ROUTING OPEN SHOP satisfy the triangle inequality and exploit Lemma 2.2 and Observation 2.3.

3 Schedule Completion Theorem

In this section, we present a theorem that allows us to complete *partial schedules*, which do not necessarily assign a start point to each job on each machine, into feasible schedules.

In the following, we consider only ROUTING OPEN SHOP with unit processing times and say that a machine M_q processes a job J_i at time $S(J_i, M_q)$ if it processes job J_i in the time interval $[S(J_i, M_q), S(J_i, M_q) + 1)$. We use $S(J_i, M_q) = \bot$ to denote that the processing time of job J_i on machine M_q is undefined.

Definition 3.1 (Partial schedule). A *partial schedule* with respect to given routes $(R_{M_q})_{M_q \in \mathcal{M}}$ of length at most L is a *partial* function $S \colon \mathcal{J} \times \mathcal{M} \to \mathbb{N}$ satisfying Definition 1.1(i–iii) for those jobs $J_i, J_j \in \mathcal{J}$ and machines $M_q, M_r \in \mathcal{M}$ for which $S(J_i, M_q) \neq \bot$ and $S(J_j, M_r) \neq \bot$. For a partial schedule $S \colon \mathcal{J} \times \mathcal{M} \to \mathbb{N}$, we introduce the following terminology:

$\mathcal{J}^S_{M_q} := \{J_i \in \mathcal{J} \mid S(J_i, M_q) = \bot\}$ is the set of jobs that lack processing by machine M_q,

$\mathcal{M}^S_{J_i} := \{M_q \in \mathcal{M} \mid S(J_i, M_q) = \bot\}$ is the set of machines that job J_i lacks processing of (note that $J_i \in \mathcal{J}^S_{M_q}$ if and only if $M_q \in \mathcal{M}^S_{J_i}$),

$T^S_{J_i} := \{t \leq L \mid \exists M_q \in \mathcal{M} : S(J_i, M_q) = t\}$ is the set of time units where job J_i is being processed,

$T^S_{M_q} := \{t \leq L \mid \exists J_i \in \mathcal{J} : S(J_i, M_q) = t\}$ is the set of time units where machine M_q is processing,

$T^{R_{M_q}}_v := \{t \leq L \mid \text{there is a stay } (a_i, v, b_i) \text{ on } R_{M_q} \text{ such that } a_i \leq t < b_i\}$ are the time units where M_q stays in a vertex $v \in V$, and

$\mathcal{J}_v := \{J_i \in \mathcal{J} \mid \mathcal{L}(J_i) = v\}$ is the set of jobs in vertex $v \in V$ of $\mathcal{G}$.

The schedule completion theorem will allow us to turn any *completable* partial schedule into a feasible schedule. Intuitively, a schedule is completable if a machine has enough "free time" in each vertex to process all yet unprocessed jobs and to wait for other machines in the vertex to free their jobs.

Definition 3.2 (Completable schedule). Let $(R_{M_q})_{M_q \in \mathcal{M}}$ be a family of routes and, for each vertex $v \in V$, let $\bigcup_{s=1}^{g_v} \mathcal{M}^s_v := \mathcal{M}$ be a partition of machines such that, for any two machines $M_q \in \mathcal{M}^s_v$ and $M_r \in \mathcal{M}^t_v$ with $s \neq t$, one has $T^{R_{M_q}}_v \cap T^{R_{M_r}}_v = \emptyset$.

A partial schedule $S \colon \mathcal{J} \times \mathcal{M} \to \mathbb{N}$ with respect to $(R_{M_q})_{M_q \in \mathcal{M}}$ is *completable* if, for each vertex $v \in V$, each $1 \leq s \leq g_v$, each machine $M_q \in \mathcal{M}^s_v$, and each job $J_i \in \mathcal{J}^S_{M_q} \cap \mathcal{J}_v$, it holds that

$$|T^{R_{M_q}}_v \setminus (T^S_{J_i} \cup T^S_{M_q})| \geq \max \begin{cases} \max\limits_{M_r \in \mathcal{M}^s_v} |\mathcal{J}^S_{M_r} \cap \mathcal{J}_v|, \\ \max\limits_{J_j \in \mathcal{J}_v} |\mathcal{M}^S_{J_j} \cap \mathcal{M}^s_v|. \end{cases} \tag{3.1}$$

Example 3.3. Let $(R_{M_q})_{M_q \in \mathcal{M}}$ be routes such that all machines are in the same vertex at the same time, that is, $T_v^{R_{M_q}} = T_v^{R_{M_r}}$ for all vertices $v \in V$ and machines $M_q, M_r \in \mathcal{M}$. Moreover, assume that each machine $M_q \in \mathcal{M}$ stays in each vertex $v \in V$ at least $\max\{|\mathcal{J}_v|, |\mathcal{M}|\}$ time, that is, $|T_v^{R_{M_q}}| \geq \max\{|\mathcal{J}_v|, |\mathcal{M}|\}$. Then the empty schedule is completable and, by the following schedule completion theorem, there is a feasible schedule with respect to the routes $(R_{M_q})_{M_q \in \mathcal{M}}$.

Theorem 3.4 (Schedule completion theorem). *If a partial schedule $S: \mathcal{J} \times \mathcal{M} \to \mathbb{N}$ with respect to routes $(R_{M_q})_{M_q \in \mathcal{M}}$ is completable, then there is a feasible schedule $S' \supseteq S$ with respect to the routes $(R_{M_q})_{M_q \in \mathcal{M}}$ and it can be computed in time polynomial in $|\mathcal{J}| + |\mathcal{M}| + |V| + \sum_{v \in V, M_q \in \mathcal{M}} |T_v^{R_{M_q}}|$.*

We prove Theorem 3.4 using Galvin's theorem about properly list-coloring the edges of bipartite graphs [17,33].

Definition 3.5 (Proper edge coloring, list chromatic index). A *proper edge coloring* of a graph $G = (V, E)$ is a coloring $C: E \to \mathbb{N}$ of the edges of G such that $C(e_1) \neq C(e_2)$ if $e_1 \cap e_2 \neq \emptyset$.

A graph $G = (V, E)$ is k-edge-choosable if, for every family $\{L_e \subseteq \mathbb{N} \mid e \in E\}$ satisfying $|L_e| \geq k$ for all $e \in E$, G allows for a proper edge coloring $C: E \to \mathbb{N}$ with $C(e) \in L_e$. The *list chromatic index* $\chi'_\ell(G)$ of G is the least integer k such that G is k-edge-choosable.

Theorem 3.6 (Galvin [17]). *For any bipartite multigraph G, it holds that $\chi'_\ell(G) = \Delta(G)$, where $\Delta(G)$ is the maximum degree of G.*

Moreover, given a bipartite multigraph $G = (V, E)$ and, for each edge $e \in E$, a set $L_e \subseteq \mathbb{N}$ with $|L_e| \geq \Delta(G)$, a proper edge coloring $C: E \to \mathbb{N}$ with $C(e) \in L_e$ is computable in polynomial time.

Before Galvin [17] proved Theorem 3.6, its special case with $G = K_{n,n}$ being a complete bipartite graph was known as Dinitz' conjecture. A self-contained proof of Theorem 3.6 was later given by Slivnik [33], who also pointed out the polynomial-time computability of the coloring. We now use Theorem 3.6 to prove Theorem 3.4.

Proof (of Theorem 3.4). Let $B = (\mathcal{J} \cup \mathcal{M}, X)$ be a bipartite graph with an edge $\{J_i, M_q\} \in X$ if and only if $S(J_i, M_q) = \perp$ for $J_i \in \mathcal{J}$ and $M_q \in \mathcal{M}$. We compute a proper edge coloring C of B such that, for each edge $\{J_i, M_q\} \in X$, we have

$$C(\{J_i, M_q\}) \in T_{\mathcal{L}(J_i)}^{R_{M_q}} \setminus (T_{J_i}^S \cup T_{M_q}^S) \tag{3.2}$$

$$\text{and define } S'(J_i, M_q) := \begin{cases} C(\{J_i, M_q\}) & \text{if } \{J_i, M_q\} \in X \text{ and} \\ S(J_i, M_q) & \text{otherwise.} \end{cases}$$

It remains to show that (1) the edge coloring C is computable in time polynomial in $|\mathcal{J}| + |\mathcal{M}| + |V| + \sum_{v \in V, M_q \in \mathcal{M}} |T_v^{R_{M_q}}|$ and that (2) S' is a feasible schedule.

(1) We obtain the proper edge coloring C by independently computing, for each induced subgraph $B_{vs} := B[\mathcal{J}_v \cup \mathcal{M}_v^s]$ for all $v \in V$ and $1 \le s \le g_v$, a proper edge coloring C_{vs} satisfying (3.2). To this end, observe that the maximum degree of B_{vs} is

$$\Delta := \max \begin{cases} \max_{M_r \in \mathcal{M}_v^s} |\mathcal{J}_{M_r}^S \cap \mathcal{J}_v|, \\ \max_{J_j \in \mathcal{J}_v} |\mathcal{M}_{J_j}^S \cap \mathcal{M}_v^s|. \end{cases}$$

By Theorem 3.6, if, for each edge $e \in X$, we have a list L_e of colors with $|L_e| \ge \Delta$, then B_{vs} has a proper edge coloring C_{vs} with $C_{vs}(e) \in L_e$ for each edge e of B_{vs}. For each edge $\{J_i, M_q\}$ of B_{vs}, we choose

$$L_{\{J_i, M_q\}} := T_v^{R_{M_q}} \setminus (T_{J_i}^S \cup T_{M_q}^S).$$

Since S is completable (Definition 3.2), we have $|L_e| \ge \Delta$ for each edge e of B_{vs}. Thus, B_{vs} admits a proper edge coloring C_{vs} satisfying (3.2).

We now let $C := \bigcup_{v \in V, 1 \le s \le g_v} C_{vs}$. This is a proper edge coloring for the bipartite graph B since, for edges e_{vs} of B_{vs} and e_{wt} of B_{wt} with $v \ne w$ or $s \ne t$, we have $L_{e_{vs}} \cap L_{e_{wt}} = \emptyset$: for any vertex $v \in V$ and machines $M_q \in \mathcal{M}_v^s, M_r \in \mathcal{M}_v^t$ with $s \ne t$, one has $T_v^{R_{M_q}} \cap T_v^{R_{M_r}} = \emptyset$, and for any machine $M_q \in \mathcal{M}$ and $v \ne w \in V$, one has $T_v^{R_{M_q}} \cap T_w^{R_{M_q}} = \emptyset$. Moreover, C satisfies (3.2) since each C_{vs} for $v \in V$ and $1 \le s \le g_v$ satisfies (3.2).

Regarding the running time, it is clear that, for each $v \in V$ and $1 \le s \le g_v$, the bipartite graph B_{vs} and the sets $L_{\{J_j, M_q\}}$ of allowed colors for each edge $\{J_j, M_q\}$ are computable in time polynomial in $|\mathcal{J}| + |\mathcal{M}| + |T_v^{R_{M_q}}|$.[1] Moreover, by Theorem 3.6, the sought edge coloring C_{vs} for each B_{vs} is computable in time polynomial in $|B_{vs}| + \sum_{e \in E(B_{vs})} |L_e|$.

(2) We first show that S' is a schedule. For each job $J_i \in \mathcal{J}$ and each machine $M_q \in \mathcal{M}$ we have $S(J_i, M_q) \ne \bot$ or $\{J_i, M_q\} \in X$. Thus, $S'(J_i, M_q) = S(J_i, M_q) \ne \bot$ or $S'(J_i, M_q) = C(\{J_i, M_q\}) \ne \bot$ and S' is a schedule. We show that S' is feasible.

First, let $J_i \in \mathcal{J}$ be a job and $M_q, M_r \in \mathcal{M}$ be distinct machines. We show that $S'(J_i, M_q) \ne S'(J_i, M_r)$. We distinguish three cases. If $S(J_i, M_q) \ne \bot$ and $S(J_i, M_r) \ne \bot$, then $S'(J_i, M_q) = S(J_i, M_q) \ne S(J_i, M_r) = S'(J_i, M_r)$. If $S(J_i, M_q) = \bot = S(J_i, M_r)$, then $S'(J_i, M_q) = C(\{J_i, M_q\}) \ne C(\{J_i, M_r\}) = S'(J_i, M_r)$. Finally, if $S(J_i, M_q) = \bot$ and $S(J_i, M_r) \ne \bot$, then $S'(J_i, M_q) \ne S'(J_i, M_r)$ since $S'(J_i, M_q) = C(\{J_i, M_q\}) \in T_v^{R_{M_q}} \setminus (T_{J_i}^S \cup T_{M_q}^S)$ and $S(J_i, M_r) \in T_{J_i}^S$.

Now, let $J_i, J_j \in \mathcal{J}$ be distinct jobs and $M_q \in \mathcal{M}$ be a machine. We show $S'(J_i, M_q) \ne S'(J_j, M_q)$. We distinguish three cases. If $S(J_i, M_q) \ne \bot$ and $S(J_j, M_q) \ne \bot$, then $S'(J_i, M_q) = S(J_i, M_q) \ne S(J_j, M_q) = S'(J_j, M_q)$. If $S(J_i, M_q) = \bot = S(J_j, M_q)$, then $S'(J_i, M_q) = C(\{J_i, M_q\}) \ne C(\{J_j, M_q\}) = S'(J_j, M_q)$. Finally, if $S(J_i, M_q) = \bot$ and $S(J_j, M_q) \ne \bot$, then $S'(J_i, M_q) \ne S'(J_j, M_q)$ since $S'(J_i, M_q) = C(\{J_i, M_q\}) \in T_v^{R_{M_q}} \setminus (T_{J_i}^S \cup T_{M_q}^S)$ and $S'(J_j, M_q) \in T_{M_q}^S$. $\qquad \square$

---

[1] We abstain from a more detailed running time analysis since no such analysis is available for the forthcoming application of Theorem 3.6 (yet).

4   Upper and Lower Bounds

In this section, we show lower and upper bounds on the lengths of optimal solutions to ROUTING OPEN SHOP with unit processing times. These will be exploited in our fixed-parameter algorithm and make first steps towards approximation algorithms.

We assume ROUTING OPEN SHOP instances to be preprocessed to satisfy the triangle inequality. By Lemma 2.1, this does not change the length of optimal schedules. However, it ensures that the minimum cost of a cycle visiting each vertex of the graph $\mathcal{G} = (V, E)$ with travel times $c \colon E \to \mathbb{N}$ *at most once* coincides with the minimum cost of a cycle doing so *exactly once* [32], that is, of a *Hamiltonian cycle*.

A simple lower bound is given by the fact that, in view of Observation 2.3, all machines have to visit each vertex at least once and to process $|\mathcal{J}|$ jobs.

Observation 4.1. Let H be a minimum-cost Hamiltonian cycle in the graph $\mathcal{G} = (V, E)$ with metric travel times $c \colon E \to \mathbb{N}$. Then, any feasible schedule has length at least $c(H) + |\mathcal{J}|$.

A trivial upper bound can be given by letting the machines work sequentially.

Observation 4.2. Given a Hamiltonian cycle H for the graph $\mathcal{G} = (V, E)$ with travel times $c \colon E \to \mathbb{N}$, a feasible schedule of length $c(H) + |\mathcal{J}| + |\mathcal{M}| - 1$ is computable in $O(|\mathcal{J}| \cdot |\mathcal{M}| + |V|)$ time.

This bound can be improved if $c(H) + 1 \leq |\mathcal{M}| \leq |\mathcal{J}|$ or $c(H) + 1 \leq |\mathcal{J}| \leq |\mathcal{M}|$:

Proposition 4.3. *Given a Hamiltonian cycle H for the graph $\mathcal{G} = (V, E)$ with travel times $c \colon E \to \mathbb{N}$, a feasible schedule of length $2c(H) + \max\{|\mathcal{J}|, |\mathcal{M}|\}$ is computable in $O(|\mathcal{J}|^2 + |\mathcal{M}| + |V|)$ time.*

We next study for which instances one gets an upper bound that matches the lower bound from Observation 4.1. In Example 3.3, we have already seen that arbitrary machine routes that stay in each vertex v at least $\max\{|\mathcal{J}_v|, |\mathcal{M}|\}$ time can be completed into a feasible schedule. We therefore distinguish vertices v for which staying $|\mathcal{J}_v|$ time is both necessary and sufficient.

Definition 4.4 (Criticality of vertices). For a vertex $v \in V$, we denote by

$$k(v) := \max\{0, |\mathcal{M}| - |\mathcal{J}_v|\} \text{ the } \textit{criticality of } v, \text{ and by}$$
$$K := \sum_{v \in V} k(v) \text{ the } \textit{total criticality}.$$

A vertex $v \in V$ is *critical* if $k(v) > 0$, that is, if $|\mathcal{J}_v| < |\mathcal{M}|$.

Proposition 4.5. *Given a Hamiltonian cycle H for the graph $\mathcal{G} = (V, E)$ with travel times $c \colon E \to \mathbb{N}$, a feasible schedule of length at most $c(H) + |\mathcal{J}| + K$ can be computed in polynomial time.*

Proof. Let $H = (v_1, v_2, \ldots, v_{|V|})$ and $v_{|V|+1} := v_1$. Without loss of generality, assume that $v_1 = v^*$, where v^* is the depot vertex. Each machine $M_q \in \mathcal{M}$ uses the same route R of $|V| + 1$ stays $(a_1, v_1, b_1), \ldots, (a_{|V|+1}, v_{|V|+1}, b_{|V|+1})$, where

$$a_1 := 0, \qquad\qquad b_{|V|+1} := a_{|V|+1},$$
$$a_{i+1} := b_i + c(v_i, v_{i+1}), \qquad b_i := a_i + |\mathcal{J}_{v_i}| + k(v_i), \quad \text{for } i \in \{1, \ldots, |V|\}.$$

Each stay (a_i, v_i, a_i) lasts $|\mathcal{J}_{v_i}| + k(v_i) = \max\{|\mathcal{J}_{v_i}|, |\mathcal{M}|\}$ time. By Theorem 3.4, the empty schedule S is completable into a feasible schedule S' with respect to the route R for each machine and S' is computable in time polynomial in $|\mathcal{J}| + |\mathcal{M}| + \sum_{v \in V} |T_v^R| \in O(|\mathcal{J}| + |\mathcal{M}| + K) \subseteq O(|\mathcal{J}| + |\mathcal{M}| + |V| \cdot |\mathcal{M}|)$. Finally, the route R has length

$$b_{|V|+1} = a_{|V|+1} = b_{|V|} + c(v_{|V|}, v_1) + \sum_{i=1}^{|V|}(|\mathcal{J}_{v_i}| + k(v_i)) = c(H) + |\mathcal{J}| + K. \qquad \square$$

Combining Observation 4.1 and Proposition 4.5 and that a minimum-cost Hamiltonian cycle is computable in $O(2^{|V|} \cdot |V|^2)$ time using the algorithm of Bellman [4], Held and Karp [24], we obtain a first fixed-parameter tractability result:

Corollary 4.6. ROUTING OPEN SHOP *with unit processing times is fixed-parameter tractable parameterized by* $|V|$ *if there are no critical vertices.*

Corollary 4.6 makes clear that, given the schedule completion theorem, critical vertices are the main obstacle for solving ROUTING OPEN SHOP with unit processing times: while staying $|\mathcal{J}_v|$ time in a noncritical vertex $v \in V$ is both necessary and sufficient, staying in critical vertices $|\mathcal{M}|$ time is sufficient, but not necessary. Indeed, as shown in Fig. 1, in the presence of critical vertices, there might not even be optimal schedules in which the machines travel along Hamiltonian cycles.

S	M_1	M_2	M_3	M_4	M_5	M_6	M_7
J_1	0	1	2	6	7	8	3
J_2	2	3	7	4	5	6	1
J_3	5	6	4	1	2	3	7

Fig. 1. On the left: a graph with one job in each vertex, travel times as denoted on the edges, and the depot being J_1. On the right: a schedule S of length 9 to process these jobs on seven machines. Note that machine M_7 does not travel along a Hamiltonian cycle, but along route J_1, J_2, J_1, J_3, J_1. One can show that any schedule in which machines travel along Hamiltonian cycles has length at least 10.

5   Fixed-Parameter Algorithm

In this section, we present a fixed-parameter algorithm for ROUTING OPEN SHOP with unit processing times, which is our main algorithmic result:

Theorem 5.1. ROUTING OPEN SHOP *with unit processing times is solvable in* $2^{O(|V||\mathcal{M}|^2 \log |V||\mathcal{M}|)} \cdot poly(|\mathcal{J}|)$ *time.*

The outline of the algorithm for Theorem 5.1 is as follows: in Sect. 5.1, we use the schedule completion Theorem 3.4 to show that the routes of a minimum-length schedule comply with one of $2^{O(|V||\mathcal{M}|^2 \log |V||\mathcal{M}|)}$ *pre-schedules*, which determines the sequence of vertices that each machine stays in, the durations of stays in critical vertices, and the time offsets between stays in critical vertices.

In Sect. 5.2, we use integer linear programming to compute, for each pre-schedule, shortest complying routes so that each machine stays in each non-critical vertex v for at least $|\mathcal{J}_v|$ time. The schedule for noncritical vertices is then implied by the schedule completion Theorem 3.4, whereas we compute the schedule for critical vertices using brute force.

5.1   Enumerating Pre-schedules

One can show that the routes of a minimum-length schedule comply with some pre-schedule:

Definition 5.2 (Pre-schedule). A *pre-stay* is a triple $(M_q, v, \sigma) \in M \times V \times \{1, \dots, |V||\mathcal{M}| + 2\}$, intuitively meaning that machine $M_q \in \mathcal{M}$ has its σ-th stay in vertex $v \in V$. We call $T = ((M_{q_i}, v_i, \sigma_i))_{i=1}^s$ a *pre-stay sequence* if,

(i) for each $M_q \in \mathcal{M}$, the σ_i with $q_i = q$ increase in steps of one for increasing i.

Machine routes $(R_{M_q})_{M_q \in \mathcal{M}}$, where $R_{M_q} = ((a_k^q, w_k^q, b_k^q))_{k=1}^{t_q}$, *comply* with a pre-stay sequence if

(ii) route R_{M_q} has a stay (a_k^q, w_k^q, b_k^q) if and only if (M_q, w_k^q, k) is in T and,
(iii) for pre-stays (M_{q_i}, v_i, σ_i) and (M_{q_j}, v_j, σ_j) with $i < j$, one has $a_{\sigma_i}^{q_i} \leq a_{\sigma_j}^{q_j}$.

Let $\mathcal{K} := \{i \leq s \mid v_i \text{ is critical}\}$ be the indices of pre-stays in critical vertices of T. A *length assignment* is a map $A\colon \mathcal{K} \to \{0, \dots, 2|\mathcal{M}| - 1\}$. Machine routes $(R_{M_q})_{M_q \in \mathcal{M}}$ *comply* with a length assignment A if,

(iv) for each pre-stay (M_{q_i}, v_i, σ_i) on T with $i \in \mathcal{K}$, one has $b_{\sigma_i}^{q_i} - a_{\sigma_i}^{q_i} = A(i)$.

A *displacement* is a map $D\colon \mathcal{K} \to \{0, \dots, 2|\mathcal{M}|\}$. The machine routes $(R_{M_q})_{M_q \in \mathcal{M}}$ *comply* with a displacement D if

(v) for two pre-stays (M_{q_i}, v_i, σ_i) and (M_{q_j}, v_j, σ_j) with $i, j \in \mathcal{K}$ and $k \notin \mathcal{K}$ for all k with $i < k < j$, one has

$$a_{\sigma_j}^{q_j} \geq a_{\sigma_i}^{q_i} + 2|\mathcal{M}| \qquad \text{if } D(j) = 2|\mathcal{M}| \text{ and}$$
$$a_{\sigma_j}^{q_j} = a_{\sigma_i}^{q_i} + D(j) \qquad \text{if } D(j) < 2|\mathcal{M}|.$$

We call (T, A, D) a *pre-schedule* and say that machine routes *comply* with (T, A, D) if they comply with each of T, A, and D, that is, (i)–(v) hold.

We show that an optimal solution for ROUTING OPEN SHOP with unit processing times can be found by solving instances of the following problem:

Problem 5.3.
Input: An instance I of ROUTING OPEN SHOP with unit processing times, a pre-schedule (T, A, D), and a natural number L.
Task: Compute a schedule whose machine routes $(R_{M_q})_{M_q \in \mathcal{M}}$ have length at most L and comply with (T, A, D), if such a schedule exists.

Proposition 5.4. *For a* ROUTING OPEN SHOP *instance I with unit processing times there is a set $\mathcal{I}$ of $2^{O(|V||\mathcal{M}|^2 \log |V||\mathcal{M}|)}$ instances of Problem 5.3 such that*

(i) if some instance $(I, (T, A, D), L) \in \mathcal{I}$ has a solution S, then S is a schedule of length at most L for I and

(ii) there is a minimum-length schedule S for I such that S is a solution for at least one instance $(I, (T, A, D), L) \in \mathcal{I}$, where L is the length of S.

Moreover, the set $\mathcal{I}$ can be generated in $2^{O(|V||\mathcal{M}|^2 \log |V||\mathcal{M}|)} \cdot \text{poly}(|\mathcal{J}|)$ time.

Having Proposition 5.4, for proving Theorem 5.1, it remains to solve Problem 5.3 in $2^{O(|V||\mathcal{M}|^2 \log |V||\mathcal{M}|)} \cdot \text{poly}(|\mathcal{J}|)$ time since a shortest schedule for an instance I of ROUTING OPEN SHOP with unit processing times can be found by solving the instances $(I, (T, A, D), L) \in \mathcal{I}$ for increasing L. The proof of Proposition 5.4 is based on proving that there are at most $2^{O(|V||\mathcal{M}|^2 \log |V||\mathcal{M}|)}$ pre-schedules and the following two lemmas.

Lemma 5.5. *Each of the routes $(R_{M_q})_{M_q \in \mathcal{M}}$ of an optimal schedule consists of at most $|V||\mathcal{M}| + 2$ stays.*

Proof. Let H be a minimum-cost Hamiltonian cycle for the graph $\mathcal{G}$ with travel times $c: E \to \mathbb{N}$. Let $M_q \in \mathcal{M}$ be an arbitrary machine. It has to stay in all vertices and return to the depot, that is, its tour R_{M_q} has at least $|V| + 1$ stays. Moreover, by Observation 4.1, its length is at least $c(H) + |\mathcal{J}|$. By Lemma 2.2, each additional stay has length at least one.

   Thus, if R_{M_q} had more than $|V| + K + 1$ stays, where K is the total critically of vertices in the input instance (cf. Definition 4.4), then it would have length at least $c(H) + |\mathcal{J}| + K + 1$, contradicting the optimality of the schedule by Proposition 4.5. Thus, the number of stays on R_{M_q} is at most

$$|V| + K + 1 = |V| + \sum_{v \in V} \max\{0, |\mathcal{M}| - |\mathcal{J}_v|\} + 1 \le |V||\mathcal{M}| + 2$$

since, by Observation 2.3, only for the depot v^* one might have $\mathcal{J}_{v^*} = \emptyset$.   □

Lemma 5.5 implies that there is a pre-stay sequence that the routes $(R_{M_q})_{M_q \in \mathcal{M}}$ of an optimal schedule comply with. The following lemma implies that there are also length assignments and displacements that $(R_{M_q})_{M_q \in \mathcal{M}}$ comply with. For the notation used in Lemma 5.6, recall Definition 3.1.

Lemma 5.6. For each feasible schedule S with respect to routes $(R_{M_q})_{M_q \in \mathcal{M}}$, there is a feasible schedule S' of the same length with respect to routes $(R'_{M_q})_{M_q \in \mathcal{M}}$ such that $|\mathcal{T}_v^{R'_{M_q}}| \leq \max\{|\mathcal{J}_v|, |\mathcal{M}|\} + |\mathcal{M}| - 1$ for each vertex $v \in V$.

Proof. For each machine $M_q \in \mathcal{M}$, construct the route R'_{M_q} from the route $R_{M_q} = ((a_k, v_k, b_k))_{k=1}^s$ as follows:

1. If $|\mathcal{T}_v^{R_{M_q}}| \leq \max\{|\mathcal{J}_v|, |\mathcal{M}|\} + |\mathcal{M}| - 1$, then $R'_{M_q} := R_{M_q}$,
2. Otherwise, let $R'_{M_q} := ((a'_1, v_1, b'_1))_{k=1}^s$, where $a_i \leq a'_i \leq b'_i \leq b_i$ for $1 \leq i \leq s$ are chosen arbitrarily with $|\mathcal{T}_v^{R'_{M_q}}| = \max\{|\mathcal{J}_v|, |\mathcal{M}|\} + |\mathcal{M}| - 1$.

Denote by $\overline{\mathcal{M}} := \{M_q \in \mathcal{M} \mid R_{M_q} \neq R'_{M_q}\}$ the set of machines whose tours have been altered. If $\overline{\mathcal{M}} = \emptyset$, then there is nothing to prove. Henceforth, assume $\overline{\mathcal{M}} \neq \emptyset$. Then, S might not be a feasible schedule for the routes $(R'_{M_q})_{M_q \in \mathcal{M}}$ but

$$S^*(J_i, M_q) := \begin{cases} \bot & \text{if } M_q \in \overline{\mathcal{M}}, \\ S(J_i, M_q) & \text{otherwise} \end{cases}$$

is a partial schedule for the routes $(R'_{M_q})_{M_q \in \mathcal{M}}$ since the machines in $\overline{\mathcal{M}}$ do not process any jobs in S^*. We show that S^* is completable with respect to $(R'_{M_q})_{M_q \in \mathcal{M}}$ in terms of Definition 3.2.

To this end, choose an arbitrary vertex $v \in V$ and an arbitrary machine $M_q \in \mathcal{M}$ with some unprocessed job $J_i \in \mathcal{J}_v^{S^*}$. Then, $M_q \in \overline{\mathcal{M}}$, since only machines in $\overline{\mathcal{M}}$ have unprocessed jobs in S^*. Moreover, $|\mathcal{T}_{J_i}^{S^*}| \leq |\mathcal{M}| - 1$, since at least the machine M_q does not process J_i. Finally $\mathcal{T}_{M_q}^{S^*} = \emptyset$ since M_q does not process any jobs in S^*. Thus,

$$|\mathcal{T}_v^{R'_{M_q}} \setminus (\mathcal{T}_{J_i}^{S^*} \cup \mathcal{T}_{M_q}^{S^*})| \geq \max\{|\mathcal{J}_v|, |\mathcal{M}|\} + |\mathcal{M}| - 1 - (|\mathcal{M}| - 1)$$
$$= \max\{|\mathcal{J}_v|, |\mathcal{M}|\}$$

and Theorem 3.4 shows how to complete S^* into a feasible schedule S' for the routes $(R'_{M_q})_{M_q \in \mathcal{M}}$. $\square$

Remark 5.7. Lemma 5.6 gives an upper bound of $\max\{|\mathcal{J}_v|, |\mathcal{M}|\} + |\mathcal{M}| - 1$ on the total amount of time that each machine stays in a vertex v in an optimal schedule. Note that neither Example 3.3 nor Proposition 4.5 give such an upper bound: these show that, in order to obtain a feasible schedule, it is sufficient that each machine stays in each vertex v for *at least* $\max\{|\mathcal{J}_v|, |\mathcal{M}|\}$ time. They do not exclude that, in an optimal schedule, a machine might stay in a vertex significantly longer in order to enable other machines to process their jobs faster.

5.2 Computing Routes and Completing the Schedule

In this section, we provide the last missing ingredient for our fixed-parameter algorithm for ROUTING OPEN SHOP with unit processing times:

Proposition 5.8. *Problem 5.3 is solvable* $2^{O(|V||\mathcal{M}|^2 \log |V||\mathcal{M}|)} \cdot poly(|\mathcal{J}|)$ *time.*

By Proposition 5.4, this proves Theorem 5.1. The key to our algorithm for Proposition 5.8 is the following lemma.

Lemma 5.9. *Let* $(I, (T, A, D), L)$ *be an instance of Problem 5.3 that has a solution. Then, for arbitrary routes* $(R_{M_q})_{M_q \in \mathcal{M}}$ *of length* L *complying with* (T, A, D) *and satisfying* $|\mathcal{T}_v^{R_{M_q}}| \geq |\mathcal{J}_v|$ *for each non-critical vertex* $v \in V$,

(i) *there is a partial schedule* S *with respect to* $(R_{M_q})_{M_q \in \mathcal{M}}$ *such that* $S(J_i, M_q) \neq \perp$ *if and only if* $\mathcal{L}(J_i)$ *is critical,*

(ii) *any such partial schedule is completable with respect to* $(R_{M_q})_{M_q \in \mathcal{M}}$.

Lemma 5.9 shows that, to solve Problem 5.3, it is sufficient to compute routes $(R_{M_q})_{M_q \in \mathcal{M}}$ of length L that comply with a given pre-schedule (T, A, D) and stay in each uncritical vertex v for at least $|\mathcal{J}_v|$ units of time. If no such routes are found, then the instance of Problem 5.3 has no schedule of length L since *any* feasible schedule has to spend at least $|\mathcal{J}_v|$ units of time in each vertex v. If such routes are found, then a feasible schedule with respect to them can be computed independently using the schedule completion Theorem 3.4 for noncritical vertices by Lemma 5.9(ii) and using brute force for critical vertices:

Lemma 5.10. *Let* $(I, (T, A, D), L)$ *be an instance of Problem 5.3 and* $(R_{M_q})_{M_q \in \mathcal{M}}$ *be arbitrary routes complying with* (T, A, D).

If there is a partial schedule S *for* I *that satisfies Lemma 5.9(i), then we can find it in* $2^{O(|V||\mathcal{M}|^2 \log |\mathcal{M}|)} \cdot poly(|\mathcal{J}|)$ *time.*

Proof. Observe that, in total, there are at most $|V| \cdot |\mathcal{M}|$ jobs in critical vertices. Thus, we determine $S(J_i, M_q)$ for at most $|V| \cdot |\mathcal{M}|^2$ pairs $(J_i, M_q) \in \mathcal{J} \times \mathcal{M}$. By Lemma 5.6, each machine can process all of its jobs in a critical vertex staying there no longer than $2|\mathcal{M}| - 1$ units of time. Thus, for each of the at most $|V| \cdot |\mathcal{M}|^2$ pairs $(J_i, M_q) \in \mathcal{J} \times \mathcal{M}$, we enumerate all possibilities of choosing $S(J_i, M_q)$ among the smallest $2|\mathcal{M}| - 1$ numbers in $\mathcal{T}_{\mathcal{L}(J_i)}^{R_{M_q}}$. There are $(2|\mathcal{M}| - 1)^{|V| \cdot |\mathcal{M}|^2} \in 2^{O(|V||\mathcal{M}|^2 \log |\mathcal{M}|)}$ possibilities to do so.   □

Finally, we compute the routes required by Lemma 5.9 by testing the feasibility of an integer linear program with $O(|\mathcal{M}| \cdot (|V||\mathcal{M}| + 2))$ variables and constraints, which, by Lenstra's theorem below, works in $2^{O(|V||\mathcal{M}|^2 \log |V||\mathcal{M}|)}$ time. Together with Lemma 5.10 and Theorem 3.4, this completes the proof of Proposition 5.8.

Theorem 5.11. (Lenstra [28]; see also Kannan [26]). *A feasible solution to an integer linear program with* p *variables and* m *constraints is computable in* $p^{O(p)} \cdot poly(m)$ *time, if such a feasible solution exists.*

6   Conclusion

We have proved the schedule completion Theorem 3.4 and used it for a fixed-parameter algorithm for ROUTING OPEN SHOP with unit processing times. Precisely, we used it to prove upper bounds on various parameters of optimal schedules. This suggests that Theorem 3.4 will be likewise beneficial for approximation algorithms. Indeed, our Sect. 4 makes first steps into this direction.

A natural direction for future research is determining the parameterized complexity of ROUTING OPEN SHOP with unit processing times parameterized by the number $|V|$ of vertices. Even the question whether the problem is polynomial-time solvable for constant $|V|$ is open, yet we showed fixed-parameter tractability in the absence of critical vertices (Corollary 4.6). Finally, it would be desirable to find a fast polynomial-time algorithm for finding the coloring whose existence is witnessed by Galvin's theorem (Theorem 3.6).

References

1. Allahverdi, A., Ng, C.T., Cheng, T.C.E., Kovalyov, M.Y.: A survey of scheduling problems with setup times or costs. Eur. J. Oper. Res. **187**(3), 985–1032 (2008)
2. Averbakh, I., Berman, O., Chernykh, I.: A $\frac{6}{5}$-approximation algorithm for the two-machine routing open-shop problem on a two-node network. Eur. J. Oper. Res. **166**(1), 3–24 (2005)
3. Averbakh, I., Berman, O., Chernykh, I.: The routing open-shop problem on a network: Complexity and approximation. Eur. J. Oper. Res. **173**(2), 531–539 (2006)
4. Bellman, R.: Dynamic programming treatment of the Travelling Salesman Problem. J. ACM **9**(1), 61–63 (1962)
5. van Bevern, R., Chen, J., Hüffner, F., Kratsch, S., Talmon, N., Woeginger, G.J.: Approximability and parameterized complexity of multicover by c-intervals. Inform. Process. Lett. **115**(10), 744–749 (2015a)
6. van Bevern, R., Komusiewicz, C., Sorge, M.: Approximation algorithms for mixed, windy, and capacitated arc routing problems. In: Proceedings of the 15th ATMOS. OASIcs, vol. 48. Schloss Dagstuhl-Leibniz-Zentrum für Informatik (2015b)
7. van Bevern, R., Mnich, M., Niedermeier, R., Weller, M.: Interval scheduling and colorful independent sets. J. Sched. **18**, 449–469 (2015)
8. van Bevern, R., Niedermeier, R., Sorge, M., Weller, M.: Complexity of arc routing problems. In: Arc Routing: Problems, Methods, and Applications. SIAM (2014)
9. van Bevern, R., Niedermeier, R., Suchý, O.: A parameterized complexity view on non-preemptively scheduling interval-constrained jobs: few machines, small looseness, and small slack. J. Sched. (in press, 2016). doi:10.1007/s10951-016-0478-9
10. Bodlaender, H.L., Fellows, M.R.: W[2]-hardness of precedence constrained k-processor scheduling. Oper. Res. Lett. **18**(2), 93–97 (1995)
11. Chernykh, I., Kononov, A., Sevastyanov, S.: Efficient approximation algorithms for the routing open shop problem. Comput. Oper. Res. **40**(3), 841–847 (2013)
12. Cygan, M., Fomin, F.V., Kowalik, L., Lokshtanov, D., Marx, D., Pilipczuk, M., Pilipczuk, M., Saurabh, S.: Parameterized Algorithms. Springer, Heidelberg (2015)
13. Dorn, F., Moser, H., Niedermeier, R., Weller, M.: Efficient algorithms for Eulerian Extension and Rural Postman. SIAM J. Discrete Math. **27**(1), 75–94 (2013)
14. Downey, R.G., Fellows, M.R.: Fundamentals of Parameterized Complexity. Texts in Computer Science. Springer, London (2013)
15. Fellows, M.R., McCartin, C.: On the parametric complexity of schedules to minimize tardy tasks. Theor. Comput. Sci. **298**(2), 317–324 (2003)
16. Flum, J., Grohe, M.: Parameterized Complexity Theory. Texts in Theoretical Computer Science. An EATCS Series. Springer, Heidelberg (2006)
17. Galvin, F.: The list chromatic index of a bipartite multigraph. J. Comb. Theory B **63**(1), 153–158 (1995)

18. Gonzalez, T., Sahni, S.: Open shop scheduling to minimize finish time. J. ACM **23**(4), 665–679 (1976)
19. Gutin, G., Jones, M., Sheng, B.: Parameterized complexity of the k-Arc Chinese postman problem. In: Schulz, A.S., Wagner, D. (eds.) ESA 2014. LNCS, vol. 8737, pp. 530–541. Springer, Heidelberg (2014)
20. Gutin, G., Jones, M., Wahlström, M.: Structural parameterizations of the mixed Chinese postman problem. In: Bansal, N., Finocchi, I. (eds.) ESA 2015. LNCS, vol. 9294, pp. 668–679. Springer, Heidelberg (2015)
21. Gutin, G., Muciaccia, G., Yeo, A.: Parameterized complexity of k-Chinese Postman Problem. Theor. Comput. Sci. **513**, 124–128 (2013)
22. Gutin, G., Wahlström, M., Yeo, A.: Parameterized Rural Postman and Conjoining Bipartite Matching problems (2014). arXiv:1308.2599v4
23. Halldórsson, M.M., Karlsson, R.K.: Strip graphs: recognition and scheduling. In: Fomin, F.V. (ed.) WG 2006. LNCS, vol. 4271, pp. 137–146. Springer, Heidelberg (2006)
24. Held, M., Karp, R.M.: A dynamic programming approach to sequencing problems. J. SIAM **10**(1), 196–210 (1962)
25. Hermelin, D., Kubitza, J.-M., Shabtay, D., Talmon, N., Woeginger, G.: Scheduling two competing agents when one agent has significantly fewer jobs. In: Proceedings of the 10th IPEC. LIPIcs, vol. 43, pp. 55–65. Schloss Dagstuhl-Leibniz-Zentrum für Informatik (2015)
26. Kannan, R.: Minkowski's convex body theorem and integer programming. Math. Oper. Res. **12**(3), 415–440 (1987)
27. Kononov, A.: $O(\log n)$-approximation for the routing open shop problem. RAIRO-Oper. Res. **49**(2), 383–391 (2015)
28. Lenstra, H.W.: Integer programming with a fixed number of variables. Math. Oper. Res. **8**(4), 538–548 (1983)
29. Mnich, M., Wiese, A.: Scheduling and fixed-parameter tractability. Math. Program. **154**(1–2), 533–562 (2015)
30. Niedermeier, R.: Invitation to Fixed-Parameter Algorithms. Oxford University Press, Oxford (2006)
31. Pyatkin, A.V., Chernykh, I.D.: The Open Shop problem with routing in a two-node network and allowed preemption (in Russian). Diskretnyj Analiz i Issledovaniye Operatsij **19**(3), 65–78 (2012). English translation in J. Appl. Ind. Math. **6**(3), 346–354
32. Serdyukov, A.I.: On some extremal by-passes in graphs (in Russian). Upravlyayemyye sistemy **17**, 76–79 (1978). zbMATH 0475.90080
33. Slivnik, T.: Short proof of Galvin's Theorem on the list-chromatic index of a bipartite multigraph. Comb. Probab. Comput. **5**, 91–94 (1996)
34. Sorge, M., van Bevern, R., Niedermeier, R., Weller, M.: A new view on Rural Postman based on Eulerian Extension and Matching. J. Discrete Alg. **16**, 12–33 (2012)
35. Yu, W., Liu, Z., Wang, L., Fan, T.: Routing open shop and flow shop scheduling problems. Eur. J. Oper. Res. **213**(1), 24–36 (2011)
36. Zhu, X., Wilhelm, W.E.: Scheduling and lot sizing with sequence-dependent setup: A literature review. IIE Trans. **38**(11), 987–1007 (2006)

Max-Closed Semilinear Constraint Satisfaction

Manuel Bodirsky and Marcello Mamino$^{(\boxtimes)}$

Institut für Algebra, TU Dresden, 01062 Dresden, Germany
marcello.mamino@tu-dresden.de

Abstract. A semilinear relation $S \subseteq \mathbb{Q}^n$ is *max-closed* if it is preserved by taking the componentwise maximum. The constraint satisfaction problem for max-closed semilinear constraints is at least as hard as determining the winner in Mean Payoff Games, a notorious problem of open computational complexity. Mean Payoff Games are known to be in NP∩co-NP, which is not known for max-closed semilinear constraints. Semilinear relations that are max-closed and additionally closed under translations have been called *tropically convex* in the literature. One of our main results is a new duality for open tropically convex relations, which puts the CSP for tropically convex semilinear constraints in general into NP ∩ co-NP. This extends the corresponding complexity result for scheduling under and-or precedence constraints, or equivalently the max-atoms problem. To this end, we present a characterization of max-closed semilinear relations in terms of syntactically restricted first-order logic, and another characterization in terms of a finite set of relations L that allow primitive positive definitions of all other relations in the class. We also present a subclass of max-closed constraints where the CSP is in P; this class generalizes the class of max-closed constraints over finite domains, and the feasibility problem for max-closed linear inequalities. Finally, we show that the class of max-closed semilinear constraints is *maximal* in the sense that as soon as a single relation that is not max-closed is added to L, the CSP becomes NP-hard.

1 Introduction

A relation $R \subseteq \mathbb{Q}^n$ is *semilinear* if R has a first-order definition in $(\mathbb{Q}; +, \leq, 1)$; equivalently, R is a finite union of finite intersections of (open or closed) linear half spaces; see Ferrante and Rackoff [15]. In this article we study the computational complexity of constraint satisfaction problems with semilinear constraints. Informally, a constraint satisfaction problem (CSP) is the problem of deciding whether a given finite set of constraints has a common solution. It has been a fruitful approach to study the computational complexity of CSPs depending on the type of constraints allowed in the input.

Formally, we fix a set D, a set of relation symbols $\tau = \{R_1, R_2, \dots\}$, and a τ-structure $\Gamma = (D; R_1^\Gamma, R_2^\Gamma, \dots)$ where $R_i^\Gamma \subseteq D^{k_i}$ is a relation over D of arity k_i. For finite τ the computational problem CSP(Γ) is defined as follows:

INSTANCE: a finite set of formal variables $x_1, \dots, x_n$, and a finite set of expressions of the form $R(x_{i_1}, \dots, x_{i_k})$ with $R \in \tau$.

© Springer International Publishing Switzerland 2016
A.S. Kulikov and G.J. Woeginger (Eds.): CSR 2016, LNCS 9691, pp. 88–101, 2016.
DOI: 10.1007/978-3-319-34171-2_7

QUESTION: is there an assignment $x_1^s, \ldots, x_n^s \in D$ such that $(x_{i_1}^s, \ldots, x_{i_k}^s) \in R^\Gamma$ for all constraints of the form $R(x_{i_1}, \ldots, x_{i_k})$ in the input?

When the domain of Γ is the set of rational numbers $\mathbb{Q}$, and all relations of Γ are semilinear, we say that Γ is semilinear. It is possible, and sometimes essential, to also define $\mathrm{CSP}(\Gamma)$ when the signature τ is infinite. However, in this situation it is important to discuss how the symbols from τ are represented in the input of the CSP. For instance, in the semilinear setting, the most natural choice is often to assume that the relations are given as quantifier-free formulas, say in disjunctive normal form, with coefficients in binary.

A famous example of a computational problem that can be formulated as $\mathrm{CSP}(\Gamma)$ for a semilinear structure Γ is the feasibility problem for linear programming: this CSP is well-known to be in P [25]. Here, it is natural to assume that the relations in the input are represented as linear inequalities with coefficients in binary. It follows that the CSP for *any* semilinear Γ (represented, say, in DNF) is in NP, because we can non-deterministically select a disjunct from the representation of each of the given constraints, and then verify in polynomial time whether the obtained set of linear inequalities is satisfiable.

We would like to systematically study the computational complexity of $\mathrm{CSP}(\Gamma)$ for all semilinear structures Γ. This is a very ambitious goal. Several partial results are known [5,6,8,23,24]. Let us also mention that it is easy to find for every structure Δ with a finite domain a semilinear structure Γ so that $\mathrm{CSP}(\Delta)$ and $\mathrm{CSP}(\Gamma)$ are the same computational problem. But already the complexity classification of CSPs for finite structures is open [14].

Even worse, there are concrete semilinear structures whose CSP has an open computational complexity. An important example of this type is the max-atoms problem [4], which is the CSP for the semilinear structure Γ that contains all ternary relations of the form

$$M_c \overset{\text{def}}{=} \left\{ (x_1, x_2, x_3) \mid x_1 + c \leq \max(x_2, x_3) \right\}$$

where $c \in \mathbb{Q}$ is represented in binary. It is an open problem whether the max-atoms problem is in P, but it is known to be polynomial-time equivalent to determining the winner in mean payoff games [27], which is in $\mathrm{NP} \cap \mathrm{co\text{-}NP}$. Note that here the assumption that c is represented in binary is important: when c is represented in unary, or when we drop all but a finite number of relations in Γ, the resulting problem is in P.

An important tool to study the computational complexity of $\mathrm{CSP}(\Gamma)$ is the concept of *primitive positive definability*. A primitive positive formula is a first-order formula of the form $\exists x_1, \ldots, x_n (\psi_1 \wedge \cdots \wedge \psi_m)$ where ψ are atomic formulas (in primitive positive formulas, no disjunction, negation, and universal quantification is allowed). Jeavons, Gyssens, and Cohen [21] showed that the CSP for expansions of Γ by finitely many primitive positive definable relations is polynomial-time reducible to $\mathrm{CSP}(\Gamma)$.

Primitive positive definability in Γ can be studied using the *polymorphisms* of Γ, which are a multi-variate generalization of endomorphisms. We say that $f \colon \Gamma^k \to \Gamma$ is a polymorphism of a τ-structure Γ if

$$(f(a_1^1, \ldots, a_1^k), \ldots, f(a_m^1, \ldots, a_m^k)) \in R^\Gamma$$

for all $R \in \tau$ and $(a_1^1, \ldots, a_m^1), \ldots, (a_1^k, \ldots, a_m^k) \in R^\Gamma$. For finite structures Γ, a relation R is primitive positive definable in Γ *if and only if* R is preserved by all polymorphisms of Γ. And indeed, the *tractability conjecture* of Bulatov, Jeavons, and Krokhin [11] in a reformulation due to Kozik and Barto [3] states that $CSP(\Gamma)$ is in P if and only if Γ has a polymorphism f which is *cyclic*, this is, has arity $n \geq 2$ and satisfies $\forall x_1, \ldots, x_n \; f(x_1, \ldots, x_n) = f(x_2, \ldots, x_n, x_1)$.

Polymorphisms are also relevant when Γ is infinite. For example, a semilinear relation is convex if and only if it has the cyclic polymorphism $(x, y) \mapsto (x+y)/2$, hence this polymorphism identifies the semilinear CSPs which are tractable by linear programming. Indeed, when Γ has a cyclic polymorphism, and assuming the tractability conjecture, Γ cannot interpret[1] primitively positively any hard finite-domain CSP, which is the standard way of proving that a CSP is NP-hard.

A fundamental cyclic operation is the maximum operation, $(x, y) \mapsto \max(x, y)$. The constraints for the max-atoms problem are examples of semilinear relations that are not convex, but that have max as a polymorphism; we also say that they are *max-closed*. When a *finite* structure is max-closed, with respect to some ordering of the domain, then the CSP for this structure is known to be in P [20]. The complexity of the CSP for max-closed *semilinear* constraints, on the other hand, is open.

2   Results

We show that the CSP for semilinear max-closed relations that are *translation-invariant*, that is, have the polymorphism $x \mapsto x+c$ for all $c \in \mathbb{Q}$, is in NP $\cap$ co-NP (Sect. 5). Such relations have been called *tropically convex* in the literature [13][2]. This class is a non-trivial extension of the max-atoms problem (for instance it contains relations such as $x \leq (y + z)/2$), and it is not covered by the known reduction to mean payoff games [1,2,27]. Indeed, it is open whether the CSP for tropically convex semilinear relations can be reduced to mean payoff games (in fact, Zwick and Paterson [30] believe that mean payoff games are "strictly easier" than simple stochastic games, which reduce to our problem via the results presented in Sect. 4). The containment in NP$\cap$co-NP can be slightly extended to the CSP for the structure that includes additionally all relations $x = c$ for $c \in \mathbb{Q}$ (represented in binary). It follows from our results (Corollary 8) that the class of semilinear tropically convex sets is the smallest class of semilinear sets that has the same polymorphisms as the max atoms constraints $x \leq \max(y, z) + c$ with $c \in \mathbb{Q}$.

In our proof, we first present a characterization of max-closed semilinear relations in terms of syntactically restricted first-order logic (Sect. 3). We show

---

[1]   Interpretations in the sense of model theory; we refer to Hodges [19] since we do not need this concept further.

[2]   The original definition of tropical convexity is for the dual situation, considering min instead of max.

that a semilinear relation is max-closed if and only if it can be defined by a *semilinear Horn formula*, which we define as a finite conjunction of *semilinear Horn clauses*, that is, finite disjunctions of the form

$$\bigvee_{i=1}^{m} \bar{a}_i^\top \bar{x} \succ_i c_i$$

where

1. $\bar{a}_1, \ldots, \bar{a}_m \in \mathbb{Q}^n$ and there is a $k \le n$ such that $\bar{a}_{i,j} \ge 0$ for all i and $j \ne k$,
2. $\bar{x} = (x_1, \ldots, x_n)$ is a vector of variables,
3. $\succ_i \in \{\ge, >\}$ are strict or non-strict inequalities, and
4. $c_1, \ldots, c_m \in \mathbb{Q}$ are coefficients.

Example 1. The ternary relation M_c from the max-atoms problem can be defined by the semilinear Horn clause $x_2 - x_1 \ge c \lor x_3 - x_1 \ge c$.

Example 2. A linear inequality $a_1 x_1 + \cdots + a_n x_n \ge c$ is max-closed if and only if at most one of $a_1 \ldots a_n$ is negative.

Example 3. Conjunctions of implications of the form

$$(x_1 \le c_1 \land \cdots \land x_n \le c_n) \Rightarrow x_i < c_0 \tag{1}$$

are max-closed since such an implication is equivalent to the semilinear Horn clause

$$(-x_i > -c_0) \lor \bigvee_i x_i > c_i .$$

It has been shown by Jeavons and Cooper [20] that over finite ordered domains, a relation is max-closed[3] if and only if it can be defined by finite conjunctions of implications of the form (1). Over infinite domains, this is no longer true, as demonstrated by the relations in the previous examples.

We also show that the classes $\mathcal{C}$ of max-closed semilinear relations and $\mathcal{C}_t$ of tropically convex semilinear relations are *finitely generated* in the sense of universal algebra, this is, there exists a finite subset L_0 of $\mathcal{C}$ (resp. L_t of $\mathcal{C}_t$) that can primitively positively define all other relations in $\mathcal{C}$ (resp. $\mathcal{C}_t$). The primitive positive definitions in our finite bases can even be computed efficiently from semilinear Horn formulas with coefficients in binary, hence this infinite set of relations and *any* finite basis (either in the case of $\mathcal{C}$ or $\mathcal{C}_t$) have CSPs of polynomially related complexity. This is in sharp contrast to the situation for max-atoms.

Our proof of the containment in NP ∩ co-NP is based on a duality for open tropically convex semilinear sets, which extends the duality that has been

---

[3] Also the results in Jeavons and Cooper [20] have been formulated in the dual situation for min instead of max.

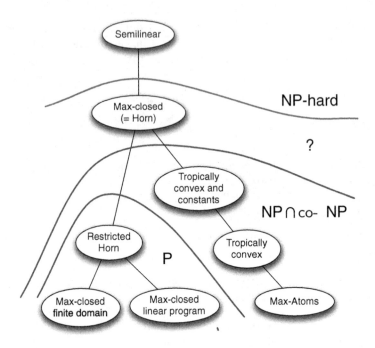

Fig. 1. An overview of our results

observed for the max-atoms problem in [18]. To prove the duality, we translate instances of our problem into a condition on associated mean payoff stochastic games (also called limiting average payoff stochastic games; see [16] for a general reference), and then exploit the symmetry implicit in the definition of such games. The connections between the max-atoms problem, mean payoff deterministic games, and tropical polytopes have been explored extensively in the computer science literature. However, the theory of stochastic games introduces profound changes over the deterministic setting, and employs additional nontrivial techniques. Even though this field is active since the 70s, to the best of our knowledge, no application of its results to semilinear feasibility problems has been published yet. Note that solving mean payoff stochastic games is in NP ∩ co-NP, as a consequence of the existence of optimal positional strategies; see [17,26]. However, we cannot use this fact directly, because stochastic games only relate to a subset of tropically convex sets. We conclude our argument combining the duality with semilinear geometry techniques. Interestingly, at several places in our proofs, we need to replace $\mathbb{Q}$ with appropriate non-Archimedean structures.

Our next result is the identification of a class of max-closed semilinear relations whose CSP can be solved in polynomial time (Sect. 6). This class consists of the semilinear relations that can be defined by *restricted Horn clauses*, which are semilinear Horn clauses satisfying the additional condition that there are $k \leq n$ and $l \leq m$ such that $\bar{a}_{i,j} \geq 0$ for all $(i,j) \neq (k,l)$.

Finally, we observe that for every relation R that is not max-closed, the problem $\mathrm{CSP}(\Gamma_0, R)$ is NP-hard.

Caveat: All proofs are *sketches*, full details in the full paper [7].

3   A Syntactic Characterization of Max-Closure

In this section, the letter F will denote an ordered field: for technical reasons we need to work in this slightly more general setting.

Definition 4. *A semilinear Horn clause is called* closed *if all inequalities are non-strict. We say that $X \subset F^n$ is a* basic max-closed set *if it is the graph of a semilinear Horn clause, i.e. there is $k \leq n$ such that X can be written as a finite union*

$$X = \bigcup_i \left\{ (x_1, \ldots, x_n) \mid a_{i,1}x_1 + \cdots + a_{i,n}x_n \succ_i c_i \right\}$$

where $\succ_i$ can be either $>$ or $\geq$, and $a_{i,j} \geq 0$ for all i and all $j \neq k$. We say that X is basic closed max-closed *if it is the graph of a closed semilinear Horn clause.*

Theorem 5

1. *Let $X \subset F^n$ be a semilinear set. Then X is max-closed if and only if it is a finite intersection of basic max-closed sets.*
2. *Let $X \subset F^n$ be a closed semilinear set. Then X is max-closed if and only if it is a finite intersection of basic closed max-closed sets.*
3. *$X \subset \mathbb{Q}^n$ is primitive positive definable in $\Gamma_0 = (\mathbb{Q}; <, 1, -1, S_1, S_2, M_0)$ where*

$$S_1 = \left\{ (x, y) \mid 2x \leq y \right\} \qquad S_2 = \left\{ (x, y, z) \mid x \leq y + z \right\}$$
$$M_0 = \left\{ (x, y, z) \mid x \leq y \vee x \leq z \right\}$$

   if and only if X is semilinear and max-closed.
4. *$X \subset \mathbb{Q}^n$ is primitively positively definable in $\Gamma_0' = (\mathbb{Q}; 1, -1, S_1, S_2, M_0)$ if and only if X is semilinear closed and max-closed.*

Definition 6. *We say that $X \subset F^n$ is a* basic tropically convex set *if it is a basic max-closed set*

$$X = \bigcup_i \left\{ (x_1, \ldots, x_n) \mid a_{i,1}x_1 + \cdots + a_{i,n}x_n \succ_i c_i \right\}$$

as in Definition 4 and $\sum_j a_{i,j} = 0$ for all i.

Theorem 7

1. *Let $X \subset F^n$ be a semilinear set. Then X is tropically convex if and only if it is a finite intersection of basic tropically convex sets.*

2. $X \subset \mathbb{Q}^n$ *is primitive positive definable in* $\Gamma_t = (\mathbb{Q}; <, T_1, T_{-1}, S_3, M_0)$ *where*

$$T_{\pm 1} = \{(x, y) \mid x \le y \pm 1\} \qquad S_3 = \left\{ (x, y, z) \mid x \le \frac{y + z}{2} \right\}$$

if and only if X is semilinear and tropically convex.

Corollary 8. *A semilinear set $X \subset \mathbb{Q}^n$ is tropically convex if and only if it is preserved by every polymorphism that preserves the max-atoms language (i.e. all sets of the form $\{(x, y, z) \mid x \le \max(y, z) + c\}$ for $c \in \mathbb{Q}$).*

Proof. Translations and maximum are polymorphisms of max-atoms, so one direction is trivial. For the converse, by Theorem 7 (2), it suffices to prove that the relations $<$, T_1, T_{-1}, S_3, and M_0 are preserved by all polymorphisms of max-atoms. This is immediate for T_1, T_{-1}, and M_0 since they have primitive positive definitions over max-atoms. The relation $<$, on the other hand, is an ascending union of max-atoms constraints $\bigcup_{c \in \mathbb{Q}, c > 0} \{(x, y) \mid x \le y - c\}$ and S_3 is an intersection of max-atoms constraints $\bigcap_{c \in \mathbb{Q}} \{(x, y, z) \mid z \le \max(y + c, z - c)\}$. It's well known that ascending unions and arbitrary intersections of primitive positive definable sets are preserved by polymorphisms [28]. □

The following observation is important in view of its implications on the complexity of the constraint satisfaction problems.

Observation 9. *Given a max-closed (resp. tropically convex) set X written as a finite intersection of basic max-closed (resp. basic tropically convex) sets with the constants represented in binary, we can compute in polynomial time a primitive positive definition of X in the structure Γ_0 (resp. Γ_t).*

In this extended abstract, we will omit the proof of the existence of a finite basis, i.e. Theorem 5 (3 and 4) and the proof of Theorem 7.

Definition 10. *We say that $x \in X \subset F^n$ is of type k in X, with $k = 1, \ldots, n$, if*

$$x - Q_k \stackrel{\text{def}}{=} \{x - y \mid y \in Q_k\} \subset X$$

where

$$Q_k = \{y \in (F^{\ge 0})^n \mid y_k = 0\}$$

Observe that if X is the complement of a max-closed set, then every point of X is of type k in X for at least one k (Figure 2).

Proof (Proof of Theorem 5 (2)). The *if* part is immediate, hence we concentrate on the *only if*. Let $\bar{X}$ be $F^n \setminus X$. Consider the subsets $\bar{X}_1, \ldots, \bar{X}_n$ of the points in $\bar{X}$ of type $1, \ldots, n$ respectively. Since X is max-closed, each point of $\bar{X}$ is of type k for some $k = 1, \ldots, n$, hence $\bar{X} = \bar{X}_1 \cup \cdots \cup \bar{X}_n$ moreover, each of these sets is open and semilinear. By standard techniques, we can further split each of the sets $\bar{X}_i$ into a finite union $\bar{X}_i = \bar{X}_{i,1} \cup \cdots \cup \bar{X}_{i,m_i}$ of convex open semilinear sets. Now, by definition, $\bar{X}_i = \bar{X}_i - Q_i$ (i.e. $\{x - y \mid x \in \bar{X}_i, y \in Q_i\}$), hence

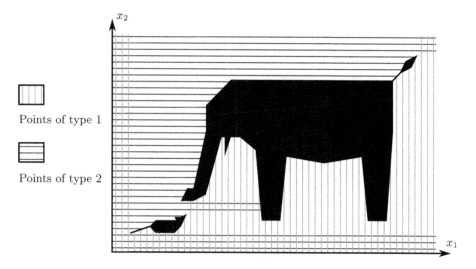

Fig. 2. A max-closed set in $\mathbb{Q}^2$

we can replace the sets $\bar{X}_{i,j}$ with $\tilde{X}_{i,j} = \bar{X}_{i,j} - Q_i$. The sets $\tilde{X}_{i,j}$ are convex, hence they are intersections of half spaces (see [12, Corollary 4.9] and also [29]). Therefore, for each i, j, the set $X_{i,j} = F^n \setminus \tilde{X}_{i,j}$ is a union of half spaces, we can conclude that it is semilinear Horn observing that only the coefficient of x_i can be negative, by the construction of $\tilde{X}_{i,j}$. □

Proof (Proof of Theorem 5—$(2)\to(1)$). We apply (2) to the field of formal Laurent series $F((\epsilon))$ with coefficients in F. The semilinear set $X \subset F^n$ has a unique extension X^* to $F((\epsilon))^n$, which is the set defined by the same formula that defines X. First we modify the extension X^* of X locally to obtain a closed semilinear max-closed set $\tilde{X} \subset F((\epsilon))^n$ such that $\tilde{X} \cap F^n = X$. To this aim, decompose X^* into relatively open convex facets. Consider for each facet its ϵ-*interior*, defined as the points of the facet that are at least ϵ far from its relative boundary. Finally let $\tilde{X}$ be the closure under max of the ϵ-interiors.

By (2), $\tilde{X}$ is an intersection of basic max-closed sets. The coefficients appearing in the intersection are, in general, elements of $F((\epsilon))$. We need to to replace these coefficients with elements of F. More precisely, we have to rewrite terms of the form

$$a_1 x_1 + \cdots + a_n x_n - c \geq 0 \qquad\qquad (\star)$$

where $a_1, \ldots, a_n, c \in F((\epsilon))$ and $x_1, \ldots, x_n$ range over F as positive boolean combinations of F-linear terms of the form

$$a_1' x_1 + \cdots + a_n' x_n - c' \succ 0$$

with $a_1', \ldots, a_n', c' \in F$. Moreover, we know that at most one of the a_i, say a_1, is negative, and we must make sure that all our a_i' except at most a_1' are non-negative. Without loss of generality, we can divide the term $(\star)$ by the absolute value of its largest coefficient, as a result we can assume that the leading term of $(\star)$ has coefficients in F. Now, $(\star)$ is equivalent to

$$\text{leading term} > 0 \ \vee \ \big(\text{leading term} \geq 0 \ \wedge \ \text{remaining terms} > 0\big)$$

This forms the basis of an inductive procedure. To ensure termination, at each step, we divide the remaining terms again by the largest coefficient, so the number of coefficients not in F decreases. The requirement that the coefficients of $x_2, \ldots, x_n$ must be positive is not necessarily preserved by the remaining terms, however if some coefficient $a_i = (a_i)_0 + \epsilon(a_i)_{>0}$ has $(a_i)_{>0}$ negative, then the leading term $(a_i)_0$ of the same coefficient must be positive, hence we can force the positivity condition by adding to the remaining terms a suitable multiple of the leading terms. $\qquad\square$

4   A Duality for Max-Plus-Average Inequalities

Let $\mathcal{O}_n$ be the class of functions mapping $\big(\mathbb{Q} \cup \{+\infty\}\big)^n$ to $\mathbb{Q} \cup \{+\infty\}$ of either of the following forms

$$(x_1, \ldots, x_n) \mapsto \max(x_{j_1} + k_1, \ldots, x_{j_m} + k_m)$$
$$(x_1, \ldots, x_n) \mapsto \min(x_{j_1} + k_1, \ldots, x_{j_m} + k_m) \qquad (\star\star)$$
$$(x_1, \ldots, x_n) \mapsto \frac{\alpha_1 x_{j_1} + \cdots + \alpha_m x_{j_m}}{\alpha_1 + \cdots + \alpha_m} + k$$

where $k, k_i \in \mathbb{Q}$ and $\alpha_i \in \mathbb{Q}^{>0}$.

For any given vector of operators $\bar{o} \in \mathcal{O}_n^n$ we consider the following satisfiability problems: the *primal* $P(\bar{o})$ and the *dual* $D(\bar{o})$

$$P(\bar{o}): \begin{cases} \bar{x} \in \mathbb{Q}^n \\ \bar{x} < \bar{o}(\bar{x}) \end{cases} \qquad D(\bar{o}): \begin{cases} \bar{y} \in (\mathbb{Q} \cup \{+\infty\})^n \setminus \{+\infty\}^n \\ \bar{y} \geq \bar{o}(\bar{y}) \end{cases}$$

where $<$ and $\geq$ are meant to hold component-wise.

Theorem 11. *For any $\bar{o} \in \mathcal{O}_n^n$ one and only one of the problems $P(\bar{o})$ and $D(\bar{o})$ is satisfiable.*

For the proof of Theorem 11, we make use of zero-sum stochastic games with perfect information, in the flavours known as the discounted and the limiting average payoff. A stochastic game is played by two players, MAX and MIN, moving a token along the edges of a directed graph G. Each vertex v of G is either assigned to one of the players, that moves when the token is on v, or it is a stochastic vertex. Each edge e of G has a payoff $\text{po}(e) \in \mathbb{Q}$. The out-edges of a stochastic vertex have also a probability $\text{pr}(e) \in \mathbb{Q}$ of being taken when exiting that vertex. We assume that all vertices have at least one out-edge, so a play never ends.

Let $e_1, e_2, \ldots$ be the edges traversed during a play p of the game G. The discounted payoff $\mathbf{v}_\beta(p)$ of p with discounting factor $\beta \in [0, 1[$ and the limiting average payoff $\mathbf{v}_1$ are

$$\mathbf{v}_\beta(p) \stackrel{\text{def}}{=} (1 - \beta) \sum_{i=1}^{\infty} \mathrm{po}(e_i)\beta^{i-1} \qquad \mathbf{v}_1(p) \stackrel{\text{def}}{=} \liminf_{T \to \infty} \frac{1}{T} \sum_{i=1}^{T} \mathrm{po}(e_i)$$

In such formulas we suppress the dependency on G, to ease the notation. Clearly the objective of MAX is to maximize the payoff, that of MIN is to minimize it.

Our argument is based on the fact below (see [16] Theorem 6.3.7 plus Theorem 6.3.5 and its proof). An alternate approach would have been to use the normal form described in [9]. In fact, probably, the duality and the existence of a normal form in the sense of [9] imply each other. However we obtain our result via a different method.

Observation 12. *For any stochastic game G with perfect information:*

1. *Both players possess positional strategies π_{MAX} and π_{MIN} which are optimal for the limiting average payoff and for all discount factors β sufficiently close to 1. We denote by $\mathbf{v}_\beta(v)$ the expected $\mathbf{v}_\beta$ when the game is started from vertex v and both players follow their optimal strategies.*
2. *Calling $\mathbf{v}_\beta = (\mathbf{v}_\beta(v))_{v \in G}$ the value vector of G with discount factor β, we have that the value vector for the limiting average payoff can be written as $\mathbf{v}_1 = \lim_{\beta \to 1} \mathbf{v}_\beta$.*
3. *The vector $\mathbf{v}_\beta$ can be written as a power series in $(1 - \beta)$.*

We map each vector of operators $\bar{o} \in \mathcal{O}_n^n$ to a stochastic game $G_{\bar{o}}$ with n vertices $\{v_1, \ldots, v_n\}$ corresponding to the n components of $\bar{o}$. Each vertex v_i is assigned to MAX, MIN, or is stochastic according to whether $\bar{o}_i$ is max, min, or a weighted average. When the variable x_j appears in o_i, we put an edge between v_i and v_j. In this case, by $(\star\star)$, the operator o_i must be of the form

$$(x_1, \ldots, x_n) \mapsto \max(\ldots, x_j + k, \ldots)$$
$$(x_1, \ldots, x_n) \mapsto \min(\ldots, x_j + k, \ldots)$$
$$(x_1, \ldots, x_n) \mapsto \frac{\cdots + \alpha x_j + \cdots}{\cdots + \alpha + \cdots} + k$$

and we assign payoff k to the edge $v_i \to v_j$. The probabilities are the weights for the weighted average operators.

Lemma 13. *Let $\bar{o} \in \mathcal{O}_n^n$ be a vector of operators, and let $\mathbf{v}_1$ denote the value vector of the game $G_{\bar{o}}$ with the limiting average payoff. The problem $P(\bar{o})$ is satisfiable if and only if $\mathbf{v}_1(v_i) > 0$ for all vertices v_i of $G_{\bar{o}}$.*

Proof. (*if* direction) Contrary to the intuition, the mean payoff value vector $\mathbf{v}_1$ of $G_{\bar{o}}$ is not a solution of $P(\bar{o})$, nor can a solution be computed from the value vector, as exemplified by the case of ergodic games [10].

Using Fact 12 (3), let $\bar{a}_i \in \mathbb{R}^n$ for $i = 1, \ldots, \infty$ be the coefficients of the series representing the discounted payoff value vector of $G_{\bar{o}}$

$$\mathbf{v}_\beta = \sum_{i=0}^{\infty} \bar{a}_i (1 - \beta)^i$$

For N denoting a (large) real number, define $\bar{x}_N \in \mathbb{R}^n$ by

$$(\bar{x}_N)_j = N(\bar{a}_0)_j + (\bar{a}_1)_j$$

We claim that, for large enough N, the vector $\bar{x}_N$ satisfies $P(\bar{o})$.

The claim can be verified writing the limit discount equation for $G_{\bar{o}}$, which $\mathbf{v}_\beta$ must satisfy, and then truncating the series expansion on both sides to the second term.

(*only if* direction) Fix a solution $\bar{x}$ of $P(\bar{o})$. For each pair of vertices v_i and v_j of $G_{\bar{o}}$ replace the payoff $\mathrm{po}(v_i, v_j)$ with $\mathrm{po}(v_i, v_j) + \bar{x}_j - \bar{x}_i$. It is easy to check that this new game is equivalent to $G_{\bar{o}}$, because in any given play the corrections $\bar{x}_j - \bar{x}_i$ form a telescopic sum, then the player MAX can always avoid negative payoffs. $\square$

Lemma 14. *Let $\bar{o} \in \mathcal{O}_n^n$ be a vector of operators, and let $\mathbf{v}_1$ denote the value vector of the game $G_{\bar{o}}$ with the limiting average payoff. The problem $D(\bar{o})$ is satisfiable if and only if $\mathbf{v}_1(v_i) \leq 0$ for some vertex v_i of $G_{\bar{o}}$.*

Theorem 11 follows immediately from Lemmas 13 and 14.

5 Complexity of Tropically Convex CSPs

In this section, we will apply our duality to tropically convex constraint satisfaction problems. By Theorem 7, we know that the tropically convex relations are precisely those primitively positively definable in the structure $\Gamma_t = (\mathbb{Q}; <, T_1, T_{-1}, S_3, M_0)$. The CSP of Γ_t subsumes max-atoms (see [4]), but is more general than it.

Theorem 15. *The problem $\mathrm{CSP}(\Gamma_t)$ is in $\mathsf{NP} \cap \mathsf{co\text{-}NP}$.*

We would like to stress that, instead of the finite constraint language of Γ_t, we could have chosen to work with basic tropically convex sets (in the sense of Sect. 3) encoded with the constants expressed in binary. In view of Observation 9 this choice is immaterial. We begin with a corollary of Theorem 11 for non-strict inequalities.

Corollary 16. *For any vector of operators $\bar{o} \in \mathcal{O}_n^n$ we consider the following satisfiability problems: the primal $P'(\bar{o})$ and the dual $D'(\bar{o})$*

$$P'(\bar{o}): \begin{cases} \bar{x} \in \mathbb{Q}^n \\ \bar{x} \leq \bar{o}(\bar{x}) \end{cases} \qquad D'(\bar{o}): \begin{cases} \bar{y} \in (\mathbb{Q} \cup \{+\infty\})^n \setminus \{+\infty\}^n \\ \bar{y} > \bar{o}(\bar{y}) \end{cases}$$

where $\leq$ and $>$ are meant to hold component-wise, and we stipulate that $+\infty > +\infty$. Then one and only one of the problems $P'(\bar{o})$ and $D'(\bar{o})$ is satisfiable.

To prove Theorem 15 we need this technical statement.

Definition 17. *Consider a quantifier free semilinear formula $\phi(t, \bar{x})$ with rational coefficients, where t denotes a variable and $\bar{x}$ denotes a tuple of variables. We say that $\phi(t, \bar{x})$ is satisfiable in $0+$ if*

$$\exists t_0 > 0 \; \forall t \in \left]0, t_0\right] \; \exists \bar{x} \; \phi(t, \bar{x})$$

Lemma 18. *The problem, given ϕ as in Definition 17 with coefficients encoded in binary, of deciding whether ϕ is satisfiable in $0+$ is in* NP.

Proof (Proof of Theorem 15). First we replace each strict inequality $A < B$ in the input by $\exists \epsilon > 0 \; A \le B - \epsilon$. Hence we can apply Corollary 16, so we get that our problem is not satisfiable if and only if for all $\epsilon > 0$ some $D'(\bar{o}_\epsilon)$ is satisfiable. We then conclude by Lemma 18.   □

6   Tractable and Intractable Cases

We present an algorithm that tests satisfiability of a given restricted Horn formula Φ. Recall that each restricted Horn clause has at most one literal which contains a variable with a negative coefficient. We call this literal the *positive literal* of the clause, and all other literals the *negative literals*.

Solve(Φ)
Do
      Let Ψ be the clauses in Φ that contain at most one literal.
      If Ψ is unsatisfiable then return *unsatisfiable*.
      For all negative literals ϕ in clauses from Φ
         If $\Psi \wedge \phi$ is unsatisfiable, then Ψ implies $\neg\phi$:
            remove ϕ from all clauses in Φ.
Loop until no literal has been removed
Return *satisfiable*.

For testing whether conjunctions of literals are satisfiable, we use a polynomial time algorithm for linear program feasibility see [22,25]. Since the algorithm always removes false literals, it is clear that if the algorithm returns *unsatisfiable*, then Φ is indeed unsatisfiable. Suppose now that we are in the final step of the algorithm and the procedure returns *satisfiable*. Then for each negative literal ϕ of Φ the set $\Psi \wedge \phi$ has a solution. The maximum of these solutions must be a satisfying assignment for Φ.

Proposition 19. *Let $R \subseteq \mathbb{Q}^n$ an n-ary relation that is not max-closed. Then* $\mathrm{CSP}(\Gamma_0, R)$ *is* NP-hard.

Proof. Follows from [20, Theorems 6.5 and 6.6].   □

Acknowledgements. Both authors have received funding from the European Research Council under the European Community's Seventh Framework Programme (FP7/2007-2013 Grant Agreement no. 257039), and the German Research Foundation (DFG, project number 622397).

References

1. Akian, M., Gaubert, S., Guterman, A.: Tropical polyhedra are equivalent to mean payoff games. Int. Algebra Comput. **22**(1), 43 (2012). 125001
2. Atserias, A., Maneva, E.: Mean-payoff games and propositional proofs. Inf. Comput. **209**(4), 664–691 (2011). In: Abramsky, S., Gavoille, C., Kirchner, C., Meyer auf der Heide, F., Spirakis, P.G. (eds.) ICALP 2010, Part I. LNCS, vol. 6198, pp. 102–113. Springer, Heidelberg (2010)
3. Barto, L., Kozik, M.: Absorbing subalgebras, cyclic terms and the constraint satisfaction problem. Logical Methods Comput. Sci. **8**(1:07), 1–26 (2012)
4. Bezem, M., Nieuwenhuis, R., Rodríguez-Carbonell, E.: The max-atom problem and its relevance. In: Cervesato, I., Veith, H., Voronkov, A. (eds.) LPAR 2008. LNCS, vol. 5330, pp. 47–61. Springer, Heidelberg (2008)
5. Bodirsky, M., Jonsson, P., von Oertzen, T.: Semilinear program feasibility. In: Albers, S., Marchetti-Spaccamela, A., Matias, Y., Nikoletseas, S., Thomas, W. (eds.) ICALP 2009, Part II. LNCS, vol. 5556, pp. 79–90. Springer, Heidelberg (2009)
6. Bodirsky, M., Jonsson, P., von Oertzen, T.: Essential convexity and complexity of semi-algebraic constraints. Logical Methods Comput. Sci. **8**(4), 1–25 (2012). An extended abstract about a subset of the results has been published under the title Semilinear Program Feasibility at ICALP 2010
7. Bodirsky, M., Mamino, M.: Max-closed semilinear constraint satisfaction (2015). arXiv:1506.04184
8. Bodirsky, M., Martin, B., Mottet, A.: Constraint satisfaction problems over the integers with successor. In: Halldórsson, M.M., Iwama, K., Kobayashi, N., Speckmann, B. (eds.) ICALP 2015, Part I. LNCS, vol. 9134, pp. 256–267. Springer, Heidelberg (2015)
9. Boros, E., Elbassioni, K., Gurvich, V., Makino, K.: Every stochastic game with perfect information admits a canonical form. RRR-09-2009, RUTCOR, Rutgers University (2009)
10. Boros, E., Elbassioni, K., Gurvich, V., Makino, K.: A pumping algorithm for ergodic stochastic mean payoff games with perfect information. In: Eisenbrand, F., Shepherd, F.B. (eds.) IPCO 2010. LNCS, vol. 6080, pp. 341–354. Springer, Heidelberg (2010)
11. Bulatov, A.A., Krokhin, A.A., Jeavons, P.G.: Classifying the complexity of constraints using finite algebras. SIAM J. Comput. **34**, 720–742 (2005)
12. Andradas, C., Rubio, R., Vélez, M.P.: An algorithm for convexity of semilinear sets over ordered fields. Real Algebraic and Analytic Geometry Preprint Server, No. 12
13. Develin, M., Sturmfels, B.: Tropical convexity. Doc. Math. **9**, 1–27 (2004)
14. Feder, T., Vardi, M.Y.: The computational structure of monotone monadic SNP and constraint satisfaction: a study through Datalog and group theory. SIAM J. Comput. **28**, 57–104 (1999)
15. Ferrante, J., Rackoff, C.: A decision procedure for the first order theory of real addition with order. SIAM J. Comput. **4**(1), 69–76 (1975)
16. Filar, J., Vrieze, K.: Competitive Markov Decision Processes. Springer, New York (1996)
17. Gillette, D.: Stochastic games with zero probabilities. Contrib. Theor. Games **3**, 179–187 (1957)

18. Grigoriev, D., Podolskii, V.V.: Tropical effective primary and dual nullstellensätze. In: 32nd International Symposium on Theoretical Aspects of Computer Science, STACS 2015, Garching, Germany, 4–7 March 2015, pp. 379–391 (2015)
19. Hodges, W.: A Shorter Model Theory. Cambridge University Press, Cambridge (1997)
20. Jeavons, P.G., Cooper, M.C.: Tractable constraints on ordered domains. Artif. Intell. **79**(2), 327–339 (1995)
21. Jeavons, P., Cohen, D., Gyssens, M.: Closure properties of constraints. J. ACM **44**(4), 527–548 (1997)
22. Jonsson, P., Bäckström, C.: A unifying approach to temporal constraint reasoning. Artif. Intell. **102**(1), 143–155 (1998)
23. Jonsson, P., Lööw, T.: Computation complexity of linear constraints over the integers. Artif. Intell. **195**, 44–62 (2013)
24. Jonsson, P., Thapper, J.: Constraint satisfaction and semilinear expansions of addition over the rationals and the reals (2015). arXiv:1506.00479
25. Khachiyan, L.: A polynomial algorithm in linear programming. Dokl. Akad. Nauk SSSR **244**, 1093–1097 (1979)
26. Liggett, T.M., Lippman, S.A.: Stochastic games with perfect information and time average payoff. SIAM Rev. **11**(4), 604–607 (1969)
27. Möhring, R.H., Skutella, M., Stork, F.: Scheduling with and/or precedence constraints. SIAM J. Comput. **33**(2), 393–415 (2004)
28. Pöschel, R.: A general galois theory for operations and relations and concrete characterization of related algebraic structures. Technical report of Akademie der Wissenschaften der DDR (1980)
29. Scowcroft, P.: A representation of convex semilinear sets. Algebra Univers. **62**(2–3), 289–327 (2009)
30. Zwick, U., Paterson, M.: The complexity of mean payoff games on graphs. Theor. Comput. Sci. **158**(1&2), 343–359 (1996)

Computing and Listing st-Paths in Public Transportation Networks

Kateřina Böhmová[1], Matúš Mihalák[2], Tobias Pröger[1(✉)],
Gustavo Sacomoto[3,4], and Marie-France Sagot[3,4]

[1] Institut für Theoretische Informatik, ETH Zürich, Zürich, Switzerland
tobias.proeger@inf.ethz.ch
[2] Department of Knowledge Engineering,
Maastricht University, Maastricht, The Netherlands
[3] INRIA Grenoble Rhône-Alpes, Montbonnot-Saint-Martin, France
[4] UMR CNRS 5558 – LBBE, Université Lyon 1, Lyon, France

Abstract. Given a set of directed paths (called *lines*) L, a *public transportation network* is a directed graph $G_L = (V_L, A_L)$ which contains exactly the vertices and arcs of every line $l \in L$. An st-route is a pair (π, γ) where $\gamma = \langle l_1, \dots, l_h \rangle$ is a line sequence and π is an st-path in G_L which is the concatenation of subpaths of the lines $l_1, \dots, l_h$, in this order. Given a threshold β, we present an algorithm for listing all st-paths π for which a route (π, γ) with $|\gamma| \leq \beta$ exists, and we show that the running time of this algorithm is polynomial with respect to the input and the output size. We also present an algorithm for listing all *line sequences* γ with $|\gamma| \leq \beta$ for which a route (π, γ) exists, and show how to speed it up using preprocessing. Moreover, we show that for the problem of finding an st-route (π, γ) that minimizes the number of different lines in γ, even computing an $o(\log |V|)$-approximation is NP-hard.

1 Introduction

Motivation. Given a public transportation network (in the following called *transit network*) and two locations s and t, a common goal is to find a fastest route from s to t, i.e. an st-route whose travel time is minimum among all st-routes. A fundamental feature of any public transportation information system is to provide, given s, t and a target arrival time t_A, a fastest st-route that reaches t no later than at time t_A. This task can be solved by computing a shortest path in an auxiliary graph that also reflects time [16]. However, if delays occur in the network (which often happens in reality), then the goal of computing a *robust* st-route that is likely to reach t on time, naturally arises.

The problem of finding robust routes received much attention in the literature (for a survey, see, e.g., [1]). Recently, Böhmova et al. [4,5] proposed the following two-stage approach for computing robust routes. In the first step, all st-routes ignoring time are listed explicitly, and only after that, timetables and historic traffic data are incorporated to evaluate the robustness of each possible route.

A.S. Kulikov and G.J. Woeginger (Eds.): CSR 2016, LNCS 9691, pp. 102–116, 2016.
DOI: 10.1007/978-3-319-34171-2_8

However, from a practical point of view it is undesirable to list *all* possible st-routes, for two reasons: (1) the number of listed routes might be huge, leading to a non-satisfactory running time, and (2) many routes might be inacceptable for the user, e.g., because they use many more transfers than necessary. Having a huge number of transfers is not only uncomfortable, but usually also has a negative impact on the robustness of routes, since each transfer bears a risk of missing the next connection when vehicles are delayed.

Our Contribution. The main contribution of the present paper are three algorithms that list all st-routes for which the number of transfers does not exceed a given threshold β. The running time of our algorithms are polynomial with respect to the sum of the input and the output size. As a subroutine of this algorithm we need to compute a route with a minimum number of transfers which is known to be solvable efficiently [16]. However, we show that finding a route with a minimum number of *different* lines cannot be approximated within $(1 - \varepsilon) \ln n$ for any $\varepsilon > 0$ unless NP = P.

We note that for bus networks it is reasonable to consider directed networks (instead of undirected ones), because real-world transportation networks (such as the one in the city of Barcelona) may contain one-way streets in which buses can only operate in a single direction.

Related Work. Listing combinatorial objects (such as paths, cycles, spanning trees, etc.) in graphs is a widely studied field in computer science (see, e.g., [2]). The currently fastest algorithm for listing all st-paths in directed graphs was presented by Johnson [13] in 1975 and runs in time $\mathcal{O}((n + m)(\kappa + 1))$ where n and m are the number of vertices and arcs, respectively, and κ is the number of all st-paths (i.e., the size of the output). For undirected graphs, an optimal algorithm was presented by Birmelé et al. [3]. A related problem is the K-shortest path problem, which asks, for a given constant K, to compute the first K distinct shortest st-paths. Yen [19] and Lawler [15] studied this problem for directed graphs. Their algorithm uses Dijkstra's algorithm [8] and can be implemented to run in time $\mathcal{O}(K(nm + n^2 \log n))$ using Fibonacci heaps [11]. For undirected graphs, Katoh et al. [14] proposed an algorithm with running time $\mathcal{O}(K(m + n \log n))$. Eppstein [10] gave an $\mathcal{O}(K + m + n \log n)$ algorithm for listing the first K distinct shortest st-*walks*, i.e., paths in which vertices are allowed to appear more than once. Recently, Rizzi et al. [17] studied a different parameterization of the K shortest path problem where they ask to list all st-paths with length at most α for a given α. The difference to the classical K shortest path problem is that the lengths (instead of the overall number) of the paths output is bounded. Thus, depending on the value of α, K might be exponential in the input size. The running time of the proposed algorithm coincides with the running time of the algorithm of Yen and Lawler for directed graphs, and with the running time of the algorithm of Katoh et al. for undirected graphs. However, the algorithm of Rizzi et al. uses only $\mathcal{O}(n + m)$ space which is linear in the input size. All these algorithms cannot directly be used for our listing problem, since we have the additional constraint to list only paths for which a route of length at most β exists, and since lines can share multiple transfers. A more detailed explanation is given in Sect. 4.

2   Preliminaries

Mathematical Preliminaries. Let $G = (V, A)$ be a directed graph. A *walk* in G is a sequence of vertices $\langle v_0, \ldots, v_k \rangle$ such that $(v_{i-1}, v_i) \in A$ for all $i \in [1, k]$. For a walk $w = \langle v_0, \ldots, v_k \rangle$ and a vertex $v \in V$, we write $v \in w$ if and only if there exists an index $i \in [0, k]$ such that $v = v_i$. Analogously, for a walk $w = \langle v_0, \ldots, v_k \rangle$ and an arc $a = (u, v) \in A$, we write $a \in w$ if and only if there exists an index $i \in [1, k]$ such that $u = v_{i-1}$ and $v = v_i$. The length of a walk $w = \langle v_0, \ldots, v_k \rangle$ is k, the number of arcs in the walk, and is denoted by $|w|$. A walk w of length $|w| = 0$ is called *degenerate*, and *non-degenerate* otherwise. For two walks $w_1 = \langle u_0, \ldots, u_k \rangle$ and $w_2 = \langle v_0, \ldots, v_l \rangle$ with $u_k = v_0$, $w_1 \cdot w_2$ denotes the *concatenation* $\langle u_0, \ldots, u_k = v_0, \ldots, v_l \rangle$ of w_1 and w_2. A *path* is a walk $\pi = \langle v_0, \ldots, v_k \rangle$ such that $v_i \neq v_j$ for all $i \neq j$ in $[0, k]$, i.e. a path is a walk without crossings. Given a path $\pi = \langle v_0, \ldots, v_k \rangle$, every contiguous subsequence $\pi' = \langle v_i, \ldots, v_j \rangle$ is called a *subpath* of π. A path $\pi = \langle s = v_0, v_1, \ldots, v_{k-1}, v_k = t \rangle$ is called an *st-path*. For a vertex $v \in V$, let $N_G^-(v)$ denote the out-neighborhood of v. Given two integers i, j, we define the function δ_{ij} (*Kronecker delta*) as 1 if $i = j$ and 0 if $i \neq j$.

Lines and Transit Networks. Given a set of non-degenerate paths (called *lines*) L, the *transit network induced by* L is the graph $G_L = (V_L, A_L)$ where V_L contains exactly the vertices v for which L contains a line l with $v \in l$, and A_L contains exactly the arcs a for which L contains a line l with $a \in l$. This definition is similar to the definition of the *station graph* in [18], and it does not include travel times or timetables since we are only interested in the structure of the network. The modeling differs from classical graph-based models like the *time-expanded* or the *time-dependent* model which incorporate travel times explicitly by adding additional vertices or cost functions in the arcs, respectively (see, e.g., [6,16] for more information on these models). However, for finding robust routes with the approach in [5], the above definition is sufficient since travel times are integrated at a later stage. In the following, let $M_L = \sum_{l \in L} |l|$ denote the sum of the lengths of all lines. In the rest of this paper, we omit the index L from V_L, A_L and M_L to simplify the notation.

Given a path $\pi = \langle v_0, \ldots, v_k \rangle$ in G_L and a sequence of lines $\gamma = \langle l_1, \ldots, l_h \rangle$, we say that the pair (π, γ) is a *route* if π is equal to the concatenation of non-degenerate subpaths $\pi_1, \ldots, \pi_h$ of the lines $l_1, \ldots, l_h$, in this order. Notice that a line might occur multiple times in γ (see Fig. 1); however, we assume that any two consecutive lines in γ are different. For every $i \in \{1, \ldots, h-1\}$, we say that a *line change* between the lines l_i and l_{i+1} occurs. The *length* of the route (π, γ) is $|\gamma|$, i.e. the number of line changes plus one. Given two vertices $u, v \in V$, a *uv-route* is a route (π, γ) such that π is a *uv*-path. A minimum *uv*-route has smallest length among all *uv*-routes in G_L, and we define the *L-distance* $d_L(u, v)$ from u to v as the length of a minimum *uv*-route. For a path π and a line $l \in L$, let $l - \pi$ be the union of (possibly degenerate) paths that we obtain after removing every vertex $v \in \pi$ and its adjacent arcs from l (see Fig. 1). For simplicity, we also call each of these unions of paths a *line*, although they might be disconnected

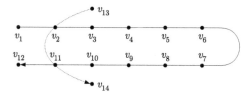

Fig. 1. A transit network with one-way streets induced by a line $l_1 = \langle v_1, \ldots, v_{12} \rangle$ (solid) and a line $l_2 = \langle v_{13}, v_2, v_{11}, v_{14} \rangle$ (dotted). To travel from $s = v_1$ to $t = v_{12}$, it is reasonable to use l_1 until v_2, after that use l_2 from v_2 to v_{11} and from there use l_1 again. We have $l_1 - l_2 = \langle v_1, v_3, v_4, \ldots, v_{10}, v_{12} \rangle$, and $l_2 - l_1 = \langle v_{13}, v_{14} \rangle$.

and/or degenerated. However, we note that all algorithms in this paper also work for disconnected and/or degenerate lines. Given a path π and a set L of lines, let $L - \pi = \{l - \pi \mid l \in L\}$ denote the set of all lines in which every vertex from π has been removed. Analogously to our previous definitions, given a path π and a graph G, we define $G - \pi$ as the graph from which every vertex $v \in \pi$ and its adjacent arcs have been removed.

Problems. An algorithm that systematically lists all or a specified subset of solutions of a combinatorial optimization problem is called a *listing algorithm*. The *delay* of a listing algorithm is the maximum of the time elapsed until the first solution is output and the times elapsed between any two consecutive solutions are output [12,17].

Problem 1 (Finding a minimum st-route). Given a transit network $G_L = (V, A)$ and two vertices $s, t \in V$, find a minimum route from s to t.

Problem 2 (Finding an st-route with a minimum number of different lines). Given a transit network $G_L = (V, A)$ and two vertices $s, t \in V$, find a route from s to t that uses a minimum number of different lines from L.

Notice that, although Problems 1 and 2 sound similar, they are in general not equivalent. Figure 2 shows an example for a transit network in which the optimal solutions of the problems differ.

A natural listing problem is to list all possible st-routes. However, this formulation has the disadvantage that the number of possible solutions is huge, and that there might exist many redundant solutions since a path π can give rise to multiple distinct routes (e.g., if some arc of π is shared by two lines) and vice versa. Moreover, from a practical point of view, also routes that contain many line changes are undesirable. Thus, we formulate the following two listing problems.

Problem 3 (Listing β-bounded st-paths). Given a transit network $G_L = (V, A)$, two vertices $s, t \in V$, and $\beta \in \mathbb{N}$, output all st-paths π such that there exists at least one route (π, γ) with length at most β.

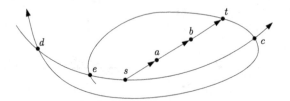

Fig. 2. A transit network induced by the lines $l_1 = \langle s, a \rangle$, $l_2 = \langle a, b \rangle$, $l_3 = \langle b, t \rangle$, $l_4 = \langle d, e, s, c \rangle$ and $l_5 = \langle e, t, c, d \rangle$. The route $r_1 = (\langle s, a, b, t \rangle, \langle l_1, l_2, l_3 \rangle)$ is an optimal solution for Problem 1. It uses three different lines and two transfers. However the optimal solution for Problem 2 is the route $r_2 = (\langle s, c, d, e, t \rangle, \langle l_4, l_5, l_4, l_5 \rangle)$ which uses only two different lines but three transfers.

Problem 4 (Listing β-bounded line sequences). Given a transit network $G_L = (V, A)$, two vertices $s, t \in V$, and $\beta \in \mathbb{N}$, output all line sequences γ such that there exists at least one route (π, γ) with length at most β.

3   Finding an Optimal Solution

In this section we discuss solutions to the Problems 1 and 2. As a preliminary observation we show that for *undirected* lines (i.e., undirected connected graphs where every vertex has degree 2 or smaller) and *undirected* transit networks, the problems are equivalent and can be solved in time $\Theta(M)$. Essentially they are easy because lines can always be traveled in both directions. Of course, this does not hold in the case of directed graphs (see Fig. 2). While Problem 1 can be solved in time $\Theta(M)$ using Dial's (implementation of Dijkstra's) algorithm [7] on an auxiliary graph similar to the one presented in [16], Problem 2 turns out to be NP-hard to approximate.

Theorem 1. *If all lines in L are undirected and G_L is the undirected induced transit network, then Problems 1 and 2 coincide and can be solved in time $\Theta(M)$ where $M = \sum_{l \in L} |l|$ is the input size.*

Proof. Let $r = (\pi, \gamma)$ with $\pi = (\pi_1, \ldots, \pi_h)$ and $\gamma = (l_1, \ldots, l_h)$ be an optimal solution to Problem 2. We first show that there always exists an optimal solution $\bar{r} = (\bar{\pi}, \bar{\gamma})$ that uses every line in $\bar{\gamma}$ exactly once. Suppose that some line l occurred multiple times in γ. Let i be the smallest index such that $l_i = l$, and let j be the largest index such that $l_j = l$. Let v be the first vertex on π_i (i.e., the first vertex on the subpath served by the first occurrence of l), and let w be the last vertex on π_j (i.e., the last vertex on the subpath served by the last occurrence of l). Let π_{sv} be the subpath of π starting in s and ending in v, π_{vw} be a subpath of l from v to w, and π_{wt} be the subpath of π starting in w and ending in t. The route $r' = (\pi', \gamma')$ with $\pi' = \pi_{sv} \cdot \pi_{vw} \cdot \pi_{wt}$ and $\gamma' = (l_1, \ldots, l_{i-1}, l, l_{j+1}, \ldots, l_h)$ is still an st-route, it uses the line l exactly once, and overall it does not use more different lines than r does. Thus, repeating the above argument for every line l

that occurs multiple times, we obtain a route $\bar{r} = (\bar{\pi}, \bar{\gamma})$ which uses every line in $\bar{\gamma}$ exactly once and which is still an optimal solution to Problem 2.

The above argument can also be applied to show that every optimal solution (π, γ) to Problem 1 uses every line in γ exactly once. Now it easy to see that Problem 1 has a solution with exactly k line changes if and only if Problem 2 has a solution with exactly $k+1$ different lines. Therefore, Problems 1 and 2 are equivalent. They can efficiently be solved as follows. For a given transit network $G_L = (V, A)$, consider the vertex-line incidence graph $G' = (V \cup L, A')$ where

$$A' = \{\{v, l\} \mid v \in V \wedge l \in L \wedge \text{line } l \text{ contains vertex } v\}. \tag{1}$$

Breadth-first search can be used to find a shortest st-path $\langle s, l_1, v_1, \ldots, v_{k-1}, l_k, t \rangle$ in G'. Let $\gamma = (l_1, \ldots, l_k)$ be the sequence of lines in this path. Now we use a simple greedy strategy to find a path π in the transit network G_L such that π is the concatenation of subpaths of $l_1, \ldots, l_k$: we start in s, follow l_1 in an arbitrary direction until we find the vertex v_1; if v_1 is not found, we traverse l_1 in the opposite direction until we find v_1. From v_1 we search v_2 on line l_2, and continue correspondingly until we reach t on line l_k. Now the pair (π, γ) is a route with a minimum number of transfers (and, with a minimum number of different lines).

We have $|V \cup L| \in \mathcal{O}(M)$ and $|A'| \in \Theta(M)$, thus the breadth-first search runs in time $\Theta(M)$. Furthermore, G' can be constructed from G_L in time $\Theta(M)$. Thus, for undirected lines and undirected transit networks, Problems 1 and 2 can be solved in time $\Theta(M)$. □

To solve Problem 1 for a *directed* transit network $G_L = (V, A)$, one can construct a weighted auxiliary graph $\Gamma[G_L] = (V[\Gamma], A[\Gamma])$ such that $V \subseteq V[\Gamma]$, and for any two vertices $s, t \in V$ the cost of a shortest st-path in $\Gamma[G_L]$ is exactly $d_L(s, t)$. For a given vertex $v \in V$, let $L_v \subseteq L$ be the set of all lines that contain v. We add every vertex $v \in V$ to $V[\Gamma]$. Additionally, for every vertex $v \in V$ and every line $l \in L_v$, we create a new vertex v_l and add it to $V[\Gamma]$. The set $A[\Gamma]$ contains three different types of arcs:

(1) For every arc $a = (u, v)$ in a line l, we create a *traveling* arc (u_l, v_l) with cost 0. These arcs are used for traveling along a line l.
(2) For every vertex v and every line $l \in L_v$, we create a *boarding* arc (v, v_l) with cost 1. These arcs are used to board the line l at vertex v.
(3) For every vertex v and every line $l \in L_v$, we create a *leaving* arc (v_l, v) with cost 0. These arcs are used to leave the line l at vertex v.

This construction is a simplified version of the *realistic time-expanded graph* for the *Minimum Number of Transfers Problem* described in [16]. We nevertheless describe and analyse it explicitly because it will be used as a subroutine in the listing algorithms in Sect. 4, and the details of the construction are important for the running time analysis of our listing algorithms.

Theorem 2. *Problem 1 is solvable in time* $\Theta(M)$ *where* $M = \sum_{l \in L} |l|$ *is the input size.*

$X = \{x_1, x_2, x_3, x_4, x_5\}$
$S_1 = \{x_2, x_4, x_5\}$

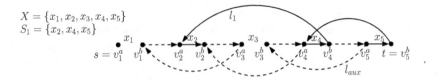

Fig. 3. The correspondence between a set $S_1 \subseteq X$ and a line l_i of the transit network.

Proof. Let $G_L = (V, A)$ be a transit network and $s, t \in V$ be arbitrary. We compute the graph $\Gamma[G_L]$ and run Dial's algorithm [7] on the vertex s. Let π_{st} be a shortest st-path in $\Gamma[G_L]$. It is easy to see that the cost of π_{st} is exactly $d_L(s, t)$ [16]. Furthermore, π_{st} induces an st-path in G_L by replacing every traveling arc (v_l, w_l) by (v, w), and ignoring the arcs of the other two types [18]. Analogously the line sequence can be extracted from π_{st} by considering the lines l of all boarding arcs (v, v_l) in π_{st} (or, alternatively, by considering the lines l of all leaving arcs (v_l, v) in π_{st}).

For every vertex v served by a line l, $\Gamma[G_L]$ contains at most two vertices (namely, v_l and v), thus we have $|V[\Gamma]| \in \mathcal{O}(M)$. Furthermore, $A[\Gamma]$ contains every arc a of every line, and exactly two additional arcs for every vertex v_l. Thus we obtain $|A[\Gamma]| \in \mathcal{O}(M)$. Since the largest arc weight is $C = 1$ and Dial's algorithm runs in time $\mathcal{O}(|V[\Gamma]|C + |A[\Gamma]|)$, Problem 1 can be solved in time $\mathcal{O}(M)$.   □

In contrast to the previous Theorem, we will show now that finding a route with a minimum number of *different* lines is NP-hard to approximate.

Theorem 3. *Problem 2 is NP-hard to approximate within* $(1 - \varepsilon) \ln n$ *for any* $\varepsilon > 0$ *unless NP = P.*

Proof. We construct an approximation preserving reduction from SETCOVER. The reduction is similar to the one presented in [20] for the minimum-color path problem. Given an instance $I = (X, \mathcal{S})$ of SETCOVER, where $X = \{x_1, \ldots, x_n\}$ is the ground set, and $\mathcal{S} = \{S_1, \ldots, S_m\}$ is a family of subsets of X, the goal is to find a minimum cardinality subset $\mathcal{S}' \subseteq \mathcal{S}$ such that the union of the sets in $\mathcal{S}'$ contains all elements from X.

We construct from I a set of lines L that induces a transit network $G_L = (V, A)$ as follows. See Fig. 3 along with the construction. The set L consists of $m+1$ lines and induces $2n$ vertices. The vertex set $V = \{v_1^a, v_1^b, v_2^a, v_2^b, \ldots, v_n^a, v_n^b\}$ contains two vertices v_i^a and v_i^b for each element x_i of the ground set X. Let $V^O = \langle v_1^a, v_1^b, \ldots, v_n^a, v_n^b \rangle$ be the order naturally defined by V. The set of lines $L = \{l_1, \ldots, l_m, l_{aux}\}$ contains one line for each set in $\mathcal{S}$, plus one auxiliary line l_{aux}. For a set $S_i \in \mathcal{S}$, consider the set of vertices that correspond to the elements in S_i and order them according to V^O to obtain $\langle v_{i_1}^a, v_{i_1}^b, v_{i_2}^a, v_{i_2}^b, \ldots, v_{i_r}^a, v_{i_r}^b \rangle$. Now we define the line l_i as $\langle v_{i_r}^a, v_{i_r}^b, v_{i_{(r-1)}}^a, v_{i_{(r-1)}}^b, \ldots, v_{i_1}^a, v_{i_1}^b \rangle$. The auxiliary line l_{aux} is defined as $\langle v_{n-1}^b, v_n^a, v_{n-2}^b, v_{n-1}^a, \ldots, v_1^b, v_2^a \rangle$. Observe that the set of arcs A induced by L contains two types of arcs. First, there are arcs of the form (v_i^a, v_i^b) or in the form (v_i^b, v_{i+1}^a) for some $i \in [1, n]$. These are the only arcs in A whose direction agrees with the order V^O, and we refer to them as *forward* arcs. Second, for all the other arcs $(u, v) \in A$ we have $u > v$ with respect to the order V^O, and we refer to these arcs as *backward* arcs. We note that every line l_i is constructed so that the forward arcs of l_i correspond to those elements of X that are contained in S_i, and the backward arcs connect the forward arcs, in the order opposite to V^O, thus making the lines connected. The auxiliary line l_{aux} consists of all the forward arcs in the form (v_i^b, v_{i+1}^a), that are again connected in the opposite order by backward arcs.

Now, for $s = v_1^a$ and $t = v_n^b$, we show that an st-route with a minimum number of different lines in the given transit network G_L provides a minimum SETCOVER for I, and vice versa. Since t is after s in the order V^O, and the only forward arcs in G_L are of the form (v_i^a, v_i^b) or (v_i^b, v_{i+1}^a) for some i, it follows that any route from s to t in G_L goes via all the vertices, in the order V^O. Thus, for each st-route $r = (\pi, \gamma)$, there exists an st-route $r' = (\pi', \gamma')$ which does not use any additional lines to those used in r, but contains no backward arc. That is, γ' is a subsequence of γ, and $\pi' = \langle v_1^a, v_1^b, v_2^a, v_2^b, \ldots, v_n^a, v_n^b \rangle$. In particular, there exists an st-route that minimizes the number of different lines, and its path is $\langle v_1^a, v_1^b, v_2^a, v_2^b, \ldots, v_n^a, v_n^b \rangle$. Clearly, l_{aux} must be used in every st-route, as it represents the only way to reach v_{i+1}^a from v_i^b. Now, if a line l_i is used in the st-route r, all the forward arcs in l_i correspond to the arcs (v_i^a, v_i^b) of the path in r and in this way the line l_i "covers" these arcs. Since there is a one to one mapping between the lines $l_1, \ldots, l_m$ and the sets in $\mathcal{S}$, by finding an st-route with $k+1$ different lines, one finds a solution of size k to the original SETCOVER. Similarly each solution of size k to the original SETCOVER can be mapped to an st-route with $k + 1$ lines. Thus our reduction is approximation preserving, and based on the inapproximability of SETCOVER [9] this concludes the proof. □

4   Listing All Solutions

Motivation. In [5], the authors describe an algorithm for Problem 4 whose worst-case running time might be exponential in β, independently of κ, the number of listed line sequences. A naïve approach for solving Problem 3 is to use this algorithm to generate all feasible line sequences γ and then to compute the corresponding paths (there might be more than one) for each feasible γ. However, this approach does not only have the disadvantage of a possibly huge running time, also for every path π there might be many line sequences γ such that (π, γ) is route in G_L. Since we want to output every path π at most once, we would need to store $\Omega(\kappa)$ many paths.

Another straightforward idea to solve Problem 3 might be to construct an auxiliary graph from G_L and then use one the well-known algorithms for listing paths, e.g., Yen's algorithm. For example, one could create a directed graph $G_X = (V, A_X)$ where A_X contains an arc between v and w if there exists a line that visits v before w. Any path between s and t in G_X of length at most β

induces an st-path in G_L. However, as before, exponentially many paths in G_X might correspond to one path in G_L which again might lead to an exponential gap between the running time and the sum of the input and the output size.

Improved Idea for Problem 3. Let $\mathcal{P}_{st}^{\beta}(L)$ denote the set of all st-paths π such that there exists a route (π, γ) with length at most β in the transit network G_L. To obtain a polynomial delay algorithm that uses only $\mathcal{O}(M)$ space, we use the so-called *binary partition method* described in [3,17]: The transit network G_L is traversed in a depth first search fashion starting from s, and the solution space $\mathcal{P}_{st}^{\beta}(L)$ is recursively partitioned at every call until it contains exactly one solution (i.e., one path) that is then output.

When the algorithm considers a partial su-path π_{su}, we first check whether $u = t$. In that case, π_{su} is output. Otherwise, we compute the graph G' that is the transit network G_L from which all vertices (and all adjacent edges) in π_{su} are removed. To bound the running time of the algorithm we maintain the invariant that the current partition (i.e., the paths in $\mathcal{P}_{st}^{\beta}(L)$ with prefix π_{su}) contains at least one solution. More concretely, we require that G' contains at least one ut-path π_{ut} that extends π_{su} so that $\pi_{su} \cdot \pi_{ut} \in \mathcal{P}_{st}^{\beta}(L)$. The idea behind this algorithm is similar to the one in [17] for listing all α-bounded paths; here, however, new ideas to maintain the invariant are necessary because our objective is to list only paths π for which a length-bounded route (π, γ) in G_L exists (instead of listing all paths whose length itself is bounded).

Checking Whether to Recurse or Not. Let π_{su} be the su-path that the algorithm currently considers, $L' = L - \pi_{su}$, $G' = G_L - \pi_{su} = G_{L'}$, and $v \in N_{G_L}^-(u) \cap G'$, i.e., v is a neighbor of u that is not contained in π_{su}. We recursively continue on $\pi_{su} \cdot (u, v)$ only if the invariant (I) is satisfied, i.e., if $\mathcal{P}_{st}^{\beta}(L)$ contains a path with prefix $\pi_{su} \cdot (u, v)$.

Let $d_{G_L}(\pi_{su}, (u, v), l_i)$ be the length of a minimum route $(\pi_{su} \cdot (u, v), \gamma)$ in G_L such that l_i is the last line of γ. Let $d_{G'}^{L'}(v, t, l_j)$ be the L'-distance from v to t in G' such that l_j is the first line used. For a vertex $v \in V$, let $L_v \subseteq L$ be the set of all lines that contain an outgoing arc from v. Analogously, for an arc $(u, v) \in A$, let $L_{(u,v)}$ be the set of all lines that contain (u, v). Now, the set $\mathcal{P}_{st}^{\beta}(L)$ contains a path with prefix $\pi_{su} \cdot (u, v)$ if and only if

$$\min \left\{ d_{G_L}(\pi_{su}, (u, v), l_i) - \delta_{ij} + d_{G'}^{L'}(v, t, l_j) \mid l_i \in L_{(u,v)} \text{ and } l_j \in L_v \right\} \le \beta. \quad (2)$$

Basically, $\min\{d_{G_L}(\pi_{su}, (u, v), l_i) - \delta_{ij} + d_{G'}^{L'}(v, t, l_j) \mid l_i \in L_{(u,v)} \text{ and } l_j \in L_v\}$ is the length of the minimum route that has prefix $\pi_{su} \cdot (u, v)$.

Computing $d_{G_L}(\pi_{su}, (u, v), l_i)$ and $d_{G'}^{L'}(v, t, l_j)$. We can use the solution for Problem 1 to compute the values $d_{G_L}(\pi_{su}, (u, v), l_i)$ and $d_{G'}^{L'}(v, t, l_j)$. The values $d_{G_L}(\pi_{su}, (u, v), l_i)$ need to be computed only for arcs $(u, v) \in A$ with $v \notin \pi_{su}$ (i.e., only for arcs from u to a vertex $v \in N_{G_L}^-(u) \cap G'$), and only for lines $l_i \in L_{(u,v)}$. Consider the graph G'' that contains every arc from π_{su} and every arc $(u, v) \in A$ with $v \notin \pi_{su}$, and that contains exactly the vertices incident to these arcs.

Now we compute $H = \Gamma[G'']$ and run Dial's algorithm on the vertex s. For every $v \in N_{G_L}^-(u) \cap G'$ and every line $l_i \in L_{(u,v)}$, the length of a shortest path in H from s to v_{l_i} is exactly $d_{G_L}(\pi_{su}, (u,v), l_i)$. For computing $d_{G'}^{L'}(v, t, l_j)$, we can consider the L'-distances from t in the reverse graph G'^R (with all the arcs and lines in L' reversed). Considering G' instead of G_L ensures that lines do not use vertices that have been deleted in previous recursive calls of the algorithm. Thus we compute $\Gamma[G'^R]$ and run Dial's algorithm on the vertex t. Then, the length of a shortest path in $\Gamma[G'^R]$ from t to v_{l_j} is exactly $d_{G'}^{L'}(v, t, l_j)$.

Algorithm. Algorithm 1 shows the details of the aforementioned approach. To limit the space consumption of the algorithm, we do not pass the graph G' as a parameter to the recursive calls, but compute it at the beginning of each recursive call from the current prefix π_{su}. For the same reason, we do not perform the recursive calls immediately in step 8, but first create a list $V_R \subseteq V$ of vertices for which the invariant (I) is satisfied, and only then recurse on $(v, \pi_{su} \cdot (u,v))$ for every $v \in V_R$. To list all paths in $\mathcal{P}_{st}^\beta$, we invoke LISTPATHS$(s, \langle s \rangle)$.

Algorithm 1. LISTPATHS(u, π_{su})

---

1 **if** $u = t$ **then** OUTPUT(π_{su}); **return**
2 $L' \leftarrow L - \pi_{su}$; $G' \leftarrow G_L - \pi_{su}$
3 Compute $d_{G_L}(\pi_{su}, (u,v), l_i)$ for each $v \in N_{G_L}^-(u) \cap G'$ and $l_i \in L_{(u,v)}$
4 Compute $d_{G'}^{L'}(v, t, l_j)$ for each $v \in N_{G_L}^-(u) \cap G'$ and $l_j \in L_v$
5 $V_R \leftarrow \emptyset$
6 **for** $v \in N_{G_L}^-(u) \cap G'$ **do**
7 $\quad$ $d \leftarrow \min\{d_{G_L}(\pi_{su}, (u,v), l_i) + d_{G'}^{L'}(v, t, l_j) - \delta_{ij} \mid i \in L_{(u,v)} \text{ and } l_j \in L_v\}$
8 $\quad$ **if** $d \leq \beta$ **then** $V_R \leftarrow V_R \cup \{v\}$
9 **for** $v \in V_R$ **do**
10 $\quad$ LISTPATHS$(v, \pi_{su} \cdot (u,v))$

Theorem 4. *Algorithm 1 has delay $\mathcal{O}(nM)$, where n is the number of vertices in G_L and $M = \sum_{l \in L} |l|$ is the input size. The total time complexity is $\mathcal{O}(nM \cdot \kappa)$, where κ is the number of returned solutions. Moreover, the space complexity is $\mathcal{O}(M)$.*

Proof. We first analyse the cost of a given call to the algorithm without including the cost of the recursive calls performed inside. Theorem 2 states that steps 3 and 4 can be performed in time $\mathcal{O}(M)$. We will now show that steps 6–8 can be implemented in time $\mathcal{O}(M)$. Notice that for a fixed prefix π_{su} and a fixed vertex $v \in N_{G_L}^-(u) \cap G'$, for computing the minimum in step 7, we need to consider only the values $d_{G_L}(\pi_{su}, (u,v), l_i)$ that are minimum among all $d_{G_L}(\pi_{su}, (u,v), \cdot)$, and only the values $d_{G'}^{L'}(v, t, l_j)$ that are minimum among all $d_{G'}^{L'}(v, t, \cdot)$. Let $\Lambda_v \subseteq L_{(u,v)}$ be the list of all lines l_i for which $d_{G_L}(\pi_{su}, (u, v), l_i)$ is minimum among all $d_{G_L}(\pi_{su}, (u, v), \cdot)$. Analogously, let $\Lambda'_v \subseteq L_v$ be the list of all lines l_j for which $d_{G'}^{L'}(v, t, l_j)$ is minimum among all $d_{G'}^{L'}(v, t, \cdot)$. Let

$$\mu_v = \min \left\{ d_{G_L}(\pi_{su}, (u, v), l_i) \mid l_i \in \Lambda_v \right\} \tag{3}$$

$$\mu'_v = \min \left\{ d_{G'}^{L'}(v, t, l_j) \mid l_j \in \Lambda'_v \right\} \tag{4}$$

be the minimum values of $d_{G_L}(\pi_{su}, (u, v), \cdot)$ and $d_{G'}^{L'}(v, t, \cdot)$, respectively. Both values as well as the lists Λ_v and Λ'_v can be computed in steps 3 and 4, and their computation only takes overall time $\mathcal{O}(M)$. Now the expression in step 7 evaluates to $\mu_v + \mu'_v$ if $\Lambda_v \cap \Lambda'_v = \emptyset$, and to $\mu_v + \mu'_v - 1$ otherwise. Assuming that Λ_v and Λ'_v are ordered ascendingly by the index of the contained lines l_i, it can easily be checked with $|\Lambda_v| + |\Lambda'_v| \leq |L_{(u,v)}| + |L_v|$ many comparisons if their intersection is empty or not. Using this method, each of the values $d_{G_L}(\pi_{su}, \cdot, \cdot)$ and $d_{G'}^{L'}(\cdot, t, \cdot)$ is accessed exactly once (when computing Λ_v and Λ'_v), and since each of these values has a unique corresponding vertex in the graphs H and $\Gamma[G'^R]$, there exist at most $\mathcal{O}(M)$ many such values. Thus, the running time of the steps 6–8 is bounded by $\mathcal{O}(M)$ which is also an upper bound on the running time of Algorithm 1 (ignoring the recursive calls).

We now look at the structure of the recursion tree. The height of the recursion tree is bounded by n, since at every level of the recursion tree a new vertex is added to the current partial solution and any solution has at most n vertices. Solutions are output in the leaves. Since the length of a path between any two leaves in the recursion tree is at most $2n$, the delay is in $\mathcal{O}(nM)$.

For analysing the space complexity, observe that L', G' and the values $d_{G_L}(\pi_{su}, (u, v), l_i)$ and $d_{G'}^{L'}(v, t, l_j)$ can be removed from the memory after step 8 since they are not needed any more. Thus, we only need to store the lists V_R between the recursive calls. Consider a path in the recursion tree, and for each recursive call i, let u^i be the vertex u and V_R^i be the list V_R of the i-th recursive call. Since V_R^i contains only vertices adjacent to u^i and u^i is never being considered again in *any* succeeding recursive call $j > i$, we have

$$\sum_i |V_R^i| \leq |A_L|, \tag{5}$$

which proves the space complexity of $\mathcal{O}(M)$.   □

A More Efficient Solution to Problem 4. We can now combine the ideas from above with the ideas in [5] to develop a polynomial delay listing algorithm for Problem 4. We first compute the values $d_{G_L}^L(v, t, l)$ for every vertex v and every line l that contains v (using the solution to Problem 1). After that, assume that a partial line sequence $\gamma = \langle l_1, \dots, l_k \rangle$, $k \leq \beta$, was already computed and that u is the earliest vertex on l_k which can be reached from l_{k-1} among all possible st-routes (π, γ); see Fig. 4 for an example. If l_k visits t after u, then we output the line sequence γ. We also have to check whether γ can be extended. Figure 1 shows why this can be reasonable even if t is reachable via l_k. We compute a set L' of possible line candidates l that can be reached from l_k after u, and also the transfer vertices v_l by which l is reached as early as possible (i.e., there is no vertex v'_l visited by l_k after u that is visited by l before v_l). This can be done by considering the successors of u on l_k with increasing distance from u, and keeping track of the optimal transfer from l_k. Now, for each $l \in L'$, we check whether $d^L_{G_L}(v_l, t, l) \leq \beta - k$. In such a case there exists a $v_l t$-route starting with the line l, and extending γ by l gives a route of length at most β. Otherwise, there either is no $v_l t$-route starting with l, or it uses too many transfers.

Algorithm 2 shows the details. To solve Problem 4, it is sufficient to invoke LISTLINESEQUENCES$(s, \langle l \rangle)$ for every $l \in L_s$ where $d^L_{G_L}(s, t, l) \leq \beta$. The idea of extending the partial line sequence step-by-step and how to find the optimal transfer vertex is similar to [5]; here, however, we only extend γ by l' if this definitely leads to a solution that is output. As the following theorem shows, this guarantees a polynomial delay.

Algorithm 2. LISTLINESEQUENCES$(u, \langle l_1, \ldots, l_k \rangle)$

---

1   Compute $d^L_{G_L}(v, t, l)$ for each $v \in G_L$ and $l \in L$
2   **for** $l \in L$ **do** $v_l \leftarrow \infty$
3   **for each** *successor v of u on l_k in increasing order* **do**
4   **if** $v = t$ **then** OUTPUT$(\langle l_1, \ldots, l_k \rangle)$
5   **for** $l \in L_v$ **do**
6        $\lfloor$ **if** $v_l = \infty$ **or** l *visits v earlier than* v_l **then** $v_l \leftarrow v$

7   **for** $l \in L$ **do**
8   **if** $v_l \neq \infty$ **and** $d^L_{G_L}(v_l, t, l) \leq \beta - k$ **then**
9        $\lfloor$ LISTLINESEQUENCES$(v_l, \langle l_1, \ldots, l_k, l \rangle)$

Theorem 5. *Algorithm 2 has delay $\mathcal{O}(\beta M)$, and its total time complexity is $\mathcal{O}(\beta M \cdot \kappa)$, where $M = \sum_{l \in L} |l|$ is the input size and κ is the number of returned solutions. Moreover, the space complexity is $\mathcal{O}(M)$.*

Proof. As before, step 1 can be computed using $\mathcal{O}(M)$ operations and requires $\mathcal{O}(M)$ space. Steps 3–6 can also be implemented to run in time $\mathcal{O}(M)$, as every step 6 takes only constant time and is performed at most once for every vertex v and every line l containing v. Since the recursive calls in step 9 are only performed if it is guaranteed to output a solution, we observe that the height of the recursion tree is bounded by $\mathcal{O}(\beta)$, hence the delay of the algorithm is $\mathcal{O}(\beta M)$. The time complexity of $\mathcal{O}(\beta M \cdot \kappa)$ immediately follows. $\square$

As in [5], we assumed in the above running time analysis that the test whether a given vertex v is visited by a line l can be performed in constant time using

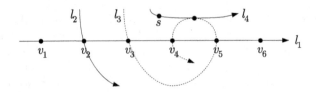

Fig. 4. The earliest transfer from l_3 to l_1 is v_3. However, the earliest transfer using the line sequence $\langle l_4, l_3, l_1 \rangle$ is v_4. We have $f(l_2, l_1, v_1) = v_2$ and $f(l_2, l_1, v_k) = \infty$ for every $k \neq 1$. We have $f(l_3, l_1, v_5) = f(l_3, l_1, v_6) = \infty$, $f(l_3, l_1, v_3) = f(l_3, l_1, v_4) = v_5$ and $f(l_3, l_1, v_1) = f(l_3, l_1, v_2) = v_3$. Moreover, $f(l_3, l_4, v_k) = \infty$ for every k.

suitable hash tables. The same is true for the test whether a line visits a vertex v earlier than some other vertex w. Moreover, when the algorithm is invoked with the parameters $\gamma = \langle l_1, \ldots, l_k \rangle$ and u, and if additionally t is visited by l_k after u, then in practice the successors of t on l_k do not have to be visited any more in step 3 of the Algorithm 2, hence the current call of the algorithm can terminate after γ is output in step 4.

A Faster Algorithm with Preprocessing. Although the enumerating all length-bounded routes is already reasonably fast for urban transportation networks like the one in Zürich [4], an overall running time of $\mathcal{O}(\beta M \cdot \kappa)$ is undesirable from a practical point of view. Algorithm 2 has delay $\mathcal{O}(\beta M)$ for two reasons: (1) initially we compute the values $d_{G_L}^L(\cdot, t, \cdot)$, and (2) for every partial line sequence $\langle l_1, \ldots, l_k \rangle$ we investigate all possible transfer vertices from l_k to other lines to find the optimal one for every line. Issue (1) can easily be solved by computing the values $d_{G_L}^L(v, t, l)$ for every $v, t \in V$ and every line $l \in L$ in advance and then storing them. Since for every t there are at most $\mathcal{O}(M)$ many values $d_{G_L}^L(\cdot, t, \cdot)$ and all of them can be computed in time $\mathcal{O}(M)$, we need overall time $\mathcal{O}(M|V|)$. To solve issue (2), we precompute for every line l, every vertex v on l and every line $l' \neq l$ the vertex $f(l, l', v)$ which is visited by l after v and by which l' is reached as early as possible (on l'). If no such a vertex exists, we set $f(l, l', v) = \infty$. For every line $l = \langle v_1, \ldots, v_k \rangle \in L$, the values $f(l, \cdot, \cdot)$ can be computed as follows. We consider the vertices $v_k, \ldots, v_1$ in this order. We set $f(l, l', v_k) = \infty$ for every $l' \in L$. After that, considering a vertex v_i, we set

$$f(l, l', v_i) = \begin{cases} v_{i+1} & \text{if } l' \in L_{v_{i+1}} \text{ and } f(l, l', v_{i+1}) = \infty \\ v_{i+1} & \text{if } l' \in L_{v_{i+1}} \text{ and } l' \text{ visits } v_{i+1} \text{ before } f(l, l', v_{i+1}) \\ f(l, l', v_{i+1}) & \text{otherwise} \end{cases}$$

(6)

Since the computation of each entry requires only constant time, the values $f(l, l', v)$ can be computed using time and space $\mathcal{O}(M|L|)$. Hence, for preprocessing time and space $\mathcal{O}(M(|V|+|L|))$ suffice. Now, however, st-route listing queries can be performed much faster using the following algorithm.

Algorithm 3. LISTLINESEQUENCES$(u, \langle l_1, \ldots, l_k \rangle)$

---

1 **if** l_k *visits* t *after* u **then** OUTPUT$(\langle l_1, \ldots, l_k \rangle)$
2 **for** $l \in L$ **do**
3   $\quad v_l \leftarrow f(l_k, l, u)$
4   $\quad$ **if** $v_l \neq \infty$ **and** $d^L_{G_L}(v_l, t, l) \leq \beta - k$ **then**
5   $\quad \quad$ LISTLINESEQUENCES$(v_l, \langle l_1, \ldots, l_k, l \rangle)$

Theorem 6. *The values* $d^L_{G_L}(v, t, l)$ *and* $f(l, l', v)$ *can be precomputed using time and space* $\mathcal{O}(M(|V| + |L|))$. *Assuming that these values have been precomputed, Algorithm 3 has delay* $\mathcal{O}(\beta|L|)$, *and its total time complexity is* $\mathcal{O}(\beta|L| \cdot \kappa)$, *where* $|L|$ *is the number of lines and* κ *is the number of returned solutions.*

Proof. The straightforward proof is similar to the proof of Theorem 5. $\square$

To see the speedup, remember that Algorithm 2 has a delay of $\mathcal{O}(\beta M)$ while Algorithm 3 has a delay of only $\mathcal{O}(\beta|L|)$. In real networks, $M = \sum_{l \in L} |l|$ is usually way larger than $|L|$ is.

Acknowledgements. We thank the anonymous reviewers for pointing out how the running times of our listing algorithms can be improved by a factor of $\Theta(\log M)$. Furthermore we thank Peter Widmayer for many helpful discussions. This work has been partially supported by the Swiss National Science Foundation (SNF) under the grant number 200021 138117/1, and by the EU FP7/2007-2013 (DG CONNECT.H5-Smart Cities and Sustainability), under grant agreement no. 288094 (project eCOMPASS). Kateřina Böhmová is a recipient of a Google Europe Fellowship in Optimization Algorithms, and this research is supported in part by this Google Fellowship. Gustavo Sacomoto is a recipient of a grant from the European Research Council under the European Community's Seventh Framework Programme (FP7/2007-2013)/ERC grant agreement n° [247073]10 SISYPHE.

References

1. Bast, H., Delling, D., Goldberg, A.V., Müller-Hannemann, M., Pajor, T., Sanders, P., Wagner, D., Werneck, R.F.: Route planning in transportation networks. CoRR abs/1504.05140 (2015)
2. Bezem, G., Leeuwen, J.V.: Enumeration in graphs. Technical report RUU-CS-87-07, Utrecht University (1987)
3. Birmelé, E., Ferreira, R.A., Grossi, R., Marino, A., Pisanti, N., Rizzi, R., Sacomoto, G.: Optimal listing of cycles and st-paths in undirected graphs. In: SODA 2013, pp. 1884–1896 (2013)
4. Böhmová, K., Mihalák, M., Neubert, P., Pröger, T., Widmayer, P.: Robust routing in urban public transportation: evaluating strategies that learn from the past. In: ATMOS 2015, pp. 68–81 (2015)

5. Böhmová, K., Mihalák, M., Pröger, T., Šrámek, R., Widmayer, P.: Robust routing in urban public transportation: how to find reliable journeys based on past observations. In: ATMOS 2013, pp. 27–41 (2013)
6. Brodal, G.S., Jacob, R.: Time-dependent networks as models to achieve fast exact time-table queries. Electr. Notes Theor. Comput. Sci. **92**, 3–15 (2004)
7. Dial, R.B.: Algorithm 360: shortest-path forest with topological ordering. Commun. ACM **12**(11), 632–633 (1969)
8. Dijkstra, E.W.: A note on two problems in connexion with graphs. Numer. Math. **1**(1), 269–271 (1959)
9. Dinur, I., Steurer, D.: Analytical approach to parallel repetition. In: STOC 2014, pp. 624–633 (2014)
10. Eppstein, D.: Finding the k shortest paths. SIAM J. Comput. **28**(2), 652–673 (1998)
11. Fredman, M.L., Tarjan, R.E.: Fibonacci heaps and their uses in improved network optimization algorithms. J. ACM **34**(3), 596–615 (1987)
12. Johnson, D.S., Papadimitriou, C.H., Yannakakis, M.: On generating all maximal independent sets. Inf. Process. Lett. **27**(3), 119–123 (1988)
13. Johnson, D.B.: Finding all the elementary circuits of a directed graph. SIAM J. Comput. **4**(1), 77–84 (1975)
14. Katoh, N., Ibaraki, T., Mine, H.: An efficient algorithm for k shortest simple paths. Networks **12**(4), 411–427 (1982)
15. Lawler, E.L.: A procedure for computing the k best solutions to discrete optimization problems and its application to the shortest path problem. Mgmt. Sci. **18**, 401–405 (1972)
16. Müller-Hannemann, M., Schulz, F., Wagner, D., Zaroliagis, C.D.: Timetable information: models and algorithms. In: Geraets, F., Kroon, L., et al. (eds.) ATMOS 2004, Part I. LNCS, vol. 4359, pp. 67–90. Springer, Heidelberg (2007)
17. Rizzi, R., Sacomoto, G., Sagot, M.-F.: Efficiently listing bounded length st-paths. In: Jan, K., Miller, M., Froncek, D. (eds.) IWOCA 2014. LNCS, vol. 8986, pp. 318–329. Springer, Heidelberg (2015)
18. Schulz, F., Wagner, D., Zaroliagis, C.D.: Using multi-level graphs for timetable information in railway systems. In: Mount, D.M., Stein, C. (eds.) ALENEX 2002. LNCS, vol. 2409, pp. 43–59. Springer, Heidelberg (2002)
19. Yen, J.Y.: Finding the k shortest loopless paths in a network. Mgmt. Sci. **17**, 712–716 (1971)
20. Yuan, S., Varma, S., Jue, J.P.: Minimum-color path problems for reliability in mesh networks. In: INFOCOM 2005, pp. 2658–2669 (2005)

Compositional Design of Stochastic Timed Automata

Patricia Bouyer[1], Thomas Brihaye[2], Pierre Carlier[1,2(✉)], and Quentin Menet[2]

[1] LSV, CNRS, ENS Cachan, Université Paris-Saclay, Cachan, France
Carlier@lsv.fr
[2] Université de Mons, Mons, Belgium

Abstract. In this paper, we study the model of stochastic timed automata and we target the definition of adequate composition operators that will allow a compositional approach to the design of stochastic systems with hard real-time constraints. This paper achieves the first step towards that goal. Firstly, we define a parallel composition operator that (we prove) corresponds to the interleaving semantics for that model; we give conditions over probability distributions, which ensure that the operator is well-defined; and we exhibit problematic behaviours when this condition is not satisfied. We furthermore identify a large and natural subclass which is closed under parallel composition. Secondly, we define a bisimulation notion which naturally extends that for continuous-time Markov chains. Finally, we importantly show that the defined bisimulation is a congruence w.r.t. the parallel composition, which is an expected property for a proper modular approach to system design.

1 Introduction

Compositional design and compositional verification are two crucial aspects of the development of computerised systems for which correctness needs to be guaranteed or quantified. It is indeed convenient and natural to model separately each component of a system and model their interaction, and it is easier and probably less error-prone than to model at once the complete system.

In the last twenty years a huge effort has been made to design expressive models, with the aim to faithfully represent computerised systems. This is for instance the case of systems with real-time constraints for which the model of timed automata [1,2] is successfully used. Many applications like communication protocols require models integrating both real-time constraints and randomised aspects (see e.g. [25]), which requires the development of specific models. Recently, a model of stochastic timed automata (STA) has been proposed as a natural extension of timed automata with stochastic delays and stochastic edge choices (see [8] for a survey of the results so far concerning this model). Advantages of the

The first and the third authors are supported by ERC project EQualIS. The second author is partly supported by FP7-EU project Cassting. The fourth author was a postdoctoral researcher at the Belgian National Fund for Scientific Research (FNRS).

A.S. Kulikov and G.J. Woeginger (Eds.): CSR 2016, LNCS 9691, pp. 117–130, 2016.
DOI: 10.1007/978-3-319-34171-2_9

STA model are twofold: (i) it is based on the well-understood and powerful model of timed automata, allowing to express hard real-time constraints like deadlines (unlike for the widely used model of *continuous-time Markov chains* (CTMCs in short)); (ii) it enjoys nice decidability properties (see [8,9]). On the other hand, there is no obvious way of designing in a compositional manner a complex system using this model.

In this paper we are inspired by the approach of [24], and we target the definition of (parallel) composition operators allowing for a component-based modelling framework for STA. This paper achieves the first steps towards that goal:

1. We define a parallel composition operator that (we prove) corresponds to the interleaving semantics for that model; we give conditions over families of distributions over delays, which ensure that the operator is well-defined; we exhibit problematic behaviours when this condition is not satisfied. We furthermore identify a class of such well-behaving STA that is closed under parallel composition. Note that this class of well-behaving systems encompasses the class of CTMCs.
2. We define a bisimulation notion which naturally extends that for CTMCs [5,6,17], and we importantly show that the bisimulation is a congruence w.r.t. parallel composition; this is an expected property for a proper modular approach to system design.

The next step will be to extend the current composition operator with some synchronisation between components. For CTMCs, this has required much effort over the years to come up with a satisfactory solution, yielding for instance the model of *interactive Markov chains* (IMCs) [21,22]. We believe we will benefit a lot from this solution and plan to follow a similar approach for STA; we leave it as further work (the current work focuses on races between components and establishes all useful properties at the level of STA).

Related Works. We do not list all works concerned with the verification of stochastic real-time systems, but will focus on those interested in compositional design. The first natural related work is that on interactive Markov chains (IMCs in short) [21,22], which extend CTMCs with interaction, and for which compositional verification methods have been investigated [13,23]. However in this model, only soft real-time constraints can be evaluated (that is, they may not be always satisfied by the system, but their likelihood is then quantified), and the model cannot evolve differently, depending on constraints over clocks. Our ultimate goal is to extend the elegant approach of IMCs to a model based on timed automata.

Other related approaches are based on process algebras (note that originally IMCs presented as a process algebra as well [21]). There have been several proposals, among which the IGSMP calculus [12], whose semantics is given as generalised semi-Markov processes (GSMPs); and the stochastic process algebra ♠ [15,16], whose semantics is given as ♠-stochastic timed automata (we write ♠-STA). Our model very much compares to the latter, so we will briefly describe it. In such a system, when a clock variable is activated, it is sampled according to a predefined distribution, and then it acts as a countdown timer: when time elapses, the clock variables decrease down to 0. Transitions can be fired once all clocks specified on the transition have reached value 0. First notice that both STA and ◊-STA allow to express hard real-time constraints, e.g. strict deadlines to be satisfied by the system (which is not the case of CTMCs or IMCs). Then the ◊-STA model is at the basis of several modelling languages like Modest [10] and comes with several notions of bisimulations with nice congruence properties, and with a complete equational theory. It is interesting to mention as well that ◊-STA allow for infinitely many states and clock variables, whereas STA do not (they have been defined on top of timed automata, with desirable decidability properties in mind). Similarly to ◊-STA, STA extend (finite-state and finite-variable) GSMPs,[1] but for different reasons: ◊-STA allows for fixed-delay events and non-determinism, whereas STA allows for more intricate timing constraints and branchings.[2] Finally, it is worth mentioning the modelling language Modest [10], whose semantics is given as a very general notion of stochastic timed automata (we call them Modest-STA), which comes with an interesting tool suite [19,20], and which encompasses all the models we have mentioned. STA in general, and the subclass that is closed under parallel composition while enjoying decidability properties, can be viewed as a fragment of Modest-STA.

The full version of this work and detailed proofs are given in [11].

2   Stochastic Timed Automata

In this section, we recall the notion of *timed automaton* [2], and that of *stochastic timed automaton* [8]. Let $X = \{x_1, \ldots, x_n\}$ be a finite set of real-valued variables called *clocks*. A *clock valuation* over X is a mapping $\nu : X \to \mathbb{R}_+$ where $\mathbb{R}_+$ is the set of nonnegative real numbers. We write $\mathbb{R}_+^X$ for the set of clock valuations over X. If $\nu \in \mathbb{R}_+^X$, we write ν_i for $\nu(x_i)$ and we then denote ν by $(\nu_1, \ldots, \nu_n)$. If $\tau \in \mathbb{R}_+$, we write $\nu + \tau$ for the clock valuation defined by $(\nu_1 + \tau, \ldots, \nu_n + \tau)$. If $Y \in 2^X$ (the power set of X), $[Y \leftarrow 0]\nu$ is the valuation that assigns to x, 0 if $x \in Y$ and $\nu(x)$ otherwise. A *guard*[3] over X is a finite conjunction of expressions of the form $x_i \sim c$ where $c \in \mathbb{N}$ and $\sim \in \{<, >\}$. We denote by $\mathcal{G}(X)$ the set of guards over X. We write $\nu \models g$ if ν satisfies g, which is defined in a natural way.

Definition 1. *A timed automaton (TA in short) is a tuple $\mathcal{A} = (L, L_0, X, E, \mathsf{AP}, \mathcal{L})$ where: (i) L is a finite set of locations, (ii) $L_0 \subseteq L$ is a set of initial locations, (iii) X is a finite set of clocks, (iv) $E \subseteq L \times \mathcal{G}(X) \times 2^X \times L$ is a finite set of edges, (v) AP is a set of atomic propositions and (vi) $\mathcal{L} : L \to 2^{\mathsf{AP}}$ is a labelling function.*

---

[1] This can be seen using the residual-time semantics given in [14,18].

[2] Somehow, the clock behaviour in GSMPs and in ◊-STA is that of countdown timers (which can be seen as event-predicting clocks of [3]), which is not as rich as general clocks in standard timed automata.

[3] We restrict to open guards for technical reasons due to stochastic aspects.

The semantics of a TA is a labelled timed transition system $T_A = (Q, Q_0, \mathbb{R}_+ \times E, \rightarrow, \mathsf{AP}, \mathcal{L})$ where $Q = L \times \mathbb{R}_+^X$ is the set of states, $Q_0 = L_0 \times \mathbf{0}_X$ is the set of initial states (valuation $\mathbf{0}_X$ assigns 0 to each clock), $\mathcal{L} : Q \rightarrow 2^{\mathsf{AP}}$ labels each state $q = (l, \nu) \in Q$ by $\mathcal{L}(l)$ and $\rightarrow \subseteq Q \times (\mathbb{R}_+ \times E) \times Q$ is the transition relation defined as follows: if $e = (l, g, Y, l') \in E$ and $\tau \in \mathbb{R}_+$, then we have $(l, \nu) \xrightarrow{\tau, e} (l', \nu')$ if $(\nu + \tau) \models g$ and $\nu' = [Y \leftarrow 0](\nu + \tau)$. If $q = (\ell, \nu)$, for every $\tau \geq 0$, $q + \tau$ denotes $(\ell, \nu + \tau)$. A *finite* (resp. *infinite*) run ρ is a finite (resp. infinite) sequence $\rho = q_1 \xrightarrow{\tau_1, e_1} q_2 \xrightarrow{\tau_2, e_2} \ldots$. Given $q \in Q$, we write $\mathsf{Runs}(\mathcal{A}, q)$ for the set of infinite runs in $\mathcal{A}$ from q. Given $q \in Q$ and $e \in E$ we define $I(q, e) = \{\tau \in \mathbb{R}_+ \mid \exists q' \in Q \text{ s.t. } q \xrightarrow{\tau, e} q'\}$ and $I(q) = \bigcup_{e \in E} I(q, e)$.

We now define the notion of *stochastic timed automaton* [8], by equipping every state of a TA with probabity measures over both delays and edges.

Definition 2. *A stochastic timed automaton (STA in short) is a tuple $\mathcal{A} = (L, L_0, X, E, AP, \mathcal{L}, (\mu_q, p_q)_{q \in L \times \mathbb{R}_+^X})$ where $(L, L_0, X, E, AP, \mathcal{L})$ is a timed automaton and for every $q = (l, \nu) \in L \times \mathbb{R}_+^X$,*

 (i) *μ_q is a probability distribution over $I(q)$ and p_q is a probability distribution over E such that for each $e = (l, g, Y, l') \in E$, $p_q(e) > 0$ iff $\nu \models g$,*
 (ii) *μ_q is equivalent to the restriction of the Lebesgue measure on $I(q)$,[4] and*
 (iii) *for each edge e, the function $p_{q+\bullet}(e) : \mathbb{R}_+ \rightarrow [0, 1]$ that assigns to each $t \geq 0$ the value $p_{q+t}(e)$, is measurable.*

We fix $\mathcal{A}$ a STA, with the notations of the definition. We let $Q = L \times \mathbb{R}_+^X$ be the set of states of $\mathcal{A}$, and pick $q \in Q$. We aim at defining a probability distribution $\mathbb{P}_\mathcal{A}$ over $\mathsf{Runs}(\mathcal{A}, q)$. Let $e_1, \ldots, e_k$ be edges of $\mathcal{A}$, and $\mathcal{C} \subseteq \mathbb{R}_+^k$ be a Borel set. The *(constrained) symbolic path* starting from q and determined by $e_1, \ldots, e_k$ and $\mathcal{C}$ is the following set of finite runs: $\pi_\mathcal{C}(q, e_1, \ldots, e_k) = \{\rho = q \xrightarrow{\tau_1, e_1} q_1 \cdots \xrightarrow{\tau_k, e_k} q_k \mid (\tau_1, \ldots, \tau_k) \in \mathcal{C}\}$. Given a symbolic path π, we define the cylinder generated by π as the subset $\mathsf{Cyl}(\pi)$ of $\mathsf{Runs}(\mathcal{A}, q)$ containing all runs ρ with a prefix ρ' in π.

We inductively define a measure over the set of symbolic paths as follows:

$$\mathbb{P}_\mathcal{A}(\pi_\mathcal{C}(q, e_1, \ldots, e_k)) = \int_{t_1 \in I(q, e_1)} p_{q+t_1}(e_1)\, \mathbb{P}_\mathcal{A}(\pi_{\mathcal{C}_{[\tau_1/t_1]}}(q_{t_1}, e_2, \ldots, e_k))\, \mathrm{d}\mu_q(t_1),$$

where for every $t_1 \geq 0$, q_{t_1} is such that $q \xrightarrow{t_1, e_1} q_{t_1}$ and $\mathcal{C}_{[\tau_1/t_1]}$ replaces variable τ_1 by t_1 in $\mathcal{C}$; we initialise with $\mathbb{P}_\mathcal{A}(\pi(q)) = 1$. The formula for $\mathbb{P}_\mathcal{A}$ relies on the fact that the probability of taking transition e_1 at time t_1 coincides with the probability of waiting t_1 time units and then choosing e_1 among the enabled transitions, i.e. $p_{q+t_1}(e_1)\, \mathrm{d}\mu_q(t_1)$. Now, one can extend $\mathbb{P}_\mathcal{A}$ to the cylinders by $\mathbb{P}_\mathcal{A}(\mathsf{Cyl}(\pi)) = \mathbb{P}_\mathcal{A}(\pi)$, where π is a symbolic path. Using some extension theorem as Carathéodory's theorem, we can extend $\mathbb{P}_\mathcal{A}$ in a unique way to the σ-algebra generated by the cylinders starting in q, which we denote $\Omega_\mathcal{A}^q$.

---

[4] Two measures μ and ν on the same measurable space are equivalent whenever for every measurable set A, $\mu(A) > 0$ iff $\nu(A) > 0$.

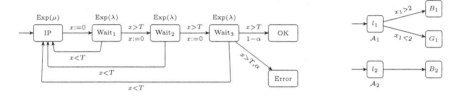

Fig. 1. The IPv4 Zeroconf STA.   **Fig. 2.** $\mathcal{A}_1 \notin$ CSTA.

Proposition 1 ([8]). *Let* $\mathcal{A} = (L, L_0, X, E, \mathsf{AP}, \mathcal{L}, (\mu_q, p_q)_{q \in L \times \mathbb{R}_+^X})$ *be a STA. For every state* $q \in Q$, $\mathbb{P}_\mathcal{A}$ *is a probability measure over* $(\mathrm{Runs}(\mathcal{A}, q), \Omega_\mathcal{A}^q)$.

Remark 1. Among others, the set of Zeno runs is measurable in $\Omega_\mathcal{A}^q$;[5] writing $\mathcal{C}_{M,k}$ for $\{(\tau_1, \ldots, \tau_k) \in \mathbb{R}_+^k \mid \tau_1 + \ldots + \tau_k \leq M\}$ it is indeed expressible as follows:

$$\bigcup_{M \in \mathbb{N}} \bigcap_{k \in \mathbb{N}_0} \bigcup_{(e_1, \ldots, e_k) \in E^k} \mathrm{Cyl}(\pi_{\mathcal{C}_{M,k}}(q, e_1, \ldots, e_k)).$$

Remark 2. A CTMC can be viewed as a STA with trivial guards on transitions and exponential distributions over delays.

We now give an example of STA.

Example 1. We model the IPv4 Zeroconf protocol using STA as done in [8] (see Fig. 1). This protocol aims at configuring IP addresses in a local network of appliances. When a new appliance is plugged, it selects an IP address at random, and broadcasts several probe messages to the network to know whether this address is already used or not. If it receives in a bounded delay an answer from the network informing that the IP is already used, then a new IP address is chosen. It may be the case that messages get lost, in which case there is an error. In [7], a simple model for the IPv4 Zeroconf protocol is given as a discrete-time Markov chain, which abstracts away timing constraints. In Fig. 1, we model the protocol as a STA with a single clock x, and exponential distributions (of parameters μ and λ) and this allows us to explicitly express the delay bound.

Discussion on the Model. STA have been defined and studied in a series of papers from 2007, with a complete journal version published as [8]. They can be used for modelling systems with stochastic aspects and real-time constraints (they are based on the standard model of timed automata [2] and extend the model of CTMCs) and are amenable to automatic verification. The class of *almost-surely fair STA*[6] is of particular interest. Indeed:

---

[5] We recall that a run $\rho = q \xrightarrow{\tau_1, e_1} q_1 \xrightarrow{\tau_2, e_2} \ldots$ is *Zeno* if $\sum_{i \geq 1} \tau_i < +\infty$.

[6] A STA is said almost-surely fair whenever $\mathbb{P}_\mathcal{A}(\mathsf{fair}) = 1$, where a run is fair if and only if (roughly speaking) any edge enabled infinitely often is taken infinitely often.

Theorem 1 ([8]). *The almost-sure model-checking problem is decidable for the class of almost-surely fair STA, with regards to ω-regular properties or properties given as deterministic timed automata.*

There exists surprisingly simple examples of STA which are not almost-surely fair (see for example [8, Fig. 9]), but large classes of STA have been identified in [8], that are almost-surely fair (they include single-clock STA and (weak-)reactive STA). Deciding whether a STA is almost-surely fair is an open problem

The approach adopted so far for modelling and verifying is monolithic. We target modular design of STA and describe a class of STA in which composition can safely be applied.

3   Parallel Composition of Stochastic Timed Automata

Compositional design is desirable for building computerised systems. Inspired by the approach of [24], we first define a parallel composition operator for STA, which corresponds to an interleaving semantics. This operator involves complex behaviours that are due to races between components. We therefore give conditions under which STA can be safely composed.

Remark 3. As already mentioned earlier, we focus here on an interleaving parallel composition operator between STA, and study the races between components. Extension to a parallel composition operator with some synchronisation is part of our future work, and we plan to adopt the idea of interactive Markov chains [21,22], which extend CTMCs with interactive actions, for the purpose of synchronisation.

3.1   Definition of the Parallel Composition

We consider two STA $\mathcal{A}_i = (L_i, L_0^{(i)}, X_i, E_i, \mathsf{AP}_i, \mathcal{L}_i, (\mu_q^{(i)}, p_q^{(i)})_{q \in L_i \times \mathbb{R}_+^{X_i}})$ for $i = 1, 2$ with $X_1 \cap X_2 = \emptyset$, and we first recall the standard (interleaving) parallel composition for the underlying TA. It is the TA $(L, L_0, X, E, \mathsf{AP}, \mathcal{L})$ where $L = L_1 \times L_2$, $L_0 = L_0^{(1)} \times L_0^{(2)}$, $X = X_1 \cup X_2$, $\mathsf{AP} = \mathsf{AP}_1 \cup \mathsf{AP}_2$, $\mathcal{L} : L \to 2^{\mathsf{AP}}$ is such that $\mathcal{L}((l_1, l_2)) = \mathcal{L}_1(l_1) \cup \mathcal{L}_2(l_2)$ and where $E = E_{1,\bullet} \cup E_{\bullet,2}$ with $E_{1,\bullet} = \{((l_1, l_2), g, Y, (l_1', l_2)) \mid (l_1, g, Y, l_1') \in E_1, \ l_2 \in L_2\}$.

Back to the STA, the parallel composition $\mathcal{A}_1 \parallel \mathcal{A}_2$ has as underlying TA the one above; it remains to equip each state $q = (q_1, q_2) \in Q_1 \times Q_2$ with probability distributions over both delays and edges, with the following constraints:

- distributions over delays from state (q_1, q_2) should reflect a *race* between the two components $\mathcal{A}_1$ and $\mathcal{A}_2$ from respectively states q_1 and q_2;
- distributions over edges should be state-based (or memoryless), that is, should not depend on how long has been waited before taking that edge, or which other actions have been done meanwhile by other components;

– globally, the product-automaton should correspond to the interleaving of $\mathcal{A}_1$ and $\mathcal{A}_2$, which we express as follows: given a property φ_1 that only concerns $\mathcal{A}_1$ and a property φ_2 that only concerns $\mathcal{A}_2$, $\mathbb{P}_{\mathcal{A}_1 \| \mathcal{A}_2}(\varphi_1 \wedge \varphi_2) = \mathbb{P}_{\mathcal{A}_1}(\varphi_1) \cdot \mathbb{P}_{\mathcal{A}_2}(\varphi_2)$.

Example 2 will illustrate the intricacy of getting these conditions satisfied.

Let $\mathcal{A}$ be a STA and let $q = (l, \nu) \in Q$ be a state of $\mathcal{A}$. We write f_q for the density function of μ_q w.r.t. the Lebesgue measure. We write F_q for the cumulative function associated to f_q.

We now define a first class of STA, called CSTA, which is suitable to define a parallel composition. We say that a STA $\mathcal{A}$ is in CSTA if:

(A) for every state q of $\mathcal{A}$, the density function associated with μ_q, denoted by f_q, is continuous everywhere on $\mathbb{R}_+$ except in a finite number of points, and
(B) the family of probability distributions $(\mu_q)_{q \in Q}$ is *weakly-memoryless*, i.e. for every $t, t' \geq 0$, $\mathbb{P}_{\mathcal{A}}(\mathbb{X}_q \geq t + t' \mid \mathbb{X}_q \geq t) = \mathbb{P}_{\mathcal{A}}(\mathbb{X}_{q+t} \geq t')$, where $\mathbb{X}_q$ (resp. $\mathbb{X}_{q+t}$) is a random variable with density function f_q (resp. f_{q+t}).

This second condition is a consistency condition between states which belong to the same 'time-elapsing fiber', that is, sets of the form $F = \{q + t \mid t \in \mathbb{R} \text{ and } q + t \in Q\}$. Indeed, $\mathbb{X}_q$ (resp. $\mathbb{X}_{q+t}$) represents the delay after which we leave state q (resp. $q + t$) via an edge. Hence if q_0 is the minimal (for time-elapsing) element of F, then for every $q = q_0 + t \in F$, the law of $\mathbb{X}_q$ has to be equal to the law of $\mathbb{X}_{q_0}$ conditioned by the fact that t time units have already passed. The distribution in q_0 can be taken arbitrary (satisfying condition (A)), and distributions for $q \in F$ can then be inferred.

Condition (B) can equivalently be written as: for every $t, t' \geq 0$,

$$f_q(t + t') = (1 - F_q(t))f_{q+t}(t') \tag{1}$$

Remark 4. Let q_0 be an initial element of a fiber, we can check that for instance,

– if $I(q_0)$ is a bounded subset of $\mathbb{R}_+$ and if μ_{q_0} is a uniform distribution over $I(q_0)$, then for every $t \in \mathbb{R}_+$, (B) imposes that μ_{q_0+t} is also uniform over $I(q_0 + t)$;
– similarly, if $I(q_0) = \mathbb{R}_+$, and if μ_{q_0} is an exponential distribution with parameter λ (denoted $\mathrm{Exp}(\lambda)$), then for every $t \in \mathbb{R}_+$, (B) imposes that μ_{q_0+t} is also an $\mathrm{Exp}(\lambda)$-distribution. This corresponds to the classical memoryless property assumed in CTMCs.

We can now explain how to build the probability distributions associated with a state $q = (q_1, q_2)$ of $\mathcal{A}_1 \| \mathcal{A}_2$. Since we leave state $q = (q_1, q_2)$ as soon as we leave q_1 or q_2, we naturally define the distribution over the delays from q as the minimum of the distributions over delays from q_1 and q_2. Under hypothesis (A) for the distributions from q_1 and q_2, one can show that the density function f_q for the minimum satisfies $f_q(t) = f_{q_1}(t)(1 - F_{q_2}(t)) + f_{q_2}(t)(1 - F_{q_1}(t))$ almost-surely for every $t \geq 0$ (w.r.t. the Lebesgue measure).

In order to define the probability distribution p_q over the enabled edges in q, one could consider that from state q, both systems $\mathcal{A}_1$ and $\mathcal{A}_2$ are in a race to win the next edge, i.e. $\mathcal{A}_1$ wins the race if the first edge taken from q is in E_1. Hence, given $t \in I(q)$, and an edge $e \in E_1$ enabled in $q + t$, one would like that $p_{q+t}(e) = w_q^1(t)p_{q_1+t}(e)$ where $w_q^1(t)$ is the probability that, starting from q, $\mathcal{A}_1$ wins the race knowing that it was won after a delay of t time units. This can be formalized, and under hypothesis (A) for f_{q_1} and f_{q_2}, we can show that if $f_q(t) \neq 0$, then $w_q^1(t) = \frac{f_{q_1}(t)(1-F_{q_2}(t))}{f_q(t)}$ almost-surely.

Definition 3. *Let $\mathcal{A}_i = (L_i, L_0^{(i)}, X_i, E_i, AP_i, \mathcal{L}_i, (\mu_q^{(i)}, p_q^{(i)})_{q \in L_i \times \mathbb{R}_+^{X_i}})$ for $i = 1, 2$ be two STA. We say that $\mathcal{A}_1$ and $\mathcal{A}_2$ are composable if $\mathcal{A}_1$ and $\mathcal{A}_2$ are in CSTA and $X_1 \cap X_2 = \emptyset$. In that case, we define the parallel composition of $\mathcal{A}_1$ and $\mathcal{A}_2$ as the STA $\mathcal{A}_1 \parallel \mathcal{A}_2 = (L, L_0, X, E, AP, \mathcal{L}, (\mu_q, p_q)_{q \in L \times \mathbb{R}_+^X})$, where for any state $q = (q_1, q_2)$ of $\mathcal{A}_1 \parallel \mathcal{A}_2$,*

(i) $(L, L_0, X, E, AP, \mathcal{L})$ is the composition of the underlying TA $\mathcal{A}_1$ and $\mathcal{A}_2$,
(ii) μ_q is defined by its density function $f_q = f_{q_1}(1 - F_{q_2}) + f_{q_2}(1 - F_{q_1})$, and
(iii) for any $t \in I(q)$, p_{q+t} is defined as follows:

$$p_{q+t}(e) = \mathbb{1}_{E_1}(e)w_q^1(t)p_{q_1+t}(e) + \mathbb{1}_{E_2}(e)w_q^2(t)p_{q_2+t}(e)$$

for every $e \in E$, where $w_q^i = \frac{f_{q_i}}{f_q}(1 - F_{q_{3-i}})$ on $I(q)$, for $i = 1, 2$.

3.2   Properties of the Parallel Composition

We are now ready to prove that this parallel composition operator satisfies all the expected properties. We assume the notations of Definition 3. First:

Lemma 1. *The distributions μ_q and p_q are well-defined, and the STA $\mathcal{A}_1 \parallel \mathcal{A}_2$ belongs to the class CSTA.*

We now give an example of a family of probability measures that do not satisfy hypothesis (B), which yields undesirable properties in the parallel composition.

Example 2 (Counter-example for condition (B)). We consider the single-clock STA $\mathcal{A}_1$ depicted in Fig. 2. We assume μ_{q_1} is an exponential distribution of parameter λ_1 (resp. λ_1') if $q_1 = (l_1, \nu_1)$ with $\nu_1 < 1$ (resp. $\nu_1 \geq 1$), and with $\lambda_1 \neq \lambda_1'$. Then for each $\nu_1 \in [0, 1[$, μ_{q_1} does not satisfy hypothesis (B). We then compose $\mathcal{A}_1$ with the STA $\mathcal{A}_2$. Each state $q_2 = (l_2, \nu_2)$ is equipped with an exponential distribution of parameter $\lambda_2 = \lambda_1'$ over the delays. It can be shown that the probability to reach B_1 in $\mathcal{A}_1$ corresponds to the probability to reach (B_1, B_2) in $\mathcal{A}_1 \parallel \mathcal{A}_2$ iff $\ln(\lambda_1) - \ln(\lambda_2) = \lambda_1 - \lambda_2$, which is not true in general.

Example 3. In order to illustrate the notion of composition, we composed two independent copies of the STA modelling the IPv4 Zeroconf protocol (see Example 1). Part of the composed STA is depicted in Fig. 3.

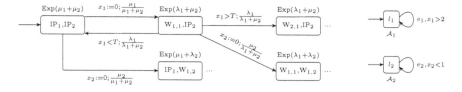

Fig. 3. The product of two STA modelling the IPv4 Zeroconf   **Fig. 4.** $\mathcal{A}_2$ is Zeno

It remains to identify when the parallel composition really coincides with an interleaving semantics. This is in general not true, as already shown in Example 2 (which does not satisfy Condition (B)), and witnessed further by Example 4 below (which satisfies both conditions (A) and (B)).

Example 4. We consider the STA $\mathcal{A}_1$ and $\mathcal{A}_2$ of Fig. 4, equipped resp. with an Exp(λ)-distribution and a uniform distribution. Let $q = (q_1, q_2)$ be a state of $\mathcal{A}_1 \parallel \mathcal{A}_2$, with $q_i = (l_i, 0)$. One can easily check that $\mathbb{P}_{\mathcal{A}_1\parallel\mathcal{A}_2}(q \rightarrow^* \xrightarrow{e_1}) = 0$ while $\mathbb{P}_{\mathcal{A}_1}(\mathrm{Cyl}(\pi(q_1, e_1))) = 1$ which contradicts the independence property we expect. One can notice that $\mathcal{A}_2$ is Zeno with probability 1.

Hence we define a subclass CSTA* of CSTA; $\mathcal{A} \in$ CSTA will be in CSTA* if:

(C) $\mathcal{A}$ is almost-surely non-Zeno.

Remark 5. Hypothesis (C) is not too restrictive since Zeno runs can be seen as faulty behaviours (they perform infinitely many actions in a finite amount of time, which is not realistic). We will see that hypothesis (C) is sufficient (together with (A) and (B)) to show that the parallel composition really coincides with an interleaving semantics. Note that condition (C) can be decided in various subclasses of STA [8].

We give some more notations. Let $\mathcal{A}$ be a STA and let φ be a property for $\mathcal{A}$. Given a state q, we say that φ is measurable from q if the set of runs starting from q satisfying φ is in $\Omega_{\mathcal{A}}^q$; we write this set $\{q \models \varphi\}$. Now let $\mathcal{A}_1$ and $\mathcal{A}_2$ be two composable STA. For $i = 1, 2$, we write ι_i for the natural projection of Runs($\mathcal{A}_1 \parallel \mathcal{A}_2, (q_1, q_2)$) onto Runs($\mathcal{A}_i, q_i$), and given a measurable property φ_i in $\mathcal{A}_i$ from q_i, we write $\{(q_1, q_2) \models \widetilde{\varphi}_i\}$ for the set $\iota_i^{-1}(\{q_i \models \varphi_i\})$. The following theorem states that the defined parallel composition is indeed interleaving.

Theorem 2. *Let $\mathcal{A}_1, \mathcal{A}_2 \in$ CSTA* be composable. Then $\mathcal{A}_1 \parallel \mathcal{A}_2 \in$ CSTA*. Moreover, for every state $q = (q_1, q_2)$ of $\mathcal{A}_1 \parallel \mathcal{A}_2$, for every properties φ_1 measurable in $\mathcal{A}_1$ from q_1 and φ_2 measurable in $\mathcal{A}_2$ from q_2, we have*

$$\mathbb{P}_{\mathcal{A}_1\parallel\mathcal{A}_2}(\{q \models \widetilde{\varphi}_1\} \cap \{q \models \widetilde{\varphi}_2\}) = \mathbb{P}_{\mathcal{A}_1}(\{q_1 \models \varphi_1\}) \cdot \mathbb{P}_{\mathcal{A}_2}(\{q_2 \models \varphi_2\}). \quad (2)$$

Proof (Sketch). Given $\mathcal{A}_1$ and $\mathcal{A}_2$ in CSTA*, thanks to Lemma 1, it suffices to prove that $\mathcal{A}_1 \parallel \mathcal{A}_2$ is almost-surely non-Zeno. This will be ensured by (2) and the fact that *non-Zeroness* is a measurable property.

The important first step to prove (2) consists in showing that, given an edge e_1 of $\mathcal{A}_1$, the probability in $\mathcal{A}_1 \parallel \mathcal{A}_2$ that e_1 is the first edge from $\mathcal{A}_1$ (with possibly edges from $\mathcal{A}_2$ taken before) performed from $q = (q_1, q_2)$ in a given set of delays Δ corresponds to the probability in $\mathcal{A}_1$ that e_1 is the first edge performed from q_1 in the same set of delays Δ. In order to do so, hypothesis (B) is crucial. The rest of the proof is long and technical but does not contain major difficulties. □

Remark 6. Note that an almost-surely non-Zeno STA $\mathcal{A}$ equipped with uniform or exponential distributions such that it satisfies conditions (A) and (B) (i.e. as in Remark 4), it holds that $\mathcal{A}$ is in CSTA*. As said before, we have large classes of STA that are almost-surely fair. For (weak-)reactive STA, it holds that they are almost-surely non-Zeno. Equipping them with uniform or exponential distributions as in Remark 4) make them also composable.

4   Bisimulation and Congruence

In this section, we define a notion of bisimulation for STA which naturally extends that for CTMCs [4,6,17]. We importantly show that the defined bisimulation is a congruence w.r.t. parallel composition: this means that, in a complex system, a component can be replaced by an equivalent one without affecting the global behaviour of the system.

4.1   Bisimulation

To define a bisimulation relation between STA, we are inspired by the approach of [17], which considers continuous-time Markov processes (CTMPs) – CTMPs generalize CTMCs to general continuous state-spaces; this definition of bisimulation that is given for CTMPs can be adapted to our context (note however that STA cannot be seen as particular CTMPs).

We first define some notions. A subset $P \subseteq \mathbb{R}^n$ is a *polyhedral set* if it is defined by a (finite) boolean combination of constraints of the form $A_1 x \leq b_1$ or $A_2 x < b_2$, where $x = (x_1, \ldots, x_n)$ is a variable, $A_1 \in \mathbb{R}^{m_1 \times n}$, $b_1 \in \mathbb{R}^{m_1}$, $A_2 \in \mathbb{R}^{m_2 \times n}$ and $b_2 \in \mathbb{R}^{m_2}$.

Let $\mathcal{A}$ be a STA, Q be its set of states, and $P(Q) = \{\cup_{l \in L}\{l\} \times C_l \mid \forall l \in L, C_l$ polyhedral set of $\mathbb{R}_+^n\}$ where n is the number of clocks of $\mathcal{A}$. The set $P(Q)$ is a proper subset of the Borel σ-algebra over $L \times \mathbb{R}_+^n$, which is closed by projection (contrary to the Borel σ-algebra). We then define the *closure of* $\mathcal{R}$ w.r.t. polyhedral sets, and we write $\mathrm{pcl}(\mathcal{R})$ as the following set $\mathrm{pcl}(\mathcal{R}) = \{A \in P(Q) \mid (a \in A \wedge a\mathcal{R}b) \Rightarrow b \in A\}$. One can notice that $\mathrm{pcl}(\mathcal{R})$ corresponds to the set of all polyhedral unions of equivalence classes. Given two equivalence relations $\mathcal{R}$ and $\mathcal{R}'$ over S we say that $\mathcal{R}'$ is *coarser* than $\mathcal{R}$ or that $\mathcal{R}$ is *finer* than $\mathcal{R}'$ if $\mathcal{R} \subseteq \mathcal{R}'$.

Definition 4. *Let* $\mathcal{A} = (L, L_0, X, E, AP, \mathcal{L}, (\mu_q, p_q)_{q \in L \times \mathbb{R}_+^X})$ *be a STA. An equivalence relation* $\mathcal{R}$ *over* $Q = L \times \mathbb{R}_+^X$ *is a bisimulation for* $\mathcal{A}$ *if for all*

$q, q' \in Q$ with qRq': (i) $\mathcal{L}(q) = \mathcal{L}(q')$, and (ii) for every $I \in \mathcal{B}(\mathbb{R}_+)$, for every $C \in \mathrm{pcl}(\mathcal{R})$,

$$\mathbb{P}_{\mathcal{A}}(\{q \models^{I,E} C\}) = \mathbb{P}_{\mathcal{A}}(\{q' \models^{I,E} C\}),$$

where $\{q \models^{I,E} C\}$ stands for $\{\rho \in \mathrm{Runs}(\mathcal{A}, q) \mid \exists \tau \in I, \exists e \in E, \rho = q \xrightarrow{\tau, e} q_1 \to \cdots \wedge q_1 \in C\}$. States q and q' are bisimilar (written $q \sim q'$) if there is a bisimulation that contains (q, q').

Given $q \in Q$, $I \in \mathcal{B}(\mathbb{R}_+)$ and $C \in \mathrm{pcl}(\mathcal{R})$ the value $\mathbb{P}_{\mathcal{A}}(\{q \models^{I,E} C\})$ can be expressed:

$$\mathbb{P}_{\mathcal{A}}(\{q \models^{I,E} C\}) = \int_{t \in I} P_{q+t}(C) f_q(t) \, \mathrm{d}t$$

where the value $P_{q+t}(C)$ corresponds to the probability to reach instantaneously C from state $q+t$. Formally: $P_{q+t}(C) = \sum_{l' \in L} \sum_{e \in E_{l'}} p_{q+t}(e) \mathbb{1}_{C_{l'}(e,\nu)}(t)$ for each $t \geq 0$ and each $C \in \mathrm{pcl}(\mathcal{R})$, where, given $l' \in L$, $E_{l'}$ is the set of edges with target l', and given $e = (l, g, Y, l')$, $C_{l'}(e, \nu) = \{t \in \mathbb{R}_+ \mid [Y \leftarrow 0](\nu + t) \in C_{l'}\}$. It can be shown that for every $t \geq 0$, P_{q+t} is a probability measure over Q.

Also, given a STA $\mathcal{A}$, one can show that $\sim$ is the coarsest bisimulation for $\mathcal{A}$.

The above natural definition enjoys the following very nice characterization, which shows that our definition is conservative w.r.t. bisimulation over CTMCs [4,6].

Proposition 2. Let $\mathcal{A}$ be a STA and let $\mathcal{R}$ be a bisimulation for $\mathcal{A}$. Then for all $q, q' \in Q$, qRq' if and only if (i) $\mathcal{L}(q) = \mathcal{L}(q')$, (ii) $\mu_q = \mu_{q'}$, and (iii) for every $C \in \mathrm{pcl}(\mathcal{R})$, $P_{q+t}(C) = P_{q'+t}(C)$ almost-surely for every $t \geq 0$.

Proof (Sketch). Point (i) is obvious, and points (ii) and (iii) come from the fact that qRq' if for each $C \in \mathrm{pcl}(\mathcal{R})$ and for each $I \in \mathcal{B}(\mathbb{R}_+)$,

$$\int_{t \in I} P_{q+t}(C) f_q(t) \, \mathrm{d}t = \int_{t \in I} P_{q'+t}(C) f_{q'}(t) \, \mathrm{d}t.$$

With $C = L \times \mathcal{B}(\mathbb{R}^n_+)$, where n is the number of clocks, we get that $P_{q+t}(C) = 1$ and thus $f_q = f_{q'}$ almost-surely, i.e. $\mu_q = \mu'_q$. It can then be easily shown that point (iii) holds. □

We now illustrate the notion of bisimulation on a simple example.

Example 5. Let us consider the simple STA $\mathcal{A}$ with two clocks on Fig. 5. We assume exponential distributions with parameter λ for every state at l_1 or l_2, and from a state of the form $q = (l_0, (\nu_1, \nu_2))$ with $\nu_1 < 1$ or $\nu_2 < 1$, $I(q) = [0, 1 - \min(\nu_1, \nu_2)[$ and so we can equip q with a uniform distribution on the interval $I(q)$ for the delays.

The coarsest bisimulation $\sim$ can easily be computed and is shown on the right part of Fig. 5: at location l_0, it is described by the following equivalence classes, for each $\nu \in [0, 1[$: $A_\nu = \{l_0\} \times (\{(\nu_1, \nu) \mid \nu_1 \geq \nu\} \cup \{(\nu, \nu_2) \mid \nu_2 \geq \nu\})$.

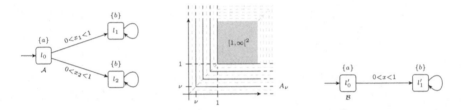

Fig. 5. A simple example for bisimulation.   **Fig. 6.** $\mathcal{B}$ is bisimilar to $\mathcal{A}$.

We extend the previous notion of bisimulation to two STA in a standard way (see [7]), by considering the union of the two STA, and a bisimulation relation between the initial states. If $\mathcal{A}_1$ and $\mathcal{A}_2$ are two STA, we write $\mathcal{A}_1 \sim \mathcal{A}_2$ when the two STA are bisimilar.

Example 6. Let us consider the one-clock STA $\mathcal{B}$ (Fig. 6). Assuming that we have the same probability distributions as STA $\mathcal{A}$ of Fig. 5, it can be easily established that $\mathcal{B} \sim \mathcal{A}$ by noticing that for each $\nu \in [0, 1[$, (l'_0, ν) is bisimilar to each state of A_ν.

4.2 Congruence

One of the main objectives of defining behavioural equivalences is to aim at modular design and proof of correctness. This is only possible if bisimulation is a *congruence w.r.t. parallel composition*, that is, if $\mathcal{A}_1 \sim \mathcal{A}_2$, then for every $\mathcal{B}$, $\mathcal{A}_1 \parallel \mathcal{B} \sim \mathcal{A}_2 \parallel \mathcal{B}$. We first prove the following natural lemma which is a key point for proving the congruence of the bisimulation w.r.t. parallel composition. Though very intuitive, the result is surprisingly quite technical to prove.

Lemma 2. *Let $\mathcal{A}, \mathcal{B} \in \mathrm{CSTA}^*$ with sets of states resp. Q_A and Q_B. If $\mathcal{R}$ is a bisimulation for $\mathcal{A}$ then the equivalence relation $\mathcal{R}'$ over $Q_A \times Q_B$ defined by $\mathcal{R}' = \{((q_1, q), (q_2, q)) \mid q_1 \mathcal{R} q_2 \text{ and } q \in Q_B\}$, is a bisimulation for $\mathcal{A} \parallel \mathcal{B}$.*

We can now state the main result of this section:

Theorem 3. *Bisimulation is a congruence w.r.t. parallel composition. That is: if $\mathcal{A}_1$, $\mathcal{A}_2$ and $\mathcal{B}$ are three STA in CSTA^*, if $\mathcal{A}_1 \sim \mathcal{A}_2$ then $\mathcal{A}_1 \parallel \mathcal{B} \sim \mathcal{A}_2 \parallel \mathcal{B}$.*

5 Conclusion

In this paper we have described a formal framework for compositional design of stochastic timed automata. We have established properties that should be satisfied by distributions over delays for well-defined parallel composition between components. We have proposed a natural notion of bisimulation and proven that it is a congruence w.r.t. parallel composition. We have also identified a subclass of STA which is closed under parallel composition.

We plan to extend our current work to so-called *interactive* STA (following [21,22]): the idea will be to add non-guarded interactive synchronizing events which take priority over delays when they are enabled. We hope that a parallel composition with synchronisation can be nicely defined in that setting, and that the model will enjoy nice properties as is the case in this paper.

There are many other plans for the future:

- Following the approach of [5,17], we would like to give a logical characterization of the bisimulation using (a subset of) CSL;
- We would like to be able, given a STA, to compute a small quotient automaton that would allow reduce the size of the system;
- All algorithms that have been developed so far for analyzing STA require a unique STA describing the system under analysis; we target the development of compositional verification (or approximation) methods, as it is done for instance for interactive Markov chains [13,23]. We would then like to see how this performs in practice.

References

1. Alur, R., Dill, D.: Automata for modeling real-time systems. In: Paterson, M.S. (ed.) Automata, Languages and Programming. LNCS, vol. 443, pp. 322–335. Springer, Heidelberg (1990)
2. Alur, R., Dill, D.: A theory of timed automata. Theor. Comput. Sci. **126**(2), 183–235 (1994)
3. Alur, R., Fix, L., Henzinger, T.A.: A determinizable class of timed automata. In: Dill, D.L. (ed.) Computer Aided Verification. LNCS, vol. 818, pp. 1–13. Springer, Heidelberg (1994)
4. Baier, C., Haverkort, B., Hermanns, H., Katoen, J.-P.: Model-checking algorithms for continuous-time Markov chains. IEEE Trans. Softw. Eng. **29**(7), 524–541 (2003)
5. Baier, C., Hermanns, H., Katoen, J.-P., Wolf, V.: Comparative branching-time semantics for Markov chains. Inf. Comput. **200**, 149–214 (2005)
6. Baier, C., Hermanns, H., Katoen, J.-P., Wolf, V.: Bisimulation and simulation relations for Markov chains. In: Proceedings of the Workshop Essays on Algebraic Process Calculi, vol. 162. ENTCS, pp. 73–78 (2006)
7. Baier, C., Katoen, J.-P.: Principles of Model Checking. MIT Press, Cambridge (2008)
8. Bertrand, N., Bouyer, P., Brihaye, T., Menet, Q., Baier, Ch., Größer, M., Jurdziński, M.: Stochastic timed automata. Logical Methods Comput. Sci. **10**(4), 1–73 (2014)
9. Bertrand, N., Bouyer, P., Brihaye, Th., Markey, N.: Quantitative model-checking of one-clock timed automata under probabilistic semantics. In: Proceedings of 5th International Conference on Quantitative Evaluation of Systems (QEST 2008). IEEE Computer Society Press (2008)
10. Bohnenkamp, H., D'Argenio, P., Hermanns, H., Katoen, J.-P.: MODEST: a compositional modeling formalism for hard and softly timed systems. IEEE Trans. Softw. Eng. **32**(10), 812–830 (2006)

11. Bouyer, P., Brihaye, T., Carlier, P., Menet, Q.: Compositional design of stochastic timed automata. Research Report LSV-15-06, Laboratoire Spécification et Vérification, ENS Cachan, France, 51 pages, December 2015
12. Bravetti, M., Gorrieri, R.: The theory of interactive generalized semi-Markov processes. Theor. Comput. Sci. **282**(1), 5–32 (2002)
13. Brázdil, T., Hermanns, H., Krcál, J., Kretínský, J., Rehák, V.: Verification of open interactive Markov chains. In: Proceedings of the 31st Conference on Foundations of Software Technology and Theoretical Computer Science (FSTTCS 2012), vol. 18. LIPIcs, pp. 474–485. Springer (2012)
14. Brázdil, T., Krčál, J., Křetínský, J., Řehák, V.: Fixed-delay events in generalized semi-Markov processes revisited. In: Katoen, J.-P., König, B. (eds.) CONCUR 2011. LNCS, vol. 6901, pp. 140–155. Springer, Heidelberg (2011)
15. D'Argenio, P., Katoen, J.-P.: A theory of stochastic systems Part I: Stochastic automata. Inf. Comput. **203**(1), 1–38 (2005)
16. D'Argenio, P., Katoen, J.-P.: A theory of stochastic systems part II: Process algebra. Inf. Comput. **203**(1), 39–74 (2005)
17. Desharnais, J., Panangaden, P.: Continuous stochastic logic characterizes bisimulation of continuous-time Markov processes. J. Logic Algebraic Program. **56**, 99–115 (2003)
18. Glynn, P.W.: A GSMP formalism for discrete event systems. Proc. IEEE **77**(1), 14–23 (1989)
19. Hartmanns, A.: Modest - a unified language for quantitative models. In: Proceedings of the Forum on Specification and Design Languages (FDL 2012), pp. 44–51. IEEE Computer Society Press (2012)
20. Hartmanns, A., Hermanns, H.: The modest toolset: an integrated environment for quantitative modelling and verification. In: Ábrahám, E., Havelund, K. (eds.) TACAS 2014 (ETAPS). LNCS, vol. 8413, pp. 593–598. Springer, Heidelberg (2014)
21. Hermanns, H.: Interactive Markov Chains: The Quest for Quantified Quality. LNCS, vol. 2428. Springer, Heidelberg (2002)
22. Hermanns, H., Katoen, J.-P.: The how and why of interactive Markov chains. In: de Boer, F.S., Bonsangue, M.M., Hallerstede, S., Leuschel, M. (eds.) FMCO 2009. LNCS, vol. 6286, pp. 311–337. Springer, Heidelberg (2010)
23. Hermanns, H., Krčál, J., Křetínský, J.: Compositional verification and optimization of interactive Markov chains. In: D'Argenio, P.R., Melgratti, H. (eds.) CONCUR 2013 – Concurrency Theory. LNCS, vol. 8052, pp. 364–379. Springer, Heidelberg (2013)
24. Hermanns, H., Zhang, L.: From concurrency models to numbers - performance and dependability. In: Software and Systems Safety - Specification and Verification, vol. 30. NATO Science for Peace and Security Series, pp. 182–210. IOS Press (2011)
25. Stoelinga, M.: Fun with FireWire: a comparative study of formal verification methods applied to the IEEE 1394 root contention protocol. Formal Aspects Comput. **14**(3), 328–337 (2003)

Online Bounded Analysis

Joan Boyar[1], Leah Epstein[2], Lene M. Favrholdt[1], Kim S. Larsen[1(✉)],
and Asaf Levin[3]

[1] Department of Mathematics and Computer Science,
University of Southern Denmark, Odense, Denmark
{joan,lenem,kslarsen}@imada.sdu.dk
[2] Department of Mathematics, University of Haifa, Haifa, Israel
lea@math.haifa.ac.il
[3] Faculty of IE&M, The Technion, Haifa, Israel
levinas@ie.technion.ac.il

Abstract. Though competitive analysis is often a very good tool for the analysis of online algorithms, sometimes it does not give any insight and sometimes it gives counter-intuitive results. Much work has gone into exploring other performance measures, in particular targeted at what seems to be the core problem with competitive analysis: the comparison of the performance of an online algorithm is made to a too powerful adversary. We consider a new approach to restricting the power of the adversary, by requiring that when judging a given online algorithm, the optimal offline algorithm must perform as well as the online algorithm, not just on the entire final request sequence, but also on any prefix of that sequence. This is limiting the adversary's usual advantage of being able to exploit that it knows the sequence is continuing beyond the current request. Through a collection of online problems, including machine scheduling, bin packing, dual bin packing, and seat reservation, we investigate the significance of this particular offline advantage.

1 Introduction

An *online problem* is an optimization problem where requests from a request sequence I are given one at a time, and for each request an irrevocable decision must be made for that request before the next request is revealed. For a minimization problem, the goal is to minimize some cost function, and if ALG is an online algorithm, we let $\text{ALG}(I)$ denote this cost on the request sequence I. Similarly, for a maximization problem, the goal is to maximize some value function (a.k.a. profit), and if ALG is an online algorithm, we let $\text{ALG}(I)$ denote this value on the request sequence I.

1.1 Performance Measures

Competitive analysis [27,34] is the most common tool for comparing online algorithms. For a minimization problem, an online algorithm is *c-competitive* if there

---

Supported in part by the Danish Council for Independent Research, Natural Sciences, and the Villum Foundation.

A.S. Kulikov and G.J. Woeginger (Eds.): CSR 2016, LNCS 9691, pp. 131–145, 2016.
DOI: 10.1007/978-3-319-34171-2_10

exists a constant α such that for all input sequences I, $\text{ALG}(I) \leq c\text{OPT}(I) + \alpha$. Here, OPT denotes an optimal offline algorithm. The (asymptotic) *competitive ratio* of ALG is the infimum over all such c. Similarly, for a maximization problem, an online algorithm is *c-competitive* if there exists a constant α such that for all input sequences I, $\text{ALG}(I) \geq c\text{OPT}(I) - \alpha$. Again, OPT denotes an optimal offline algorithm. The (asymptotic) *competitive ratio* of ALG is the supremum over all such c. In both cases, if the inequality can be established using $\alpha = 0$, we refer to the result as being *strict* (some authors use the terms *absolute* or *strong*). Note that for maximization problems, we use the convention of competitive ratios smaller than 1.

For many online problems, competitive analysis gives useful and meaningful results. However, researchers also realized from the very beginning that this is not always the case: Sometimes competitive analysis does not give any insight and sometimes it even gives counter-intuitive results, in that it points to the worse of two algorithms as the better one. A recent list of examples with references can be found in [21]. Much work has gone into exploring other performance measures, in particular targeted at what seems to be the core problem with competitive analysis that the comparison of the performance of an online algorithm is made to a too powerful adversary, controlling an optimal offline algorithm.

Four main techniques for addressing this have been employed, sometimes in combination. We discuss these ideas below. No chronological order is implied by the order the techniques are presented in. First, one could completely eliminate the optimal offline algorithm by comparing algorithms to each other directly. Measures taking this approach include max/max analysis [8], relative worst order analysis [12], bijective and average analysis [3], and relative interval analysis [20]. Second, one could limit the resources of the optimal offline algorithm, or correspondingly increase the resources of the online algorithm, as is done in extra resource analysis [31,34]. Thus, the offline algorithm's knowledge of the future is counter-acted by requiring that it solves a harder version of the problem than the online algorithm. Alternatively, the online algorithm could be given limited knowledge of the future in terms of some form of look-ahead, as has been done for paging. In those set-ups, one assumes that the online algorithm can see a fixed number ℓ of future requests, though it varies whether it is simply the next ℓ requests, or, for instance, the next ℓ expensive requests [36], the next ℓ new requests [16], or the next ℓ distinct requests [1]. Third, one could limit the adversary's control over exactly which sequence is being used to give the bound by grouping sequences and/or considering the expected value over some set as has been done with the statistical adversary [33], diffuse adversary [30], random order analysis [29], worst order analysis [12], Markov model [28], and distributional adversary [25].

Finally, one could limit the adversary's choice of sequences it is allowed to present to the online algorithm. An early approach to this, which at the same time addressed issues of locality of reference, was the access graph model [9], where a graph defines which requests are allowed to follow each other. Another locality of reference approach was taken in [2], limiting the maximum number of different requests allowed within some fixed-sized sliding window. Both of these models were targeted at the paging problem, and the techniques are not meant to be generally applicable to online algorithm analysis. A resource-based approach is taken in [14], where only sequences that could be fully accommodated given some resource are considered, eliminating some pathological worst-case sequences. A generalization of this, where the competitive ratio is found in the limit, appears in [13,15]. All of these approaches are aimed at removing pathological sequences from consideration such that the worst-case (or, in principle, expected case) behavior is taken over a smaller and more realistic set of sequences, thereby obtaining results corresponding better with observed behavior in practice. A similar concept for scheduling problems is the "known-OPT" model, where the cost of an optimal offline solution is known in advance [6]. Finally, loose competitive analysis [37] allows for a a set of sequences, asymptotically smaller than the whole infinite set of input sequences, to be disregarded, while the remaining sequences should either be c-competitive or have small cost. In this way, infrequent pathological as well as unimportant (due to low cost) sequences can be eliminated.

1.2   Online Bounded Analysis

Much work can be done in all of these four categories. In this paper, we consider a new approach to restricting the power of the adversary that does not really fit into any of the known categories. Given an online algorithm, we require that the optimal offline algorithm perform as well as the online algorithm, not just on the entire final request sequence, but also on any prefix of that sequence. In essence, this is limiting the adversary's usual advantage of being able to exploit that it knows the sequence is continuing beyond the current request, without completely eliminating this advantage. Since the core of the problem of the adversary's strength is its knowledge of the future, is seems natural to try to limit that advantage directly.

This new measure is generally applicable to online problems, since it is only based on the objective function. Comparing with other measures, it is a new element that the behavioral restriction imposed on the optimal offline algorithm is determined by the online algorithm, which is the reason we name this technique *online bounded analysis*. It is adaptive in the sense that online algorithms attempting non-optimal behavior face increasingly harder conditions from the adversary the farther the online algorithm goes in the direction of non-optimality (on prefixes). The measure judges greediness more positively than does competitive analysis, since making greedy choices limits the adversary's options more, so the focus shifts towards the quality of a range of greedy or near-greedy decisions.

Behavioral restrictions on the optimal offline algorithm have been seen before, as in [17], where it is used as a tool to arrive at the final result. Here they first show a $O(1)$-competitive result against an offline algorithm restricted to, among other things, using shortest remaining processing time for job selection. Later they show that this gives rise to a schedule at most three times as bad as for an unrestricted offline algorithm. Thus, the end goal is the usual competitive ratio, and the restriction employed in the process is problem specific.

1.3   Our New Measure

If I is an input sequence for some optimization problem and A is a deterministic online algorithm for this problem, we let $A(I)$ denote the objective function value returned by A on the input sequence I.

We let OPT_A denote the offline algorithm which is optimal under the restriction that it can never be worse than A on any prefix of an input sequence, i.e., for all sequences I, and all prefixes I' of I, for a minimization problem $\text{OPT}_A(I') \leq A(I')$ (for a maximization problem, $\text{OPT}_A(I') \geq A(I')$), and no algorithm with that property is strictly better than OPT_A on any sequence. We say that OPT_A is the *online bounded optimal solution* (for A).

For a minimization problem, if for some constant, c, it holds for all sequences I that $A(I) \leq c\text{OPT}_A(I)$, then we say that A has an *online bounded ratio* of at most c. The online bounded ratio of A is the infimum over all such c. Similarly, for a maximization problem, if for some constant, c, it holds for all sequences I that $A(I) \geq c\text{OPT}_A(I)$, then we say that A has an *online bounded ratio* of at least c. The online bounded ratio of A is the supremum over all such c. Note that we use the convention that an online bounded ratio for a minimization problem is at least 1, while this ratio for a maximization problem is at most 1.

1.4   Results

Through a collection of online problems, including machine scheduling, bin packing, dual bin packing, and seat reservation, we investigate the workings of online bounded analysis. As is apparent from the large collection of measures that have been defined, there is not any one measure which is best for everything. With our approach, we try to learn more about the nature of online problems, greediness, and robustness. As a first approach, we study this new idea in the simplest possible setting, but many measures combine ideas, so in future work, it would be natural to investigate this basic idea in combination with elements from other measures.

First, we observe that some results from competitive analysis carry over. Then we note that some problem characteristics imply that a greedy algorithm is optimal.

For machine scheduling, we obtain the following results. For minimizing makespan on $m \geq 2$ identical machines, we get an online bounded ratio of $2 - \frac{1}{m-1}$ for GREEDY. Though this is smaller than the competitive ratio of $2 - \frac{1}{m}$ [26], it is a comparable result, demonstrating that non-greedy behavior is not the key to the adversary performing better by a factor close to two for large m. For two uniformly related machines, we prove that GREEDY has online bounded ratio 1. This is consistent with competitive ratio results, where GREEDY has been proven optimal [18,24]. For the case where the faster machine is at least ϕ (the golden ratio) times faster than the slower machine, competitive analysis finds that GREEDY and FAST, the algorithm that only uses the faster machine, are equally good. Using relative worst order analysis, GREEDY is deemed the better algorithm [23], which seems reasonable since GREEDY is never worse on any sequence than FAST, and sometimes better. We also obtain this positive distinction, establishing the online bounded ratios 1 and $\frac{s+1}{s}$ (if the faster machine is s times faster than the slower one) for GREEDY and FAST, respectively.

For the Santa Claus machine scheduling problem [7], we prove that GREEDY is optimal for identical machines with respect to the online bounded ratio. For two related machines with speed ratio s, we present an algorithm with an online bounded ratio better than $\frac{1}{s}$ and show that no online algorithm has a higher online bounded ratio. For this problem, it is known that the best possible competitive ratio for identical machines is $\frac{1}{m}$, and the best possible competitive ratio for two related machines is $\frac{1}{s+1}$ [5, 22, 35].

For classic bin packing, we show that any Any-Fit algorithm has an online bounded ratio of at least $\frac{3}{2}$. We observe that for bin covering, the best online bounded ratio is equal to the best competitive ratio [19]. For these problems, asymptotic measures are used. We show a connection between results concerning the competitive ratio on accommodating sequences and the online bounded ratio. For dual bin packing (namely, the multiple knapsack problem with equal capacity knapsacks and unit weights items), we show that the online bounded ratio is the same as the competitive ratio on accommodating sequences (that is, sequences where OPT packs all items) for a large class of algorithms including First-Fit, Best-Fit, and Worst-Fit. It then follows from results in [13] that any algorithm in this class has an online bounded ratio of at least $\frac{1}{2}$. Furthermore, the online bounded ratio of First-Fit and Best-Fit is $\frac{5}{8}$, and that of Worst-Fit is $\frac{1}{2}$. We also note that, for any dual bin packing algorithm, an upper bound on the competitive ratio on accommodating sequences is also an upper bound on the online bounded ratio. Using a result from [13], this implies that any (possibly randomized) algorithm has an online bounded ratio of at most $\frac{6}{7}$.

For seat reservation, we have preliminary results, and conjecture that results are similar to machine scheduling for identical machines, in that ratios similar to but slightly better than those obtained using competitive analysis can be established.

We found that the new measure sometimes leads to the same results as the standard competitive ratio, and in some cases it leads to a competitive ratio of 1. However, there are problem variants for which we obtain an intermediate value, which confirms the relevance of our approach.

The proofs which have been omitted can be found in the full paper [10].

2   Online Bounded Analysis

Before considering concrete problems, we discuss some generic properties.

2.1   Measure Properties

The online bounded ratio of an algorithm is never further away from 1 than the competitive ratio, since the online algorithm's performance is being compared to a (possibly) restricted "optimal" algorithm.

Since algorithms are compared with different optimal algorithms, one might be concerned that two algorithms, A and O, could have online bounded ratio 1, and yet one algorithm could do better on some sequences than the other. This is not possible. To see this, consider some sequence I and assume that A does better than O on I, yet both algorithms have online bounded ratio 1.

If their online bounded ratio is 1, there is no point where one algorithm makes a decision which changes the objective value more than the other does, since the adversary could end the sequence there and the one algorithm would not have online bounded ratio 1. Thus, both algorithms have the same objective function value at all points, so they always compete against the same adversary. If algorithm A performs better than algorithm O on I, then algorithm O does not have online bounded ratio 1.

For some problems, such as paging, OPT_A is the same as OPT under competitive analysis for all algorithms A, because OPT's behavior on any sequence is also optimal on any prefix of that sequence. Thus, the competitive analysis results for paging and similar problems also hold with this measure, giving the same online bounded ratio as competitive ratio.

2.2   Greedy is Sometimes Optimal

It is sometimes the case that there is one natural greedy algorithm that always has a unique greedy choice in each step. In such situations, the greedy algorithm is optimal with respect to this measure, having online bounded ratio 1. For example, consider the weighted matching in a graph where the edges arrive in an online fashion (the edge-arrival model) and the algorithm in each step decides if the current edge is added to the matching or discarded. Here, the greedy algorithm, denoted by GREEDY, adds the current edge if adding the edge will keep the solution feasible (that is, its two end-vertices are still exposed by the matching that the algorithm created so far) and the weight of the edge is strictly positive. Note that indeed the online bounded ratio of GREEDY is 1, as the solution constructed by $\text{OPT}_{\text{GREEDY}}$ must coincide with the solution created by GREEDY. The last claim follows by a trivial induction on the number of edges considered so far by both GREEDY and $\text{OPT}_{\text{GREEDY}}$. If GREEDY adds the current edge, then by the definition of $\text{OPT}_{\text{GREEDY}}$, we conclude that $\text{OPT}_{\text{GREEDY}}$ adds the current edge. If GREEDY discards the current edge because at least one of its end-vertices is matched, then $\text{OPT}_{\text{GREEDY}}$ cannot add the current edge either (using the induction assumption). Last, if GREEDY discards the current edge since its weight is non-positive, then we can remove the edge from the bounded optimal solution, $\text{OPT}_{\text{GREEDY}}$, if it was added (removing it from $\text{OPT}_{\text{GREEDY}}$ will not affect the future behavior of $\text{OPT}_{\text{GREEDY}}$ since $\text{OPT}_{\text{GREEDY}}$ must accept an edge whenever GREEDY does). Similar proofs hold in other cases when there is a unique greedy choice for OPT in each step. Note that for the weighted matching problem where vertices arrive in an online fashion and when a vertex arrives the edge set connecting this vertex to earlier vertices is revealed with their weights (the vertex-arrival model), the standard lower bound for weighted matching holds as can be seen in the following lower bound construction.

The first three vertices arrive in the order $1, 2, 3$ and when vertex 3 arrives, two edges $\{1, 3\}, \{2, 3\}$ are revealed each of which has weight of 1 (vertices 1 and 2 are not connected). At this point, an online algorithm with a finite online bounded ratio must add one of these edges to the matching. Then, either 1 or 2 are matched in the current solution, and in the last step, vertex 4 arrives with an edge of weight M connecting 4 to the vertex among 1 and 2 that was matched by the algorithm. Observe that when vertex 3 arrives, the algorithm adds an edge to the matching while the bounded optimal solution can add the other edge, and this will allow the bounded optimal solution to add the last edge as well.

The argument for the optimality of GREEDY for the weighted matching problem in the edge-arrival model clearly holds if all weights are 1 also. This unweighted matching problem in the edge-arrival model is an example of a maximization problem in the online complexity class Asymmetric Online Covering (AOC) [11]:

Definition 1. *An online accept-reject problem, P, is in* Asymmetric Online Covering (AOC) *if, for the set Y of requests accepted:*

For minimization (maximization) problems, the objective value of Y is $|Y|$ if Y is feasible and ∞ $(-\infty)$ otherwise, and any superset (subset) of a feasible solution is feasible.

For all maximization problems in the class AOC, there is an obvious greedy algorithm, GREEDY, which accepts a request whenever acceptance maintains feasibility. The argument above showing that the online bounded ratio of GREEDY is 1 for the weighted matching problem in the edge-arrival model generalizes to all maximization problems in AOC.

Theorem 1. *For any maximization problem in AOC, the online bounded ratio of GREEDY is 1. Thus, GREEDY is optimal according to online bounded analysis for Online Independent Set in the vertex-arrival model, Unweighted Matching in the edge-arrival model, and Online Disjoint Path Allocation where requests are paths.*

Note that this does not hold for all minimization problems in AOC. For example, Cycle Finding in the vertex-arrival model, the problem of accepting as few vertices as possible, but accepting enough so that there is a cycle in the induced subgraph accepted, is AOC-Complete. However, consider the first vertex requested in a graph with only one cycle. GREEDY is forced to accept it, since the vertex could be part of the unique cycle, but $\text{OPT}_{\text{GREEDY}}$ will reject the vertex if it is not in that cycle.

However, there are online bounded optimal greedy algorithms for minimization problems in AOC, such as Vertex Cover, which are *complements* of maximization problems in AOC (Independent Set in the case of Vertex Cover). By complement, we mean that set S is a maximal feasible set in the maximization problem if and only if the requests not in S are a feasible solution for the minimization problem. The greedy algorithm in the case of these minimization problems would be the algorithm that accepts exactly those requests that GREEDY for the complementary maximization problem rejects.

3   Machine Scheduling: Makespan

We consider the load balancing problem of minimizing makespan for online job scheduling on m identical machines without preemption, and analyze the classic greedy algorithm (also known as list scheduling). At any point, GREEDY schedules the next job on a least loaded machine. Since the machines are identical, ties can be resolved arbitrarily without loss of generality. It is known that the competitive ratio of GREEDY is $2 - \frac{1}{m}$ [26]. With the more restricted optimal algorithm, we get a smaller value of $2 - \frac{1}{m-1}$ as the online bounded ratio of GREEDY.

Lemma 1. *For the problem of minimizing makespan for online job scheduling on m identical machines, GREEDY has online bounded ratio of at most $2 - \frac{1}{m-1}$.*

Proof. Consider a sequence I. Let j be the first job in I that is completed at the final makespan of GREEDY, and assume that it has size w. Let t and s be the starting times of j in $\mathrm{OPT_{GREEDY}}$ and GREEDY, respectively, and let ℓ and ℓ' be the makespans of $\mathrm{OPT_{GREEDY}}$ and GREEDY, respectively, just before the arrival of j. Let V denote the total size of the jobs in I just before j arrives.

We have the following inequalities: $\mathrm{OPT_{GREEDY}} \geq t + w$ and $\mathrm{OPT_{GREEDY}} \geq \ell$. In addition, since, just before j arrived, the machine where $\mathrm{OPT_{GREEDY}}$ placed j had load t and the other machines had load at most ℓ, $V \leq t + (m-1)\ell$. Since $m - 1 \geq 1$, $V \leq (m-1)(t + \ell)$.

Because GREEDY placed j on its least loaded machines, all machines had load at least s before j arrived. At least one machine had load ℓ', so $V \geq (m-1)s + \ell'$. By the definition of online bounded analysis, $\ell \leq \ell'$. Thus, $V \geq (m-1)s + \ell$. Combining the upper and lower bounds on V gives $(m-1)s \leq (m-1)t + (m-2)\ell$ and $s \leq t + \frac{m-2}{m-1}\ell$. We now bound GREEDY's makespan:

$$\mathrm{GREEDY}(I) = s + w = (s - t) + (t + w)$$
$$\leq \left(\tfrac{m-2}{m-1}\right) \cdot \ell + \mathrm{OPT_{GREEDY}}(I) \leq \left(2 - \tfrac{1}{m-1}\right) \mathrm{OPT_{GREEDY}}(I)$$

$\square$

Lemma 2. *For the problem of minimizing makespan for online job scheduling on m identical machines, GREEDY has online bounded ratio of at least $2 - \frac{1}{m-1}$.*

By Lemmas 1 and 2 we find the following.

Theorem 2. *For the problem of minimizing makespan for online job scheduling on m identical machines, GREEDY has online bounded ratio $2 - \frac{1}{m-1}$.*

Most interestingly, Theorem 2 establishes the existence of an online algorithm, GREEDY, for makespan minimization on two identical machines with an online bounded ratio of 1. Next, we generalize this last result to the case of two uniformly related machines. Note that for two uniformly related machines we can assume that machine number 1 is strictly faster than machine number 2, and the two speeds are $s > 1$ and 1.

We define GREEDY as the algorithm that assigns the current job to the machine such that adding the job there results in a solution of a smaller makespan breaking ties in favor of assigning the job to the slower machine (that is, to machine number 2). If an algorithm breaks ties in favor of assigning the job to the faster machine (let this algorithm be called GREEDY'), then its online bounded ratio is strictly above 1, as the following example implies. The first job has size $s - 1$ (and it is assigned to machine 1), and the second job has size 1 (and assigning it to any machine will result in the current makespan 1). The first job must be assigned to machine 1 by OPT$_{\text{GREEDY}'}$, and it assigns the second job to the second machine. A third job of size $s + 1$ arrives. This job is assigned to the first machine by OPT$_{\text{GREEDY}'}$, obtaining a makespan of 2. GREEDY' will have a makespan of at least $\min\{2 + 1/s, s + 1\} > 2$ as $s > 1$.

Theorem 3. *For the problem of minimizing makespan for online job scheduling on two uniformly related machines, GREEDY has online bounded ratio 1.*

We now consider the algorithm FAST that simply schedules all jobs on the faster machine. In contrast to GREEDY, FAST does not have an online bounded ratio of 1. This also contrasts with competitive analysis, since FAST has an optimal competitive ratio for $s \geq \phi$, where $\phi = \frac{1+\sqrt{5}}{2} \approx 1.618$.

Theorem 4. *For two related machines with speed ratio s, FAST has an online bounded ratio of $\frac{s+1}{s}$.*

By Theorem 2, the result of Theorem 3 cannot be extended to three or more identical machines for GREEDY. We conclude this section by proving that such a generalization is impossible, not only for GREEDY, but for any deterministic online algorithm.

Theorem 5. *Let $m \geq 3$. For the problem of minimizing makespan for online job scheduling on m identical machines, any deterministic online algorithm A has online bounded ratio of at least $\frac{4}{3}$.*

An obvious next step would be to try to match the general lower bound of $\frac{4}{3}$ by designing an algorithm that places each job on the most loaded machine where the bound of $\frac{4}{3}$ would not be violated. However, even for $m = 3$, this would not work, as seen by the input sequence $I = \langle \frac{3}{4}, \frac{1}{4}, \frac{5}{12}, \frac{1}{6}, \frac{7}{12}, \frac{5}{6} \rangle$. The algorithm would combine the first two jobs on one machine and the following two on another machine. Since the optimal makespan at this point is $\frac{3}{4}$, the algorithm will schedule the fifth job on the third machine. When the last job arrives, all machines have a load of at least $\frac{7}{12}$, resulting in a makespan of $\frac{17}{12} > 1.4$. Note that I can be scheduled such that each machine has a load of exactly 1. Since the algorithm has a makespan of 1 already after the second job, the online bounded restriction is actually no restriction on OPT for this sequence.

4   Machine Scheduling: Santa Claus

In contrast to makespan, the objective in Santa Claus scheduling is to maximize the minimum load. Traditionally, the algorithm GREEDY for this problem assigns any new job to a machine achieving a minimum load in the schedule that was created up to the time just before the job is added to the solution (breaking ties arbitrarily). For identical machines, this algorithm is equivalent to the greedy algorithm for makespan minimization. Unlike the makespan minimization problem, where this algorithm has online bounded ratio of 1 only for two identical machines, here we show that GREEDY has an online bounded ratio of 1 for any number of identical machines.

Theorem 6. *For the Santa Claus problem on m identical machines, GREEDY has online bounded ratio 1.*

Proof. Let a configuration be a multi-set of the current loads on all of the machines, i.e., without any annotation of which machine is which. As long as $\mathrm{OPT_{GREEDY}}$ also assigns each job to a machine with minimum load, the configurations of GREEDY and $\mathrm{OPT_{GREEDY}}$ are identical.

Consider the first time $\mathrm{OPT_{GREEDY}}$ does something different from GREEDY. If, when that job j arrives, there is a unique machine with minimum load, $\mathrm{OPT_{GREEDY}}$ would have a worse objective value than GREEDY after placing j, so, by definition of online bounded analysis, this cannot happen. Now consider the situation where $k \geq 2$ machines have minimum load. Then, after processing j, GREEDY has $k - 1$ machines with minimum load, whereas $\mathrm{OPT_{GREEDY}}$ has k. In that case, no more than $k - 2$ further jobs can be given. This is seen as follows: If $k - 1$ jobs were given, GREEDY would place one on each of its $k - 1$ machines with minimum load, and, thus, raise the minimum. $\mathrm{OPT_{GREEDY}}$, on the other hand, would not be able to raise (at this step) the minimum of all of its k machines with minimum load, and would therefore not be optimal; a contradiction.

Thus, $\mathrm{OPT_{GREEDY}}$ can only have a different configuration than GREEDY after GREEDY (and $\mathrm{OPT_{GREEDY}}$) have obtained their final (and identical) objective value, and so, the online bounded ratio of GREEDY is 1.   □

Next, we show that unlike the makespan minimization problem, for which there is an online algorithm with online bounded ratio of 1 even for the case of two uniformly related machines (Theorem 3), such a result is impossible for the Santa Claus problem on two uniformly related machines.

Theorem 7. *For the Santa Claus problem on two uniformly related machines with speed ratio s, no deterministic online algorithm has an online bounded ratio larger than $\frac{1}{s}$.*

Proof. For any online algorithm A, we consider a setting of two uniformly related machines with speeds 1 and s. The input consists of exactly two jobs. After the first job is assigned by A, the objective function value remains zero, and only if the algorithm assigns the two jobs to distinct machines, will it have a positive objective function value. Thus, when there are only two jobs, OPT_A is simply the optimal solution for the instance. The first job is of size 1. If A assigns the job to the machine of speed s, then the next job is of size s. At this point OPT_A has value 1 (by assigning the first job to the slower machine and the second to the faster machine), but A has either zero value (if both jobs are assigned to the faster machine) or a value of $\frac{1}{s}$. In the second case where A assigns the first job (of size 1) to the slower machine of speed 1, the second job has size $\frac{1}{s}$. At this point OPT_A has value $\frac{1}{s}$ (by assigning the first job to the faster machine and the second to the slower machine), but A has either zero value (if both jobs are assigned to the slower machine) or a value of $\frac{1/s}{s}$.   □

Interestingly, the online bounded ratio of the following simple algorithm matches this bound. The algorithm G assigns each job to the least loaded machine. While for identical machines, this algorithm and GREEDY are equivalent, for related machines this is not the case. The same algorithm is the one that achieves the best possible competitive ratio $\frac{1}{s+1}$ [22].

Theorem 8. *For the Santa Claus problem on two uniformly related machines with speed ratio s, the online bounded ratio of G is $\frac{1}{s}$.*

5   Classic Bin Packing and Bin Covering

In classic bin packing, the input is a sequence of items of sizes s, $0 < s \leq 1$, that should be packed in as few bins of size 1 as possible. We say that a bin is open if at least one item has been placed in the bin. An Any-Fit algorithm is an algorithm that never opens a new bin if the current item fits in a bin that is already open. In this section, we use the *asymptotic online bounded ratio*. Thus, we allow for an additive constant, exactly as with the asymptotic (non-strict) competitive ratio.

Theorem 9. *Any Any-Fit algorithm has an online bounded ratio of at least $\frac{3}{2}$.*

In classic bin covering, the input is as in bin packing, and the goal is to assign items to bins so as to maximize the number of bins whose total assigned size is at least 1. For this problem, it is known that a simple greedy algorithm (which assigns all items to the active bin until the total size assigned to it becomes 1 or larger, and then it moves to the next bin and defines it as active) has the best possible competitive ratio $\frac{1}{2}$. The negative result [19] is proven using inputs where the first batch of items consists of a large number of very small items, and it is followed by a set of large identical items of sizes close to 1 (where the exact size is selected based on the actions of the algorithm). The total size of the very small items is strictly below 1, so as long as large items were not presented yet, the value of any algorithm is zero. An optimal offline solution packs the very small items such that packing every large item results in a bin whose contents have a total size of exactly 1. Thus, no algorithm can perform better on any prefix, and this construction shows that the online bounded ratio is at most $\frac{1}{2}$.

6   Dual Bin Packing

As in the previous section, we use the asymptotic online bounded ratio here. Dual bin packing is like the classic bin packing problem, except that there is only a limited number, n, of bins and the goal is to pack as many items in these n bins as possible. Known results concerning the competitive ratio on accommodating sequences can be used to obtain results for the online bounded ratio.

In general, accommodating sequences [14,15] are defined to be those sequences for which OPT does not get a better result by having more resources. For the dual bin packing problem, accommodating sequences are sequences of items that can be fully accommodated in the n bins, i.e., OPT packs all items.

We show that, for a large class of algorithms for dual bin packing containing First-Fit and Best-Fit, the online bounded ratio is the same as the competitive ratio on accommodating sequences. To show that this does not hold for all algorithms, we also give an example of a $\frac{2}{3}$-competitive algorithm on accommodating sequences that has an online bounded ratio of 0.

Dual bin packing is an example of a problem in a larger class of problems which includes the seat reservation problem discussed below. A problem is an *accept/reject accommodating problem* if algorithms can only accept or reject requests, the goal is accept as many requests as possible, and the accommodating sequences are those where OPT accepts all requests.

Theorem 10. *For any online algorithm* ALG *for any accept/reject accommodating problem, the competitive ratio of* ALG *on accommodating sequences is equal to the online bounded ratio of* ALG *on accommodating sequences.*

Note that this result applies to all algorithms for dual bin packing. Since any accommodating sequence is also a valid adversarial sequence for the case with no restrictions on the sequences, we obtain the following corollary of Theorem 10.

Corollary 1. *For any online algorithm* ALG *for any accept/reject accommodating problem, any upper bound on the competitive ratio of* ALG *on accommodating sequences is also an upper bound on the online bounded ratio of* ALG.

A *fair* algorithm for dual bin packing is an algorithm that never rejects an item that it could fit in a bin. A *rejection-invariant* algorithm is an algorithm that does not change its behavior based on rejected items.

Theorem 11. *For any fair, rejection-invariant dual bin packing algorithm* ALG, *the online bounded ratio of* ALG *equals the competitive ratio of* ALG *on accommodating sequences.*

One algorithm which is fair and rejection-invariant is First-Fit, which packs each item in the first bin it fits in (and rejects it if no such bin exists). Another example of a fair, rejection-invariant algorithm is Best-Fit, which packs each item in a most full bin that can accommodate it. Worst-Fit is the algorithm that packs each item in a most empty bin.

Corollary 2. *Best-Fit and First-Fit have online bounded ratios of $\frac{5}{8}$. Worst-Fit has an online bounded ratio of $\frac{1}{2}$.*

Corollary 3. *Any fair, rejection-invariant dual bin packing algorithm has an online bounded ratio of at least $\frac{1}{2}$. Any (possibly randomized) dual bin packing algorithm has an online bounded ratio of at most $\frac{6}{7}$.*

The algorithm Unfair-First-Fit (UFF) defined in [4] is designed to work well on accommodating sequences. Whenever an item larger than $\frac{1}{2}$ arrives, UFF rejects the item unless it will bring the number of accepted items below $\frac{2}{3}$ of the total number of items that are accepted by an optimal solution of the prefix of items given so far (for an accommodating sequence this is the number of items in the prefix). The competitive ratio of UFF on accommodating sequences is $\frac{2}{3}$ [4]. We show that, in contrast to Theorem 11, UFF has an online bounded ratio of 0.

Theorem 12. *Unfair-First-Fit has an online bounded ratio of 0.*

7   Unit Price Seat Reservation

Since even the unit price seat reservation problem has a terrible competitive ratio, depending on the number of stations, this problem has often been studied using the competitive ratio on accommodating sequences, which for the seat reservation problem restricts the input sequences considered to those where OPT could have accepted all of the requests. By Theorem 10, for accommodating sequences, the competitive ratio and the online bounded ratio are identical.

The unit price seat reservation problem has competitive ratio $\Theta(1/k)$, where k is the number of stations. This does not change for the online bounded ratio, even though, both the original proof, showing that no deterministic fair online algorithm (that does not reject an interval if it is possible to accept it) for the unit price problem is more than $\frac{8}{k+5}$-competitive [14], and the proof improving this to $\frac{4}{k-2\sqrt{k-1}+4}$ [32], used an optimal offline algorithm which rejected some requests before the online algorithm did. The main ideas in these proofs was that the adversary could give small request intervals which OPT could place differently from the algorithm, allowing it to reject some long intervals and still be fair. Rejecting long intervals allowed it to accept many short intervals which the algorithm was forced to reject. By using small intervals involving only the last few stations, one can arrange that the online algorithm has to reject intervals early. Then, giving nearly the same sequence as for the $\frac{8}{k+5}$ bound, using two fewer stations, OPT can still reject the same long intervals and do just as badly asymptotically. Note that we reuse the $[k-3, k-2)$ intervals.

Theorem 13. *No deterministic fair online algorithm for the unit price seat reservation problem has an online bounded ratio of more than $\frac{11}{k+7}$.*

Using a similar proof, one can show that the online bounded ratios of First-Fit and Best-Fit are at least $\frac{5}{k+1}$. The major difference is that in the first part,

First-Fit and Best-Fit each reject $n/2$ intervals, so in the second part, O can also reject $n/2$ intervals. Since any fair online algorithm for the unit price problem is $2/k$-competitive, any fair online algorithm for the unit price problem has an online bound ratio of at least $2/k$.

References

1. Albers, S.: On the influence of lookahead in competitive paging algorithms. Algorithmica **18**, 283–305 (1997)
2. Albers, S., Favrholdt, L.M., Giel, O.: On paging with locality of reference. In: 34th Annual ACM Symposium on the Theory of Computing (STOC), pp. 258–267 (2002)
3. Angelopoulos, S., Dorrigiv, R., López-Ortiz, A.: On the separation and equivalence of paging strategies. In: 18th ACM-SIAM Symposium on Discrete Algorithms (SODA), pp. 229–237 (2007)
4. Azar, Y., Boyar, J., Epstein, L., Favrholdt, L.M., Larsen, K.S., Nielsen, M.N.: Fair versus unrestricted bin packing. Algorithmica **34**(2), 181–196 (2002)
5. Azar, Y., Epstein, L.: On-line machine covering. J. Sched. **1**(2), 67–77 (1998)
6. Azar, Y., Regev, O.: On-line bin-stretching. Theoret. Comput. Sci. **268**(1), 17–41 (2001)
7. Bansal, N., Sviridenko, M.: The Santa Claus problem. In: 38th Annual ACM Symposium on the Theory of Computing (STOC), pp. 31–40 (2006)
8. Ben-David, S., Borodin, A.: A new measure for the study of on-line algorithms. Algorithmica **11**(1), 73–91 (1994)
9. Borodin, A., Irani, S., Raghavan, P., Schieber, B.: Competitive paging with locality of reference. J. Comput. Syst. Sci. **50**(2), 244–258 (1995)
10. Boyar, J., Epstein, L., Favrholdt, L.M., Larsen, K.S., Levin, A.: Online bounded analysis. CoRR, abs/1602.06708 (2016)
11. Boyar, J., Favrholdt, L., Mikkelsen, J., Kudahl, C.: Advice complexity for a class of online problems. In: 32nd International Symposium on Theoretical Aspects of Computer Science (STACS). Leibniz International Proceedings in Informatics, vol. 30, pp. 116–129 (2015)
12. Boyar, J., Favrholdt, L.M.: The relative worst order ratio for on-line algorithms. ACM Trans. Algorithms **3**(2), 24 (2007). article 22
13. Boyar, J., Favrholdt, L.M., Larsen, K.S., Nielsen, M.N.: Extending the accommodating function. Acta Informatica **40**(1), 3–35 (2003)
14. Boyar, J., Larsen, K.: The seat reservation problem. Algorithmica **25**, 403–417 (1999)
15. Boyar, J., Larsen, K.S., Nielsen, M.N.: The accommodating function–a generalization of the competitive ratio. SIAM J. Comput. **31**(1), 233–258 (2001)
16. Breslauer, D.: On competitive on-line paging with lookahead. Theoret. Comput. Sci. **209**(1–2), 365–375 (1998)
17. Chan, S.-H., Lam, T.-W., Lee, L.-K., Liu, C.-M., Ting, H.-F.: Sleep management on multiple machines for energy and flow time. In: Aceto, L., Henzinger, M., Sgall, J. (eds.) ICALP 2011, Part I. LNCS, vol. 6755, pp. 219–231. Springer, Heidelberg (2011)
18. Cho, Y., Sahni, S.: Bounds for list schedules on uniform processors. SIAM J. Comput. **9**(1), 91–103 (1980)

19. Csirik, J., Totik, V.: On-line algorithms for a dual version of bin packing. Discrete Appl. Math. **21**, 163–167 (1988)
20. Dorrigiv, R., López-Ortiz, A., Munro, J.I.: On the relative dominance of paging algorithms. Theoret. Comput. Sci. **410**, 3694–3701 (2009)
21. Ehmsen, M.R., Kohrt, J.S., Larsen, K.S.: List factoring and relative worst order analysis. Algorithmica **66**(2), 287–309 (2013)
22. Epstein, L.: Tight bounds for bandwidth allocation on two links. Discrete Appl. Math. **148**(2), 181–188 (2005)
23. Epstein, L., Favrholdt, L.M., Kohrt, J.S.: Separating online scheduling algorithms with the relative worst order ratio. J. Comb. Optim. **12**(4), 363–386 (2006)
24. Epstein, L., Noga, J., Seiden, S.S., Sgall, J., Woeginger, G.J.: Randomized online scheduling on two uniform machines. In: Tenth Annual ACM-SIAM Symposium on Discrete Algorithms (SODA), pp. 317–326 (1999)
25. Giannakopoulos, Y., Koutsoupias, E.: Competitive analysis of maintaining frequent items of a stream. Theoret. Comput. Sci. **562**, 23–32 (2015)
26. Graham, R.L.: Bounds for certain multiprocessing anomalies. Bell Syst. Tech. J. **45**, 1563–1581 (1966)
27. Karlin, A.R., Manasse, M.S., Rudolph, L., Sleator, D.D.: Competitive snoopy caching. Algorithmica **3**, 79–119 (1988)
28. Karlin, A.R., Phillips, S.J., Raghavan, P.: Markov paging. SIAM J. Comput. **30**(3), 906–922 (2000)
29. Kenyon, C.: Best-fit bin-packing with random order. In: 7th ACM-SIAM Symposium on Discrete Algorithms (SODA), pp. 359–364 (1996)
30. Koutsoupias, E., Papadimitriou, C.H.: Beyond competitive analysis. In: 35th Annual Symposium on Foundations of Computer Science (FOCS), pp. 394–400 (1994)
31. Koutsoupias, E., Papadimitriou, C.H.: Beyond competitive analysis. SIAM J. Comput. **30**(1), 300–317 (2000)
32. Miyazaki, S., Okamoto, K.: Improving the competitive ratios of the seat reservation problem. In: Calude, C.S., Sassone, V. (eds.) TCS 2010. IFIP AICT, vol. 323, pp. 328–339. Springer, Heidelberg (2010)
33. Raghavan, P.: A statistical adversary for on-line algorithms. In: On-Line Algorithms. Series in Discrete Mathematics and Theoretical Computer Science, vol. 7, pp. 79–83. American Mathematical Society (1992)
34. Sleator, D.D., Tarjan, R.E.: Amortized efficiency of list update and paging rules. Commun. ACM **28**(2), 202–208 (1985)
35. Woeginger, G.J.: A polynomial-time approximation scheme for maximizing the minimum machine completion time. Oper. Res. Lett. **20**(4), 149–154 (1997)
36. Young, N.: Competitive paging and dual-guided algorithms for weighted caching and matching (thesis). Technical Report CS-TR-348-91, Computer Science Department, Princeton University (1991)
37. Young, N.E.: The k-server dual and loose competitiveness for paging. Algorithmica **11**, 525–541 (1994)

Affine Computation and Affine Automaton

Alejandro Díaz-Caro[1][(✉)] and Abuzer Yakaryılmaz[2][(✉)]

[1] Universidad Nacional de Quilmes, Roque Sáenz Peña 352,
B1876BXD Bernal, Buenos Aires, Argentina
alejandro.diaz-caro@unq.edu.ar
[2] National Laboratory for Scientific Computing, Petrópolis, RJ 25651-075, Brazil
abuzer@lncc.br

Abstract. We introduce a quantum-like classical computational concept, called affine computation, as a generalization of probabilistic computation. After giving the basics of affine computation, we define affine finite automata (AfA) and compare it with quantum and probabilistic finite automata (QFA and PFA, respectively) with respect to three basic language recognition modes. We show that, in the cases of bounded and unbounded error, AfAs are more powerful than QFAs and PFAs, and, in the case of nondeterministic computation, AfAs are more powerful than PFAs but equivalent to QFAs.

1   Introduction

Using negative amplitudes, allowing interference between states and configurations, is one of the fundamental properties of quantum computation that does not exist in classical computation. Therefore, it is interesting to define a quantum-like classical system allowing to use negative values. However, both quantum and probabilistic systems are linear and it seems not possible to define a classical linear computational systems using negative values (see also the discussions regarding fantasy quantum mechanics in [1]). On the other hand, it is possible to define such a system *almost linearly*, as we do in this paper.

A probabilistic state is a l_1-norm 1 vector defined on non-negative real numbers, also called a stochastic vector. A probabilistic operator is a linear operator mapping probabilistic states to probabilistic states, which is also called a stochastic matrix. Equivalently, a matrix is stochastic if each of its columns is a probabilistic state. Similarly, a quantum state is a l_2-norm 1 vector defined over complex numbers. A quantum operator is a linear operator mapping quantum

The arXiv number is 1602.04732 [4].

Díaz-Caro was partially supported by STIC-AmSud project 16STIC04 FoQCoSS.
Yakaryılmaz was partially supported by CAPES with grant 88881.030338/2013-01 and some parts of the work were done while Yakaryılmaz was visiting Buenos Aires in July 2015 to give a lecture at ECI2015 (Escuela de Ciencias Informáticas 2015, Departamento de Computación, Facultad de Ciencias Exactas y Naturales, Universidad de Buenos Aires), partially supported by CELFI, Ministerio de Ciencia, Tecnología e Innovación Productiva.

A.S. Kulikov and G.J. Woeginger (Eds.): CSR 2016, LNCS 9691, pp. 146–160, 2016.
DOI: 10.1007/978-3-319-34171-2_11

states to quantum states, which is also called a unitary matrix. Equivalently, a matrix is unitary if each of its columns (also each row) is a quantum state.

Our aim is to define a new system that (1) is a generalization of probabilistic system, (2) can have negative values, (3) evolves linearly, and (4) is defined in a simple way like the probabilistic and quantum systems. When working on non-negative real numbers, l_1-norm is the same as the summation of all entries. So, by replacing "l_1-norm 1" condition with "summation to 1" condition, we can obtain a new system that allows negative values in its states. The linear operators preserving the summation of vectors are barycentric-preserving, also called *affine transformations*, which can give the name of our new system: *affine system*. Thus, the state of an affine system is called an *affine state*, the entries of which sum to 1. Moreover, a matrix is an affine transformation if each of its columns is an affine state. It is clear that an affine system is a probabilistic one if negative values are not used. Thus, the new affine system satisfies all four conditions above.

The only renaming detail is how to get information from the system. For this purpose, we define an operator similar to the measurement operator in quantum computation that projects the computation into the computational basis. It is intuitive that the "weights" of negative and positive values should be same if their magnitudes are the same. Moreover, each state should be observed with the probability calculated based on the value of its magnitude. Therefore, we normalize each magnitude (since the summation of all magnitudes can be bigger than 1) and each normalized magnitude gives us the probability of "observing" the corresponding state. We call this operator as *weighting operator*.

In the paper, we give the basics of affine systems in detail and start to investigate affine computation by defining the affine finite automaton (AfA) (due to the simplicity of automata models). Then, we compare it with probabilistic finite automata (PFAs) and quantum finite automata (QFAs) with respect to the basic language recognition modes. We show that, in the cases of bounded and unbounded error, AfAs are more powerful than QFAs and PFAs, and, in the case of nondeterministic computation, AfAs are more powerful than PFAs but equivalent to QFAs. Our results are also the evidence that although an AfA has a finite number of basis states, it can store more information. This is why we use small "f" in the abbreviation of AfA.

Throughout the paper, we focus on the finite dimensional systems. In Sect. 2, we give the basics of probabilistic and quantum systems. In Sect. 3, we describe the basics of affine systems. Then, we give the definitions of classical and quantum finite automata in Sect. 4. The definition of affine finite automaton is given in Sect. 5. Our results are given in Sect. 6. We close the paper with Sect. 7.

2   Probabilistic and Quantum Systems

A *probabilistic system* has a finite number of states, say $E = \{e_1, \ldots, e_n\}$ $(n > 0)$, called *deterministic states* of the system. At any moment, the system can be in a probabilistic distribution of these states: $v = (p_1 \quad p_2 \quad \cdots \quad p_n)^T$, where p_j

represents the probability of system being in state e_j $(1 \leq j \leq n)$. Here v is called a *probabilistic state*, which is a stochastic (column) vector, i.e.

$$0 \leq p_i \leq 1 \text{ and } \sum_{i=1}^{n} p_i = 1.$$

It is clear that v is a vector in $\mathbb{R}^n$ and all the deterministic states form the standard basis of $\mathbb{R}^n$. Moreover, all the probabilistic states form a simplex in $\mathbb{R}^n$, represented by linear equation $x_1 + x_2 + \cdots + x_n = 1$ whose variables satisfy $0 \leq x_j \leq 1$ $(1 \leq j \leq n)$.

The system evolves from a probabilistic state to another one by a linear operator: $v' = Av$, where A is an $n \times n$ matrix and $A[k,j]$ represents the probability of going from e_j to e_k $(1 \leq j, k \leq n)$. Since v' is a probabilistic state and so a stochastic vector, A is a (left) stochastic matrix, each column of which is a stochastic vector. Assume that the system is in v_0 at the beginning, and A_t is the probabilistic operator at the t-th time step $(t = 1, 2, \ldots)$. Then, the evolution of the system is as follows: $v_t = A_t A_{t-1} \cdots A_1 v_0$. At the t-th step, the probability of observing the j-th state is $v_t[j]$.

A quantum system is a *non-trivial linear* generalization of a probabilistic one, which forms a Hilbert space (a complex vector space with inner product). A basis of the Hilbert space, say $\mathcal{H}^n$, can be seen as the set of "deterministic states" of the system. Unless otherwise is specified, the standard basis is used: $B = \{|q_1\rangle, \ldots, |q_n\rangle\}$, where each $|q_j\rangle$ is a zero vector except the j-th entry, which is 1. Remark that $\mathcal{H}^n = span\{|q_1\rangle, \ldots, |q_n\rangle\}$. At any moment, the system can be in a linear combination of basis states:

$$|v\rangle = \begin{pmatrix} \alpha_1 \\ \alpha_2 \\ \vdots \\ \alpha_n \end{pmatrix} \in \mathcal{H}^n,$$

where $\alpha_j \in \mathbb{C}$ is called the *amplitude* of the system being in state $|q_j\rangle$. Moreover, the value $|\alpha_j|^2$ represents the *probability* of the system being in state $|q_j\rangle$. We call $|v\rangle$ the *(pure) quantum state* of the system, which is a norm-1 (column) vector: $\sqrt{\langle v|v\rangle} = 1 \Leftrightarrow \langle v|v\rangle = 1 \Leftrightarrow \sum_{j=1}^{n} |\alpha_j|^2 = 1$. Remark that all the quantum states from a sphere in $\mathbb{C}^n$, i.e. $x_1^2 + x_2^2 + \cdots + x_n^2 = 1$.

Similar to the probabilistic case, the system evolves from a quantum state to another one by a linear operator: $|v'\rangle = U|v\rangle$, where U is an $n \times n$ matrix and $U[k,j]$ represents the transition amplitude of going from $|q_j\rangle$ to $|q_k\rangle$ $(1 \leq j, k \leq n)$. Since $|v'\rangle$ is a quantum state and so a norm-1 vector, U is a unitary matrix, the columns/rows of which form an orthonormal set. Moreover, $U^{-1} = U^{\dagger}$.

To retrieve information from a quantum system, we apply measurement operators. In its simplest form, when in quantum state $|v\rangle$, we can make a measurement in the computation basis and then we can observe $|q_j\rangle$ with probability $p_j = |\alpha_j|^2$ and so the new state becomes $|q_j\rangle$ (if $p_i > 0$). We can also split the set B into m disjoint subsets: $B = B_1 \cup \cdots \cup B_m$ and $B_j \cap B_k = \emptyset$ for $1 \leq j \neq k \leq m$.

Based on this classification, $\mathcal{H}^n$ is split into m pairwise orthogonal subspaces: $\mathcal{H}^n = \mathcal{H}_1 \oplus \cdots \oplus \mathcal{H}_m$ where $\mathcal{H}_j = span\{|q\rangle \mid |q\rangle \in B_j\}$. We can design a projective measurement operator P to force the system to be observed in one of these subspaces, i.e.

$$P = \left\{ P_1, \ldots, P_m \mid P_j = \sum_{|q\rangle \in B_j} |q\rangle\langle q| \text{ and } 1 \leq j \leq m \right\},$$

where P_j is a zero-one projective matrix that projects any quantum state to $\mathcal{H}_j^n$. More formally, $|v\rangle = |\widetilde{v_1}\rangle \oplus \cdots \oplus |\widetilde{v_m}\rangle$, $|\widetilde{v_j}\rangle = P_j|v\rangle \in \mathcal{H}_j^n$. Each P_j is called a *projective operator* and the index is called a *measurement outcome*. Then, the probability of observing the outcome "j" is calculated as $p_j = \langle \widetilde{v_j} | \widetilde{v_j} \rangle$. If it is observed ($p_j > 0$), then the new state is obtained by normalizing $|\widetilde{v_j}\rangle$, which is called *unnormalized (quantum) state*, $|v_j\rangle = \frac{|\widetilde{v_j}\rangle}{\sqrt{p_j}}$.

From a mathematical point of view, any quantum system defined on $\mathcal{H}^n$ can be simulated by a quantum system straightforwardly defined on $\mathbb{R}^{2n}$ (e.g. [8]). Therefore, we can say that the main distinguishing property of quantum systems is using negative amplitudes rather than using complex numbers.

After making projective measurements, for example, the quantum system can be in a mixture of pure quantum states, i.e.

$$\left\{ (p_j, |v_j\rangle) \mid 1 \leq j \leq m, \sum_{j=1}^{m} p_j = 1 \right\}.$$

We can represent such a mixture as a single mathematical object called *density matrix*, an $(n \times n)$-dimensional matrix: $\rho = \sum_{j=1}^{m} p_j |v_j\rangle\langle v_j|$, which is called the *mixed state* of the system. A nice property of ρ is that the k-th diagonal entry represents the probability of the system of being in the state $|q_k\rangle$, i.e. $Tr(\rho) = 1$.

It is clear that unitary operators are not the generalizations of stochastic operators. However, by interacting a quantum system with an auxiliary system, more general quantum operators can be applied on the main quantum system. They are called *superoperators*.[1] Formally, a superoperator $\mathcal{E}$ is composed by a finite number of operation elements $\{E_j \mid 1 \leq j \leq m\}$, where $m > 0$, satisfying that $\sum_{j=1}^{m} E_j^\dagger E_j = I$. When $\mathcal{E}$ is applied to the mixed state ρ, the new mixed state is obtained as

$$\rho' = \mathcal{E}(\rho) = \sum_{j=1}^{m} E_j \rho E_j^\dagger.$$

In fact, a superoperator includes a measurement and the indices of operation elements can be seen as the outcomes of the measurement(s). When $\mathcal{E}$ is applied to pure state $|v\rangle$, we can obtain up to m new pure states. The probability of

---

[1] A superoperator can also be obtained by applying a series of unitary and measurements operators where the next unitary operator is selected with respect to the last measurement outcome.

observing the outcome of "j", say p_j, calculated as $p_j = \langle \tilde{v}_j | \tilde{v}_j \rangle$, $\quad |\tilde{v}_j\rangle = E_j|v\rangle$, where $|\tilde{v}_j\rangle$ is called an unnormalized state vector if it is not a zero vector. If the outcome "j" is observed ($p_j > 0$), then the new state becomes, $|v_j\rangle = \frac{|\tilde{v}_j\rangle}{\sqrt{p_j}}$. Remark that using unnormalized state vectors sometimes make the calculations easier since the probabilities can be calculated directly from them.

If we apply the projective measurement $P = \{P_j \mid 1 \le j \le m\}$ to the mixed state ρ, where $m > 0$, the probability of observing the outcome j, say p_j, and the new state, say ρ_j, is calculated as follows:

$$\tilde{\rho}_j = P_j \rho P_j, \quad p_j = Tr(\tilde{\rho}_j), \quad \text{and} \quad \rho_j = \frac{\tilde{\rho}_j}{p_j} (\text{if } p_j > 0).$$

The reader may ask how a quantum system can be a linear generalization of a probabilistic system. We omit the details here but any probabilistic operator can be implemented by a superoperator. Moreover, a mixed-state can be represented as a single column vector, and each superoperator can be represented as a single matrix. Then, all computations can be represented linearly. We refer the reader to [12,14,17] for the details.

3   Affine Systems

Inspired from quantum systems, we define the finite-dimensional affine system (AfS) as a *non-linear* generalization of a probabilistic system by allowing to use negative "probabilities". Let $E = \{e_1, \ldots, e_n\}$ be the set of basis states, which are the deterministic states of an n-dimensional probabilistic system. Any affine state is a linear combination of E

$$v = \begin{pmatrix} a_1 \\ a_2 \\ \vdots \\ a_n \end{pmatrix}$$

such that each entry can be an arbitrary real number but the summation of all entries must be 1:

$$\sum_{i=1}^{n} a_i = 1.$$

So, any probabilistic state, a stochastic column vector, is an affine state. However, on contrary to a probabilistic state, an affine state can contain negative values. Moreover, all the affine states form a surface in $\mathbb{R}^n$, i.e. $x_1 + x_2 + \cdots + x_n = 1$.

Both, probabilistic and quantum states, form finite objects (simplex and sphere, respectively). For example, in $\mathbb{R}^2$, all the probabilistic states form the line $x + y = 1$ on $(\mathbb{R}^+ \cup \{0\})^2$ with length $\sqrt{2}$ and all the quantum states form the unit circle with length 2π. On the other hand, affine states form infinite objects (plane). In $\mathbb{R}^2$, all the affine states form the infinite line $x + y = 1$. Therefore, it seems that, with the same dimension, affine systems can store more information.

In this paper, we provide some evidences to this interpretation. On the other hand, affine systems might not be comparable with quantum systems due to the fact of forming different geometrical objects (e.g. line versus circle).

Any affine transformation is a linear operator, that is, a mapping between affine states. We can easily show that any matrix is an affine operator if and only if for each column, the summation of all entries is equal to 1. The evolution of the system is as follows: when in affine state v, the new affine state v' is obtained by $v' = Av$, where A is the affine transformation such that $A[j, k]$ represents the transition value from e_k to e_j.

In quantum computation, the sign of the amplitudes does not matter when making a measurement. We follow the same idea for affine systems. More precisely, the *magnitude* of an affine state is the l_1-norm of the state:

$$|v| = |a_1| + |a_2| + \cdots + |a_n| \geq 1.$$

Then, we can say that the probability (*weight*) of observing the j-th state is $\frac{|a_j|}{|v|}$, where $1 \leq j \leq n$. To retrieve this information, we use an operator (possible non-linear) called *weighting operator*, which can be seen as a counterpart of the measurements in the computational basis for quantum systems. Therefore, we can make a weighting in the basis E and the system collapses into a single deterministic state.

One may ask whether we can use a weighting operator similar to a projective measurement. Assume that the system is in the following affine state

$$v = \begin{pmatrix} 1 \\ -1 \\ 1 \end{pmatrix}$$

and we make weighting based on the separation $\{e_1\}$ and $\{e_2, e_3\}$. Then, we can observe the system in the first state with weight $\frac{1}{3}$ and in the second and third states with weight $\frac{2}{3}$. But, in the latter case, the new state is not an affine state since the summation of entries will always be zero whatever normalization factor is used. Therefore, once we make a weighting, the system must collapse to a single state. On the other hand, one may still define an *affine system with extended weighting* by allowing this kind of weighting with the assumption that if the new state has a zero summation, then the system terminates, i.e. no further evolution can occur. Such kind of assumptions may be used cleverly to gain some computational power.

One may also define an affine state as a l_1-norm 1 vector on the real numbers and require that each new state is normalized after each linear affine operator. A straightforward calculation shows that the weighting results will be exactly the same as the previous definition, so both systems are equivalent. However, this time the overall evolution operator, a linear affine operator followed by normalization, is not linear. With respect to this new definition, say *normalized affine systems*, all the affine states form finite objects: $|x_1| + |x_2| + \cdots + |x_n| = 1$. It is, for example, a square on $\mathbb{R}^2$: $|x| + |y| = 1$. One could see this square as an approximation of the unit circle but remark that we cannot use unitary operators as affine operators directly. On the other hand, we may define a more general model by replacing linear affine operators with arbitrary linear operators. We call this system *general affine systems* or *general normalized affine systems*. In this paper, we focus only on the standard definition where the states are vectors with a barycentric sum to 1, and the transformations are affine operators preserving such barycenters.

4   Classical and Quantum Automata

Unless otherwise specified, we denote the input alphabet as Σ, not containing the left end-marker ¢ and the right end-marker \$. The set of all the strings generated on Σ is denoted by Σ^*. We define $\widetilde{\Sigma} = \Sigma \cup \{\text{¢}, \$\}$ and $\tilde{w} = \text{¢}w\$$ for any string $w \in \Sigma^*$. For any given string $w \in \Sigma^*$, $|w|$ is the length of the string, $|w|_\sigma$ is the number of occurrences of the symbol σ in w, and w_j is the j-th symbol of w.

For a given machine/automaton M, $f_M(w)$ denotes the accepting probability (value) of M on the string w.

A probabilistic finite automaton (PFA) [10] P is 5-tuple

$$P = (E, \Sigma, \{A_\sigma \mid \sigma \in \widetilde{\Sigma}\}, e_s, E_a),$$

where E is the set of deterministic states, $e_s \in E$ is the starting state, $E_a \subseteq E$ is the set of accepting state(s), and A_σ is the stochastic transition matrix for the symbol $\sigma \in \widetilde{\Sigma}$. Let $w \in \Sigma^*$ be the given input. The input is read as $\tilde{w}$ from left to right, symbol by symbol. After reading the j-th symbol, the probabilistic state is $v_j = A_{\tilde{w}_j} v_{j-1} = A_{\tilde{w}_j} A_{\tilde{w}_{j-1}} \cdots A_{\tilde{w}_1} v_0$, where $v_0 = e_s$ and $1 \leq j \leq |\tilde{w}|$. The final state is denoted $v_f = v_{|\tilde{w}|}$. The accepting probability of P on w is calculated as $f_P(w) = \sum_{e_k \in E_a} v_f[k]$.

A quantum finite automaton (QFA) [2] M is a 5-tuple

$$M = (Q, \Sigma, \{\mathcal{E}_\sigma \mid \sigma \in \widetilde{\Sigma}\}, q_s, Q_a),$$

where Q is the set of basis states, $\mathcal{E}_\sigma$ is the transition superoperator for symbol σ, q_s is the starting state, and $Q_a \subseteq Q$ is the set of accepting states. For a given input $w \in \Sigma^*$, the computation of M on w is traced as $\rho_j = \mathcal{E}_{\tilde{w}_j}(\rho_{j-1})$, where $\rho_0 = |q_s\rangle\langle q_s|$ and $1 \leq j \leq |\tilde{w}|$. The final state is denoted $\rho_f = \rho_{|\tilde{w}|}$. The accepting probability of M on w is calculated as $f_M(w) = \sum_{q_j \in A_a} \rho_f[j,j]$.

If we restrict the entries of the transitions matrices of a PFA to zeros and ones, we obtain a deterministic finite automaton (DFA). A DFA is always in a single state during the computation and the input is accepted if and only if the computation ends in an accepting state. A language is said to be recognized by a DFA (then called regular [11]) if and only if any member of the language is accepted by the DFA. The class of regular languages are denoted by REG.

Let $\lambda \in [0,1)$ be a real number. A language L is said to be recognized by a PFA P with cutpoint λ if and only if

$$L = \{w \in \Sigma^* \mid f_P(w) > \lambda\}.$$

Any language recognized by a PFA with a cutpoint is called stochastic language [10] and the class of stochastic languages are denoted by SL, a superset of REG. A language is said to be recognized by a PFA P with unbounded error if L or the complement of L is recognized by P with cutpoint [17]. (Remark that it is still not known whether SL is closed under complement operation.)

As a special case, if $\lambda = 0$, the PFA is also called a nondeterministic finite automaton (NFA). Any language recognized by a NFA is also regular.

A language L is said to be recognized by P with isolated cutpoint λ if and only if there exists a positive real number δ such that (1) $f_P(w) \geq \lambda + \delta$ for any $w \in L$ and (2) $f_P(w) \leq \lambda - \delta$ for any $w \notin L$. When the cutpoint is required to be isolated, PFAs are not more powerful than DFAs: Any language recognized by a PFA with isolated cutpoint is regular [10].

Recognition with isolated cutpoint can also be formulated as recognition with bounded error. Let $\epsilon \in [0, \frac{1}{2})$. A language L is said to be recognized by a PFA P with error bound ϵ if and only if (1) $f_P(w) \geq 1 - \epsilon$ for any $w \in L$ and (2) $f_P(w) \leq \epsilon$ for any $w \notin L$. As a further restriction of bounded error, if $f_P(w) = 1$ for any $w \in L$, then it is called negative one-sided error bound, and, if $f_P(w) = 0$ for any $w \notin L$, then it is called positive one-sided error bound. If the error bound is not specified, it is said that L is recognized by P with [negative/positive one-sided] bounded error.

A language L is called exclusive stochastic language [9] if and only if there exists a PFA P and a cutpoint $\lambda \in [0, 1]$ such that $L = \{w \in \Sigma^* \mid f_P(w) \neq \lambda\}$. The class of exclusive stochastic languages is denoted by $\mathsf{SL}^{\neq}$. Its complement class is denoted by $\mathsf{SL}^{=}$ ($L \in \mathsf{SL}^{\neq} \leftrightarrow \overline{L} \in \mathsf{SL}^{=}$). Note that for any language in $\mathsf{SL}^{\neq}$ we can pick any cutpoint between 0 and 1 but not 0 or 1 since when fixing the cutpoint to 0 or 1, we can recognize only regular languages. Note that both $\mathsf{SL}^{\neq}$ and $\mathsf{SL}^{=}$ are supersets of REG (but it is still open whether REG is a proper subset of $\mathsf{SL}^{\neq} \cap \mathsf{SL}^{=}$).

In the case of QFAs, they recognize all and only regular languages with bounded-error [7] and stochastic languages with cutpoint [15,17]. However, their nondeterministic versions (NQFAs) are more powerful: NQAL, the class of languages defined by NQFAs (QFAs with cutpoint 0), is identical to $\mathsf{SL}^{\neq}$ [16].

5   Affine Finite Automaton

Now we define the affine finite automaton (AfA). An AfA M is a 5-tuple

$$M = (E, \Sigma, \{A_\sigma \mid \sigma \in \widetilde{\Sigma}\}, e_s, E_a),$$

where all the components are the same as that of PFA except that A_σ is an affine transformation matrix. Let $w \in \Sigma^*$ be the given input. After reading the whole input, a weighting operator is applied and the weights of the accepting states determine the accepting probability of M on w, i.e.

$$f_M(w) = \sum_{e_k \in E_a} \frac{|v_f[k]|}{|v_f|} \in [0, 1].$$

The languages recognized by AfAs are defined similarly to PFAs and QFAs. Any language recognized by an AfA with cutpoint is called *affine language*. The class of affine languages is denoted AfL. Any language recognized by an AfA with cutpoint 0 (called nondeterministic AfA (NAfA)) is called *nondeterministic affine language*. The related class is denoted NAfL. A language is called exclusive affine language if and only if there exists an AfA M and a cutpoint $\lambda \in [0,1]$ such that $L = \{w \in \Sigma^* \mid f_M(w) \neq \lambda\}$. The class of exclusive affine languages is denoted AfL$^{\neq}$ and its complement class is denoted AfL$^{=}$. Any language recognized by an AfA with bounded error is called *bounded affine language*. The related class is denoted BAfL. If it is a positive one-sided error (all non-members are accepted with value 0), then the related class is denoted BAfL^0, and, if it is a negative one (all members are accepted with value 1), then the related class is denoted BAfL^1. Note that if $L \in$ BAfL^0, then $\overline{L} \in$ BAfL^1, and vice versa. Any language recognized by an AfA with zero-error is called *exact affine language* and the related class is denoted EAfL.

6   Main Results

We present our results under three subsections.

6.1   Bounded-Error Languages

We start with a 2-state AfA, say M_1, for the language EQ $= \{w \in \{a,b\}^* \mid |w|_a = |w|_b\}$. Let $E = \{e_1, e_2\}$ be the set of states, where e_1 is the initial and only accepting state. None of the end-markers is used (or the related operators are the identity). At the beginning, the initial affine state is $v_0 = \begin{pmatrix} 1 \\ 0 \end{pmatrix}$. When reading symbols a and b, the following operators are applied:

$$A_a = \begin{pmatrix} 2 & 0 \\ -1 & 1 \end{pmatrix} \quad A_b = \begin{pmatrix} \frac{1}{2} & 0 \\ \frac{1}{2} & 1 \end{pmatrix},$$

respectively. Then, the value of the first entry of the affine state is multiplied by 2 for each a and by $\frac{1}{2}$ for each b, and so, the second entry takes the value of "1 minus the value of the first entry", i.e. if M reads m as and n bs, then the new affine state is $\begin{pmatrix} 2^{m-n} \\ 1 - 2^{m-n} \end{pmatrix}$. That is, for any member, the final affine state is $v_f = \begin{pmatrix} 1 \\ 0 \end{pmatrix}$ and so the input is accepted with value 1. For any non-member, the final state can be one of the followings

$$\cdots, \begin{pmatrix} 8 \\ -7 \end{pmatrix}, \begin{pmatrix} 4 \\ -3 \end{pmatrix}, \begin{pmatrix} 2 \\ -1 \end{pmatrix}, \begin{pmatrix} \frac{1}{2} \\ \frac{1}{2} \end{pmatrix}, \begin{pmatrix} \frac{1}{4} \\ \frac{3}{4} \end{pmatrix}, \begin{pmatrix} \frac{1}{8} \\ \frac{7}{8} \end{pmatrix}, \cdots.$$

Thus, the maximum accepting value is obtained when $v_f = \begin{pmatrix} 2 \\ -1 \end{pmatrix}$, which gives the accepting value $\frac{|2|}{|2|+|-1|} = \frac{2}{3}$. Therefore, we can say that the language EQ can be recognized by the AfA M_1 with isolated cutpoint $\frac{5}{6}$ (the isolation gap is $\frac{1}{6}$). Since it is a nonregular language, we can follow that AfAs can recognize more languages than PFAs and QFAs with isolated cutpoints (bounded error).

By using 3 states, we can also design an AfA $M_2(x)$ recognizing EQ with better error bounds, where $x \geq 1$: $M_2(x) = \{\{e_1, e_2, e_3\}, \{a, b\}, \{A_a, A_b\}, e_1, \{e_1\}\}$, where $A_a = \begin{pmatrix} 1 & 0 & 0 \\ x & 1 & 0 \\ -x & 0 & 1 \end{pmatrix}$ and $A_b = \begin{pmatrix} 1 & 0 & 0 \\ -x & 1 & 0 \\ x & 0 & 1 \end{pmatrix}$. The initial affine state is $v_0 = (1, 0, 0)$ and after reading m as and n bs, the affine state will be $\begin{pmatrix} 1 \\ (m-n)x \\ (n-m)x \end{pmatrix}$.

Then, the accepting value will be 1 if $m = n$, and, $\frac{1}{2x|m-n|+1}$ if $m \neq n$. Notice that it is at most $\frac{1}{2x+1}$ if $m \neq n$. Thus, by picking larger x, we can get smaller error bound.

Theorem 1. REG $\subsetneq$ BAfL^1 *and* REG $\subsetneq$ BAfL$^0 \subseteq$ NAfL.

The knowledgable readers can notice that in the algorithm $M_2(x)$, we actually implement a blind counter [5][2]. Therefore, by using more states, we can implement more than one blind counter.

Corollary 1. *Any language recognized by a deterministic multi-blind-counter automaton is in* BAfL^1.

Since AfA is a generalization of PFA, we can also obtain the following result.

Theorem 2. *Any language recognized by a probabilistic multi-blind-counter automaton with bounded-error is in* BAfL.

6.2   Cutpoint Languages

Lapiņš [6] showed that the language LAPIŅŠ $= \{a^m b^n c^p \mid m^4 > n^2 > p > 0\}$ is nonstochastic and it is not in SL. It is clear that the following language is also nonstochastic: LAPIŅŠ$' = \{w \in \{a, b, c\}^* \mid |w|_a^4 > |w|_b^2 > |w|_c\}$ or equivalently LAPIŅŠ$' = \{w \in \{a, b, c\}^* \mid |w|_a^2 > |w|_b \text{ and } |w|_b^2 > |w|_c\}$.

---

[2] A counter is blind if its status (whether its value is zero or not) cannot be accessible during the computation. A multi-blind-counter finite automaton is an automaton having $k > 0$ blind counter(s) such that in each transition it can update the value(s) of its counter(s) but never access the status of any counter. Moreover, an input can be accepted by such automaton only if the value of every counter is zero at the end of the computation.

Theorem 3. *The language* LAPINŠ′ *is recognized by an AfA with cutpoint* $\frac{1}{2}$.

Proof. We start with a basic observation about AfAs. Let $\begin{pmatrix} m \\ 1-m \end{pmatrix}$ be an affine state, where m is an integer. Then:

- If $m > 0$, $|m| > |1 - m| = m - 1$ since $m - 1$ is closer to 0.
- If $m \leq 0$, $|m| = -m < |1 - m| = -m + 1$ since $-m$ is closer to 0.

So, if the first state is an accepting state and the second one is not, then we can determine whether $m > 0$ with cutpoint $\frac{1}{2}$, which can be algorithmically useful.

For a given input $w \in \{a, b, c\}^*$, we can easily encode $|w|_a$, $|w|_b$, and $|w|_c$ into the values of some states. Our aim is to determine $|w|_a^2 > |w|_b$ and $|w|_b^2 > |w|_c$. Even though PFAs and QFAs can make the similar encodings, they can make only a single comparison. Here, we show that AfAs can make both compressions. First we present some encoding integer matrices. If we apply matrix $\begin{pmatrix} 1 & 0 \\ 1 & 1 \end{pmatrix}$ m times to $\begin{pmatrix} 1 \\ 0 \end{pmatrix}$, the value of the second entry becomes m:

$$\begin{pmatrix} 1 & 0 \\ 1 & 1 \end{pmatrix}\begin{pmatrix} 1 \\ 0 \end{pmatrix} = \begin{pmatrix} 1 \\ 1 \end{pmatrix} \text{ and } \begin{pmatrix} 1 & 0 \\ 1 & 1 \end{pmatrix}\begin{pmatrix} 1 \\ x \end{pmatrix} = \begin{pmatrix} 1 \\ x + 1 \end{pmatrix} \Rightarrow \begin{pmatrix} 1 & 0 \\ 1 & 1 \end{pmatrix}^m \begin{pmatrix} 1 \\ 0 \end{pmatrix} = \begin{pmatrix} 1 \\ m \end{pmatrix}.$$

For obtaining m^2, we can use the following initial state and matrix:

$$\begin{pmatrix} 1 & 0 & 0 \\ 2 & 1 & 0 \\ 0 & 1 & 1 \end{pmatrix}\begin{pmatrix} 1 \\ 0 \\ 0 \end{pmatrix} = \begin{pmatrix} 1 \\ 1 \\ 0 \end{pmatrix} \text{ and } \begin{pmatrix} 1 & 0 & 0 \\ 2 & 1 & 0 \\ 0 & 1 & 1 \end{pmatrix}\begin{pmatrix} 1 \\ 2x - 1 \\ (x-1)^2 \end{pmatrix} = \begin{pmatrix} 1 \\ 2x + 1 \\ x^2 \end{pmatrix}$$

$$\Rightarrow \begin{pmatrix} 1 & 0 & 0 \\ 2 & 1 & 0 \\ 0 & 1 & 1 \end{pmatrix}^m \begin{pmatrix} 1 \\ 0 \\ 0 \end{pmatrix} = \begin{pmatrix} 1 \\ 2m - 1 \\ m^2 \end{pmatrix}.$$

We can easily embed such matrices into affine operators (by using some additional states) and then we can obtain the value like $|w|_a$ and $|w|_a^2$ as the values of some states. If required, the appropriate initial states can be prepared on the left end-marker. Moreover, on the right end-marker, we can make some basic arithmetic operations by using a combination of more than one affine operators. Furthermore, we can easily tensor two AfA and obtain a single AfA that indeed can simulate both machines in parallel.

Let $|w|_a = x$, $|w|_b = y$, and $|w|_c = z$, and, M_1 and M_2 be two AfAs that respectively have the following final states after reading w:

$$v_f(M_1) = \begin{pmatrix} x^2 \\ y \\ 1 - x^2 - y \\ 0 \\ \vdots \\ 0 \end{pmatrix} \text{ and } v_f(M_2) = \begin{pmatrix} y^2 - z \\ 1 - y^2 + z \\ 0 \\ \vdots \\ 0 \end{pmatrix}.$$

If we tensor both machines and apply an affine transformation to arrange the values in a certain way, the final state will be

$$v_f = \left(x^2(y^2 - z) \quad x^2(1 - y^2 + z) \quad y \quad \frac{1-T}{2} \quad \frac{1-T}{2} \quad 0 \quad \cdots \quad 0 \right)^T,$$

where T is the summation of the first three entries. We select the first and fourth states as accepting states. Then, the difference between the accepting and the remaining values is $\Delta = x^2(|y^2 - z| - |1 - y^2 + z|) - y$. Remark that $\delta = |y^2 - z| - |1 - y^2 + z|$ is either 1 or -1.

- If w is a member, then $\Delta = x^2(1) - y$, which is greater than 0.
- If w is not a member, then we have different cases.
  - $x^2 \leq y$: Δ will be either $x^2 - y$ or $-x^2 - y$ and in both case it is equal to zero or less than zero.
  - $x^2 > y$ but $y^2 \leq z$: Δ will be $-x^2 - y$ and so less than zero.

Thus, the final AfA can recognize LAPINŠ' with cutpoint $\frac{1}{2}$.   □

Since AfAs can recognize a nonstochastic language with cutpoint, they are more powerful than PFAs and QFAs with cutpoint (and also with unbounded-error).

Corollary 2. SL $\subsetneq$ AfL.

6.3   Nondeterministic Languages

Now, we show that NAfAs are equivalent to NQFAs.

Lemma 1. SL$^{\neq} \subseteq$ NAfL.

Proof. Let L be a language in SL$^{\neq}$. Then, there exists an n-state PFA P such that $L = \{w \in \Sigma^* \mid f_P(w) \neq \frac{1}{2}\}$, where $n > 0$. Let $A_\$$ be the transition matrix for the right end-marker and $v_f(w)$ be the final probabilistic state for the input $w \in \Sigma^*$. We can trivially design a probabilistic transition matrix $A'_\$$ such that the first and second entries of the probabilistic state $v'_f(w) = A'_\$ v_f(w)$ are $1 - f_P(w)$ and $f_P(w)$, respectively, and the others are zeros. Let $A''_\$$ be the following affine operator:

$$\begin{pmatrix} 1 & -1 & \mathbf{0} \\ 0 & 2 & \mathbf{0} \\ \mathbf{0} & \mathbf{0} & \mathbf{I} \end{pmatrix}.$$

Then, the first and second entries of $v''_f = A''_\$ v'_f$ are $1 - 2f_P(w)$ and $2f_P(w)$, respectively, and the others are zeros. So, based on P, we can design an AfA M by making at most two modifications: (i) the single accepting state of M is the first one and (ii) the affine operator for the right end-marker is $A''_\$ A'_\$ A_\$$. Then, if $f_P(w) = \frac{1}{2}$ if and only if $f_M(w) = 0$. That is, $L \in$ NAfL.   □

Lemma 2. NAfL $\subseteq$ NQAL.

Proof. Let $L \in$ NAfL. Then, there exists an AfA $M = (E, \Sigma, \{A_\sigma \mid \sigma \in \tilde{\Sigma}, e_s, E_a\})$ such that $w \in L$ if and only if $f_M(w) > 0$ for any input $w \in \Sigma^*$. Now, we design a nondeterministic QFA $M' = \{E \cup F, \Sigma, \{\mathcal{E}_\sigma \mid \sigma \in \Sigma\}, e_s, E_a\}$ for language L, where F is a set of finite states and $E \cap F = \emptyset$.

We provide a simulation of M by M'. The idea is to trace the computation of M through a single pure state. Let $w \in \Sigma^*$ be the input string. The initial affine state is $v_0 = e_s$ and the initial quantum state is $|v_0\rangle = |e_s\rangle$. Assume that each superoperator has $k > 0$ operation elements.

A superoperator can map a pure state to more than one pure state. Therefore, the computation of M' can be also traced/shown as a tree, say T_w. We build the tree level by level. The root is the initial state. For the first level, we apply $\mathcal{E}_\textcent$ to the initial state and obtain k vectors:

$$|\widetilde{v_{(j)}}\rangle = E_{\textcent, j}|v_0\rangle, \quad 1 \leq j \leq k,$$

some of which are unnormalized pure states and maybe the others are zero vectors. We connect all these vectors to the root. For the second level, we apply $\mathcal{E}_{\tilde{w}_2}$ to each vectors on the first level. Although it is clear that zero vectors can always be mapped to zero vectors, we keep them for simplicity. From the node corresponding $|\widetilde{v_{(j)}}\rangle$, we obtain the following children:

$$|\widetilde{v_{(j,j')}}\rangle = E_{\tilde{w}_2, j'}|\widetilde{v_{(j)}}\rangle, \quad 1 \leq j' \leq k.$$

We continue in this way (by increasing the indices of vectors by one in each level) and at the end, we obtain $k^{|\tilde{w}|}$ vectors at the leafs, some of which are unnormalized pure states. The indices of the vectors at the leafs are from $(1, \ldots, 1)$ to $(k, \ldots, k)$. Remark that $|\widetilde{v_{(1,\ldots,1)}}\rangle$ is calculated as

$$|\widetilde{v_{(1,\ldots,1)}}\rangle = E_{\tilde{w}_{|\tilde{w}|}, 1} E_{\tilde{w}_{|\tilde{w}|-1}, 1} \cdots E_{\tilde{w}_1, 1}|v_0\rangle,$$

where all the operation elements are the ones having index of 1. Remark that if α is a value of an accepting state in one of these pure states, then its contribution to the total accepting probability will be $|\alpha|^2$.

This tree facilities to describe our simulation. Each superoperator $\mathcal{E}_\sigma = \{E_{\sigma,1}, \ldots, E_{\sigma,k}\}$ is defined based on A_σ. Among the others, $E_{\sigma,1}$ is the special one that keeps the transitions of A_σ and all the others exist for making $\mathcal{E}_\sigma$ a valid operator. The details of $E_{\sigma,1}$ and the other operation elements of $\mathcal{E}_\sigma$ are as follows:

$$E_{\sigma,1} = \frac{1}{l_\sigma}\left(\begin{array}{c|c} A_\sigma & 0 \\ \hline 0 & I \end{array}\right) \quad \text{and} \quad E_{\sigma,j} = \frac{1}{l_\sigma}\left(\begin{array}{c|c} 0 & 0 \\ \hline * & * \end{array}\right), \quad (2 \leq j \leq k)$$

where $l_\sigma \geq 1$ is a normalization factor and the parts denoted by "*" can be arbitrary filled to make $\mathcal{E}_\sigma$ a valid operator. (Note that there have already been some methods to fill the parts denoted by "*" in a straightforward way [16,17].)

The Hilbert space of M' can be decomposed into two orthogonal subspaces: $\mathcal{H}_e = span\{|e\rangle \mid e \in E\}$ and $\mathcal{H}_f = span\{|f\rangle \mid f \in F\}$. So, any pure state $|v\rangle$ can be decomposed as $|v\rangle = |v_e\rangle \oplus |v_f\rangle$, where $|v_e\rangle \in \mathcal{H}_e$ and $|v_f\rangle \in \mathcal{H}_f$. It is clear that any $E_{\sigma,1}$ ($\sigma \in \widetilde{\Sigma}$) keeps the vector inside of the subspaces: $E_{\sigma,1} : \mathcal{H}_e \to \mathcal{H}_e$ and $E_{\sigma,1} : \mathcal{H}_f \to \mathcal{H}_f$. Then, $E_{\sigma,1}$ maps $|v\rangle = |v_e\rangle \oplus |v_f\rangle$ to $\frac{1}{l_\sigma} A_\sigma |v_e\rangle \oplus \frac{1}{l_\sigma} |v_f\rangle$. Therefore, when $E_{\sigma,1}$ is applied, the part of computation in $\mathcal{H}_f$ never affects the part in $\mathcal{H}_e$.

All the other operational elements map any vector inside $\mathcal{H}_f$ and so they never affect the part in $\mathcal{H}_e$. Remark that any pure state lies in $\mathcal{H}_f$ never produce an accepting probability since the set of accepting states are a subset of E.

Now, we have enough details to show why our simulation works. When considering all leaves of T_w, only $|\widetilde{v_{(1,...,1)}}\rangle$ lies in $\mathcal{H}_e$ and all the others lie in $\mathcal{H}_f$. Then, the accepting probability can be produced only from $|\widetilde{v_{(1,...,1)}}\rangle$, the value of which can be straightforwardly calculated as

$$|\widetilde{v_{(1,...,1)}}\rangle = \frac{1}{l_w}(v_f, *, \ldots, *), \quad l_w = \prod_{j=1}^{|\tilde{w}|} l_{\tilde{w}_j},$$

where "$*$" are some values of the states in F. It is clear that $f_M(w) = 0$ if and only if $f_{M'}(w) = 0$.

Remark that each superoperator can have a different number of operation elements and this does not change our simulation. Moreover, the size of F can be arbitrary. If it is small, then we need to use more operation elements and if it is big enough, then we can use less operation elements.   □

Theorem 4. NAfL = NQAL.

Proof. The equality follows from the fact that $\mathsf{SL}^{\neq} = \mathsf{NQAL}$ [16] and the previous two lemmas: $\mathsf{NQAL} \subseteq \mathsf{SL}^{\neq} \subseteq \mathsf{NAfL} \subseteq \mathsf{NQAL}$ □

7   Concluding Remarks

We introduce affine computation as a generalization of probabilistic computation by allowing to use negative "probabilities". After giving the basics of the new system, we define affine finite automaton and compare it with probabilistic and quantum finite automata. We show that our new automaton model is more powerful than the probabilistic and quantum ones in bounded- and unbounded-error language recognitions and equivalent to quantum one in nondeterministic language recognition mode. After the paper was accepted, we also showed that exclusive affine languages form a superset of exclusive quantum and stochastic languages [4]. Moreover, some other new results on computational power and succinctness of AfAs were recently obtained in [3,13]. These are only the initial results. We believe that the further investigations on the affine computational models can provide new insights on using negative transition values.

Acknowledgements. We thank Marcos Villagra for his very helpful comments. We also thank the anonymous reviewers for their very helpful comments.

References

1. Aaronson, S.: Quantum computing, postselection, and probabilistic polynomial-time. Proc. Roy. Soc. A **461**(2063), 3473–3482 (2005)
2. Ambainis, A., Yakaryılmaz, A.: Automata and quantum computing. Technical report arXiv:1507.01988 (2015)
3. Belovs, A., Montoya, J.A., Yakaryılmaz, A.: Can one quantum bit separate any pair of words with zero-error? Technical report arXiv:1602.07967 (2016)
4. Díaz-Caro, A., Yakaryılmaz, A.: Affine computation and affine automaton. Technical report arXiv:1602.04732 (2016)
5. Greibach, S.A.: Remarks on blind and partially blind one-way multicounter machines. Theor. Comput. Sci. **7**, 311–324 (1978)
6. Lapins, J.: On nonstochastic languages obtained as the union and intersection of stochastic languages. Avtom. Vychisl. Tekh. **4**, 6–13 (1974). Russian
7. Li, L., Qiu, D., Zou, X., Li, L., Wu, L., Mateus, P.: Characterizations of one-way general quantum finite automata. Theoret. Comput. Sci. **419**, 73–91 (2012)
8. Moore, C., Crutchfield, J.P.: Quantum automata and quantum grammars. Theoret. Comput. Sci. **237**(1–2), 275–306 (2000)
9. Paz, A.: Introduction to Probabilistic Automata. Academic Press, New York (1971)
10. Rabin, M.O.: Probabilistic automata. Inf. Control **6**, 230–243 (1963)
11. Rabin, M.O., Scott, D.: Finite automata and their decision problems. Scott. IBM J. Res. Dev. **3**, 114–125 (1959)
12. Say, A.C.C., Yakaryılmaz, A.: Quantum finite automata: a modern introduction. In: Calude, C.S., Freivalds, R., Kazuo, I. (eds.) Computing with New Resources. LNCS, vol. 8808, pp. 208–222. Springer, Heidelberg (2014)
13. Villagra, M., Yakaryılmaz, A.: Language recognition power and succintness of affine automata. Technical report arXiv:1602.05432 (2016)
14. Watrous, J.: Quantum computational complexity. In: Encyclopedia of Complexity and System Science. Springer, Also available at (2009). arXiv:0804.3401
15. Yakaryılmaz, A., Say, A.C.C.: Languages recognized with unbounded error by quantum finite automata. In: Frid, A., Morozov, A., Rybalchenko, A., Wagner, K.W. (eds.) CSR 2009. LNCS, vol. 5675, pp. 356–367. Springer, Heidelberg (2009)
16. Yakaryılmaz, A., Say, A.C.C.: Languages recognized by nondeterministic quantum finite automata. Quantum Inf. Comput. **10**(9&10), 747–770 (2010)
17. Yakaryılmaz, A., Say, A.C.C.: Unbounded-error quantum computation with small space bounds. Inf. Comput. **279**(6), 873–892 (2011)

On Approximating (Connected) 2-Edge Dominating Set by a Tree

Toshihiro Fujito$^{(\boxtimes)}$ and Tomoaki Shimoda

Department of Computer Science and Engineering,
Toyohashi University of Technology, Toyohashi 441-8580, Japan
`fujito@cs.tut.ac.jp`

Abstract. The *edge dominating set* problem (EDS) is to compute a minimum edge set such that every edge is dominated by some edge in it. This paper considers a variant of EDS with extensions of multiple and connected dominations combined. In the *b*-EDS problem, each edge needs to be dominated *b* times. CONNECTED EDS requires an edge dominating set to be connected while it has to form a tree in TREE COVER. Although each of EDS, *b*-EDS, and CONNECTED EDS (or TREE COVER) has been well studied, each known to be approximable within 2 (or 8/3 for *b*-EDS in general), nothing is known when these extensions are imposed simultaneously on EDS unlike in the case of the (vertex) dominating set problem.

We consider CONNECTED 2-EDS and 2-TREE COVER (i.e., a combination of 2-EDS and TREE COVER), and present a polynomial algorithm approximating each within 2. Moreover, it will be shown that the single tree computed is no larger than twice the optimum for (not necessarily connected) 2-EDS, thus also approximating 2-EDS equally well. It also implies that 2-EDS with clustering properties can be approximated within 2 as well.

1 Introduction

In an (undirected) graph $G = (V, E)$ a vertex is said to *dominate* itself and all the vertices adjacent to it, and a vertex set $S \subseteq V$ is a *dominating set* for G if every vertex in G is dominated by some in S. The problem of computing a minimum size dominating set is called DOMINATING SET *(DS)*. Domination is one of the most fundamental concepts in graph theory with numerous applications to a variety of areas [1,10]. The two books [24,25] contain the main results and applications of domination in graphs.

Many variants of the basic concepts of domination have appeared in the literature, and two of them are relevant to the current work. One is *multiple domination*, introduced by Fink and Jacobson [18,19] with an obvious motivation of fault tolerance and/or robustness, and in fact there exist various types of multiple domination; *k-domination* requires every vertex *not* in $S \subseteq V$ to be

This work is supported in part by the Kayamori Foundation of Informational Science Advancement and JSPS KAKENHI under Grant Number 26330010.

A.S. Kulikov and G.J. Woeginger (Eds.): CSR 2016, LNCS 9691, pp. 161–176, 2016.
DOI: 10.1007/978-3-319-34171-2_12

dominated by k members of S while k-*tuple domination* requires every vertex in G to be by k members of S (A survey on combinatorial properties of theirs and other multiple dominations can be found in [8]). Another well-known variant of DS demands connectivity on dominating sets, and $S \subseteq V$ is a *connected dominating set* for G if S is a dominating set for G inducing a connected subgraph of G. The problem CONNECTED DOMINATING SET has been studied in graph theory for many years, and recently it has become a hot topic due to its application in wireless networks for virtual backbone construction [7,12,13]. Moreover, a unifying generalization of multiple and connected dominations has recently received considerable attention, leading to the concept of k-*connected m-dominating set* problem [11]. A minimum vertex set $S \subseteq V$ is sought here such that S is an m-dominating set and simultaneously the subgraph $G[S]$ induced by S is k-connected, where various approximation results are becoming available mainly for unit-disk graphs [22,27] as well as for general graphs [33,38].

In this paper we focus on the algorithmic aspects of EDGE DOMINATING SET *(EDS)*, which is yet another natural variant of DS, and its generalizations. Here, an edge is said to *dominate* itself and all the edges incident to it, and an edge set $F \subseteq E$ is an *edge dominating set (eds)* for G if every edge in G is dominated by some in F. The multiple domination variant of EDS is known as the b-EDGE DOMINATING SET *(b-EDS)* problem, and here, $F \subseteq E$ is a *b-edge dominating set (b-eds)* if *every* edge in G is dominated by b members of F (and hence, it corresponds to the b-tuple DS on line graphs). The CONNECTED EDGE DOMINATING SET problem is a connected variant of EDS, and it is also known as TREE COVER, the problem of covering all the edges in G by the vertex set of a tree, as any connected eds is necessarily cycle-free if it is minimal. Despite the fact that EDS is equivalent to the special case of DS with G restricted to line graphs, EDS seems to have its own history of being a topic of research. Since it was shown to be NP-complete by Yannakakis and Gavril [37], the computational complexity of EDS has been well explored. Computing any maximal matching is a 2-approximation to EDS since it is equivalent to the problem of computing the *minimum maximal matching* [23]. No better approximation has been found in the general case although EDS admits a PTAS for some special cases [4,26], and some nontrivial approximation lower bounds have been derived (under some likely complexity hypothesis) [9,14,31]. The parameterized complexity of EDS has also been extensively studied [6,14,15,20,36]. The general case of b-EDS is known to be approximable within 8/3 [5], but within 2 for $b \leq 3$ [21], and Connected EDS can be within 2 [2,30]. Interestingly though, to the best of our knowledge, no counterpart of the k-connected m-dominating set problem for EDS has appeared in the literature, and nothing is known for Connected b-EDS for $b \geq 2$.

1.1   Our Work

The main subject of the paper is the CONNECTED 2-EDS problem *(C2-EDS)*; that is, the problem of computing a minimum 2-eds inducing a connected subgraph. Another way of imposing the connectivity on b-EDS is to enforce a solution to be a subgraph of certain types in the spirit of Arkin et al. [2]. It is natural to define 2-TREE COVER (2-TC) as the problem of computing a smallest tree, termed a 2-*tree cover*, in connected G such that its edge set forms a 2-eds for G. Notice that a spanning tree is always a 2-tree cover (assuming G is a feasible instance) and that C2-EDS and 2-TC are not equivalent as an optimal solution for the former is not necessarily a tree.

To explain the contributions of the current work and their implications, consider the following natural integer program formulation of 2-EDS, the version where solutions are not required to be connected:

$$\min \left\{ x(E) \mid x(\delta(e)) \geq 2 \text{ and } x_e \in \{0,1\}, \forall e \in E \right\},$$

where $x(F) = \sum_{e \in F} x_e$ for $F \subseteq E$, and $\delta(e) = \{e\} \cup \{e' \in E \mid e'$ is adjacent to $e\}$ for $e \in E$. Replacing the integrality constraints by linear constraints $0 \leq x_e$ results in the LP and its dual in the following forms:

$$\text{LP:(P)} \quad \min z_P(x) = x(E) \qquad\qquad \text{LP:(D)} \quad \max z_D(y) = \sum_{e \in E} 2y_e$$

$$\text{subject to:} \quad x(\delta(e)) \geq 2, \quad \forall e \in E \qquad \text{subject to:} \quad y(\delta(e)) \leq 1, \quad \forall e \in E$$
$$x_e \geq 0, \quad \forall e \in E \qquad\qquad\qquad y_e \geq 0, \quad \forall e \in E$$

Let $\text{dual}_2(G)$ denote the optimal value of LP:(D) above for graph G. The main contribution of the paper is to present an approximation algorithm for 2-TC such that (1) it computes a tree where the edge set of which is a 2-eds for G, and (2) the tree computed is of a size no larger than $2 \cdot \text{dual}_2(G)$. Since $\text{dual}_2(G)$ lower bounds the minimum size of 2-eds, it follows that 2-TC as well as C2-EDS can be approximated within 2. It also implies in turn that 2-EDS can be approximated by a tree within 2. It is also worth pointing out that, considering the case of G being a complete graph, the gap of LP:(P) from the integral optimum for C2-EDS can be arbitrarily close to 2. As mentioned earlier, satisfying both connectivity and multiple domination has been already considered in the case of the vertex domination. It then tends to be harder to approximate, however, even for unit-disk graphs [32,34,35], than when satisfying either one only, in good contrast with the case of EDS as shown in this paper.

A vertex set $C \subseteq V$ is a *vertex cover (vc)* if every edge in G is incident to some member of C, a *connected vertex cover* if it is a vc inducing a connected subgraph, and it is called a t-*total vertex cover (t-tvc)* $(t \geq 1)$ if it is a vc such that each connected component of the subgraph induced by C has at least t vertices. Hence, if C is a t-tvc, each member of the vertex cover C belongs to a "cluster" containing at least t members of C. Having such clustering properties could be desirable or required in some applications, and variants with such properties enforced are considered in other combinatorial optimization problems as well, such as r-*gatherings* [3]. The problem t-*TVC* of computing a minimum t-tvc was introduced in [17,29], and was further studied in [16], where t-TVC was shown approximable within 2 for each $t \geq 1$ [17]. This is based on the Savage's algorithm computing a connected vc of size no larger than twice the minimum vc size [30]. Our algorithm reveals the existence of a similar phenomenon between

2-EDS and C2-EDS, and when 2-EDS with clustering properties is considered, the above result implies that it can be approximated within 2 for any cluster size t (≥ 1) (Note: No such relation is known between EDS and CEDS).

2   Preliminaries

For a vertex set $U \subseteq V$, $\Gamma(U)$ denotes the set of all the vertices in G which are adjacent to at least one vertex of U (i.e., $\Gamma(U) = \{v \in V \mid \{u, v\} \in E$ for some $u \in U\}$), and $\Gamma(u)$ means $\Gamma(\{u\})$. For a vertex $u \in V$ let $\delta(u)$ denote the set of edges incident to u. The degree of a vertex u is denoted by $\deg(u)$. An *empty* tree is meant to be one consisting of a single node (and no edge).

2.1   Gallai-Edmonds Decomposition

For any graph G denote by D the set of such vertices in G that are not covered by some maximum matchings of G. Let A denote the set of vertices in $V - D$ adjacent to at least one vertex in D, and let $C = V - A - D$. A graph G is called *factor-critical* if removal of any vertex from G results in a graph having a perfect matching. For a bipartite graph G with bipartition (X, Y), the *surplus of a set* $W \subseteq X$ is defined by $|\Gamma(W)| - |W|$, and the *surplus of the bipartite graph* G is the minimum surplus of non-empty subsets of X.

This decomposition, which can be computed in polynomial time via the Edmonds matching algorithm, provides important information concerning all the maximum matchings in G:

Theorem 1 (the Gallai-Edmonds structure theorem. See [28]).

1. *The components of the subgraph induced by D are factor-critical.*
2. *The subgraph induced by C has a perfect matching.*
3. *The bipartite graph obtained from G by deleting the vertices of C and the edges in $G[A]$ and by contracting each component of D to a single vertex has positive surplus (as viewed from A).*

Call a vertex in A an *a-node* and one in D a *d-node*. The set of a-nodes in a subgraph H of G is denoted by $A(H)$ and that of d-nodes by $D(H)$.

3   Algorithm Design Overview

One possible way to relate 2-EDS and matchings is to observe that $(1/2)y_M$ is dual feasible to LP:(D) for any matching $M \subseteq E$ and the incidence vector $y_M \in \{0, 1\}^E$ of M, which implies that the minimum 2-eds size is lower bounded by $|M|$ (or the minimum b-eds size in general lower bounded by $b|M|/2$). Therefore, any 2-eds of size bounded by $2|M|$ for some matching M is a 2-approximation to 2-EDS, and in fact such a 2-eds always exists [21]. When it has to be additionally connected, however, such a graph can be easily found where any connected 2-eds is larger than $2|M|$ even if M is a maximum matching, and we take the following measures to deal with this; (1) locate and identify subgraphs of G, with the help of the Gallai-Edmonds decomposition of G, where matching based duals do not work, and (2) devise more elaborate dual assignments for such subgraphs. The algorithm is thus designed, based on the G-E decomposition of G, to compute a forest $\mathcal{F}$ of trees in G s.t.

Property 1. $\mathcal{F}$ spans $A \cup C$ entirely and all the nontrivial components of $G[D]$.

Property 2. Every a-node has degree of at least 2 (i.e., non-leaf) in $\mathcal{F}$.

The nodes in G not spanned by $\mathcal{F}$ then are singletons in $G[D]$ only (by Property 1), and they are adjacent only to a-nodes, each of which is of degree at least 2 in $\mathcal{F}$ (by Property 2). It thus follows that a feasible 2-eds for G can be computed by minimally connecting all the trees in $\mathcal{F}$ into a single tree. In fact it is relatively easy to find a forest satisfying the properties above for the following reasons. To defer the consideration on how to span nontrivial components of $G[D]$, suppose each of them is shrunken into a single d-node, which we call a *fat* d-node, and let G' denote the resulting graph. It can be induced from Theorem 1.3 that $G'[A \cup D]$ contains a forest such that every node in A has degree 2 in it (see [28]). Thus, by additionally using a tree spanning each component of $G[C]$ (and trees spanning nontrivial components of $G[D]$ at the end), a forest satisfying both Properties 1 and 2 can be obtained.

We also need to be concerned with the quality of solutions thus computed, and to bound the approximation ratio of the algorithm, we consider additionally imposing the following two properties on $\mathcal{F}$:

Property 3. $\mathcal{F}$ is "compact" enough that all the trees can be glued together, using only extra edges (and no extra nodes).

Property 4. There exists a dual feasible $y \in \mathbb{R}^E$ s.t. $y(E[T]) \geq (|E(T)| + 1)/4$ for each $T \in \mathcal{F}$, where $E[T]$ denotes the set of edges induced by $V(T)$.

Suppose a tree T_{fin} is constructed by minimally connecting all the trees in $\mathcal{F}$, and to do so, it suffices to use $(|E(T)| + 1)$ edges per tree $T \in \mathcal{F}$ due to Property 3. The total number of edges in T_{fin} is thus no larger than

$$\sum_{T \in \mathcal{F}} (|E(T)| + 1) \leq \sum_{T \in \mathcal{F}} 4y(E[T]) \leq \sum_{e \in E} 4y_e \leq 2 \cdot \text{dual}_2(G),$$

for some dual feasible $y \in \mathbb{R}^E$, where the first inequality is due to Property 4, implying that this is a 2-approximation algorithm for 2-EDS.

Our main goal is thus to devise a way to compute a forest $\mathcal{F}$ satisfying Properties 3 and 4 on top of 1 and 2. It is noted here that Property 3 follows easily when T (or the component T spans) has a perfect matching M, by assigning $1/4$ to y_e for each edge $e \in M$. For this and some other reasons to be explained later, it is not so hard to deal with the components in $G[C]$ as well as the nontrivial components in $G[D]$, and much of the work will be spent in computing trees in $G'[A \cup D]$.

3.1   Two Types of Trees

To ensure Property 2 in building trees in $G'[A \cup D]$, a path (u, v, w) of length 2 with an a-node v in the middle will be taken as an "unit-edge", which we call a 2-*edge*. A tree composed of 2-edges only, with any 2-edge (u, v, w) treated as an "edge" $\{u, w\}$, is called a 2-*tree*.

Basic Trees. A 2-edge (u, v, w) is called a (dad)-*edge* when both of u and w are d-nodes. A tree consisting of (dad)-edges only is called *basic*. Clearly, for any basic tree T, every a-node is of degree 2 in T, and all the leaves of T are d-nodes.

A basic tree T is *maximal* if no node of T is adjacent to an a-node outside of T; that is, if $\Gamma(V(T)) \setminus V(T) \subseteq D$. It will be shown that maximal basic trees can be efficiently computed in $G'[A \cup D]$, and as will be seen, maximal basic trees constitute basic building blocks in construction of a forest in $G'[A \cup D]$.

Bridged Trees. Recall that D is an independent set in G' (as nontrivial components are shrunken to singletons) while A is not in general, and there could be edges among a-nodes. Therefore, there could be two disjoint basic trees and an edge connecting them at an a-node in each of them, for instance, and they can never be maximal basic trees. To avoid such situation we introduce an additional type of tree comprised of 2-edges.

A 2-edge (u, v, w) is called a (aad)-*edge* when one of u and w is an a-node and the other is a d-node. A path $P = (d_1, a_1, a_2, d_2)$ of length 3 is called a *bridge-path* when both of the end-nodes d_1 and d_2 are d-nodes and both of the mid-nodes a_1 and a_2 are a-nodes. A tree T constructed by starting with a bridge-path and inductively extending T by attaching either a (dad)-edge or an (aad)-edge is called a *bridged* tree, where an (aad)-edge is always attached only at an a-node of T.

As in a basic tree, an a-node cannot occur at a leaf in any bridged tree, from the way it is constructed, and hence, its degree is at least 2 (which can be larger unlike basic trees). Also observe that a bridged tree is perfectly matchable, unlike a basic tree, implying that it possesses Property 4. Once again, a bridged tree T in G is called *maximal* if no node of T is adjacent to an a-node outside of T; that is, if $\Gamma(V(T)) \setminus V(T) \subseteq D$.

4   Algorithms

4.1   Tree Construction

Let H be a connected component of $G'[A \cup D]$. We describe here how to construct within H either a maximal basic tree or a maximal bridged tree, and outline three procedures used for this purpose, basicTree(s), basicTree&Path(s), and bridgedTree(s) (More detailed and precise descriptions for them will be provided in the full version).

1. The procedure `basicTree(s)`, when called with an initial d-node s, returns a basic tree T containing s. It computes T in the DFS fashion (using a stack data structure to keep currently active d-nodes), starting with s, by repeatedly extending T by a (dad)-edge from the currently visited d-node d if there exists an a-node in $\Gamma(d) \setminus A(T)$ while backtracking to a previous d-node if no such a-nodes exist. It differs from the ordinary DFS-tree construction in how to handle a case where the addition of a new edge would introduce a cycle in T. Suppose there exists an unvisited a-node a in the neighbor of the currently visited d-node d. Let d' be another d-node adjacent to a. If $d' \notin D(T)$, adding the 2-edge (d, a, d') to T extends T to a larger basic tree. If $d' \in D(T)$, on the other hand, $T \cup \{(d, a, d')\}$ contains a cycle. Denote this cycle by B. Then, as $|A(B)| = |D(B)|$, there must exist an edge (a', d'') s.t. $a' \in A(B)$ and $d'' \notin D(B)$ due to Theorem 1.3 If $d'' \notin D(T)$ this time, by adding both (d, a, d') and (a', d'') while removing the cycle (by dropping one edge incident to a'), T can be properly extended to a larger basic tree. Certainly, it could be the case again that $d'' \in D(T)$, and hence, adding (a', d'') introduces another cycle in T. In such a case, let B denote the union of these cycles, and repeat the same argument and procedure. It would eventually find an (ad)-edge of which the d-node lies outside of T, and at that point, T can be properly extended by adding some number of (ad)-edges, on top of the initial (d, a, d'), and removing the same number of (ad)-edges within T. This part will be taken care of by procedure `processCycle`.

   The procedure `basicTree(s)` thus extends an edge from any d-node to any neighboring a-node if it is not yet a part of the tree. Therefore, it computes a basic tree T s.t. $\Gamma(D(T)) \subseteq A(T)$. It may not be maximal, however, as there could be an a-node in T adjacent to another a-node outside of T, and the next procedure `basicTree&Path(s)` computes a tree T, starting with a single d-node s, s.t. T is either a maximal basic tree or otherwise, T contains a single bridge-path joining two vertex disjoint basic trees.

2. The procedure `basicTree&Path(s)`, which is a slightly modified `basicTree(s)`, returns a tree T containing s s.t. T is either maximal basic or T contains a single bridge-path joining two vertex disjoint basic trees. It uses a flag `bridged` (initially, `false`) so that T contains at most one bridge-path. It constructs a tree T just like `basicTree(s)` does, but when trying to extend T from a current d-node d, it first checks whether a bridge-path P exists emanating from d (before the existence of a (dad)-edge incident to d), and if exists, it extends T along P, as long as `bridged` is false (that is when no bridge-path is yet contained in T). Because the extension via a bridge-path has a higher priority over that via a (dad)-edge, all the nodes in P must have been unvisited except for d when P is found, and the other end of P becomes the next active d-node (while `bridged` is set to true).

   `basicTree&Path(s)` behaves exactly same as `basicTree(s)` once `bridged` becomes true, except that we also need to modify the way to handle possible occurrences of cycles so that the middle edge (a_1, a_2) of a bridge-path $P = (d_1, a_1, a_2, d_2)$ already contained in T, is never deleted (as it would destroy the "bridgedness"). This can be easily handled by treating P as if

$P = (d_1, a_1, d_2)$ (or $P = (d_1, a_2, d_2)$ depending on the particular case).
Clearly, basicTree&Path(s) returns T such that it is either a basic tree
or it contains a single bridge-path joining two vertex disjoint basic trees.
Observe that, in the former case, no bridge-path has been found throughout
the process of extending T, which implies that no a-node of T is adjacent
to another a-node (i.e., $\Gamma(A(T)) \subseteq C \cup D$). Therefore, when T returned by
basicTree&Path(s) is basic (i.e., bridged is false), it must be a maximal
basic tree.

3. The procedure bridgedTree(s) returns a tree T containing a d-node s s.t. T
   is either a basic or a bridged tree, and maximal one in either case. It first
   calls basicTree&Path(s), which returns a tree T. If T is basic, the procedure ends here. Otherwise, it must be a tree consisting of a single bridge-
   path $P = (d_1, a_1, a_2, d_2)$ joining two vertex disjoint basic trees, T_1 and T_2.
   We extend T to a maximal bridged tree. Recall that either T_1 or T_2 is
   no longer extendable from any of its d-nodes as they are constructed by
   basicTree&Path(s), but they can possibly be from the a-nodes of them by
   (aad)-edges. Besides, it might be also possible to further grow T by extending it from each of a_1 and a_2. Therefore, we first collect all of these a-nodes
   of T in midAs, and check if any unvisited node a' exists in $\Gamma(a)$ for each
   $a \in$ midAs. If it doesn't, extension from a is complete and a is removed from
   midAs. If it does exist, on the other hand, any d-node d' adjacent to a' must
   be unvisited, and we undertake the following steps: compute a basic tree T'
   by calling basicTree(d'), extend T by attaching T' via 2-edge (a, a', d'), and
   add all the a-nodes of T' together with a' to midAs. These operations are
   repeated as long as midAs is nonempty, and when the procedure terminates,
   T must be a maximal bridged tree.

4.2   Forest Construction

Assuming again that H is a connected component of $G'[A \cup D]$, we describe
how to build a forest $\mathcal{F}_H$ in H so that all the nodes in $A(H)$ are spanned by
the trees in $\mathcal{F}_H$. A call of bridgedTree($\cdot$) returns a maximal 2-tree T, but it
may not span $A(H)$ entirely because of structural constraints imposed on T. We
therefore construct a forest $\mathcal{F}_H$ of trees, basic and bridged, in H by repeatedly
calling bridgedTree($\cdot$), so that every a-node in H becomes a part of some tree
in it, and we do so paying attention to an additional property to be imposed on
$\mathcal{F}_H$. We say that a forest $\mathcal{F}$ is *compact* if the set of vertices of the trees in it
induces a connected subgraph in G'.

If a forest $\mathcal{F}_H$ is compact, all the trees in it can be connected together using
a minimum number of edges, and such $\mathcal{F}_H$ can be easily computed by restarting
bridgedTree($\cdot$) every time from a node adjacent to an already computed tree.
algorithm buildForest(H):

1. Initialize the forest $\mathcal{F}_H$ to be empty, and set $i = 1$.
2. Pick any d-node r_1 in H.
3. While there exists a node in $A(H)$ not spanned by any tree in $\mathcal{F}_H$ do

    (a) Call `bridgedTree`(r_i), let T_i be the tree returned, and add T_i to $\mathcal{F}_H$.
    (b) Pick an unvisited r_{i+1} adjacent to some T_j for $1 \leq j \leq i$ if it exists, and increment i.
4. Output $\mathcal{F}_H$.

Let $\mathcal{F}_H = \{T_1, T_2, \cdots, T_k\}$ be the forest constructed by `buildForest`(H). It is straightforward to observe the following:

- (1) r_i belongs to T_i, and (2) there exists j $(1 \leq j < i)$ and an edge (a_{j_i}, r_i) in H s.t. the a-node a_{j_i} belongs to T_j, for all i $(1 \leq i \leq k)$. It follows that $\mathcal{F}_H$ is compact as all of the T_i's can be connected together by those $(k-1)$ edges.
- The tree T_1 is maximal within H, and when H_i denotes the graph remaining after all the vertices of earlier constructed T_j's $(1 \leq j \leq i-1)$ are removed from H, T_i is maximal within H_i, for all i $(1 \leq i \leq k)$.

Thus, any node remaining unvisited after T_i is constructed must be a d-node if it is adjacent to T_j for some $1 \leq j \leq i$, and if no such nodes exist in the last step of this algorithm, no unvisited a-node can remain in H, and hence, the while-loop terminates immediately when this occurs. It also follows from the observation above that:

Lemma 1. *Let* $\mathcal{F}_H = \{T_1, T_2, \cdots, T_k\}$ *be the forest constructed by* `buildForest` *(H). If an edge* (u, v) *exists between* T_i *and* T_j *(i.e.,* $u \in V(T_i)$ *and* $v \in V(T_j)$*),*

- *exactly one of* u *and* v *is an* a-*node and the other is a* d-*node.*
- *if* $u \in A$ *then* $i < j$*, and otherwise,* $i > j$*.*

4.3   Forest Modification

Let $\mathcal{F}_C$ denote a forest of trees each spanning a distinct component of $G[C]$, and $\mathcal{F}_H$ the forest computed by `buildForest`(H) for a connected component H of $G'[A \cup D]$. Then, the trees in $\mathcal{F}_H$'s for all the components H of $G'[A \cup D]$, together with those in $\mathcal{F}_C$ and those trees each of which spanning a nontrivial component of $G[D]$ can be seen to constitute a forest $\mathcal{F}$ satisfying Properties 1 through 3. Such $\mathcal{F}$ may not satisfy Property 4, however, and this subsection describes how to modify $\mathcal{F}_H$'s so that $\mathcal{F}$ satisfies all of them.

Clearly all the nodes in $A(H) \cup C$ are spanned by the trees in $\mathcal{F}_H$ and $\mathcal{F}_C$. Moreover, $\mathcal{F}_H$ is a compact forest as already observed. As will be treated in the next section, one can satisfy Property 4 by assigning dual values within $E[T]$ s.t.

$$y(E[T]) \geq \frac{|E(T)| + 1}{4} \quad \text{and} \quad y(\delta(u)) \leq \frac{1}{2}, \ \forall u \in V(T)$$

when T is either a tree spanning a component in $G[C]$, a bridged tree in $G'[A \cup D]$, or a basic tree in $G'[A \cup D]$ containing a fat d-node. In a case where T is a basic tree with no fat d-nodes in it, on the other hand, it may no longer be possible to assign dual values in $E[T]$ satisfying all of these bounds (just think of the case of $K_{s,s+1}$ for instance). Yet it will be shown (in the next section) how to assign dual values in $E[T]$ so that y is feasible when restricted to within $E[T]$ and $y(E[T]) \geq (|E(T)| + 1)/4$ for a basic tree T whereas $y(\delta(u))$ is not necessarily bounded by $1/2$, $\forall u \in V(T)$. A problem here is that such y may not yield a dual feasible solution when extended to a whole $y \in \mathbb{R}^E$ as there could be a node $u \in V(T)$ with $y(\delta(u)) > 1/2$ and an edge (u, v) with $v \notin V(T)$. To avoid such a situation, therefore, we consider using basic trees of a more specific type.

Let T' be a basic tree in $\mathcal{F}_H$. We say that T' is *internally closed* if none of the a-nodes of T' is adjacent to another tree in $\mathcal{F}_H \cup \mathcal{F}_C$. A basic tree $T \in \mathcal{F}_H$ containing an internally closed tree as its sub 2-tree is called *internal* whereas it is *external* if every nonempty sub 2-tree is not internally closed. Observe that

- if no inter-tree edge is incident to $A(T)$ for a basic tree $T \in \mathcal{F}_H$, T is internally closed (and hence, internal), but T could be external even if only one of such exists, and
- in the forest output by $\texttt{buildForest}(H)$, the last tree T_k constructed must be internally closed (and hence, internal as well), if it is basic and not adjacent to C, due to the property specified by Lemma 1.

Let $T_i \in \mathcal{F}_H$ be a basic tree rooted at r_i and introduce the natural ancestor-descendant relation among nodes in it for each T_i of $\mathcal{F}_H$. The following procedure tears apart T_i into parts, paste them to other trees in $\mathcal{F}_H \cup \mathcal{F}_C$ using external edges, and the only remnant of T_i is an internally closed 2-tree or otherwise, it is just a single node, which, being of no use for us, will be subsequently discarded. **algorithm** $\texttt{resolveTree}(T_i)$:

1. Given a basic tree T_i, if it is empty or internally closed, exit here.
2. Otherwise, there must exist an inter-tree edge (a, u) s.t. $a \in A(T_i)$, and either $u \in D(T')$ for $T' \in \mathcal{F}_H$ or $u \in V(T')$ for $T' \in \mathcal{F}_C$.
3. Let (d_1, a, d_2) be the 2-edge of T_i into which (a, u) is incident, and let d_1 be closer to the root of T_i than d_2.
4. Replace the 2-edge (d_1, a, d_2) of T_i by (d_1, a, u).
5. By the operation of the previous step, T_i is divided into two, and the one containing the root is immediately attached to T' via edge (a, u). The other is a sub 2-tree of T_i, which is basic by itself, and this tree is named anew as T_i rooted at d_2.
6. Recursively call $\texttt{resolveTree}(T_i)$.

Observation 1. *In case that T' referred to in Step 2 of $\texttt{resolveTree}(T_i)$ is a tree in $\mathcal{F}_H$, it must have been constructed by $\texttt{buildForest}(H)$ later than T_i because of the existence of the inter-tree edge (a, u) (Lemma 1).*

Let $\mathcal{F}_H = \{T_1, T_2, \cdots, T_k\}$ be the forest computed by $\texttt{buildForest}(H)$. We'd like to eliminate from $\mathcal{F}_H$ any external basic tree having no fat d-node, and an obvious approach is to apply $\texttt{resolveTree}(T_i)$ to such T_i's. Recall that $\mathcal{F}_H$ computed by $\texttt{buildForest}(H)$ is a compact forest, and a call of $\texttt{resolveTree}(T_i)$ could result in removal of a d-node from T_i, and hence, the compactness of $\mathcal{F}_H$ could be lost if $\texttt{resolveTree}(\cdot)$ is applied to trees in it in an arbitrary order. To avoid this, $\mathcal{F}_H$ will be modified by the next procedure, using Observation 1, by applying resolveTree(T_i) to each of external basic trees without fat d-nodes $T_i \in \mathcal{F}_H$ in the reversal of the order they were created by buildForest(H).

algorithm modifyForest(H):

1. For $i = k - 1$ downto 1 do
   (a) If T_i is a basic tree with no fat d-node in it and not internally closed, call resolveTree(T_i).
   (b) If T_i is a single node, remove it from $\mathcal{F}_H$.
2. return $\mathcal{F}_H$.

Lemma 2. *The forest $\mathcal{F}_H$ computed by* modifyForest*(H) has the following properties:*

1. *Every tree in $\mathcal{F}_H$ is either a basic tree or a bridged tree, and every basic tree is internal if it does not contain a fat d-node.*
2. *Let $\mathcal{F}'$ denote the set of trees in $\mathcal{F}_C$ involved when* resolveTree*(T_i) was applied during* modifyForest*(H). Then, $\mathcal{F}_H \cup \mathcal{F}'$ is compact.*

Proof.

1. If T_i is a basic tree without a fat d-node and not yet internally closed, then resolveTree(T_i) is called by which T_i becomes either empty or internally closed. Therefore, every basic tree remaining in $\mathcal{F}_H$ must be internal or otherwise, it must contain a fat d-node.
2. Recall r_1 denotes the root of T_1 in $\mathcal{F}_H$. To see that $\mathcal{F}_H \cup \mathcal{F}'$ is compact, it suffices to show that every node in it can reach r_1 by passing only the nodes of trees belonging to $\mathcal{F}_H \cup \mathcal{F}'$. Before running modifyForest(H), $\mathcal{F}_H$ was compact. To be more precise, every node of T_i can reach r_1 using only the nodes of $T_1, \cdots, T_i$. Consider now the run of modifyForest(H), and suppose resolveTree(T_i) is applied to T_i. At this point of time any of T_1 through T_{i-1} has been kept intact, and application of resolveTree(T_i) can affect the reachability with r_1 only for the nodes in T_i through T_k. It cannot, however, if all the nodes of T_i remain in $\mathcal{F}_H \cup \mathcal{F}'$ after application of resolveTree(T_i), and thus we consider now the case when a node u of T_i disappears from any tree of $\mathcal{F}_H \cup \mathcal{F}'$ as a result of resolveTree(T_i). It must be the case then that u was a leaf d-node of T_i before resolveTree(T_i) was applied to T_i. Then, any other node of T_i can be seen to remain reachable from r_1 even after removal of u, but so must be any node of trees T_{i+1} through T_k because no edge can exist in the first place between u and those trees (Lemma 1). □

4.4 Overall Algorithm

Now that all of the component procedures have been presented, the entire algorithm, which we refer to as C2EDS, can be described as follows:

1. Compute the Gallai-Edmonds decomposition (A, C, D) of G.
2. Compute a forest $\mathcal{F}_C$ of trees each spanning every component of $G[C]$.

3. Shrink every nontrivial component of $G[D]$ into a fat d-node and denote the resulting graph by G'.
4. For each connected component H of $G'[A \cup D]$ do
   (a) Construct a forest $\mathcal{F}_H$ by running buildForest(H).
   (b) Restructure $\mathcal{F}_H \cup \mathcal{F}_C$ by running modifyForest(H).
5. For each fat d-node d in G' do
   (a) Let H be the component of $G'[A \cup D]$ containing d, and T_d be a spanning tree of the original component of $G[D]$ shrunk into d.
   (b) If d belongs to $T \in \mathcal{F}_H \cup \mathcal{F}_C$, modify T by inserting T_d into T in place of d,
   (c) Else (i.e., d does not belong to any tree in $\mathcal{F}_H \cup \mathcal{F}_C$), add T_d to $\mathcal{F}_H$.
6. Let $\tilde{\mathcal{F}}$ be the forest of all the trees in $\mathcal{F}_C$ and $\mathcal{F}_H$'s for all the components H of $G'[A \cup D]$. Compute and output a tree by minimally connecting all the trees in $\tilde{\mathcal{F}}$.

Lemma 3. *The forest $\tilde{\mathcal{F}}$ obtained in Step 6 of C2EDS is a compact forest consisting of the following types of trees, where, in cases 1, 2 and 5, any fat d-node d in G' is replaced by a spanning tree of the original component shrunken into d:*

1. *a tree T spanning a component of $G[C]$, with possibly some number of basic trees T' attached to T, where exactly one leaf u of T' is superimposed with a node of T,*
2. *a bridged tree in $G'[A \cup D]$,*
3. *an internal basic tree in $G'[A \cup D]$ with no fat d-node in it,*
4. *a tree spanning a nontrivial component of $G[D]$,*
5. *a basic tree in $G'[A \cup D]$ containing a fat d-node.*

Proof. It is clear from Lemma 2.1 and the way $\mathcal{F}_C$ and $\mathcal{F}_H$'s are constructed in G' that $\tilde{\mathcal{F}}$ consists of only trees of those types listed above. It remains to show that $\tilde{\mathcal{F}}$ is compact in G. Shrink every spanning tree of a nontrivial component in $G[D]$ to a fat d-node in $\tilde{\mathcal{F}}$. It follows from Lemma 2.2 and the fact that $A \cup C$ is spanned by $\tilde{\mathcal{F}}$ that it is compact in G'. But then, when every fat d-node is replaced back to a spanning tree of the original component in $G[D]$, the resulting $\tilde{\mathcal{F}}$ must be compact in G. □

5   Analysis

To see that the tree T output by the algorithm C2EDS of Sect. 4.4 is indeed a connected 2-eds, it suffices to verify that each vertex of G not belonging to T is non-adjacent to a leaf of T. Observe that any vertex u of G not spanned by T must occur in D as a singleton component of $G[D]$. Thus, u is adjacent only to nodes in A, and none of them can be a leaf of T as every a-node is of degree at least 2 in T. Therefore,

Lemma 4. *The tree computed by the algorithm C2EDS is a 2-edge dominating set for G.*

The remainder of the paper is devoted to an analysis of the approximation performance of the algorithm. For a matching M in a graph H, $y \in \mathbb{R}^{E(H)}$ is called an M-*dual* if $y_e = 1/2$ for each $e \in M$ and $y_e = 0$ for each $e \notin M$. Furthermore, it is called an M-dual *respecting* u for $u \in V(H)$ if M leaves u unmatched. Easily, an M-dual y is dual feasible for any matching M and $y(\delta(u)) = 0$ if it is an M-dual respecting u. For any attachment of a basic tree T' to an existing tree T where a leaf u of T' is superimposed with a node of T, we use an M'-dual $y_{T'}$ respecting u for the dual assignment on T', where M' is any maximum matching in T' respecting u. Then, because M' matches all the nodes of T' but u, $y_{T'}$ by itself has the objective value of $|E(T')|/2$, enough for accounting for the size of T', and such an attachment can be dropped from further consideration as its dual would not interfere with the one on T when assigned as described in the proof of the next lemma. It thus suffices to show how to assign duals for those trees T listed in Lemma 3, ignoring the existence of any attachment of basic trees.

Lemma 5. *For each tree T listed below, there exists dual feasible $y \in \mathbb{R}^{E[T]}$ with $y(E[T]) \geq (|E(T)| + 1)/4$, where, in cases 1, 2 and 5, any fat d-node d in G' is replaced by a spanning tree of the original component shrunken into d:*

1. *a tree spanning a component of $G[C]$,*
2. *a bridged tree in $G'[A \cup D]$,*
3. *an internally closed basic tree in $G'[A \cup D]$ with no fat d-node in it,*
4. *a tree spanning a nontrivial component of $G[D]$,*
5. *a basic tree in $G'[A \cup D]$ containing a fat d-node.*

Moreover, $y(\delta(u)) \leq 3/4$ if u is an a-node in an internally closed basic tree in $G'[A \cup D]$ and $y(\delta(u)) \leq 1/2$ for all the other nodes u in G.

Proof. Shrunken into d is factor-critical (Theorem 1.1), and suppose d appears as a node of some tree T computed by C2EDS. When duals are assigned by an M-dual for a matching M on T, only one edge e of M is incident to d. Let u denote the unique node of B into which e is incident when d is expanded back to original B. It is assumed in what follows that dual values are assigned within B by an M_B-dual for a perfect matching M_B in $B - \{u\}$, for any of such d-nodes contained in the following trees of cases 1, 2 and 5, excluding the fat d-node d' referred to in case 5.

1. A tree spanning a component B of $G[C]$.
   Let M_B denote a perfect matching for B, and let y be an M_B-dual. Then, $y(E[T]) = |V(B)|/4 = (|E(T)| + 1)/4$ for any spanning tree T for B.
2. A bridged tree in $G'[A \cup D]$. Observe that there exists a matching in any 2-tree T of size $|E(T)|/2$, one edge per 2-edge, while any bridged tree T' has a perfect matching M in it; a matching consisting of two edges from the bridge-path and a half of the edges from each of the 2-trees attached to the bridge-path. It thus follows by considering an M-dual for T' that there exists dual feasible $y \in \mathbb{R}^{E(T')}$ such that $y(E(T')) \geq (|E(T')| + 1)/4$.

3. An internally closed basic tree in $G'[A \cup D]$ with no fat d-node in it.
   Omitted due to the space limitation.

4. A tree spanning a nontrivial component B of $G[D]$.
   Omitted due to the space limitation.

5. A basic tree containing a fat d-node.
   Suppose T is a basic tree in $G'[A \cup D]$ containing a fat d-node d'. Let y_T
   denote the M-dual where M is the maximum matching for T respecting d'.
   For the nontrivial component B' of $G[D]$ into which d' is unshrunken, let T'
   be any spanning tree for B' and let $y_{B'}$ denote the dual assignment on $E(B')$
   specified in the previous item (of "A tree spanning a nontrivial component
   B of $G[D]$"). For any fat d-node d'' contained in T other than d', consider
   the corresponding component B'' of $G[D]$ and a spanning tree T'' for B''. As
   exactly one node u in B'' is matched by M, there exists a matching M'' in
   $E(B'')$ perfectly matching all the nodes in B'' but u, and it can be seen that
   $y_T + y_{B'}$ added with the M''-dual $y_{B''}$ can remain dual feasible. It then follows
   that the total duals thus assigned within $B = G[V(T) \cup V(B') \cup V(B'')]$ is
   $|E(T)|/4 + |V(B')|/4$ added with $(|V(B'')| - 1)/4$, that is

$$y(E(B)) = \frac{|E(T)| + (|E(T')| + 1) + |E(T'')|}{4} = \frac{|E(T) \cup E(T') \cup E(T'')| + 1}{4}.$$

   This way, even if T contains more fat d-nodes in $G'[A \cup D]$, letting $\tilde{T}$ denote
   T with all of them replaced by the corresponding spanning trees for the
   components into which they become unshrunken, it can be seen there exists
   dual feasible y defined on $G[V(\tilde{T})]$ such that $y(E(B)) = (|E(\tilde{T})| + 1)/4$.   □

By Lemma 5 there is a way $y \in \mathbb{R}^{E[T]}$ to assign dual values within $E[T]$ for
each tree $T \in \tilde{\mathcal{F}}$ s.t. $y(E[T]) \geq (|E(T)|+1)/4$. Let y denote the combination of all
the dual assignments involved. It can be verified that $y(\delta(e)) \leq 1$ for each $e \in E$
as follows. Certainly, it is the case when $e \in E[T]$ for some $T \in \tilde{\mathcal{F}}$ as y restricted
to $E[T]$ is dual feasible. For $e \notin E[T]$ for any $T \in \tilde{\mathcal{F}}$, recall that $y(\delta(u)) \leq 1/2$
for all $u \in V$ unless u is an a-node in an internally closed basic tree. Since e is
not incident to an a-node of any internally closed basic tree, $y(\delta(e)) \leq 1$ holds.
Therefore, y thus determined is dual feasible as a whole, and since the final tree
output by the algorithm is constructed by minimally connecting all the trees
in the compact forest $\tilde{\mathcal{F}}$, the number of additional edges used is one less than
the number of the trees in it. The size of the tree output is thus bounded by
$4 \sum_{e \in E} y_e - 1 = 2 \cdot z_D(y) - 1 \leq 2 \cdot \mathrm{dual}_2(G) - 1$.

Theorem 2. *Given a connected graph G, the algorithm C2EDS outputs a tree,
dominating every edge at least twice, of size bounded by $2 \cdot \mathrm{dual}_2(G) - 1$.*

Corollary 1. *The algorithm C2EDS computes a tree approximating 2-EDS, C2-
EDS, and 2-TC each within a factor of 2.*

References

1. Alber, J., Betzler, N., Niedermeier, R.: Experiments on data reduction for optimal domination in networks. Ann. Oper. Res. **146**, 105–117 (2006)
2. Arkin, E., Halldórsson, M., Hassin, R.: Approximating the tree and tour covers of a graph. Inform. Process. Lett. **47**, 275–282 (1993)
3. Armon, A.: On min-max r-gatherings. Theor. Comput. Sci. **412**(7), 573–582 (2011)
4. Baker, B.: Approximation algorithms for NP-complete problems on planar graphs. J. ACM **41**, 153–180 (1994)
5. Berger, A., Fukunaga, T., Nagamochi, H., Parekh, O.: Approximability of the capacitated b-edge dominating set problem. Theor. Comput. Sci. **385**(1–3), 202–213 (2007)
6. Binkele-Raible, D., Fernau, H.: Enumerate and measure: improving parameter budget management. In: Raman, V., Saurabh, S. (eds.) IPEC 2010. LNCS, vol. 6478, pp. 38–49. Springer, Heidelberg (2010)
7. Blum, J., Ding, M., Thaeler, A., Cheng, X.: Connected dominating set in sensor networks and MANETs. In: Du, D.-Z., Pardalos, P.M. (eds.) Handbook of Combinatorial Optimization, Supplement Vol. B, pp. 329–369. Springer, New York (2005)
8. Chellali, M., Favaron, O., Hansberg, A., Volkmann, L.: k-domination and k-independence in graphs: a survey. Graphs Combin. **28**(1), 1–55 (2012)
9. Chlebík, M., Chlebíková, J.: Approximation hardness of edge dominating set problems. J. Comb. Optim. **11**(3), 279–290 (2006)
10. Cooper, C., Klasing, R., Zito, M.: Dominating sets in web graphs. In: Leonardi, S. (ed.) WAW 2004. LNCS, vol. 3243, pp. 31–43. Springer, Heidelberg (2004)
11. Dai, F., Wu, J.: On constructing k-connected k-dominating set in wireless ad hoc and sensor networks. J. Parallel Distrib. Comput. **66**(7), 947–958 (2006)
12. Du, D.-Z., Wan, P.-J.: Connected Dominating Set: Theory and Applications. Springer Optimization and Its Applications, vol. 77. Springer, New York (2013)
13. Du, H., Ding, L., Wu, W., Kim, D., Pardalos, P., Willson, J.: Connected dominating set in wireless networks. In: Pardalos, P.M., Du, D.-Z., Graham, R.L. (eds.) Handbook of Combinatorial Optimization, pp. 783–833. Springer, New York (2013)
14. Escoffier, B., Monnot, J., Paschos, V.T., Xiao, M.: New results on polynomial in approximability and fixed parameter approximability of edge dominating set. Theory Comput. Syst. **56**(2), 330–346 (2015)
15. Fernau, H.: EDGE DOMINATING SET: efficient enumeration-based exact algorithms. In: Bodlaender, H.L., Langston, M.A. (eds.) IWPEC 2006. LNCS, vol. 4169, pp. 142–153. Springer, Heidelberg (2006)
16. Fernau, H., Fomin, F.V., Philip, G., Saurabh, S.: The curse of connectivity: t-total vertex (edge) cover. In: Thai, M.T., Sahni, S. (eds.) COCOON 2010. LNCS, vol. 6196, pp. 34–43. Springer, Heidelberg (2010)
17. Fernau, H., Manlove, D.F.: Vertex and edge covers with clustering properties: complexity and algorithms. J. Discrete algorithms **7**(2), 149–167 (2009)
18. Fink, J.F., Jacobson, M.S.: n-domination in graphs. In: Graph Theory with Applications to Algorithms and Computer Science, pp. 283–300. Wiley (1985)
19. Fink, J.F., Jacobson, M.S.: On n-domination, n-dependence and forbidden subgraphs. In: Graph Theory with Applications to Algorithms and Computer Science, pp. 301–311. Wiley (1985)
20. Fomin, F.V., Gaspers, S., Saurabh, S., Stepanov, A.A.: On two techniques of combining branching and treewidth. Algorithmica **54**(2), 181–207 (2009)

21. Fujito, T.: On matchings and b-edge dominating sets: a 2-approximation algorithm for the 3-edge dominating set problem. In: Ravi, R., Gørtz, I.L. (eds.) SWAT 2014. LNCS, vol. 8503, pp. 206–216. Springer, Heidelberg (2014)
22. Gao, X., Zou, F., Kim, D., Du, D.-Z.: The latest researches on dominating problems in wireless sensor network. In: Handbook on Sensor Networks, pp. 197–226. World Scientific (2010)
23. Harary, F.: Graph Theory. Addison-Wesley, Reading (1969)
24. Haynes, T., Hedetniemi, S., Slater, P. (eds.): Domination in Graphs, Advanced Topics. Marcel Dekker, New York (1998)
25. Haynes, T., Hedetniemi, S., Slater, P.: Fundamantals of Domination in Graphs. Marcel Dekker, New York (1998)
26. Hunt III., H., Marathe, M., Radhakrishnan, V., Ravi, S., Rosenkrantz, D., Stearns, R.: A unified approach to approximation schemes for NP- and PSPACE-hard problems for geometric graphs. In: Proceedings of the Second Annual European Symposium on Algorithms, pp. 424–435 (1994)
27. Kim, D., Gao, X., Zou, F., Du, D.-Z.: Construction of fault-tolerant virtual backbones in wireless networks. In: Handbook on Security and Networks, pp. 488–509. World Scientific (2011)
28. Lovász, L., Plummer, M.: Matching Theory. North-Holland, Amsterdam (1986)
29. Małafiejski, M., Żyliński, P.: Weakly cooperative guards in grids. In: Gervasi, O., Gavrilova, M.L., Kumar, V., Laganá, A., Lee, H.P., Mun, Y., Taniar, D., Tan, C.J.K. (eds.) ICCSA 2005. LNCS, vol. 3480, pp. 647–656. Springer, Heidelberg (2005)
30. Savage, C.: Depth-first search and the vertex cover problem. Inform. Process. Lett. **14**(5), 233–235 (1982)
31. Schmied, R., Viehmann, C.: Approximating edge dominating set in dense graphs. Theoret. Comput. Sci. **414**(1), 92–99 (2012)
32. Shang, W., Wan, P., Yao, F., Hu, X.: Algorithms for minimum m-connected k-tuple dominating set problem. Theoret. Comput. Sci. **381**(13), 241–247 (2007)
33. Shi, Y., Zhang, Y., Zhang, Z., Wu, W.: A greedy algorithm for the minimum 2-connected m-fold dominating set problem. J. Comb. Optim., 1–16 (2014)
34. Thai, M.T., Zhang, N., Tiwari, R., Xu, X.: On approximation algorithms of k-connected m-dominating sets in disk graphs. Theor. Comput. Sci. **385**(13), 49–59 (2007)
35. Wu, Y., Li, Y.: Construction algorithms for k-connected m-dominating sets in wireless sensor networks. In: Proceedings of the 9th ACM International Symposium on Mobile Ad Hoc Networking and Computing, MobiHoc 2008, pp. 83–90 (2008)
36. Xiao, M., Kloks, T., Poon, S.-H.: New parameterized algorithms for the edge dominating set problem. Theor. Comput. Sci. **511**, 147–158 (2013)
37. Yannakakis, M., Gavril, F.: Edge dominating sets in graphs. SIAM J. Appl. Math. **38**(3), 364–372 (1980)
38. Zhou, J., Zhang, Z., Wu, W., Xing, K.: A greedy algorithm for the fault-tolerant connected dominating set in a general graph. J. Comb. Optim. **28**(1), 310–319 (2014)

Graph Editing to a Given Degree Sequence

Petr A. Golovach[1,2(✉)] and George B. Mertzios[3]

[1] Department of Informatics, University of Bergen, Bergen, Norway
petr.golovach@uib.no
[2] Steklov Institute of Mathematics, Russian Academy of Sciences,
St.Petersburg, Russia
[3] School of Engineering and Computing Sciences, Durham University, Durham, UK
george.mertzios@durham.ac.uk

Abstract. We investigate the parameterized complexity of the graph editing problem called EDITING TO A GRAPH WITH A GIVEN DEGREE SEQUENCE where the aim is to obtain a graph with a given degree sequence σ by at most k vertex or edge deletions and edge additions. We show that the problem is W[1]-hard when parameterized by k for any combination of the allowed editing operations. From the positive side, we show that the problem can be solved in time $2^{O(k(\Delta+k)^2)}n^2 \log n$ for n-vertex graphs, where $\Delta = \max \sigma$, i.e., the problem is FPT when parameterized by $k+\Delta$. We also show that EDITING TO A GRAPH WITH A GIVEN DEGREE SEQUENCE has a polynomial kernel when parameterized by $k + \Delta$ if only edge additions are allowed, and there is no polynomial kernel unless NP $\subseteq$ coNP/poly for all other combinations of allowed editing operations.

1 Introduction

The aim of graph editing (or graph modification) problems is to modify a given graph by applying a bounded number of permitted operations in order to satisfy a certain property. Typically, vertex deletions, edge deletions and edge additions are the considered as the permitted editing operations, but in some cases other operations like edge contractions and vertex additions are also permitted.

We are interested in graph editing problems where the aim is to obtain a graph satisfying some given degree constraints. These problems usually turn out to be NP-hard (with rare exceptions). Hence, we are interested in the parameterized complexity of such problems. Before we state our results we briefly discuss the known related (parameterized) complexity results.

Related Work. The investigation of the parameterized complexity of the editing problems with degree constraints was initiated by Moser and Thilikos in [21]

The research leading to these results has received funding from the European Research Council under the European Union's Seventh Framework Programme (FP/2007-2013)/ERC Grant Agreement n. 267959, the Government of the Russian Federation grant 14.Z50.31.0030 and by the EPSRC Grant EP/K022660/1.

A.S. Kulikov and G.J. Woeginger (Eds.): CSR 2016, LNCS 9691, pp. 177–191, 2016.
DOI: 10.1007/978-3-319-34171-2_13

and Mathieson and Szeider [20]. In particular, Mathieson and Szeider [20] considered the DEGREE CONSTRAINT EDITING problem that asks for a given graph G, nonnegative integers d and k, and a function $\delta\colon V(G) \to 2^{\{0,\ldots,d\}}$, whether G can be modified into a graph G' such that the degree $d_{G'}(v) \in \delta(v)$ for each $v \in V(G')$, by using at most k editing operations. They classified the (parameterized) complexity of the problem depending on the set of allowed editing operations. In particular, they proved that if only edge deletions and additions are permitted, then the problem can be solved in polynomial time for the case where the set of feasible degrees $|\delta(v)| = 1$ for $v \in V(G)$. Without this restriction on the size of the sets of feasible degrees, the problem is NP-hard even on subcubic planar graphs whenever only edge deletions are allowed [9] and whenever only edge additions are allowed [15]. If vertex deletions can be used, then the problem becomes NP-complete and W[1]-hard with parameter k, even if the sets of feasible degrees have size one [20]. Mathieson and Szeider [20] showed that DEGREE CONSTRAINT EDITING is FPT when parameterized by $d+k$. They also proved that the problem has a polynomial kernel in the case where only vertex and edge deletions are allowed and the sets of feasible degrees have size one. Further kernelization results were obtained by Froese, Nichterlein and Niedermeier [15]. In particular, they proved that the problem with the parameter d admits a polynomial kernel if only edge additions are permitted. They also complemented these results by showing that there is no polynomial kernel unless NP $\subseteq$ coNP/poly if only vertex or edge deletions are allowed. Golovach proved in [17] that, unless NP $\subseteq$ coNP/poly, the problem does not admit a polynomial kernel when parameterized by $d + k$ if vertex deletion and edge addition are in the list of operations, even if the sets of feasible degrees have size one. The case where the input graph is planar was considered by Dabrowski et al. in [13]. Golovach [16] introduced a variant of DEGREE CONSTRAINT EDITING in which, besides the degree restrictions, it is required that the graph obtained by editing should be connected. This variant for planar input graphs was also considered in [13].

Froese, Nichterlein and Niedermeier [15] also considered the Π-DEGREE SEQUENCE COMPLETION problem which, given a graph G, a nonnegative integer k, and a property Π of graph degree sequences, asks whether it is possible to obtain a graph G' from G by adding at most k edges such that the degree sequence of G' satisfies Π. They gave some conditions when the problem is FPT/admits a polynomial kernel when parameterized by k and the maximum degree of G. There are numerous results (see, e.g., [4,8,11,12]) about the graph editing problem where the aim is to obtain a (connected) graph whose vertices satisfy some parity restrictions on their degree. In particular, if the obtained graph is required to be a connected graph with vertices of even degree, we obtain the classical EDITING TO EULERIAN GRAPH problem (see. [4,12]).

Another variant of graph editing with degree restrictions is the DEGREE ANONYMIZATION problem, motivated by some privacy and social networks applications. A graph G is h-anonymous for a positive integer h if for any $v \in V(G)$, there are at least $h - 1$ other vertices of the same degree. DEGREE ANONYMIZATION asks, given a graph G, a nonnegative h, and a positive integer

k, whether it is possible to obtain an h-anonymous graph by at most k editing operations. The investigation of the parameterized complexity of DEGREE ANONYMIZATION was initiated by Hartung et al. [18] and Bredereck et al. [6] (see also [5,19]). In particular, Hartung et al. [18] considered the case where only edge additions are allowed. They proved that the problem is W[1]-hard when parameterized by k, but it becomes FPT and has a polynomial kernel when parameterized by the maximum degree Δ of the input graph. Bredereck et al. [6] considered vertex deletions. They proved that the problem is W[1]-hard when parameterized by $h + k$, but it is FPT when parameterized by $\Delta + h$ or by $\Delta + k$. Also the problem was investigated for the cases when vertex additions [5] and edge contractions [19] are the editing operations.

Our Results. Recall that the *degree sequence* of a graph is the nonincreasing sequence of its vertex degrees. We consider the graph editing problem where the aim is to obtain a graph with a given *degree sequence* by using the operations *vertex deletion, edge deletion,* and *edge addition*, denoted by vd, ed, and ea, respectively. Formally, the problem is stated as follows. Let $S \subseteq \{vd, ed, ea\}$.

EDITING TO A GRAPH WITH A GIVEN DEGREE SEQUENCE
Instance: A graph G, a nonincreasing sequence of nonnegative integers σ and a nonnegative integer k.
Question: Is it possible to obtain a graph G' with the degree sequence σ from G by at most k operations from S?

It is worth highlighting here the difference between this problem and the EDITING TO A GRAPH OF GIVEN DEGREES problem studied in [15,17,20]. In EDITING TO A GRAPH OF GIVEN DEGREES, a function $\delta : V(G) \to \{1, \ldots, d\}$ is given along with the input and, in the target graph G', every vertex v is required to have the *specific* degree $\delta(v)$. In contrast, in the EDITING TO A GRAPH WITH A GIVEN DEGREE SEQUENCE, only a degree sequence is given with the input and the requirement is that the target graph G' has this degree sequence, without specifying which specific vertex has which specific degree. To some extend, this problem can be seen as a generalization of the DEGREE ANONYMIZATION problem [5,6,18,19], as one can specify (as a special case) the target degree sequence in such a way that every degree appears at least h times in it.

In practical applications with respect to privacy and social networks, we might want to appropriately "smoothen" the degree sequence of a given graph in such a way that it becomes difficult to distinguish between two vertices with (initially) similar degrees. In such a setting, it does not seem very natural to specify in advance a *specific* desired degree to every *specific* vertex of the target graph. Furthermore, for anonymization purposes in the case of a social network, where the degree distribution often follows a so-called power law distribution [2], it seems more natural to identify a smaller number of vertices having all the same "high" degree, and a greater number of vertices having all the same "small" degree, in contrast to the more modest h-anonymization requirement where *every* different degree must be shared among at least h identified vertices in the target graph.

In Sect. 2, we observe that for any nonempty $S \subseteq \{vd, ed, ea\}$, EDITING TO A GRAPH WITH A GIVEN DEGREE SEQUENCE is NP-complete and W[1]-hard when parameterized by k. Therefore, we consider a stronger parameterization by $k+\Delta$, where $\Delta = \max \sigma$. In Sect. 3, we show that EDITING TO A GRAPH WITH A GIVEN DEGREE SEQUENCE is FPT when parameterized by $k+\Delta$. In fact, we obtain this result for the more general variant of the problem, where we ask whether we can obtain a graph G' with the degree sequence σ from the input graph G by at most k_{vd} vertex deletions, k_{ed} edge deletions and k_{ea} edge additions. We show that the problem can be solved in time $2^{O(k(\Delta+k)^2)}n^2 \log n$ for n-vertex graphs, where $k = k_{vd}+k_{ed}+k_{ea}$. The algorithm uses the random separation techniques introduced by Cai, Chan and Chan [7] (see also [1]). First, we construct a true biased Monte Carlo algorithm, that is, a randomized algorithm whose running time is deterministic and that always returns a correct answer when it returns a yes-answer but can return a false negative answer with a certain (small) probability. Then we explain how it can be derandomized. In Sect. 4, we show that EDITING TO A GRAPH WITH A GIVEN DEGREE SEQUENCE has a polynomial kernel when parameterized by $k+\Delta$ if $S = \{ea\}$, but for all other nonempty $S \subseteq \{vd, ed, ea\}$, there is no polynomial kernel unless $NP \subseteq coNP/poly$.

2 Basic Definitions and Preliminaries

Graphs. We consider only finite undirected graphs without loops or multiple edges. The vertex set of a graph G is denoted by $V(G)$ and the edge set is denoted by $E(G)$.

For a set of vertices $U \subseteq V(G)$, $G[U]$ denotes the subgraph of G induced by U, and by $G - U$ we denote the graph obtained from G by the removal of all the vertices of U, i.e., the subgraph of G induced by $V(G) \setminus U$. If $U = \{u\}$, we write $G - u$ instead of $G - \{u\}$. Respectively, for a set of edges $L \subseteq E(G)$, $G[L]$ is a subgraph of G induced by L, i.e., the vertex set of $G[L]$ is the set of vertices of G incident to the edges of L, and L is the set of edges of $G[L]$. For a nonempty set U, $\binom{U}{2}$ is the set of unordered pairs of elements of U. For a set of edges L, by $G - L$ we denote the graph obtained from G by the removal of all the edges of L. Respectively, for $L \subseteq \binom{V(G)}{2}$, $G + L$ is the graph obtained from G by the addition of the edges that are elements of L. If $L = \{a\}$, then for simplicity, we write $G - a$ or $G + a$.

For a vertex v, we denote by $N_G(v)$ its *(open) neighborhood*, that is, the set of vertices which are adjacent to v, and for a set $U \subseteq V(G)$, $N_G(U) = (\bigcup_{v \in U} N_G(v)) \setminus U$. The *closed neighborhood* $N_G[v] = N_G(v) \cup \{v\}$, and for a positive integer r, $N_G^r[v]$ is the set of vertices at distance at most r from v. For a set $U \subseteq V(G)$ and a positive integer r, $N_G^r[U] = \bigcup_{v \in U} N_G^r[v]$. The *degree* of a vertex v is denoted by $d_G(v) = |N_G(v)|$. The *maximum degree* $\Delta(G) = \max\{d_G(v) \mid v \in V(G)\}$.

For a graph G, we denote by $\sigma(G)$ its degree sequence. Notice that $\sigma(G)$ can be represented by the vector $\delta(G) = (\delta_0, \ldots, \delta_{\Delta(G)})$, where $\delta_i = |\{v \in V(G) \mid d_G(v) = i\}|$ for $i \in \{0, \ldots, \Delta(G)\}$. We call $\delta(G)$ the *degree vector* of G.

For a sequence $\sigma = (\sigma_1, \ldots, \sigma_n)$, we define $\delta(\sigma) = (\delta_0, \ldots, \delta_r)$, where $r = \max \sigma$ and $\delta_i = |\{\sigma_j \mid \sigma_j = i\}|$ for $i \in \{0, \ldots, r\}$. Clearly, $\delta(G) = \delta(\sigma(G))$, and the degree vector can be easily constructed from the degree sequence and vice versa. Slightly abusing notation, we write for two vectors of nonnegative integers, that $(\delta_0, \ldots, \delta_r) = (\delta'_0, \ldots, \delta'_{r'})$ for $r \leq r'$ if $\delta_i = \delta'_i$ for $i \in \{0, \ldots, r\}$ and $\delta'_i = 0$ for $i \in \{r+1, \ldots, r'\}$.

Parameterized Complexity. Parameterized complexity is a two dimensional framework for studying the computational complexity of a problem. One dimension is the input size n and another one is a parameter k. It is said that a problem is *fixed parameter tractable* (or FPT), if it can be solved in time $f(k) \cdot n^{O(1)}$ for some function f. A *kernelization* for a parameterized problem is a polynomial algorithm that maps each instance (x, k) with the input x and the parameter k to an instance (x', k') such that (i) (x, k) is a YES-instance if and only if (x', k') is a YES-instance of the problem, and (ii) $|x'| + k'$ is bounded by $f(k)$ for a computable function f. The output (x', k') is called a *kernel*. The function f is said to be a *size* of a kernel. Respectively, a kernel is *polynomial* if f is polynomial. A decidable parameterized problem is FPT if and only if it has a kernel, but it is widely believed that not all FPT problems have polynomial kernels. In particular, Bodlaender et al. [3] introduced techniques that allow to show that a parameterized problem has no polynomial kernel unless $NP \subseteq coNP/poly$. We refer to the recent books of Cygan et al. [10] and Downey and Fellows [14] for detailed introductions to parameterized complexity.

Solutions of Editing to a Graph with a Given Degree Sequence. Let (G, σ, k) be an instance of EDITING TO A GRAPH OF GIVEN DEGREE SEQUENCE. Let $U \subset V(G)$, $D \subseteq E(G - U)$ and $A \subseteq \binom{V(G) \setminus U}{2}$. We say that (U, D, A) is a *solution* for (G, σ, k), if $|U| + |D| + |A| \leq k$, and the graph $G' = G - U - D + A$ has the degree sequence σ. We also say that G' is obtained by editing with respect to (U, D, A). If vd, ed or ea is not in S, then it is assumed that $U = \emptyset$, $D = \emptyset$ or $A = \emptyset$ respectively. If $S = \{ea\}$, then instead of $(\emptyset, \emptyset, A)$ we simply write A.

We conclude this section by observing that EDITING TO A GRAPH WITH A GIVEN DEGREE SEQUENCE is hard when parameterized by k.

Theorem 1. *For any nonempty $S \subseteq \{vd, ed, ea\}$, EDITING TO A GRAPH WITH A GIVEN DEGREE SEQUENCE is NP-complete and W[1]-hard when parameterized by k.*

3   FPT-algorithm for Editing to a Graph with a Given Degree Sequence

In this section we show that EDITING TO A GRAPH WITH A GIVEN DEGREE SEQUENCE is FPT when parameterized by $k + \Delta$, where $\Delta = \max \sigma$. In fact, we obtain this result for the more general variant of the problem:

EXTENDED EDITING TO A GRAPH WITH A GIVEN DEGREE SEQUENCE
Instance: A graph G, a nonincreasing sequence of nonnegative integers σ and a nonnegative integers k_{vd}, k_{ed}, k_{ea}.
Question: Is it possible to obtain a graph G' with $\sigma(G') = \sigma$ from G by at most k_{vd} vertex deletions, k_{ed} edge deletions and k_{ea} edge additions?

Notice that we can solve EDITING TO A GRAPH WITH A GIVEN DEGREE SEQUENCE using an algorithm for EXTENDED EDITING TO A GRAPH WITH A GIVEN DEGREE SEQUENCE by trying all possible values of k_{vd}, k_{ed} and k_{ea} with $k_{vd} + k_{ed} + k_{ea} = k$.

Theorem 2. EXTENDED EDITING TO A GRAPH WITH A GIVEN DEGREE SEQUENCE *can be solved in time* $2^{O(k(\Delta+k)^2)}n^2 \log n$ *for n-vertex graphs, where* $\Delta = \max \sigma$ *and* $k = k_{vd} + k_{ed} + k_{ea}$.

Proof. First, we construct a randomized true biased Monte Carlo FPT-algorithm for EXTENDED EDITING TO A GRAPH WITH A GIVEN DEGREE SEQUENCE parameterized by $k + \Delta$ based on the random separation techniques introduced by Cai, Chan and Chan [7] (see also [1]). Then we explain how this algorithm can be derandomized.   □

Let $(G, \sigma, k_{vd}, k_{ed}, k_{ea})$ be an instance of EXTENDED EDITING TO A GRAPH WITH A GIVEN DEGREE SEQUENCE, $n = |V(G)|$.

On the first stage of the algorithm we preprocess the instance to get rid of vertices of high degree or solve the problem if we have a trivial no-instance by the following reduction rule.

Vertex Deletion Rule. If G has a vertex v with $d_G(v) > \Delta + k_{vd} + k_{ed}$, then delete v and set $k_{vd} = k_{vd} - 1$. If $k_{vd} < 0$, then stop and return a NO-answer.

To show that the rule is *safe*, i.e., by the application of the rule we either correctly solve the problem or obtain an equivalent instance, assume that $(G, \sigma, k_{vd}, k_{ed}, k_{ea})$ is a yes-instance of EXTENDED EDITING TO A GRAPH WITH A GIVEN DEGREE SEQUENCE. Let (U, D, A) be a solution. We show that if $d_G(v) > \delta + k_{vd} + k_{ed}$, then $v \in U$. To obtain a contradiction, assume that $d_G(v) > \delta + k_{vd} + k_{ed}$ but $v \notin U$. Then $d_{G'}(v) \leq \Delta$, where $G' = G - U - D + A$. It remains to observe that to decrease the degree of v by at least $k_{vd} + k_{ed} + 1$, we need at least $k_{vd} + k_{ed} + 1$ vertex or edge deletion operations; a contradiction. We conclude that if $(G, \sigma, k_{vd}, k_{ed}, k_{ea})$ is a yes-instance, then the instance obtained by the application of the rule is also a yes-instance. It is straightforward to see that if $(G', \sigma, k'_{vd}, k_{ed}, k_{ea})$ is a yes-instance of EXTENDED EDITING TO A GRAPH WITH A GIVEN DEGREE SEQUENCE obtained by the deletion of a vertex v and (U, D, A) is a solution, then $(U \cup \{v\}, D, A)$ is a solution for the original instance. Hence, the rule is safe.

We exhaustively apply the rule until we either stop and return a NO-answer or obtain an instance of the problem such that the degree of any vertex v is at most $\Delta + k$. To simplify notations, we assume that $(G, \sigma, k_{vd}, k_{ed}, k_{ea})$ is such an instance.

On the next stage of the algorithm we apply the random separation technique. We color the vertices of G independently and uniformly at random by three colors. In other words, we partition $V(G)$ into three sets R_v, Y_v and B_v (some sets could be empty), and say that the vertices of R_v are *red*, the vertices of Y_v are *yellow* and the vertices of B_v are *blue*. Then the edges of G are colored independently and uniformly at random by either *red* or *blue*. We denote by R_e the set of red and by B_e the set of blue edges respectively.

We are looking for a solution (U, D, A) of $(G, \sigma, k_{vd}, k_{ed}, k_{ea})$ such that the vertices of U are colored red, the vertices incident to the edges of A are yellow and the edges of D are red. Moreover, if X and Y are the sets of vertices incident to the edges of D and A respectively, then the vertices of $(N_G^2[U] \cup N_G[X \cup Y]) \setminus (U \cup Y)$ and the edges of $E(G) \setminus D$ incident to the vertices of $N_G[U] \cup X \cup Y$ should be blue. Informally speaking, the elements of a solution should be marked red in the case of deleted vertices and edges, and the end-vertices of added edges should be marked yellow. Then to separate the elements of a solution, we demand that the vertices and edges that are sufficiently close to it but not included in a solution should be blue. Formally, we say that a solution (U, D, A) of $(G, \sigma, k_{vd}, k_{ed}, k_{ea})$ is a *colorful* solution if there are $R_v^* \subseteq R_v$, $Y_v^* \subseteq Y_v$ and $R_e^* \subseteq R_e$ such that the following holds.

(i) $|R_v^*| \leq k_{vd}$, $|R_e^*| \leq k_{ed}$ and $|Y_v^*| \leq 2k_{ea}$.
(ii) $U = R_v^*$, $D = R_e^*$, and for any $uv \in A$, $u, v \in Y_v^*$ and $|A| \leq k_{ea}$.
(iii) If $u, v \in R_v \cup Y_v$ and $uv \in E(G)$, then either $u, v \in R_v^* \cup Y_v^*$ or $u, v \notin R_v^* \cup Y_v^*$.
(iv) If $u \in R_v \cup Y_v$ and $uv \in R_e$, then either $u \in R_v^* \cup Y_v^*$, $uv \in R_e^*$ or $u \notin R_v^* \cup Y_v^*$, $uv \notin R_e^*$.
(v) If $uv, vw \in R_e$, then either $uv, vw \in R_e^*$ or $uv, vw \notin R_e^*$.
(vi) If distinct $u, v \in R_v$ and $N_G(u) \cap N_G(v) \neq \emptyset$, then either $u, v \in R_v^*$ or $u, v \notin R_v^*$.
(vii) If $u \in R_v$ and $vw \in R_e$ for $v \in N_G(u)$, then either $u \in R_v^*$, $vw \in R_e^*$ or $u \notin R_v^*$, $vw \notin R_e^*$.

We also say that (R_v^*, Y_v^*, R_e^*) is the *base* of (U, D, A).

Our aim is to find a colorful solution if it exists. We do is by a dynamic programming algorithm based on the following properties of colorful solutions.

Let

$$L = R_e \cup \{e \in E(G) \mid e \text{ is incident to a vertex of } R_v\} \cup \{uv \in E(G) \mid u, v \in Y_v\},$$

and $H = G[L]$. Denote by $H_1, \ldots, H_s$ the components of H. Let $R_v^i = V(H_i) \cap R_e$, $Y_v^i = V(H_i) \cap Y_v$ and $R_e^i = E(H_i) \cap R_e$ for $i \in \{1, \ldots, s\}$.

Claim A. *If (U, D, A) is a colorful solution and (R_v^*, Y_v^*, R_e^*) is its base, then if H_i has a vertex of $R_v^* \cup Y_v^*$ or an edge of R_e^*, then $R_v^i \subseteq R_v^*$, $Y_v^i \subseteq Y_v^*$ and $R_e^i \subseteq R_r^*$ for $i \in \{1, \ldots, s\}$.*

Proof (of Claim A). Suppose that H_i has $u \in R_v^* \cup Y_v^*$ or $e \in R_e^*$.

If $v \in R_v^i \cup Y_v^i$, then H_i has a path $P = x_0 \ldots x_\ell$ such that $u = x_0$ or $e = x_0 x_1$, and $x_\ell = v$. By induction on ℓ, we show that $v \in R_v^*$ or $v \in Y_v^*$ respectively. If

$\ell = 1$, then the statement follows from (iii) and (iv) of the definition of a colorful solution. Suppose that $\ell > 1$. We consider three cases.

Case 1. $x_1 \in R_v \cup Y_v$. By (iii) and (iv), $x_1 \in R_v^* \cup Y_v^*$ and, because the (x_1, x_ℓ)-subpath of P has length $\ell - 1$, we conclude that $v \in R_v^*$ or $v \in Y_v^*$ by induction.

Assume from now that $x_1 \notin R_v \cup Y_v$.

Case 2. $x_0 x_1 \in R_e$. Clearly, if for the first edge e of P, $e \in R_e^*$, then $x_0 x_1 = e \in R_e^*$. Suppose that for the first vertex $u = x_0$ of P, $u \in R_v^* \cup Y_v^*$. Then by (iv), $x_0 x_1 \in R_e^*$. If $x_1 x_2 \in R_e$, then $x_1 x_2 \in R_e^*$ by (v). Since $x_1 x_2 \in R_e^*$ and the (x_1, x_ℓ)-subpath of P has length $\ell - 1$, we have that $v \in R_v^*$ or $v \in Y_v^*$ by induction. Suppose that $x_1 x_2 \notin R_e$. Then because $x_1 x_2 \in L$, $x_2 \in R_v$ and by (vii), $x_2 \in R_v^*$. If $\ell = 2$, then $x_\ell \in R_v^*$. Otherwise, as the (x_2, x_ℓ)-subpath of P has length $\ell - 2$, we have that $v \in R_v^*$ or $v \in Y_v^*$ by induction.

Case 3. $x_0 x_1 \notin R_e$. Then $u = x_0 \in R_v^* \cup Y_v^*$. Because $x_0 x_1 \in L$, $x_0 \in R_v^*$. If $x_1 x_2 \in R_e$, then $x_1 x_2 \in R_e^*$ by (vii). Since $x_1 x_2 \in R_e^*$ and the (x_1, x_ℓ)-subpath of P has length $\ell - 1$, we have that $v \in R_v^*$ or $v \in Y_v^*$ by induction. Suppose that $x_1 x_2 \notin R_e$. Then because $x_1 x_2 \in L$, $x_2 \in R_v$ and by (vi), $x_2 \in R_v^*$. If $\ell = 2$, then $x_\ell \in R_v^*$. Otherwise, as the (x_2, x_ℓ)-subpath of P has length $\ell - 2$, we have that $v \in R_v^*$ or $v \in Y_v^*$ by induction.

Suppose that $e' \in R_e^i$. Then H_i has a path $P = x_0 \ldots x_\ell$ such that $u = x_0$ or $e = x_0 x_1$, and $x_{\ell-1} x_\ell = e'$. Using the same inductive arguments as before, we obtain that $e' \in R_e^*$. ∎

By Claim A, we have that if there is a colorful solution (U, D, A), then for its base (R_v^*, Y_v^*, R_e^*), $R_v^* = \bigcup_{i \in I} R_v^i$, $Y_v^* = \bigcup_{i \in I} Y_v^i$ and $R_e^* = \bigcup_{i \in I} R_e^i$ for some set of indices $I \subseteq \{1, \ldots, s\}$.

The next property is a straightforward corollary of the definition of H.

Claim B. *For distinct* $i, j \in \{1, \ldots, s\}$, *if* $u \in V(H_i)$ *and* $v \in V(H_j)$ *are adjacent in* G, *then either* $u, v \in B_v$ *or* $(u \in Y_v^i$ *and* $v \in B_v)$ *or* $(u \in B_v$ *and* $v \in Y_v^j)$.

We construct a dynamic programming algorithm that consecutively for $i = 0, \ldots, s$, constructs the table T_i that contains the records of values of the function γ:

$$\gamma(t_{vd}, t_{ed}, t_{ea}, X, \delta) = (U, D, A, I),$$

where

(i) $t_{vd} \leq k_{vd}$, $t_{ed} \leq k_{ed}$ and $t_{ea} \leq k_{ea}$,

(ii) $X = \{d_1, \ldots, d_h\}$ is a collection (multiset) of integers, where $h \in \{1, \ldots, 2t_{ea}\}$ and $d_i \in \{0, \ldots, \Delta\}$ for $i \in \{1, \ldots, h\}$,

(iii) $\delta = (\delta_0, \ldots, \delta_r)$, where $r = \max\{\Delta, \Delta(G)\}$ and δ_i is a nonnegative integer for $i \in \{0, \ldots, r\}$,

such that (U, D, A) is a *partial solution* with the base (R_v^*, Y_v^*, R_e^*) defined by $I \subseteq \{1, \ldots, i\}$ with the following properties.

(iv) $R_v^* = \bigcup_{i \in I} R_v^i$, $Y_v^* = \bigcup_{i \in I} Y_v^i$ and $R_e^* = \bigcup_{i \in I} R_e^i$, and $t_{vd} = |R_v^*|$ and $t_{ed} = |R_e^*|$.

(v) $U = R_v^*$, $D = R_e^*$, $|A| = t_{ea}$ and for any $uv \in A$, $u, v \in Y_v^*$.

(vi) The multiset $\{d_{G'}(y) \mid y \in Y_v^*\} = X$, where $G' = G - U - D + A$.

(vii) $\delta(G') = \delta$.

In other words, t_{vd}, t_{ed} and t_{ea} are the numbers of deleted vertices, deleted edges and added edges respectively, X is the multiset of degrees of yellow vertices in the base of a partial solution, and δ is the degree vector of the graph obtained from G by the editing with respect to a partial solution. Notice that the values of γ are defined only for some $t_{vd}, t_{ed}, t_{ea}, X, \delta$ that satisfy (i)–(iii), as a partial solution with the properties (iv)–(vii) not necessarily exists, and we only keep records corresponding to the arguments $t_{vd}, t_{ed}, t_{ea}, X, \delta$ for which γ is defined.

Now we explain how we construct the tables for $i \in \{0, \ldots, s\}$.

Construction of T_0. The table T_0 contains the unique record $(0, 0, 0, \emptyset, \delta) = (\emptyset, \emptyset, \emptyset, \emptyset)$, where $\delta = \delta(G)$ (notice that the length of δ can be bigger than the length of $\delta(G)$).

Construction of T_i for $i \geq 1$. We assume that T_{i-1} is already constructed. Initially we set $T_i = T_{i-1}$. Then for each record $\gamma(t_{vd}, t_{ed}, t_{ea}, X, \delta) = (U, D, A, I)$ in T_{i-1}, we construct new records $\gamma(t'_{vd}, t'_{ed}, t'_{ea}, X', \delta') = (U', D', A')$ and put them in T_i unless T_i already contains the value $\gamma(t'_{vd}, t'_{ed}, t'_{ea}, X', \delta')$. In the last case we keep the old value.

Let $(t_{vd}, t_{ed}, t_{ea}, X, \delta) = (U, D, A, I)$ in T_{i-1}.

- If $t_{vd} + |R_v^i| > k_{vd}$ or $t_{ed} + |R_e^i| > k_{ed}$ or $t_{ea} + 2|Y_v^i| > k_{ea}$, then stop considering the record. Otherwise, let $t'_{vd} = t_{vd} + |R_v^i|$ and $t'_{ed} = t_{ed} + |R_e^i|$.
- Let $F = G - U - D + A - R_v^i - R_e^i$.
- Let $\bigcup_{j \in I} Y_v^j = \{x_1, \ldots, x_h\}$, $d_F(x_f) = d_f$ for $f \in \{1, \ldots, h\}$. Let $Y_v^i = \{y_1, \ldots, y_\ell\}$. Consider every $E_1 \subseteq \binom{Y_v^i}{2} \setminus E(F[Y_v^i])$ and $E_2 \subseteq \{x_f y_i \mid 1 \leq f \leq h, 1 \leq j \leq \ell\}$ such that $|E_1| + |E_2| \leq k_{ea} - t_{ea}$, and set $\alpha_f = |\{x_f y_j \mid x_f y_j \in E_2, 1 \leq j \leq \ell\}|$ for $f \in \{1, \ldots, h\}$ and set $\beta_j = |\{e \mid e \in E_1, e \text{ is incident to } y_j\}| + |\{x_f y_j \mid x_f y_j \in E_2, 1 \leq f \leq h\}|$ for $j \in \{1, \ldots, \ell\}$.
  - If $d_f + \alpha_f > \Delta$ for some $f \in \{1, \ldots, h\}$ or $d_F(y_j) + \beta_j > \Delta$ for some $j \in \{1, \ldots, \ell\}$, then stop considering the pair (E_1, E_2).
  - Set $t'_{ea} = t_{ea} + |E_1| + |E_2|$, $X' = \{d_1 + \alpha_1, \ldots, d_h + \alpha_h, d_F(y_1) + \beta_1, \ldots, d_F(y_\ell) + \beta_\ell\}$.
  - Let $F' = F + E_1 + E_2$. Construct $\delta' = (\delta'_0, \ldots, \delta'_r) = \delta(F')$.
  - Set $U' = U \cup R_v^i$, $D' = D \cup R_e^i$, $A' = A \cup E_1 \cup E_2$, $I' = I \cup \{i\}$, set $\gamma(t'_{vd}, t'_{ed}, t'_{ea}, X', \delta') = (U', D', A', I')$ and put the record in T_i.

We consecutively construct $T_1, \ldots, T_s$. The algorithm returns a YES-answer if T_s contains a record $(t_{vd}, t_{ed}, t_{ea}, X, \delta) = (U, D, A, I)$ for $\delta = \delta(\sigma)$ and (U, D, A) is a colorful solution in this case. Otherwise, the algorithm returns a NO-answer.

The correctness of the algorithm follows from the next claim.

Claim C. *For each $i \in \{1, \ldots, s\}$, the table T_i contains a record $\gamma(t_{vd}, t_{ed}, t_{ea}, X, \delta) = (U, D, A, I)$, if and only if there are $t_{vd}, t_{ed}, t_{ea}, X, \delta$ satisfying (i)-(iii) such that there is a partial solution (U^*, D^*, A^*) and $I^* \subseteq \{1, \ldots, i\}$ that satisfy (iv)-(vii). In particular $t_{vd}, t_{ed}, t_{ea}, X, \delta$, (U, D, A) and I satisfy (i)-(vii) if $\gamma(t_{vd}, t_{ed}, t_{ea}, X, \delta) = (U, D, A, I)$ is in T_i.*

Proof (of Claim C). We prove the claim by induction on i. It is straightforward to see that it holds for $i = 0$. Assume that $i > 0$ and the claim is fulfilled for T_{i-1}.

Suppose that a record $\gamma(t'_{vd}, t'_{ed}, t'_{ea}, X', \delta') = (U', D', A', I')$ was added in T_i. Then ether $\gamma(t'_{vd}, t'_{ed}, t'_{ea}, X', \delta') = (U', D', A', I')$ was in T_{i-1} or it was constructed for some record $(t_{vd}, t_{ed}, t_{ea}, X, \delta) = (U, D, A, I)$ from T_{i-1}. In the first case, $t'_{vd}, t'_{ed}, t'_{ea}, X', Q'$, (U', D', A') and $I' \subseteq \{1, \ldots, i\}$ satisfy (i)-(vii) by induction. Assume that $\gamma(t'_{vd}, t'_{ed}, t'_{ea}, X', \delta') = (U', D', A', I')$ was constructed for some record $(t_{vd}, t_{ed}, t_{ea}, X, Q) = (U, D, A, I)$ from T_{i-1}. Notice that $i \in I'$ in this case. Let $I = I' \setminus \{i\}$. Consider $\bigcup_{j \in I} Y_v^j = \{x_1, \ldots, x_h\}$ and $Y_v^i = \{y_1, \ldots, y_\ell\}$. By Claim B, x_f and y_j are not adjacent for $f \in \{1, \ldots, h\}$ and $j \in \{1, \ldots, \ell\}$. Then it immediately follows from the description of the algorithm that $t'_{vd}, t'_{ed}, t'_{ea}, X', \delta', (U', D', A')$ and I' satisfy (i)-(vii).

Suppose that there are $t_{vd}, t_{ed}, t_{ea}, X, \delta$ satisfying (i)-(iii) such that there is a partial solution (U^*, D^*, A^*) and $I^* \subseteq \{1, \ldots, i\}$ that satisfy (iv)-(vii). Suppose that $i \notin I^*$. Then T_{i-1} contains a record $\gamma(t_{vd}, t_{ed}, t_{ea}, X, \delta) = (U, D, A, I)$ by induction and, therefore, this record is in T_i. Assume from now that $i \in I^*$. Let $I' = I^* \setminus \{i\}$. Consider $R_v^i = \bigcup_{j \in I'} R_v^j$ and $Y_v' = \bigcup_{j \in I'} Y_v^j$. Let $E_1 = \{uv \in A \mid u, v \in T_v^i\}$ and $E_2 = \{uv \in A \mid u \in Y_v', v \in Y_v^i\}$. Define $U' = U \setminus R_v^i$, $D' = D \setminus R_e^i$ and $A' = A \setminus (E_1 \cup E_2)$. Let $t'_{vd} = |U'|$, $t_{ed} = |D'|$ and $t_{ea} = |A'|$. Consider the multiset of integers $X' = \{d_F(v) \mid v \in Y_v'\}$ and the sequence $\delta' = (\delta'_1, \ldots, \delta'_r) = \delta(F)$ for $F = G - U' - D' + A'$. We obtain that $t'_{vd}, t'_{ed}, t'_{ea}, X', \delta', (U', D', A')$ and $I' \subseteq \{1, \ldots, i-1\}$ satisfy (i)-(vii). By induction, T_{i-1} contains a record $\gamma(t'_{vd}, t'_{ed}, t'_{ea}, X', \delta') = (U'', D'', A'', I'')$. Let $Y_v' = \{x_1, \ldots, x_h\}$, $\bigcup_{j \in I''} Y_v^j = \{x'_1, \ldots, x'_h\}$ and assume that $d_F(x_f) = d_{F'}(x'_f)$ for $f \in \{1, \ldots, h\}$, where $F' = G - U'' - D'' + A''$. Consider E_2' obtained from E_2 by the replacement of every edge $x_f v$ by $x'_f v$ for $f \in \{1, \ldots, h\}$ and $v \in Y_v^i$. It remains to observe that when we consider $\gamma(t'_{vd}, t'_{ed}, t'_{ea}, X', \delta') = (U'', D'', A'', I'')$ and the pair (E_1, E_2'), we obtain $\gamma(t_{vd}, t_{ed}, t_{ea}, X, \delta) = (U, D, A, I)$ for $U = U'' \cup R_v^i$, $D = D'' \cup R_e^i$, $A = A'' \cup E_1 \cup E_2'$ and $I = I'' \cup \{i\}$.   ∎

Now we evaluate the running time of the dynamic programming algorithm. First, we upper bound the size of each table. Suppose that $\gamma(t_{vd}, t_{ed}, t_{ea}, X, \delta) = (U, D, A, I)$ is included in a table T_i. By the definition and Claim C, $\delta = \delta(G')$ for $G' = G - U - D + A$. Let $\delta = \{\delta_0, \ldots, \delta_r\}$ and $\delta(G) = (\delta'_0, \ldots, \delta'_r)$. Let $i \in \{0, \ldots, r\}$. Denote $W_i = \{v \in V(G) \mid d_G(v) = i\}$. Recall that $\delta(G) \leq \Delta + k$. If $\delta'_i > \delta_i$, then at least $\delta'_i - \delta_i$ vertices of W_i should be either deleted or get modified degrees by the editing with respect to (U, D, A). Since at most k_{vd} vertices of W_i can be deleted and we can modify degrees of at most $(k + \Delta)k_{vd} + 2(k_{ed} + k_{ea})$ vertices, $\delta'_i - \delta_i \leq (k + \Delta + 1)k_{vd} + 2(k_{ed} + k_{ea})$.

Similarly, if $\delta_i > \delta_i'$, then at least $\delta_i - \delta_i'$ vertices of $V(G) \setminus W_i$ should get modified degrees. Since we can modify degrees of at most $(k+\Delta)k_{vd}+2(k_{ed}+k_{ea})$ vertices, $\delta_i - \delta_i' \leq (k + \Delta)k_{vd} + 2(k_{ed} + k_{ea})$. We conclude that for each $i \in \{0, \ldots, r\}$,

$$\delta_i' - (k + \Delta + 1)k_{vd} + 2(k_{ed} + k_{ea}) \leq \delta_i \leq \delta_i' + (k + \Delta)k_{vd} + 2(k_{ed} + k_{ea})$$

and, therefore, there are at most $(2(k + \Delta)k_{vd} + 4(k_{ed} + k_{ea}) + 1)^r$ distinct vectors δ. Since $r = \max\{\Delta, \Delta(G)\} \leq \Delta + k$, we have $2^{O((\Delta+k)\log(\Delta+k))}$ distinct vectors δ. The number of distinct multisets X is at most $(\Delta + 1)^{2k}$ and there are at most $3(k + 1)$ possibilities for t_{vd}, t_{ed}, t_{ea}. We conclude that each T_i has $2^{O((\Delta+k)\log(\Delta+k))}$ records.

To construct a new record $\gamma(t_{vd}', t_{ed}', t_{ea}', X', \delta') = (U', D', A', I')$ from $\gamma(t_{vd}, t_{ed}, t_{ea}, X, \delta) = (U, D, A, I)$ we consider all possible choices of E_1 and E_2. Since these edges have their end-vertices in a set of size at most $2k_{ea}$ and $|E_1| + |E_2| \leq k_{ea}$, there are $2^{O(k \log k)}$ possibilities to choose E_1 and E_2. The other computations in the construction of $\gamma(t_{vd}', t_{ed}', t_{ea}', X', \delta') = (U', D', A', I')$ can be done in linear time. We have that T_i can be constructed from T_{i-1} in time $2^{O((\Delta+k)\log(\Delta+k))} \cdot n$ for $i \in \{1, \ldots, s\}$. Since $s \leq n$, the total time is $2^{O((\Delta+k)\log(\Delta+k))} \cdot n^2$.

We proved that a colorful solution can be found in time $2^{O((\Delta+k)\log(\Delta+k))} \cdot n^2$ if it exists. Clearly, any colorful solution is a solution for $(G, \sigma, k_{vd}, k_{ed}, k_{ea})$ and we can return it, but nonexistence of a colorful solution does not imply that there is no solution. Hence, to find a solution, we run the randomized algorithm N times, i.e., we consider N random colorings and try to find a colorful solution for them. If we find a solution after some run, we return it and stop. If we do not obtain a solution after N runs, we return a NO-answer. The next claim shows that it is sufficient to run the algorithm $N = 6^{2k(\Delta+k)^2}$ times.

Claim D. *There is a positive p that does not depend on the instance such that if the randomized algorithm has not found a solution for $(G, \sigma, k_{vd}, k_{ed}, k_{ea})$ after $N = 6^{2k(\Delta+k)^2}$ executions, then the probability that $(G, \sigma, k_{vd}, k_{ed}, k_{ea})$ is a no-instance is at least p.*

Proof (of Claim D). Suppose that $(G, \sigma, k_{vd}, k_{ed}, k_{ea})$ has a solution (U, D, A). Let X be the set of end-vertices of the edges of D and Y is the set of end-vertices of A. Let $W = N_G^2[U] \cup N_G[X \cup Y]$ and denote by L the set of edges incident to the vertices of $N_G[U] \cup X \cup Y$. The algorithm colors the vertices of G independently and uniformly at random by three colors and the edges are colored by two colors. Notice that if the vertices of W and the edges of L are colored correctly with respect to the solution, i.e., the vertices of U are red, the vertices of Y are yellow, all the other vertices are blue, the edges of D are red and all the other edges are blue, then (U, D, A) is a colorful solution. Hence, the algorithm can find a solution in this case.

We find a lower bound for the probability that the vertices of W and the edges of L are colored correctly with respect to the solution. Recall that $\Delta(G) \leq \Delta + k$. Hence, $|W| \leq k_{vd}(\Delta + k)^2 + 2(k_{ed} + k_{ea})(\Delta + k) \leq 2k(\Delta + k)^2$ and $|L| \leq k_{vd}(\Delta + k)^2 + 2(k_{ed} + k_{ea})(\Delta + k) \leq 2k(\Delta + k)^2$. As the vertices are colored by three colors and the edges by two, we obtain that the probability that the vertices of W and the edges of L are colored correctly with respect to the solution is at least $3^{-2k(\Delta+k)^2} \cdot 2^{-2k(\Delta+k)^2} = 6^{-2k(\Delta+k)^2}$.

The probability that the vertices of W and the edges of L are not colored correctly with respect to the solution is at most $1 - 6^{-2k(\Delta+k)^2}$, and the probability that these vertices are not colored correctly with respect to the solution for neither of $N = 6^{2k(\Delta+k)^2}$ random colorings is at most $(1 - 1/N)^N$, and the claim follows.   ∎

Claim D implies that the running time of the randomized algorithm is $2^{O(k(\Delta+k)^2)} \cdot n^2$.

The algorithm can be derandomized by standard techniques (see [1,7]) because random colorings can be replaced by the colorings induced by *universal sets*. Let m and r be positive integers, $r \leq m$. An (m, r)-*universal set* is a collection of binary vectors of length m such that for each index subset of size r, each of the 2^r possible combinations of values appears in some vector of the set. It is known that an (m, r)-universal set can be constructed in FPT-time with the parameter r. The best construction is due to Naor, Schulman and Srinivasan [22]. They obtained an (m, r)-universal set of size $2^r \cdot r^{O(\log r)} \log m$, and proved that the elements of the sets can be listed in time that is linear in the size of the set.

In our case we have $m = |V(G)| + |E(G)| \leq ((\Delta + k)/2 + 1)n$ and $r = 4k(\Delta + k)^2$, as we have to obtain the correct coloring of W and L corresponding to a solution (U, D, A). Observe that colorings induced by a universal set are binary and we use three colors. To fix it, we assume that the coloring of the vertices and edges is done in two stages. First, we color the elements of G by two colors: red and green, and then recolor the green elements by yellow or blue. By using an (m, r)-universal set of size $2^r \cdot r^{O(\log r)} \log m$, we get $4^r \cdot r^{O(\log r)} \log m$ colorings by three colors. We conclude that the running time of the derandomized algorithm is $2^{O(k(\Delta+k)^2)} \cdot n^2 \log n$.   □

4   Kernelization for Editing to a Graph with a Given Degree Sequence

In this section we show that EDITING TO A GRAPH WITH A GIVEN DEGREE SEQUENCE has a polynomial kernel when parameterized by $k + \Delta$ if $S = \{ea\}$, but for all other nonempty $S \subseteq \{vd, ed, ea\}$, there is no polynomial kernel unless NP $\subseteq$ coNP/poly.

Theorem 3. *If $S = \{ea\}$, then* EDITING TO A GRAPH WITH A GIVEN DEGREE SEQUENCE *parameterized by $k + \Delta$ has a kernel with $O(k\Delta^2)$ vertices, where $\Delta = \max \sigma$.*

Proof. Let (G, σ, k) be an instance of EDITING TO A GRAPH WITH A GIVEN DEGREE SEQUENCE and $\Delta = \max \sigma$. If $\Delta(G) > \Delta$, (G, σ, k) is a no-instance, because by edge additions it is possible only to increase degrees. Hence, we immediately stop and return a NO-answer in this case. Assume from now that

$\Delta(G) \leq \Delta$. For $i \in \{0, \ldots, \Delta\}$, denote $W_i = \{v \in V(G) \mid d_G(v) = i\}$ and $\delta_i = |W_i|$. Let $s_i = \min\{\delta_i, 2k(\Delta + 1)\}$ and let $W_i' \subseteq W_i$ be an arbitrary set of size s_i for $i \in \{0, \ldots, \Delta\}$. We consider $W = \bigcup_{i=0}^{\Delta} W_i'$ and prove the following claim.

Claim A. *If* (G, σ, k) *is a yes-instance of* EDITING TO A GRAPH WITH A GIVEN DEGREE SEQUENCE, *then there is* $A \subseteq \binom{V(G)}{2} \setminus E(G)$ *such that* $\sigma(G + A) = \sigma$, $|A| \leq k$ *and for any* $uv \in A$, $u, v \in W$.

Proof (of Claim A). Suppose that $A \subseteq \binom{V(G)}{2} \setminus E(G)$ is a solution for (G, σ, k), i.e., $\sigma(G + A) = \sigma$ and $|A| \leq k$, such that the total number of end-vertices of the edges of A in $V(G) \setminus W$ is minimum. Suppose that there is $i \in \{0, \ldots, \Delta\}$ such that at least one edge of A has at least one end-vertex in $W_i \setminus W_i'$. Clearly, $s_i = 2k(\Delta + 1)$. Denote by $\{x_1, \ldots, x_p\}$ the set of end-vertices of the edges of A in W_i and let $\{y_1, \ldots, y_q\}$ be the set of end-vertices of the edges of A in $V(G) \setminus W_i$. Since $p + q \leq 2k$, $\Delta(G) \leq \Delta$ and $s_i = 2k(\Delta + 1)$, there is a set of vertices $\{x_1', \ldots, x_p'\} \subseteq W_i'$ such that the vertices of this set are pairwise nonadjacent and are not adjacent to the vertices of $\{y_1, \ldots, y_q\}$. We construct $A' \subseteq \binom{V(G)}{2} \setminus E(G)$ by replacing every edge $x_i y_j$ by $x_i' y_j$ for $i \in \{1, \ldots, p\}$ and $j \in \{1, \ldots, q\}$, and every edge $x_i x_j$ is replaced by $x_i' x_j'$ for $i, j \in \{1, \ldots, p\}$. It is straightforward to verify that A' is a solution for (G, σ, k), but A' has less end-vertices outside W contradicting the choice of A. Hence, no edge of A has an end-vertex in $V(G) \setminus W$. ∎

If $\delta_i \leq 2k(\Delta + 1)$ for $i \in \{0, \ldots, \Delta\}$, then we return the original instance (G, σ, k) and stop, as $|V(G)| \leq 2k(\Delta + 1)^2$. From now we assume that there is $i \in \{0, \ldots, \Delta\}$ such that $\delta_i > 2k(\Delta + 1)$. We construct the graph G' as follows.

- Delete all the vertices of $V(G) \setminus W$.
- Construct $h = \Delta + 2$ new vertices $v_1, \ldots, v_h$ and join them by edges pairwise to obtain a clique.
- For any $u \in W$ such that $r = |N_G(u) \cap (V(G) \setminus W)| \geq 1$, construct edges $uv_1, \ldots, uv_r$.

Notice that $d_{G'}(v_1) \geq \ldots \geq d_{G'}(v_h) \geq \Delta + 1$ and $d_{G'}(u) = d_G(u)$ for $u \in W$. Now we consider the sequence σ and construct the sequence σ' as follows.

- The first h elements of σ' are $d_{G'}(v_1), \ldots, d_{G'}(v_h)$.
- Consider the elements of σ in their order and for each integer $i \in \{0, \ldots, \Delta\}$ that occurs j_i times in σ, add $j_i - (\delta_i - s_i)$ copies of i in σ'.

We claim that (G, σ, k) is a yes-instance of EDITING TO A GRAPH WITH A GIVEN DEGREE SEQUENCE if and only if (G', σ', k) is a yes-instance of the problem.

Suppose that (G, σ, k) is a yes-instance of EDITING TO A GRAPH WITH A GIVEN DEGREE SEQUENCE. By Claim A, it has a solution $A \subseteq \binom{V(G)}{2} \setminus E(G)$ such that for any $uv \in A$, $u, v \in W$. It is straightforward to verify that $\sigma(G' + A) = \sigma'$, i.e., A is a solution for (G', σ', k). Assume that $A \subseteq \binom{V(G')}{2} \setminus E(G)$ is a solution for (G', σ', k). Because $d_{G'}(v_1), \ldots, d_{G'}(v_h)$ are the first h elements of σ' and $d_{G'}(u) = d_G(u) \leq \Delta$ for $u \in W$, for any $uv \in A$, $u, v \in W$. Then it is straightforward to check that $\sigma(G + A) = \sigma$, i.e., A is a solution for (G, σ, k). $\square$

We complement Theorem 3 by showing that it is unlikely that EDITING TO A GRAPH WITH A GIVEN DEGREE SEQUENCE parameterized by $k + \Delta$ has a polynomial kernel for $S \neq \{ea\}$. The proof is based on the cross-composition technique introduced by Bodlaender, Jansen and Kratsch [3].

Theorem 4. *If nonempty $S \subseteq \{vd, ed, ea\}$ but $S \neq \{ea\}$, then* EDITING TO A GRAPH WITH A GIVEN DEGREE SEQUENCE *has no polynomial kernel unless* NP $\subseteq$ coNP/poly *when the problem is parameterized by $k + \Delta$ for $\Delta = \max \sigma$.*

References

1. Alon, N., Yuster, R., Zwick, U.: Color-coding. J. ACM **42**(4), 844–856 (1995)
2. Barabasi, A.L., Albert, R.: Emergence of scaling in random networks. Science **286**(5439), 509–512 (1999)
3. Bodlaender, H.L., Jansen, B.M.P., Kratsch, S.: Kernelization lower bounds by cross-composition. SIAM J. Discrete Math. **28**(1), 277–305 (2014)
4. Boesch, F.T., Suffel, C.L., Tindell, R.: The spanning subgraphs of Eulerian graphs. J. Graph Theory **1**(1), 79–84 (1977)
5. Bredereck, R., Froese, V., Hartung, S., Nichterlein, A., Niedermeier, R., Talmon, N.: The complexity of degree anonymization by vertex addition. In: Gu, Q., Hell, P., Yang, B. (eds.) AAIM 2014. LNCS, vol. 8546, pp. 44–55. Springer, Heidelberg (2014)
6. Bredereck, R., Hartung, S., Nichterlein, A., Woeginger, G.J.: The complexity of finding a large subgraph under anonymity constraints. In: Cai, L., Cheng, S.-W., Lam, T.-W. (eds.) Algorithms and Computation. LNCS, vol. 8283, pp. 152–162. Springer, Heidelberg (2013)
7. Cai, L., Chan, S.M., Chan, S.O.: Random separation: a new method for solving fixed-cardinality optimization problems. In: Bodlaender, H.L., Langston, M.A. (eds.) IWPEC 2006. LNCS, vol. 4169, pp. 239–250. Springer, Heidelberg (2006)
8. Cai, L., Yang, B.: Parameterized complexity of even/odd subgraph problems. J. Discrete Algorithms **9**(3), 231–240 (2011)
9. Cornuéjols, G.: General factors of graphs. J. Comb. Theory Ser. B **45**(2), 185–198 (1988)
10. Cygan, M., Fomin, F.V., Kowalik, L., Lokshtanov, D., Marx, D., Pilipczuk, M., Pilipczuk, M., Saurabh, S.: Parameterized Algorithms. Springer, Heidelberg (2015)
11. Cygan, M., Marx, D., Pilipczuk, M., Pilipczuk, M., Schlotter, I.: Parameterized complexity of Eulerian deletion problems. Algorithmica **68**(1), 41–61 (2014)
12. Dabrowski, K.K., Golovach, P.A., van 't Hof, P., Paulusma, D.: Editing to eulerian graphs. In: FSTTCS 2014. LIPIcs, vol. 29, pp. 97–108. Schloss Dagstuhl - Leibniz-Zentrum fuer Informatik (2014)
13. Dabrowski, K.K., Golovach, P.A., van 't Hof, P., Paulusma, D., Thilikos, D.M.: Editing to a planar graph of given degrees. In: Beklemishev, L.D., Musatov, D.V. (eds.) CSR 2015. LNCS, vol. 9139, pp. 143–156. Springer, Heidelberg (2015)
14. Downey, R.G., Fellows, M.R.: Fundamentals of Parameterized Complexity. Texts in Computer Science. Springer, Heidelberg (2013)

15. Froese, V., Nichterlein, A., Niedermeier, R.: Win-win kernelization for degree sequence completion problems. In: Ravi, R., Gørtz, I.L. (eds.) SWAT 2014. LNCS, vol. 8503, pp. 194–205. Springer, Heidelberg (2014)

16. Golovach, P.A.: Editing to a connected graph of given degrees. In: Csuhaj-Varjú, E., Dietzfelbinger, M., Ésik, Z. (eds.) MFCS 2014, Part II. LNCS, vol. 8635, pp. 324–335. Springer, Heidelberg (2014)

17. Golovach, P.A.: Editing to a graph of given degrees. Theoret. Comput. Sci. **591**, 72–84 (2015)

18. Hartung, S., Nichterlein, A., Niedermeier, R., Suchý, O.: A refined complexity analysis of degree anonymization in graphs. Inf. Comput. **243**, 249–262 (2015)

19. Hartung, S., Talmon, N.: The complexity of degree anonymization by graph contractions. In: Jain, R., Jain, S., Stephan, F. (eds.) TAMC 2015. LNCS, vol. 9076, pp. 260–271. Springer, Heidelberg (2015)

20. Mathieson, L., Szeider, S.: Editing graphs to satisfy degree constraints: A parameterized approach. J. Comput. Syst. Sci. **78**(1), 179–191 (2012)

21. Moser, H., Thilikos, D.M.: Parameterized complexity of finding regular induced subgraphs. J. Discrete Algorithms **7**(2), 181–190 (2009)

22. Naor, M., Schulman, L., Srinivasan, A.: Splitters and near-optimal derandomization. In: FOCS 1995, pp. 182–191. IEEE (1995)

Subclasses of Baxter Permutations Based on Pattern Avoidance

Shankar Balachandran and Sajin Koroth[(✉)]

Department of Computer Science and Engineering,
Indian Institute of Technology Madras, Chennai 600036, India
{shankar,sajin}@cse.iitm.ac.in

Abstract. Baxter permutations are a class of permutations which are in bijection with a class of floorplans that arise in chip design called mosaic floorplans. We study a subclass of mosaic floorplans called Hierarchical Floorplans of Order k defined from mosaic floorplans by placing certain geometric restrictions. This naturally leads to studying a subclass of Baxter permutations. This subclass of Baxter permutations are characterized by pattern avoidance. We establish a bijection, between the subclass of floorplans we study and a subclass of Baxter permutations, based on the analogy between decomposition of a floorplan into smaller blocks and *block* decomposition of permutations. Apart from the characterization, we also answer combinatorial questions on these classes. We give an algebraic generating function (but without a closed form solution) for the number of permutations, an exponential lower bound on growth rate, and a linear time algorithm for deciding membership in each subclass. Based on the recurrence relation describing the class, we also give a polynomial time algorithm for enumeration. We finally prove that Baxter permutations are closed under inverse based on an argument inspired from the geometry of the corresponding mosaic floorplans. This proof also establishes that the subclass of Baxter permutations we study are also closed under inverse. Characterizing permutations instead of the corresponding floorplans can be helpful in reasoning about the solution space and in designing efficient algorithms for floorplanning.

Keywords: Floorplanning · Pattern avoidance · Baxter permutation

1   Introduction

Baxter permutations are a well studied class of pattern avoiding permutations having real world applications. One such application is to represent floorplans in chip design. A floorplan is a rectangular dissection of a given rectangle into a finite number of indivisible rectangles using axis parallel lines. These indivisible rectangles are locations in which modules of a chip can be placed. In the floorplanning phase of chip design, relative positions of modules are decided so as to optimize cost functions like wire length, routing, area etc. Given a set of modules and an associated cost function, the floorplanning problem is to find an

© Springer International Publishing Switzerland 2016
A.S. Kulikov and G.J. Woeginger (Eds.): CSR 2016, LNCS 9691, pp. 192–206, 2016.
DOI: 10.1007/978-3-319-34171-2_14

optimal floorplan. The floorplanning problem for typical objective functions is NP-hard [8, p. 94]. Hence combinatorial search algorithms like simulated annealing [11] are used to find an optimal floorplan. The optimality of the solution and performance of such algorithms depends on the class of floorplans comprising the search space and their representation. Wong and Liu [11] were the first to use combinatorial search for solving floorplanning problems. They worked with a class of floorplans called slicing floorplans which are obtained by recursively subdividing a given rectangle into two smaller rectangles either by a horizontal or a vertical cut. The slicing floorplans correspond to a class of permutations called separable permutations [1]. Later research in this direction focused on characterizing and representing bigger classes of floorplans so that search algorithms have bigger search spaces, potentially including the optimum. One such category of floorplans is **mosaic** floorplans which are a generalization of slicing floorplans. Ackerman et al. [1] proved a bijection between mosaic floorplans and Baxter permutations. We study a subclass of mosaic floorplans obtained by some natural restrictions on mosaic floorplans. We use the bijection of Ackerman et al. [1] as a tool to characterize and answer important combinatorial problems related to this class of floorplans. For the characterization of these classes we also use characterization of a class of permutations called *simple* permutations studied by Albert and Atkinson [2].

Given a floorplan and dimensions of its basic rectangles, the area minimization problem is to decide orientation of each cell which goes into basic rectangles so as to minimize the total area of the resulting placement. This problem is NP-hard for mosaic floorplans [10], but is polynomial time for both slicing floorplans [10] and Hierarchical Floorplans of Order 5 [4]. Hence Hierarchical Floorplans of Order k is an interesting class of floorplans with provably better performance in area minimization [4] than mosaic floorplans. But the only representation of such floorplans is through a top-down representation known as hierarchical tree [4] and is known only for Hierarchical Floorplans of Order 5. Prior to this work it was not even known which floorplans with k rooms are non-sliceable and is not constructible hierarchically from mosaic floorplans of $k - 1$-rooms or less. Such a characterization is needed to extend the polynomial time area minimization algorithm based on non-dominance given in [4]. We give such a characterization and provide an efficient representation for such floorplans by generalizing generating trees to Skewed Generating Trees of Order k. We also give an exact characterization in terms of equivalent permutations.

Our main technical contributions are (i) We establish a subclass of floorplans called Hierarchical Floorplans of Order k; (ii) We characterize this subclass of floorplans using a subclass of Baxter permutations; (iii) We show that the subclass is exponential in size; (iv) We present an algorithm to check the membership status of a permutation in the subclass of Hierarchical Floorplans of Order k and (v) We present a simple proof of closure under inverse operation for Baxter permutations using the mapping between the permutations and floorplans, and the geometry of the rectangular dissection.

The remainder of the paper is organized as follows: in Sect. 2, we introduce the necessary background on floorplans and pattern avoiding permutations. In Sect. 3, we motivate and characterize the subclasses of Baxter permutations studied in this paper. Section 4 is devoted to answering interesting combinatorial problems of growth, and giving generating function on these subclasses. Section 5 gives an algorithm for membership in each class as well as for deciding given a Baxter permutation the smallest k for which it is Hierarchical Floorplans of Order k. Section 6 proves the closure of Baxter permutations under inverse. Section 7 lists some open problems.

2   Preliminaries

A floorplan is a dissection of a given rectangle by line segments which are axis parallel. The rectangles in a floorplan which do not have any other rectangle inside are called basic rectangles or rooms. For the remainder of the paper we will refer to them as rooms. A floorplan captures the relative position of the rooms via four relations defined between rooms. Given a floorplan f, the "left-of" relation denoted by L_f is defined as $(a, b) \in L_f$ if there is a vertical line segment of f going through the right edge of room a and left edge of room b or if there is a room c such that $(a, c) \in L_f$ and $(c, b) \in L_f$. When $(a, b) \in L_f$ we say that a is to the "left-of" b and is denoted by $a <_l b$. For example in the floorplan given in Fig. 1a the room labeled b is to the left of room labeled d because there is vertical segment through the right boundary of room b and left boundary of room d. Similarly for a floorplan f the "above" relation denoted by A_f is defined as $(a, b) \in A_f$ if there is a horizontal line segment of f going through the bottom edge of room a and through the top edge of room b or if there is a room c such that $(a, c) \in A_f$ and $(c, b) \in A_f$. The other two relations are inverses of these relations: "right-of" is defined as $R_f = \{(a, b) \mid (b, a) \in L_f\}$ and "below" is defined as $B_f = \{(a, b) \mid (b, a) \in A_f\}$. A cross junction in a floorplan is an intersection of two line segments such that the intersection point is not an end point of either of the line segments. A mosaic floorplan is a floorplan where there are no cross junctions. This restriction is to ensure that, in a mosaic floorplan between any two rooms, exactly one of L_f, R_f, B_f, A_f holds [1, Observation 3.3]. We denote the set of all mosaic floorplans with k rooms by M_k. The relations $X \in \{L_f, A_f, R_f, B_f\}$ can be naturally extended to that between rooms and line segments, by defining $(a, l) \in X$ if room a is supported by line segment l from the respective direction X in f. We call two mosaic floorplans f_1, f_2 equivalent if there is a bijective mapping $\psi : f_1 \rightarrow f_2$ such that $(a, b) \in X_{f_1}$ if and only if $(\psi(a), \psi(b)) \in X_{f_2}$ where $X \in \{L, R, A, B\}$, i.e. ψ preserves the relative position of rooms and line segments. For example floorplans labeled a, b in Fig. 1b are equivalent under this definition whereas a and c are not equivalent.

In this paper we study a subclass of mosaic floorplans called Hierarchical Floorplans of Order k. The subclass Hierarchical Floorplans of Order k for $k \geq 2, k \in \mathbb{N}$ (abbreviated as HFO$_k$ in the remainder of the paper) is obtained by placing the following restriction on mosaic floorplans: a mosaic floorplan is

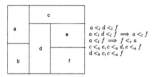

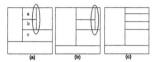

(a) ABLR relationships in a floorplan

(b) Equivalence of Floorplans - $a \equiv b$, but $a \not\equiv c$

Fig. 1. ABLR and equivalence under ABLR

HFO_k if it can be constructed using mosaic floorplans with at most k rooms by repeated application of an operation which we call *insertion*.

Definition 1 (Insertion). *Given a mosaic floorplan with k rooms $f \in M_k$ and some fixed labeling of its rooms, insertion of f by k mosaic floorplans $f_1, f_2, f_3, \ldots, f_k$ denoted by $f(f_1, \ldots, f_k)$ is the mosaic floorplan obtained by placing in f_i in ith room of f.*

Figure 2a illustrates insertion of a floorplan with two rooms by two other floorplans. In insertion, if two adjacent rooms in f (say a and b) have two segments coming from inserted floorplans f_a, f_b of same alignment (i.e., either both horizontal or both vertical) touching each other making a cross junction, then to make the resulting floorplan mosaic, one of the line segments is moved by a small $\delta > 0$ as shown in Fig. 2b. Moving a line segment by a small δ does not change the relative position of rooms. This ensures that *insertion* produces floorplans which are mosaic.

We define a mosaic floorplan f to be decomposable if there exists $k > 1$ for which there is a $g \in M_k$ and k mosaic floorplans $g_1, \ldots, g_k$ at least one of which is non trivial (i.e., has more than one room) and $f = g(g_1, \ldots, g_k)$. A mosaic floorplan is called in-decomposable if it is not decomposable.

(a) Insertion operation on floorplan

(b) Avoiding cross junction

Fig. 2. Insertion operation for mosaic floorplans

Ackerman et al. [1] established a representation for mosaic floorplans in terms of a class of pattern avoiding permutations called Baxter permutations.

Fig. 3. HFO$_2$ building blocks

The bijection is established via two algorithms, one which produces a Baxter permutation given a mosaic floorplan and another which produces a mosaic floorplan given a Baxter permutation. For explaining the results in this paper we only need the algorithm which produces a Baxter permutation π_f given a mosaic floorplan f. This algorithm has two phases, a labeling phase where every room in the mosaic floorplan f is given a unique number in $[n]$ and an extraction phase where the labels of the rooms are read off in a specific order to form a permutation $\pi_f \in S_n$. The labeling is done by successively removing the top-left room of current floorplan by sliding it out of the boundary by pulling the edge which ends at a T junction (since no cross junctions are allowed in a mosaic floorplan, for any room every edge which is within the dissected rectangle is either a horizontal segment ending in a vertical segment forming a ⊣ or is a horizontal segment on which a vertical segment ends forming a ⊥). The ith floorplan to be removed in the above process is labeled room i in the original floorplan. After the labeling phase we obtain a mosaic floorplan whose rooms are numbered from $[n]$. The permutation corresponding to the floorplan is obtained in the second phase called extraction where rooms from the bottom-left corner are successively removed by pulling the edge ending at a T junction. The ith entry of the permutation π_f is the label of the ith room removed in the extraction phase.

Figure 5a demonstrates the labeling phase and Fig. 5b demonstrates the extraction phase. If room i is labeled before room j then room i is to the *left* or *above* of room j, whereas if the room i is removed before room j, i.e., $\pi^{-1}[i] < \pi^{-1}[j]$ then room i is to the *left* of or *below* room j (see [1, Observation 3.4]). Since the permutation captures both the label and position of a room, it captures the *above, below, left* or *right* relations between rooms. Ackerman et al. (see [1, Observation 3.5]) also proved that two rooms share an edge in a mosaic floorplan f if and only if either their labels are consecutive or their positions in π_f are consecutive. For the rest of the paper we refer to this Algorithm of Ackerman et al. as FP2BP.

We now describe permutation classes which are used in this paper, including Baxter permutations mentioned earlier. For the convenience of defining pattern avoidance in permutations, we will assume that permutations are given in the one-line notation (for ex., $\pi = 3142$). A permutation $\pi \in S_n$ is said to contain a *pattern* $\sigma \in S_k$ if there are k indices $i_1, \ldots, i_k$ with $1 \leqslant i_1 < \cdots < i_k \leqslant n$ such that $\pi[i_1], \pi[i_2], \pi[i_3], \ldots, \pi[i_k]$ called *text* has the same relative ordering as σ, i.e., $\pi[i_j] < \pi[i_l]$ if and only if $\sigma_j < \sigma_l$. Note that the sub-sequence need not be formed by consecutive entries in π. If π *contains* σ it is denoted by $\sigma \leq \pi$. A permutation π *avoids* σ if it does not *contain* σ. For example $\pi = 4321$ *avoids* $\sigma = 12$ because in 4321 every number to the right of a number is smaller than itself, but π *contains* the *pattern* $\rho = 21$ because numbers at any two indices of π are in decreasing order. A permutation π is called *separable* if it *avoids* the *pattern* $\sigma_1 = 3142$ and its reverse $\sigma_2 = 2413$. Baxter permutations are a generalization of separable permutations in the following sense: they are allowed to *contain* 3142/2413 as long as any $\pi\,[i_1]\,, \pi\,[i_2]\,, \pi\,[i_3]\,, \pi\,[i_4]$ which has the same relative ordering as 3142/2413 has $|\pi\,[i_1] - \pi\,[i_4]| > 1$. For example $\pi = 41532$ is not Baxter as *text* 4153 in π matches *pattern* 3142 and the absolute difference of entry matching 3 and entry matching 2 is $4 - 3 = 1$. However $\pi = 41352$ is a Baxter permutation as the only text which matches 3142 is 4152 and the absolute difference of entries matching 3, 2 is $4 - 2 = 2$ which is greater than 1.

Another class of permutations important to this study is the class of *simple* permutations. They are a class of block in-decomposable permutations. To define this in-decomposability we need the following definition : a *block* of a permutation is a set of consecutive positions such that the values from these positions form an interval $[i, j]$ of $\mathbb{N}$. Note that the values in the block need not be in ascending order as it is in the interval corresponding to the block $[i, j]$. The notion of block in-decomposability is defined by a decomposition operation called *inflation*. We recall the definition from Sect. 2 of [2].

Definition 2 (Inflation). *Given a permutation* $\sigma \in S_k$, *inflation of* σ *by* k *permutations* $\rho_1, \rho_2, \rho_3, \ldots, \rho_k$, *denoted by* $\sigma\,(\rho_1, \ldots, \rho_k)$ *is the permutation* π *where each element* σ_i *of* σ *is replaced with a block of length* $|\rho_i|$ *whose elements have the same relative ordering as* ρ_i, *and the blocks among themselves have the same relative ordering as* σ.

For example inflation of 3124 by $21, 123, 1$ and 12 results in $\pi = 65\,123\,4\,78$ where 65 is the *block* corresponding to 21, 123 corresponds to 123, 4 corresponds to 1 and 78 corresponds to 12. If $\pi = \sigma\,(\rho_1, \ldots, \rho_k)$ then $\sigma\,(\rho_1, \ldots, \rho_k)$ is called a *block-decomposition* of π. A *block-decomposition* $\sigma\,(\rho_1, \rho_2, \ldots, \rho_k)$ is non-trivial if $\sigma \in S_k$ for $k > 1$ and at least one ρ_i is a non-singleton permutation (i.e. of more than one element). A permutation is *block-in-decomposable* if it has no non-trivial block-decomposition. Note that *inflation* on permutations as defined above is analogous to *insertion* on mosaic floorplans defined earlier.

Block in-decomposable permutations can be thought of as building blocks of all other permutations by *inflations*. Albert and Atkinson [2] studied *simple* permutations which are permutations whose only *blocks* are the trivial *blocks* (which is either a single point $\pi\,[i]$ or the whole permutation $\pi\,[1 \ldots n]$). They also defined a sub class of *simple* permutations called *exceptionally simple* permutations which are defined based on an operation called *one-point deletion*. A one-point deletion on a permutation $\pi \in S_n$ is deletion of a single element at some index i and getting a new permutation $\pi' \in S_{n-1}$ by rank ordering the remaining elements. For example one-point deletion at index 5 of 41352 gives 4135 which when rank ordered gives the permutation 3124. A permutation π is *exceptionally simple* if it is *simple* and no *one-point deletion* of π yields a *simple* permutation. Albert and Atkinson [2] characterized *exceptionally simple* permutations and proved that for any permutation $\pi \in S_n$ which is *exceptionally simple* there exists two successive *one-point* deletions which yields a *simple* permutation $\pi' \in S_{n-2}$.

3   Characterizing Hierarchical Floorplans of Order k

In this section we characterize Hierarchical Floorplans of Order k in terms of corresponding permutations using the notion of block decomposition defined earlier.

We note that this connection can be seen for a level of the hierarchy well studied in literature, namely HFO$_2$. HFO$_2$, the class of floorplans which can be built by repeated application of insertion of the two basic floorplans shown in Fig. 3 are also called slicing floorplans. Slicing floorplans are known [1] to be in bijective correspondence with *separable permutations*. Separable permutations are also the class of permutations π such that it can be obtained repeated *inflation* of 1 (the singleton permutation) by, 12 or 21. Note that both 12, 21 are *simple* permutations. Even though HFO$_2$ is well studied in literature and is known to be in bijective correspondence with separable permutations, the connection to block decomposition of permutations was not explicitly observed.

HFO$_5$ floorplans are also studied in the literature, but the only characterization till date for these floorplans is based on a discrete structure called *generating trees*. We generalize this structure for an arbitrary k in the following sense : a *generating tree* of order k is a rooted tree, where each node is labeled by an in-decomposable mosaic floorplan, say g of at most k rooms, and the number of children of a node is equal to the number of rooms in the floorplan labeling the node. The children are arranged in the order π_g^{-1} from left to right. That is the left most child corresponds to the first room to be removed in the extraction phase of FP2BP and second from left corresponds to second room to be removed and so on and so forth. The *generating tree* captures the top down application of *insertion*'s to yielding the given floorplan in the following sense : an internal node of a *generating tree* represents insertion of f - the floorplan labeling the node - by the floorplans labeling its children, $f_1, \ldots, f_k$ (ordered from left to right). There could be more than one generating tree for a floorplan owing to the fact there is ambiguity in consecutive vertical slices and in consecutive horizontal slices, as illustrated in Fig. 4. But this can be removed (proved later) by introducing two disambiguation rules called "skew". Skew rule insists that when there are multiple parallel vertical (respectively, horizontal) line segments touching the bounding box of the floorplan f, we consider only the insertion operation f_1, f_2 where f_2 is the floorplan contained to the right of (respectively, above) the first parallel line segment from left (respectively, bottom) and f_1 is the floorplan contained to the left of (respectively below) the first parallel line segment from left (respectively, bottom). Hence only the tree labeled a satisfies "skew" rule among the generating trees in Fig. 4. A generating tree satisfying "skew" rule is called *Skewed Generating Tree*.

The connection between *insertion* and block decomposition and the fact the bijection of Ackerman et al. [1] preserves this connection is the central idea

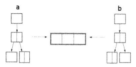

Fig. 4. Ambiguity in vertical cuts

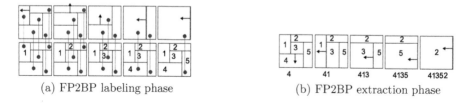

(a) FP2BP labeling phase   (b) FP2BP extraction phase

Fig. 5. FP2BP algorithm

of our paper. The following observation about the algorithm FP2BP, though not mentioned in the original paper, is not hard to see, but is useful for the characterization of HFO_k.

Lemma 1. *For a mosaic floorplan f let π_f denote the unique Baxter permutation obtained by algorithm FP2BP. If $f = g(g_1, \ldots, g_k)$ i.e., it is obtained by insertion of $g \in M_k$ by $g_1, \ldots, g_k$, then*

$$\pi_f = \pi_g\left(\pi_{g_1}, \ldots, \pi_{g_k}\right)$$

where $\pi_g\left(\pi_{g_1}, \ldots, \pi_{g_k}\right)$ denotes the permutation obtained by inflating π_g with $\pi_{g_1}, \ldots, \pi_{g_k}$.

See full version of the paper [3] for a proof. We obtain the following useful corollary from Lemma 1

Corollary 1. *A mosaic floorplan f is in-decomposable if and only if the Baxter permutation π_f corresponding to it is block in-decomposable.*

For the characterization we will also need the following connection between generating trees and block decomposition of permutations. Let T_f be a generating tree corresponding to f, satisfying the "skew" rule, then T_f captures the unique block decomposition of a permutation as defined in [2, Proposition 2]. Label every node of T_f by Baxter permutation π_{f_i} corresponding to the mosaic floorplan f_i labeling it. Mosaic floorplan g corresponding to the sub-tree rooted at f_i is obtained by the insertion of f_i by the floorplans labeling its children $f_{i_1}, \ldots, f_{i_k}$. Hence by applying Lemma 1 we get that $\pi_g = \pi_{f_i}\left(\pi_{f_{i_1}}, \ldots, \pi_{f_{i_k}}\right)$. So generating trees labeled by Baxter permutations π_{f_i} captures the block decomposition of Baxter permutation π_f corresponding to the floorplan f. Figure 6 illustrates the correspondence between *inflation* and *insertion* by showing the equivalence between *inflating* 3124 with 123, 21, 1 and 24, and *inserting*

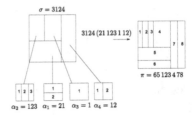

Fig. 6. Correspondence between inflation and insertion

the floorplan corresponding to 3124 with floorplans corresponding to $123, 21, 1$ and 24.

Theorem 1. *Skewed Generating Trees of Order k are in bijective correspondence with HFO_k floorplans. Moreover they capture the block decomposition of the Baxter permutation corresponding to the floorplan.*

Proof. It follows from definition of HFO_k that there is a *generating tree of order k* capturing the successive applications of insertions resulting in the final floorplan. Since HFO_k are a subclass of mosaic floorplans which are in bijective correspondence with Baxter permutations, there is unique Baxter permutation π_f corresponding to the floorplan f. Lemma 1 can now be used to prove that a *generating tree of order k* captures the block decomposition of π_f, by induction on the height of the tree. Consider the base case to be $h = 1$, i.e., the whole tree is one node labeled by an in-decomposable mosaic floorplan f and by Corollary 1, π_f is block in-decomposable. Assume that for any $h < l$, *generating trees of order k* captures the block decomposition of π_f. Take a tree of height $h = l$ corresponding to a floorplan f, and let the root node be labeled by g and children be labeled $g_1, \ldots, g_k$. By Lemma 1, $\pi_f = \pi_g(\pi_{g_1}, \ldots, \pi_{g_k})$. We can apply induction hypothesis on the children to get the decomposition of $\pi_{g_1}, \ldots, \pi_{g_k}$.

To prove the uniqueness of skewed generating trees we use the following theorem by Albert and Atkinson [2, Proposition 2] proving the uniqueness of the block-decomposition represented by skewed generating trees.

Theorem 2 [2, Proposition 2]. *For every non singleton permutation π there exists a unique simple non singleton permutation σ and permutations $\alpha_1, \ldots, \alpha_n$ such that $\pi = \sigma(\alpha_1, \ldots, \alpha_n)$. Moreover if $\sigma \neq 12, 21$ then $\alpha_1, \ldots, \alpha_n$ are also uniquely determined. If $\sigma = 12$ (respectively, 21) then α_1 and α_2 are also uniquely determined subject to the additional condition that α_1 cannot be written as $(12)[\beta, \gamma]$ (respectively as $(21)[\beta, \gamma]$)*

The proof is completed by noting that the decomposition obtained by Skewed Generating Trees of Order k satisfies the properties of the decomposition described in the above theorem. In a skewed generating tree if parent is $\sigma = 12$ (respectively, 21), then its left child cannot be 12 (respectively, 21). Hence the block-decomposition corresponding to the left child, α_1, cannot be

(12) $[\beta, \gamma]$ (respectively, (21) $[\beta, \gamma]$). Since such a decomposition is unique, the skewed generating tree also must be unique. Hence the theorem.

To characterize HFO_k in terms of pattern avoiding permutations the following insight is used: if a permutation π is Baxter then it corresponds to a mosaic floorplan. Every mosaic floorplan is HFO_k for some k. Hence for a Baxter permutation π the corresponding floorplan f_π is not HFO_k for some specific k, it will be because of existence of a node in the unique skewed generating tree corresponding to f_π, which is labeled by an in-decomposable mosaic floorplan $g \in \text{HFO}_l$ for some $l > k$. Since π is obtained by inflation of permutations including π_g corresponding to g, π will have some text which matches the *pattern* π_g because of the Lemma 2. Thus if we can figure out all the *patterns* which correspond to in-decomposable mosaic floorplans which are HFO_l for some $l > k$ then HFO_k would be all Baxter permutations which avoid those *patterns*. These insights are formalized by the following lemmas.

Lemma 2. *If* $\pi = \sigma(\rho_1, \ldots, \rho_k)$, *then* π *contains all patterns which any of* $\sigma, \rho_1, \rho_2, \ldots, \rho_k$ *contains.*

Lemma 3. *If* $\pi = \sigma(\rho_1, \ldots, \rho_k)$, *then any block in-decomposable pattern in* π *has a matching text which is completely contained in one of* $\sigma, \rho_1, \rho_2, \ldots, \rho_k$.

Refer the full version of the paper [3] for proof of the above lemma. Let f be an in-decomposable mosaic floorplan which is HFO_l for some fixed $l \in \mathbb{N}$. By Corollary 1, the permutation corresponding to f, π_f would be block in-decomposable and hence it will be a *simple* permutation of length l. It is known (see [2, Theorem 5]) that a simple permutation of length l has either a *one-point* deletion which yields another simple permutation or two *one-point* deletions giving a simple permutation. Hence by successive applications of *one-point* deletions we can reduce π_f to a *simple* permutation of length k, or an *exceptionally simple* permutation of length $k + 1$ (at which point there is no further one *one-point* deletion giving a simple permutation) for any $k < l$. Also if π' is obtained from π by a *one-point* deletion at index i, then $\pi[1, \ldots, i-1, i+1, \ldots, n]$ matches the *pattern* π'. That is π contains all patterns π' which are permutations obtained by one point deletion of π at some index. Also since pattern containment is transitive by definition, if π'' is obtained by one-point deletion of π' which in turn obtained from π by a one-point deletion, then $\pi'' \leq \pi'$ and $\pi' \leq \pi$ implies that $\pi'' \leq \pi$. From these observations we get the following characterization of HFO_k.

Theorem 3. *A mosaic floorplan f is HFO_k if and only if the permutation π_f corresponding to f (obtained by algorithm FP2BP) does not contain patterns from simple permutations of length $k + 1$ or exceptionally simple permutations of length $k + 2$.*

Proof. By Theorem 1, for any HFO_k floorplan f there is a unique Skewed Generating Trees of Order k, T_f such that it captures the block-decomposition of π_f. And in the block-decomposition of a generating tree of order k, permutations corresponding to the nodes are labeled by HFO_k permutations of length at most

k. Hence the block-decomposition of π_f contains only block in-decomposable permutations of length at most k. By Lemma 3 π_f cannot contain patterns which are block in-decomposable permutations of length strictly more than k. Thus π_f cannot contain patterns from simple permutations of length $k+1$ or from exceptionally simple permutations of length $k+2$ as they are both classes of block in-decomposable permutations of length strictly greater than k.

For the reverse direction, we prove that any mosaic floorplan which is $\mathrm{HFO}_l, l > k$ contains either a simple permutation of length $k+1$ or an exceptionally simple permutation of length $k+2$. From the fact that by definition any mosaic floorplan is HFO_j for some j and the forward direction that no HFO_k floorplan contains either a simple permutation of length $k+1$ or an exceptionally simple permutation of length $k+2$ proof is completed. Suppose if it is HFO_l for $l > 0$ then π_f would have a *text* matching a pattern $\sigma \in S_l$ which is a simple permutation. Because the generating tree T_f will have σ and so would the block decomposition of the sub-tree rooted at node σ. And by Lemma 2, π_f would also contain σ. From σ we can obtain by successive *one-point* deletions a permutation σ' which is either a simple permutation of length k or is an exceptionally simple permutation of length $k+1$. And σ' would match a *text* in π_f because π_f had a *text* matching σ and σ contains this permutation, i.e., $\sigma'' \leq \sigma \prec \pi_f \implies \sigma'' \leq \pi_f$.

From the above characterization it can be proved that the hierarchy HFO_k (it is a hierarchy because by definition $\mathrm{HFO}_i \subseteq \mathrm{HFO}_{i+1}$) is strict for $k \geqslant 7$, i.e. there is at least one floorplan which is HFO_k but is not HFO_i for any $i < k$. The natural candidates for such separation are in-decomposable mosaic floorplans on k rooms which corresponds to *simple* permutations of length k which are Baxter. It is easy to verify that for $k=5$, $\pi_5 = 41352$ is such a permutation. Note that π_5 is of the form $\pi[n-1] = n$ and $\pi[n] = 2$. From π_5 we can obtain $\pi_7 = 6413572$ by inserting 7 between 5 and 2 and appending 6 at the beginning. It can be verified that π_7 is not HFO_5. It turns out that all permutations of length at most 4 which are Baxter are also HFO_2, making HFO_5 the first odd number from where one can prove the strictness of the hierarchy. Also every HFO_6 is HFO_5, hence for even numbers separation theorem can only start from 8. Hence we prove the separation theorem for $k \geq 7$ generalizing the earlier stated idea. The generalization builds a π_{k+2} from a π_k which is an in-decomposable HFO_k having $\pi[n-1] = n$ and $\pi[n] = 2$, by setting $\pi_{k+2}[1] = n+1$, $\pi_{k+2}[i] = \pi_k[i-1], 2 \leqslant i \leqslant n$, $\pi_{k+2}[n+1] = n+2$ and $\pi_{k+2}[n+2] = 2$.

Theorem 4. *For any $k \geq 7$, there exists a floorplan f which is in HFO_{k+2} but is not in HFO_l for any $l \leq k+1$ See the full version of the paper [3] for a complete proof of the above theorem.*

4   Combinatorial Study of HFO_k

We first prove for any fixed k the existence of a rational generating function for HFO_k, which leads to a dynamic programming based algorithm for counting the number of HFO_k with n rooms. Since we have proved that the number of distinct HFO_k floorplans with n rooms is equal to the number of distinct Skewed Generating Trees of Order k with n leaves, it suffices to count such trees. Let t_n^k denote the number of distinct Skewed Generating Trees of Order k with n leaves and t_1^k represent a rectangle for any k. To provide a rational generating function for number of distinct HFO_k floorplans with n rooms, it suffices to provide one for the count t_n^k.

We describe the method for HFO_5, which can be generalized for an arbitrary k. For simplicity of notation let $t_i = t_i^5$. Skewed Generating Trees of Order 5 are labeled by simple permutation of length at most 5 which are Baxter. There are only four of them - $12, 21, 25314$ and 41352. Thus the root node of such a tree also must be labeled from one these four permutations. We obtain a recurrence by partitioning the set of Skewed Generating Trees of Order 5 into four classes decided by the label of the root. Let a_n denote the number of Skewed Generating Trees of Order 5 with n leaves whose root is labeled 12, b_n denote the number of Skewed Generating Trees of Order 5 with n leaves whose root is labeled 21, c_n denote the number of Skewed Generating Trees of Order 5 with n leaves whose root is labeled 41352 and d_n the number denote the Skewed Generating Trees of Order 5 with n leaves whose root is labeled 25314. Since these are the only in-decomposable HFO_k permutations for $k \leqslant 5$, the root (and also any internal node) has to labeled by one of these permutations. Hence we get the following recurrence for t_n^5, $t_n^5 = a_n + b_n + c_n + d_n$.

In a skewed tree if the root is labeled 12, its left child cannot be 12 but it can be 21, 41352, 25314 or a leaf node. Hence the left child of the root of a tree in a_n has to be labeled from b, c or d, but the right child has no such restriction. By definition of skewed generating trees if the root is labeled by a permutation of length l, it will have l children, such that the number of leaves of the children sum to n. Hence if root is labeled by 12, the two children will have leaves $n - i$ and i for some $i, 1 \leq i \leq n - 1$. This along with the skew rule dictates that $a_n = \sum_{i=1}^{n-1} b_{n-i}t_i + c_{n-i}t_i + d_{n-i}t_i$. Similarly if the root is 21 then its left child cannot be 21 but it can be 12, 41352, 25314 or a leaf node. But for trees whose roots are labeled 41352/25314, they can have any label for any of the five children. Hence we get, $a_n = t_{n-1}^5.1 + \sum_{i=2}^{n-1}(b_i + c_i + d_i)t_{n-i}^5$, $b_n = t_{n-1}^5.1 + \sum_{i=2}^{n-1}(a_i + c_i + d_i)t_{n-i}^5$, $c_n = \sum_{\{i,j,k,l,m \geq 1 | i+j+k+l+m=n\}} t_i^5 t_j^5 t_k^5 t_l^5 t_m^5$ and $d_n = \sum_{\{i,j,k,l,m \geq 1 | i+j+k+l+m=n\}} t_i^5 t_j^5 t_k^5 t_l^5 t_m^5$ note that $c_n = d_n$. Since a node labeled 41352/25314 ought to have five children, $c_n, d_n = 0$ for $n < 5$. Summing up a_n and b_n and using the identity $t_i^5 = a_i + b_i + c_i + d_i$ we get a recurrence for t_n^5.

Note that in a similar way recurrence relation for any HFO_k can be constructed. Even though the above recurrence failed to give us a closed form solution, it lead to a natural dynamic programming based algorithm for counting the number of HFO_k floorplans with n rooms for any fixed k. For example the recurrence for HFO_5 is given by a sixth order recurrence relation. Hence there is an $O\left(n^6\right)$ tabular algorithm computing the value of t_n using dynamic programming which recursively computes all t_i^5 for all $i < n$ and then computes t_n^5 from the recurrence relation we obtained. In general HFO_k has a recurrence relation of order k, and hence the algorithm for t_n^k would run in time $O\left(n^{k+1}\right)$ using a similar strategy.

Using the argument which proved existence of an in-decomposable HFO_k floorplan for any k, we can get a simple lower bound on the number of HFO_k floorplans with n rooms which are not HFO_j for any $j < k$. It is known [9] that the number of HFO_2 floorplans with n rooms is $\theta\left(n!\frac{(3+\sqrt{8})^n}{n^{1.5}}\right)$. If in the generating tree corresponding to an HFO_2 floorplan an in-decomposable HFO_k floorplan is inserted replacing one of the leaves (to be uniform, say the right most leaf), the resulting generating tree would be of order k and hence by Theorem 1, would correspond to an HFO_k floorplan. Hence the number of HFO_k floorplans with n rooms which are not HFO_l is at least the number of generating trees of order 2 with $n - k + 1$ leaves. And the number of generating trees of order 2 with n leaves equals the number of HFO_2 floorplans with n rooms thus giving the following exponential lower bound. For any $k \geq 7$, the number of HFO_k floorplans with n rooms which are not HFO_j for any $j < k$ is at least $\frac{(n-k)!(3+\sqrt{8})^{n-k}}{(n-k)^{1.5}}$.

5   Algorithm for Membership

For arriving at an algorithm for membership in HFO_k we note that if a given permutation is Baxter then it is HFO_k for some k. And if it is HFO_k by Theorem 1 there exits an order k generating tree corresponding to the permutation. By Theorem 1 the generating tree also captures the block decomposition of the permutation. Our algorithm identifies the block-decomposition corresponding to the generating tree of order k, level by level. And it rejects a permutation if such a reduction is not possible. The algorithm runs in time $O(n^2)$ for a fixed k. See the full version of the paper [3] for a formal description and proof of correctness.

For a fixed k one can also achieve linear time for membership owing to a new fixed parameter algorithm of Marx and Guillemot [5] which given two permutations $\sigma \in S_k$ and $\pi \in S_n$ checks if σ avoids π in time $2^{O(k^2 \log k)}n$ and a linear time algorithm for recognizing Baxter permutations by Hart and Johnson [6]. Both results [5,6] are highly non-trivial and deep. Theorem 3 guarantees that it is enough to ensure that π is Baxter and π and avoids simple permutations of length $k + 1$ and exceptionally simple permutations of length $k + 2$. Using the algorithm given by Hart and Johnson [6] we can check in linear time whether a given permutation is Baxter or not. Since there are at most $(k + 1)!$ simple permutations of length k and at most $(k+2)!$ exceptionally simple permutations of length $(k + 2)$ using the algorithm given in [5] as a sub-routine we can do the latter in $O((k + 2)!2^{c(k+2)^2 \log(k+2)}n)$ time. Since k is a fixed constant we get a linear time algorithm.

If the value k is unknown our algorithm can be used to get an $O(n^4)$ algorithm with a few modifications to find out the minimum k for which the input permutation is HFO_k. One can also obtain a context free grammar based poly-time algorithm for checking a given Baxter permutation is HFO_k, as the generating tree of an HFO_k floorplan can be thought of a parse tree of a corresponding grammar and the number of non-terminals is the number of different HFO_j floorplans with l rooms where $h, l \leq k$. See full version of the paper [3] for details.

6 Closure Properties of Baxter Permutations

Only recently it has been proved that Baxter permutations are closed under inverse [7]. The proof in [7] uses an argument based on permutations and patterns. We give a simple alternate proof of this fact using the geometrical intuition derived from mosaic floorplans. We prove that the floorplan obtained by taking a mirror image of a floorplan along the horizontal axis is a floorplan whose permutation (under the bijection of Ackerman) is the inverse of the permutation corresponding to the starting floorplan.

The intuition is that when the floorplan's mirror image about the horizontal axis is taken, it does not change the relationship between two rooms if one is to the left of the other. But if a room is below the other, it flips the relationship between the corresponding rooms. For any Baxter permutation π and two indices i, j where $i < j$, if $\pi[i] < \pi[j]$, since $\pi[i]$ appears before $\pi[j]$ by the property of the algorithm FP2BP $\pi[i]$ is to the left of $\pi[j]$ in π_f. In the inverse of π, π^{-1} indices $\pi[i]$ and $\pi[j]$ will be mapped to i and j respectively. Hence if π^{-1} is Baxter, or equivalently there is a mosaic floorplan corresponding to π^{-1}, π_f^{-1}, the rooms labeled by i and j will be such that i precedes j in the top-left deletion ordering(as $i < j$) and also in bottom left deletion ordering(as $\pi[i] < \pi[j]$). Hence i is to the left of j in π_f^{-1}. If $\pi[i] > \pi[j]$, since $\pi[i]$ appears before $\pi[j]$ by by the property of the algorithm FP2BP, $\pi[i]$ is below $\pi[j]$ in π_f. In the inverse of π, π^{-1} indices $\pi[i]$ and $\pi[j]$ will again be mapped to i and j respectively. Hence if there is a mosaic floorplan corresponding to π^{-1}, π_f^{-1}, the rooms labeled by i and j will be such that i precedes j in the top-left deletion ordering(as $i < j$) but in bottom left deletion ordering j precedes i(as $\pi[i] < \pi[j]$). Hence i is above j in π_f^{-1}. Thus mirror image about horizontal axis satisfies all these constraints on the rooms. For the formal proof of closure under inverse see the full paper. Figure 7 illustrates the link between inverse and the geometry.

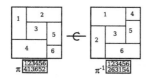

Fig. 7. Obtaining a mosaic floorplan corresponding to the inverse of a Baxter permutation

From the closure of Baxter permutations under inverse one can also get that HFO_k permutations are also closed under inverse by using the characterization based on generating trees of order k and applying the inverse Baxter permutation on each node of the tree. We also observe that there is a geometric interpretation for reverse of a Baxter permutation. It is easy to see that Baxter permutations are closed under reverse because the patterns they avoid are reverses of each other (3142/2413). We observe, without giving a proof, that for a Baxter permutation π its reverse π^r corresponds to the mosaic floorplan that is obtained by first rotating by 90° clockwise and then by taking a mirror image along the horizontal axis.

7   Open Problems

One natural open problem arising from this work is that of exact formulae for the number of HFO$_k$ floorplans. The only k for which exact count is known is $k = 2$. Our proof of closure under inverse for Baxter permutation gives rise to the following open problem. For a class of permutations characterized by pattern avoidance, like Baxter permutations, to be closed under inverse is it enough that the forbidden set of permutations defining the class is closed under inverse.

References

1. Ackerman, E., Barequet, G., Pinter, R.Y.: A bijection between permutations and floorplans, and its applications. Discrete Appl. Math. **154**(12), 1674–1684 (2006)
2. Albert, M.H., Atkinson, M.D.: Simple permutations and pattern restricted permutations. Discrete Math. **300**(1–3), 1–15 (2005)
3. Balachandran, S., Koroth, S.: Sub-families of baxter permutations based on pattern avoidance. CoRR, abs/1112.1374 (2011)
4. De Almeida, A.M., Rodrigues, R.: Trees, slices, and wheels: on the floorplan area minimization problem. Networks **41**(4), 235–244 (2003)
5. Guillemot, S., Marx, D.: Finding small patterns in permutations in linear time. In: Proceedings of the Twenty-Fifth Annual ACM-SIAM Symposium on Discrete Algorithms, SODA, Portland, Oregon, USA, pp. 82–101. SIAM, 5–7 January 2014
6. Hart, J.M.: Fast recognition of baxter permutations using syntactical and complete bipartite composite dag's. Int. J. Comput. Inf. Sci. **9**(4), 307–321 (1980)
7. Law, S., Reading, N.: The hopf algebra of diagonal rectangulations. J. Comb. Theory Ser. A **119**(3), 788–824 (2011)
8. Sait, S.M., Youssef, H.: VLSI Physical Design Automation: Theory and Practice. Lecture Notes Series on Computing. World Scientific, Singapore (1999)
9. Shen, Z.C., Chu, C.C.N.: Bounds on the number of slicing, mosaic, and general floorplans. IEEE Trans. CAD Integr. Circ. Syst. **22**(10), 1354–1361 (2003)
10. Stockmeyer, L.: Optimal orientations of cells in slicing floorplan designs. Inf. Control **57**(2), 91–101 (1983)
11. Wong, D.F., Liu, C.L.: A new algorithm for floorplan design. In: 23rd Conference on Design Automation, pp. 101–107 (1986)

On Slepian–Wolf Theorem with Interaction

Alexander Kozachinskiy[1,2(✉)]

[1] Lomonosov Moscow State University, Moscow, Russia
kozlach@mail.ru
[2] National Research University Higher School of Economics, Moscow, Russia

Abstract. In this paper we study interactive "one-shot" analogues of the classical Slepian–Wolf theorem. Alice receives a value of a random variable X, Bob receives a value of another random variable Y that is jointly distributed with X. Alice's goal is to transmit X to Bob (with some error probability ε). Instead of one-way transmission we allow them to interact. They may also use shared randomness.

We show, that for every natural r Alice can transmit X to Bob using $\left(1 + \frac{1}{r}\right) H(X|Y) + r + O(\log_2\left(\frac{1}{\varepsilon}\right))$ bits on average in $\frac{2H(X|Y)}{r} + 2$ rounds on average. Setting $r = \lceil \sqrt{H(X|Y)} \rceil$ and using a result of [2] we conclude that every one-round protocol π with information complexity I can be compressed to a (many-round) protocol with expected communication about $I + 2\sqrt{I}$ bits. This improves a result by Braverman and Rao [3], where they had $I + 5\sqrt{I}$. Further, we show (by setting $r = \lceil H(X|Y) \rceil$) how to solve this problem (transmitting X) using $2H(X|Y) + O(\log_2\left(\frac{1}{\varepsilon}\right))$ bits and 4 rounds on average. This improves a result of [4], where they had $4H(X|Y) + O(\log 1/\varepsilon)$ bits and 10 rounds on average.

In the end of the paper we discuss how many bits Alice and Bob may need to communicate on average besides $H(X|Y)$. The main question is whether the upper bounds mentioned above are tight. We provide an example of (X, Y), such that transmission of X from Alice to Bob with error probability ε requires $H(X|Y) + \Omega\left(\log_2\left(\frac{1}{\varepsilon}\right)\right)$ bits on average.

Keywords: Slepian–Wolf theorem · Communication complexity · Information complexity

1 Introduction

Assume that Alice receives a value of a random variable X and she wants to transmit that value to Bob. It is well-known [8] that Alice can do it using one message over the binary alphabet of expected length less than $H(X)+1$. Assume now that there are n independent random variables $X_1, \dots, X_n$ distributed as X, and Alice wants to transmit all $X_1, \dots, X_n$ to Bob. Another classical result from [8] states, that Alice can do it using one message of *fixed* length, namely $\approx nH(X)$, with a small probability of error.

A. Kozachinskiy—Supported by the RFBR grant 16-01-00362 and by the grant of the President of Russian Federation (MK-7312.2016.1).

A.S. Kulikov and G.J. Woeginger (Eds.): CSR 2016, LNCS 9691, pp. 207–222, 2016.
DOI: 10.1007/978-3-319-34171-2_15

One of the possible ways to generalize this problem is to provide Bob with a value of another random variable Y which is jointly distributed with X. That is, to let Bob know some partial information about X for free. This problem is the subject of the classical Slepian-Wolf Theorem [9] which asserts that if there are n independent pairs $(X_1, Y_1), \ldots, (X_n, Y_n)$, each pair distributed exactly as (X, Y), then Alice can transmit all $X_1, \ldots, X_n$ to Bob, who knows $Y_1, \ldots, Y_n$, using one message of fixed length, namely $\approx nH(X|Y)$, with a small probability of error[1] However, it turns out that a one-shot analogue of this theorem is impossible, if only one-way communication is allowed.

The situation is quite different, if we allow Alice and Bob to *interact*, that is, to send messages in both directions. In [7] Orlitsky studied this problem for the average-case communication when no error is allowed. He showed that if pair (X, Y) is uniformly distributed on it's support, then Alice may transmit X to Bob using at most

$$H(X|Y) + 3\log_2(H(X|Y) + 1) + 17$$

bits on average and 4 rounds. For the pairs (X, Y) whose support is a Cartesian product Orlitsky showed that error-less transmission of X from Alice to Bob requires $H(X)$ bits on average.

From a result of Braverman and Rao [3], it follows that for arbitrary (X, Y) it is sufficient to communicate at most

$$H(X|Y) + 5\sqrt{H(X|Y)} + O\left(\log_2\left(\frac{1}{\varepsilon}\right)\right)$$

bits on average (here ε stands for the error probability).

We show that for every positive ε and natural r there is a public-coin protocol transmitting X from Alice to Bob with error probability at most ε (for each pair of inputs) using at most

$$\left(1 + \frac{1}{r}\right) H(X|Y) + r + O\left(\log_2 \frac{1}{\varepsilon}\right)$$

bits on average in at most

$$\frac{2H(X|Y)}{r} + 2$$

rounds on average. Furthermore, there is a private-coin protocol with the same properties plus extra $O(\log\log \operatorname{supp}(X, Y))$ bits of communication. Our protocol is inspired by protocol from [1]. The idea of the protocol is essentially the same, we only apply some technical trick to reduce communication.

This improves the result of Braverman and Rao, since setting $r = \left\lceil \sqrt{H(X|Y)} \right\rceil$ above we obtain the protocol with expected communication at

---

[1] This paper is focused only on the non-symmetric version of this problem. In more general version Alice and Bob send messages to the 3rd party Charlie, who must reconstruct both random variables.

most $H(X|Y) + 2\sqrt{H(X|Y)} + O\left(\log_2\left(\frac{1}{\varepsilon}\right)\right)$. In [4], it is established a one-shot interactive analogue of the Slepian-Wolf theorem for the bounded-round communication. They showed that Alice may transmit X to Bob using at most $O(H(X|Y) + 1)$ bits and $O(1)$ rounds on average. More specifically, their protocol transmits at most $4H(X|Y) + \log_2(1/\varepsilon) + O(1)$ bits on average in 10 rounds on average. Setting $r = \lceil H(X|Y)\rceil$ above we improve this result. Indeed, we obtain the protocol with the expected length at most $2H(X|Y) + O\left(\log_2\left(\frac{1}{\varepsilon}\right)\right)$ and the expected number of rounds at most 4.

Actually, in [3] a more general result was established. It was shown there that every one-round protocol π with information complexity I can be compressed to the (many-round) protocol with expected length at most

$$\approx I + 5\sqrt{I}. \tag{1}$$

Using the result from [2], we improve Equ. 1. Namely, we show that every one-round protocol π with information complexity I can be compressed to the (many-round) protocol with expected communication length at most

$$\approx I + 2\sqrt{I}.$$

Are there random variables X, Y for which the upper bound of the form $H(X|Y) + O\left(\sqrt{H(X|Y)}\right)$ is tight? We make a step towards answering this question: we provide an example of random variables X, Y such that every public-coin communication protocol which transmits X from Alice to Bob with error probability ε (with respect to the input distribution and the protocol's randomness) must communicate at least $H(X|Y) + \Omega\left(\log_2\left(\frac{1}{\varepsilon}\right)\right)$ bits on average.

2 Definitions

2.1 Information Theory

Let X, Y be two joint distributed random variables, taking values in the finite sets, respectively, $\mathcal{X}$ and $\mathcal{Y}$.

Definition 1. *Shannon Entropy of X is defined by the formula*

$$H(X) = \sum_{x \in \mathcal{X}} \Pr[X = x] \log_2\left(\frac{1}{\Pr[X = x]}\right).$$

Definition 2. *Conditional Shannon entropy of X with respect to Y is defined by the formula:*

$$H(X|Y) = \sum_{y \in \mathcal{Y}} H(X|Y = y) \Pr[Y = y],$$

where $X|Y = y$ denotes a distribution of X, conditioned on the event $\{Y = y\}$.

If X is uniformly distributed in $\mathcal{X}$ then obviously $H(X) = \log_2(|\mathcal{X}|)$. We will also use the fact that the formula for conditional entropy may be re-written as

$$H(X|Y) = \sum_{(x,y)\in\mathcal{X}\times\mathcal{Y}} \Pr[X = x, Y = y] \log_2\left(\frac{1}{\Pr[X = x|Y = y]}\right).$$

Generalization of the Shannon entropy is Renyi entropy.

Definition 3. *Renyi entropy of X is defined by the formula*

$$H_2(X) = -\log_2\left(\sum_{x\in\mathcal{X}} \Pr[X = x]^2\right).$$

Concavity of log implies that $H(X) \geq H_2(X)$.

The mutual information of two random variables X and Y, conditioned on another random variable Z, can be defined as:

$$I(X : Y|Z) = H(X|Z) - H(X|Y, Z).$$

For the further introduction in information theory see, for example [11].

2.2 Communication Protocols

Assume that we are given jointly distributed random variables X and Y, taking values in finite sets $\mathcal{X}$ and $\mathcal{Y}$. Let R_A, R_B be a random variables, taking values in finite sets $\mathcal{R}_A$ and $\mathcal{R}_B$, such that $(X, Y), R_A, R_B$ are mutually independent.

Definition 4. *A private–coin communication protocol is a rooted binary tree, in which each non-leaf vertex is associated either with Alice or with Bob. For each non-leaf vertex v associated with Alice there is a function $f_v : \mathcal{X} \times \mathcal{R}_A \to \{0, 1\}$ and for each non-leaf vertex u associated with Bob there is a function $g_u : \mathcal{Y} \times \mathcal{R}_B \to \{0, 1\}$. For each non-leaf vertex one of the out-going edges is labeled by 0 and other is labeled by 1. Finally, for each leaf l there is a function $\phi_l : \mathcal{Y} \times \mathcal{R}_B \to \mathcal{O}$, where $\mathcal{O}$ denotes the set of all possible Bob's outputs.*

A computation according to a protocol runs as follows. Alice is given $x \in \mathcal{X}$, Bob is given $y \in \mathcal{Y}$. Assume that R_A takes a value r_a and R_B takes a value r_b. Alice and Bob start at the root of the tree. If they are in the non-leaf vertex v associated with Alice, then Alice sends $f_v(x, r_a)$ to Bob and they go by the edge labeled by $f_v(x, r_a)$. If they are in a non-leaf vertex associated with Bob then Bob sends $g_v(y, r_b)$ to Alice and they go by the edge labeled by $g_v(y, r_b)$. When they reach a leaf l Bob outputs the result $\phi_l(y, r_b)$.

A protocol is called *deterministic* if f_v, g_u and ϕ_l do not depend on the values of R_A, R_B.

A *randomized communication protocol* is a distribution over private-coin protocols with the same $\mathcal{X}$ for Alice and the same $\mathcal{Y}$ for Bob. The random variable with this distribution (*public randomness*) is denoted below by R. Before the execution starts, Alice and Bob sample R to choose the private-coin protocol to be executed.

A protocol is called *public-coin* if it is a distribution over deterministic protocols.

We distinguish between average-case communication complexity and the worst-case communication complexity. The (worst-case) communication complexity of a protocol π, denoted by $CC(\pi)$, is defined as the maximal possible depth of the leaf Alice and Bob may reach in π.

We say that protocol π communicates d bits on average (or expected length of the protocol is equal to d), if the expected depth of the leaf that Alice and Bob reach during the execution of the protocol π is equal to d, where the expectation is taken over X, Y and the protocol's randomness.

For the further introduction in Communication Complexity see [5].

3 Slepian-Wolf Theorem with Interaction

Consider the following auxiliary problem. Let A be a finite set. Assume that Alice receives an arbitrary $a \in A$ and Bob receives an arbitrary probability distribution μ on A. Alice wants to communicate a to Bob in about $\log(1/\mu(a))$ bits with small probability of error.

Lemma 1. *Let ε be a positive real and r a positive integer. There exists a public coin randomized communication protocol such that for all a in the support of μ the following hold:*

- *in the end of the communication Bob outputs $b \in A$ which is equal to a with probability at least $1 - \varepsilon$;*
- *the protocol communicates at most*

$$\log_2\left(\frac{1}{\mu(a)}\right) + \frac{\log_2\left(\frac{1}{\mu(a)}\right)}{r} + r + \log_2\left(\frac{1}{\varepsilon}\right) + 2$$

bits, regardless of the randomness.
- *the number of rounds in the protocol does not exceed*

$$\frac{2\log_2\left(\frac{1}{\mu(a)}\right)}{r} + 2.$$

Proof. Alice and Bob interpret each portion of $|A|$ consecutive bits from the public randomness source as a table of a random function $h : A \to \{0,1\}$. That is, we will think that they have access to a large enough family of mutually independent random functions of the type $A \to \{0,1\}$. Those functions will be called *hash functions* and their values *hash values* below.

The first set $k = \left\lceil \log_2\left(\frac{1}{\varepsilon}\right) \right\rceil$. Then for all $i = 0, 1 \ldots$ Bob sets:

$$S_i = \left\{ x \in A \mid \mu(x) \in (2^{-i-1}, 2^{-i}] \right\}.$$

At the beginning Alice sends k hash values of a. Then Alice and Bob work in stages numbered $1, 2 \ldots$.

On *Stage t*:

1. Alice sends r new hash values of a to Bob so that the total number of hash values of a available to Bob be $k + rt$.
2. For each $i \in \{r(t-1), \ldots, rt-1\}$ Bob computes set S'_i, which consists of all elements from S_i, which agree with all Alice's hash values.
3. If there exists $i \in \{r(t-1), \ldots, rt-1\}$ such that $S'_i \neq \varnothing$, then Bob sends 1 to Alice, outputs any element of S'_i and they terminate. Otherwise Bob sends 0 to Alice and they proceed to Stage $t+1$.

Let us at first show that the protocol terminates for all a in the support of μ. Assume that Alice has a and Bob has μ. Let $i = \left\lfloor \log_2 \left(\frac{1}{\mu(a)} \right) \right\rfloor$ so that $a \in S_i$. The protocol terminates on Stage t where

$$r(t-1) \leq i \leq rt - 1 \tag{2}$$

or earlier. Indeed all hash values of a available to Bob on Stage t coincide with hash values of some element of S_i (for instance, with those of a).

Thus Alice sends at most $k + rt$ bits to Bob and Bob sends at most t bits to Alice. The left-hand size of (2) implies that $t \leq \frac{i}{r} + 1$. Therefore Alice's communication is bounded by

$$k + rt \leq k + r \left(\frac{i}{r} + 1 \right)$$
$$= \left\lceil \log_2 \left(\frac{1}{\varepsilon} \right) \right\rceil + i + r$$
$$\leq \log_2 \left(\frac{1}{\mu(a)} \right) + r + \log_2 \left(\frac{1}{\varepsilon} \right) + 1,$$

and Bob's communication is bounded by

$$t \leq \frac{i}{r} + 1 \leq \frac{\log_2 \left(\frac{1}{\mu(a)} \right)}{r} + 1.$$

These two bounds imply that the total communication length is at most $\log_2 \left(\frac{1}{\mu(a)} \right) + \frac{\log_2 \left(\frac{1}{\mu(a)} \right)}{r} + r + \log_2 \left(\frac{1}{\varepsilon} \right) + 2$. The number of rounds equals the length of Bob's communication, multiplied by 2. Hence this number is at most $\frac{2 \log_2 \left(\frac{1}{\mu(a)} \right)}{r} + 2$. We conclude that the communication and the number of rounds are as short as required.

It remains to bound the error probability. An error may occur, if for some t a set S_i considered on Stage t has an element $b \neq a$ which agrees with hash values sent from Alice. At that time Bob has already $k + rt \geq k + i + 1$ hash values, where the inequality follows from (2). The probability that $k + i + 1$ hash values of b

coincide with those of a is 2^{-k-i-1}. Hence by union bound error probability does not exceed

$$\sum_{i=0}^{\infty} |S_i| 2^{-k-i-1} = 2^{-k} \sum_{i=0}^{\infty} |S_i| 2^{-i-1} < 2^{-k} \sum_{i=0}^{\infty} \sum_{x \in S_i} \mu(x)$$

$$= 2^{-k} \sum_{x \in A} \mu(x) = 2^{-k} = 2^{-\lceil \log_2(\frac{1}{\varepsilon}) \rceil} \le \varepsilon.$$

Theorem 1. *Let X, Y be jointly distributed random variables that take values in the finite sets $\mathcal{X}$ and $\mathcal{Y}$. Then for every positive ε and positive integer r there exists a public-coin protocol with the following properties.*

- *For every pair (x, y) from the support of (X, Y) with probability at least $1 - \varepsilon$ Bob outputs x;*
- *The expected length of communication is at most*

$$H(X|Y) + \frac{H(X|Y)}{r} + r + \log_2\left(\frac{1}{\varepsilon}\right) + 2.$$

- *The expected number of rounds is at most*

$$\frac{2H(X|Y)}{r} + 2.$$

Proof. On input x, y, Alice and Bob run protocol of Lemma 1 with $A = \mathcal{X}$, $a = x$ and μ equal to the distribution of X, conditioned on the event $Y = y$. Notice that Alice knows a and Bob knows μ.

Let us show that all the requirements are fulfilled for this protocol. The first requirement immediately follows from the first property of the protocol of Lemma 1.

From the second and the third property of the protocol of Lemma 1 it follows that for input pair x, y out protocol communicates at most:

$$\log_2\left(\frac{1}{\Pr[X = x|Y = y]}\right) + \frac{\log_2\left(\frac{1}{\Pr[X=x|Y=y]}\right)}{r} + r + \log_2\left(\frac{1}{\varepsilon}\right) + 2$$

bits in at most

$$\frac{\log_2\left(\frac{1}{\Pr[X=x|Y=y]}\right)}{r} + 2$$

rounds. Recalling that

$$H(X|Y) = \sum_{(x,y) \in \mathcal{X} \times \mathcal{Y}} \Pr[X = x, Y = y] \log_2\left(\frac{1}{\Pr[X = x|Y = y]}\right)$$

we see on average the communication and the number of rounds are as short as required.

Theorem 1 provides a trade off between the communication and the number of rounds.

- To obtain a protocol with minimal communication set $r = \left\lceil \sqrt{H(X|Y)} \right\rceil$. For such r the protocol communicates at most $H(X|Y) + 2\sqrt{H(X|Y)} + O\left(\log_2 \frac{1}{\varepsilon}\right)$ bits on average.
- To obtain a protocol with a constant number of rounds on average set, for example, $r = \lceil H(X|Y) \rceil$. For such r the protocol communicates at most $2H(X|Y) + O\left(\log_2 \frac{1}{\varepsilon}\right)$ bits on average in at most 4 rounds on average.
- In a same manner for every $\delta \in (0, 0.5)$ we can obtain a protocol with the expected communication at most $H(X|Y) + O\left(H(X|Y)^{0.5+\delta}\right)$ and the expected number of rounds at most $O\left(H(X|Y)^{0.5-\delta}\right)$.

Remark. One may wonder whether there exists a private-coin communication protocol with the same properties as the protocol of Theorem 1. Newman's theorem [6] states that every public-coin protocol can be transformed into a private-coin protocol at the expense of increasing the error probability by δ and the worst case communication by $O(\log \log |\mathcal{X} \times \mathcal{Y}| + \log 1/\delta)$ (for any positive δ). Lemma 1 provides an upper bound for the error probability and communication of our protocol for each pair of inputs. Repeating the arguments from the proof of Newman's theorem, we are able to transform the public-coin protocol of Lemma 1 into a private-coin one with the same trade off between the increase of error probability and the increase of communication length. It follows that for our problem there exists a private-coin communication protocol which errs with probability at most ε and communicates on average as many bits as the public-coin protocol from Theorem 1 plus extra $O(\log \log |\mathcal{X} \times \mathcal{Y}|)$ bits.

4   One-Round Compression

Information complexity of the protocol π with inputs (X, Y) is defined as

$$
\begin{aligned}
IC_\mu(\pi) &= I(X : \Pi|Y, R) + I(Y : \Pi|X, R) \\
&= I(X : \Pi|Y, R, R_B) + I(Y : \Pi|X, R, R_A) \\
&= I(X : \Pi, R, R_B|Y) + I(Y : \Pi, R, R_A|X),
\end{aligned}
$$

where R, R_A, R_B denote (shared, Alice's and Bob's) randomness, μ stands for the distribution of (X, Y) and Π stands for the concatenation of all bits sent in π (Π is called a *transcript*). The first term is equal to the information which Bob learns about Alice's input and the second term is equal to the information which Alice learns about Bob's input. Information complexity is an important concept in the Communication Complexity. For example, information complexity plays the crucial role in the Direct-Sum problem [10].

We will consider the special case when π is *one-round*. In this case Alice sends one message Π to Bob, then Bob outputs the result (based on his input, his randomness, and Alice's message) and the protocol terminates. Since Alice learns nothing, information complexity can be re-written as

$$I = IC_\mu(\pi) = I(X : \Pi | Y, R).$$

Our goal is to simulate a given one-round protocol π with another protocol τ which has the same input space (X, Y) and whose expected communication complexity is close to I. The new protocol τ may be many-round. The quality of simulation will be measured by the statistical distance. Statistical distance between random variables A and B, both taking values in the set V, equals

$$\delta(A, B) = \max_{U \subset V} |\Pr[A \in U] - \Pr[B \in U]|.$$

One of the main results of [3] is the following theorem.

Theorem 2. *For every one-round protocol π and for every probability distribution μ there is a public-coin protocol τ with expected length (with respect to μ and the randomness of τ) at most $I + 5\sqrt{I} + O\left(\log_2 \frac{1}{\varepsilon}\right)$ such that for each pair of inputs (x, y) after termination of τ Bob outputs a random variable Π' with $\delta\left((\Pi | X = x, Y = y), (\Pi' | X = x, Y = y)\right) \leq \varepsilon$.*

We will show that Theorem 1 and together with the main result of [2] imply that we can replace $5\sqrt{I}$ by about $2\sqrt{I}$ in this theorem. More specifically,

Theorem 3. *For every one-round protocol π and for every probability distribution μ there is a public-coin protocol τ with expected length (with respect to μ and the randomness of τ) at most*

$$I + \log_2(I + O(1)) + 2\sqrt{I + \log_2(I + O(1))} + O\left(\log_2 \frac{1}{\varepsilon}\right)$$

such that for each pair of inputs (x, y) in the protocol τ Bob outputs Π' with $\delta\left((\Pi | X = x, Y = y), (\Pi' | X = x, Y = y)\right) \leq \varepsilon$

We want to transmit Alice's message Π to Bob (who knows Y and his randomness R) in many rounds so that the expected communication length is small. By Theorem 1 this task can be solved with error ε in expected communication

$$H(\Pi | Y, R) + 2\sqrt{H(\Pi | Y, R)} + O\left(\log_2 \frac{1}{\varepsilon}\right). \tag{3}$$

Assume first that the original protocol π uses only public randomness. Then

$$I = I(X : \Pi | Y, R) = H(\Pi | Y, R) - H(\Pi | X, Y, R) = H(\Pi | Y, R).$$

Indeed, $H(\Pi | X, Y, R) = 0$, since Π is defined by X, R. Thus (3) becomes

$$I + 2\sqrt{I} + O\left(\log_2 \frac{1}{\varepsilon}\right)$$

and we are done.

In general case, when the original protocol uses private randomness, I can be much smaller than $H(\Pi|Y,R)$. Fortunately, by the following theorem from [2] we can remove private coins from the protocol with only a slight increase in information complexity.

Theorem 4. *For every one-round protocol π and for every probability distribution μ there is a one-round public-coin protocol π' with information complexity $IC_\mu(\pi') \leq I + \log_2(I + O(1))$ such that for each pairs of inputs (x,y) in the protocol π' Bob outputs Π' for which $\Pi'|X = x, Y = y$ and $\Pi|X = x, Y = y$ are identically distributed.*

Combining this theorem with our main result (Theorem 1), we obtain Theorem 3.

5   A Lower Bounds for the Average-Case Communication

Let (X, Y) be a pair of jointly distributed random variables. Assume that π is a deterministic protocol to transmit X from Alice to Bob who knows Y. Let $\pi(X, Y)$ stand for the result output by the protocol π for input pair (X, Y). We assume that for at least $1 - \varepsilon$ input pairs this result is correct:

$$\Pr[\pi(X, Y) \neq X)] \leq \varepsilon.$$

It is not hard to see that in this case the expected communication length cannot be much less than $H(X|Y)$ bits on average. Moreover, this applies for communication from Alice to Bob only.

Proposition 1. *For every deterministic protocol as above the expected communication from Alice to Bob is at least $H(X|Y) - \varepsilon \log_2 |\mathcal{X}| - 1$.*

The proof of this proposition is omitted due to space constraints.

There are random variables for which this lower bound is tight. For instance, let Y be empty and let X take the value $x \in \{0,1\}^n$ with probability $\varepsilon/2^n$ (for all such x) and let $X = $ (the empty string) with the remaining probability $1 - \varepsilon$. Then the trivial protocol with no communication solves the job with error probability ε and $H(X|Y) \approx \varepsilon \log_2 |\mathcal{X}|$.

In this section we consider the following question: are there a random variables (X, Y), for which for every public-coin communication protocol the expected communication is significantly larger than $H(X|Y)$, say close to the upper bound $H(X|Y) + 2\sqrt{H(X|Y)} + \log_2\left(\frac{1}{\varepsilon}\right)$ of Theorem 1?

Orlitsky [7] showed that if no error is allowed and the support of (X, Y) is a Cartesian product, then every deterministic protocol must communicate $H(X)$ bits on average.

Proposition 2. *Let (X, Y) be a pair of jointly distributed random variables whose support is a Cartesian product. Assume that π is a deterministic protocol, which transmits X from Alice to Bob who knows Y and*

$$\Pr[\pi(X, Y) \neq X)] = 0.$$

Then the expected length of π is at least $H(X)$.

This result can be easily generalized to the case when π is public-coin.

The main result of this section states that there are random variables (X, Y) such that transmission of X from Alice to Bob with error probability ε requires $H(X|Y) + \Omega\left(\log_2\left(\frac{1}{\varepsilon}\right)\right)$ bits on average.

The random variables X, Y are specified by two parameters, $\delta \in (0, 1/2)$ and $n \in \mathbb{N}$. Both random variables take values in $\{0, 1, \ldots, n\}$ and are distributed as follows: Y is distributed uniformly in $\{0, 1, \ldots, n\}$ and $X = Y$ with probability $1 - \delta$ and X is uniformly distributed in $\{0, 1, \ldots, n\} \setminus \{X\}$ with the remaining probability δ. That is,

$$\Pr[X = i, Y = j] = \frac{(1 - \delta)\delta_{ij} + \frac{\delta}{n}(1 - \delta_{ij})}{n + 1},$$

where δ_{ij} stands for the Kronecker's delta. Notice that X is uniformly distributed on $\{0, 1, \ldots, n\}$ as well. A straightforward calculation reveals that

$$\Pr[X = i | Y = j] = \frac{\Pr[X = i, Y = j]}{\Pr[Y = j]} = (1 - \delta - \frac{\delta}{n})\delta_{ij} + \frac{\delta}{n}$$

and

$$H(X|Y) = (1 - \delta)\log_2\left(\frac{1}{1 - \delta}\right) + \delta\log_2\left(\frac{n}{\delta}\right) = \delta\log_2 n + O(1).$$

We will think of δ as a constant, say $1/4$. For one-way protocol we are able to show that communication length must be close to $\log n$, which is about $1/\delta$ times larger than $H(X|Y)$:

Proposition 3. *Assume that π is a one-way deterministic protocol, which transmits X from Alice to Bob who knows Y and*

$$\Pr[\pi(X, Y) \neq X)] \leq \varepsilon.$$

Then the expected length of π is at least $\left(1 - \frac{\varepsilon}{\delta}\right)\log_2(n + 1) - 2$.

This result explains why the one-way one-shot analogue of the Slepian–Wolf theorem is not possible.

Proof. Let S be the number of leafs in π. For each $j \in \{0, 1, \ldots, n\}$

$$\#\{i \in \{0, 1, \ldots, n\} \mid \pi(i, j) = i\} \leq S.$$

Hence the error probability ε is at least $(n + 1 - S)\frac{\delta}{n}$. This implies that

$$S \geq n\left(1 - \frac{\varepsilon}{\delta}\right) + 1 \geq (n + 1)\left(1 - \frac{\varepsilon}{\delta}\right).$$

Let $\Pi(X)$ denote the leaf Alice and Bob reach in π (since the protocol is one-way, the leaf depends only on X). The expected length of $\Pi(X)$ is at least $H(\Pi)$ (identify each leaf with the binary string, written on the path from the root to this leaf in the protocol tree; the set of all these strings is prefix–free). Let $l_1, l_2, \ldots, l_S$ be the list of all leaves in the support of the random variable $\Pi(X)$. As X is distributed uniformly, we have

$$\Pr[\Pi = l_i] \geq \frac{1}{n+1}$$

for all i. The statement follows from

Lemma 2. *Assume that* $p_1, \ldots, p_k, q_1, \ldots, q_k \in (0,1)$ *satisfy*

$$\sum_{i=1}^{k} p_i = 1,$$

$$\forall i \in \{1, \ldots, k\} \qquad p_i \geq q_i.$$

Then

$$\sum_{i=1}^{k} p_i \log_2 \frac{1}{p_i} \geq \sum_{i=1}^{k} q_i \log_2 \frac{1}{q_i} - 2.$$

The proof of this technical lemma is omitted due to space constraints. The lemma implies that

$$H(\Pi) = \sum_{i=1}^{S} \Pr[\Pi = l_i] \log_2 \left(\frac{1}{\Pr[\Pi = l_i]} \right)$$

$$\geq \frac{S}{n+1} \log_2(n+1) - 2 \geq \left(1 - \frac{\varepsilon}{\delta}\right) \log_2(n+1) - 2.$$

The next theorem states that for any fixed δ every two-way public-coin protocol with error probability ε must communicate about $H(X|Y) + (1-\delta)\log_2(1/\varepsilon)$ bits on average.

Theorem 5. *Assume that π is a public-coin communication protocol which transmits X from Alice to Bob who knows Y and*

$$\Pr[X' \neq X] \leq \varepsilon,$$

where X' denotes the Bob's output and the probability is taken with respect to input distribution and public randomness of π. Then the expected length of π is at least

$$(1 - \delta - \delta/n) \log_2 \left(\frac{\delta}{\varepsilon + \delta/n} \right) + (\delta - 2\varepsilon) \log_2(n+1) - 2\delta.$$

The lower bound in this theorem is quite complicated and comes from its proof. To understand this bound assume that δ is a constant, say $\delta = 1/4$, and $\frac{1}{n} \leq \varepsilon \leq \frac{1}{\log_2 n}$. Then $H(X|Y) = (1/4)\log_2 n + O(1)$ and the lower bound becomes

$$\left(1 - \frac{1}{4} - \frac{1}{4n}\right)\log_2\left(\frac{\frac{1}{4}}{\varepsilon + \frac{1}{4n}}\right) + (1/4 - 2\varepsilon)\log_2(n+1) - \frac{1}{2}$$

Condition $\frac{1}{n} \leq \varepsilon$ implies that the first term is equal to

$$(3/4)\log_2\left(\frac{1}{\varepsilon}\right) - O(1).$$

Condition $\varepsilon \leq \frac{1}{\log_2 n}$ implies that the seconds term is equal to

$$(1/4)\log_2 n - O(1).$$

Therefore under these conditions the lower bound becomes

$$(1/4)\log_2 n + (3/4)\log_2\left(\frac{1}{\varepsilon}\right) - O(1) = H(X|Y) + (3/4)\log_2\left(\frac{1}{\varepsilon}\right) - O(1).$$

Proof. Let us start with the case when π is deterministic. Let $\Pi = \Pi(X,Y)$ denote the leaf Alice and Bob reach in the protocol π for input pair (X,Y). As we have seen, the expected length of communication is at least the entropy $H(\Pi(X,Y))$. Let $l_1, \ldots, l_S$ denote all the leaves in the support of the random variable $\Pi(X,Y)$. The set $\{(x,y) \mid \Pi(x,y) = l_i\}$ is a combinatorial rectangle $R_i \subset \{0,1,\ldots,n\} \times \{0,1,\ldots,n\}$. Imagine $\{0,1,\ldots,n\} \times \{0,1,\ldots,n\}$ as a table in which Alice owns columns and Bob owns rows. Let h_i be the height of R_i and w_i be the width of R_i. Let d_i stand for the number of diagonal elements in R_i (pairs of the form (j,j)). By definition of (X,Y) we have

$$\Pr[\Pi(X,Y) = l_i] = \frac{(1-\delta)d_i}{n+1} + \frac{\delta(h_i w_i - d_i)}{n(n+1)}. \tag{4}$$

The numbers $\{\Pr[\Pi(X,Y) = l_i]\}_{i=1}^{S}$ define a probability distribution over the set $\{1,2,\ldots,S\}$ and its entropy equals $H(\Pi(X,Y))$. Equation (4) represents this distribution as a weighted sum of the following distributions: $\left\{\frac{d_i}{n+1}\right\}_{i=1}^{S}$ and $\left\{\frac{h_i w_i}{(n+1)^2}\right\}_{i=1}^{S}$. That is, Eq. (4) implies that

$$\{\Pr[\Pi = l_i]\}_{i=1}^{S} = (1 - \delta - \delta/n)\left\{\frac{d_i}{n+1}\right\}_{i=1}^{S} + (\delta + \delta/n)\left\{\frac{h_i w_i}{(n+1)^2}\right\}_{i=1}^{S}.$$

Since entropy is concave, we have

$$H(\Pi) = H\left(\{\Pr[\Pi = l_i]\}_{i=1}^{S}\right)$$

$$\geq (1 - \delta - \delta/n)H\left(\left\{\frac{d_i}{n+1}\right\}_{i=1}^{S}\right) + (\delta + \delta/n)H\left(\left\{\frac{h_i w_i}{(n+1)^2}\right\}_{i=1}^{S}\right) \tag{5}$$

The lower bound of the theorem follows from lower bounds of the entropies of these distributions.

A lower bound for $H\left(\left\{\frac{d_i}{n+1}\right\}_{i=1}^{S}\right)$. In each row of R_i there is at most 1 element (x, y), for which $\pi(x, y) = x$. The rectangle R_i consists of d_i diagonal elements and hence there are at least $d_i^2 - d_i$ elements (x, y) in R_i for which $\pi(x, y) \neq x$. Summing over all i we get

$$\varepsilon \geq \sum_{i=1}^{S} \frac{\delta(d_i^2 - d_i)}{n(n+1)}$$

and thus

$$\sum_{i=1}^{S} \left(\frac{d_i}{n+1}\right)^2 \leq \frac{\varepsilon + \delta/n}{\delta}.$$

Since Renyi entropy is a lower bound for the Shannon entropy, we have

$$H\left(\left\{\frac{d_i}{n+1}\right\}_{i=1}^{S}\right) \geq \log_2\left(\frac{1}{\sum_{i=1}^{S}\left(\frac{d_i}{n+1}\right)^2}\right) \geq \log_2\left(\frac{\delta}{\varepsilon + \delta/n}\right).$$

A lower bound for $H\left(\left\{\frac{h_i w_i}{(n+1)^2}\right\}_{i=1}^{S}\right)$. In R_i, there are at most h_i good pairs (for which π works correctly). At most d_i of them has probability $\frac{1-\delta}{n+1}$. Hence

$$\Pr[\Pi = l_i, \pi(X, Y) = X] \leq \frac{(1-\delta)d_i}{n+1} + \frac{\delta(h_i - d_i)}{n(n+1)}$$

and

$$1 - \varepsilon \leq \Pr[\pi(X, Y) = X] = \sum_{i=1}^{S} \Pr[\Pi = l_i, \pi(X, Y) = X]$$

$$\leq \sum_{i=1}^{S}\left(\frac{(1-\delta)d_i}{n+1} + \frac{\delta(h_i - d_i)}{n(n+1)}\right) = 1 - \delta - \delta/n + \frac{\delta}{n(n+1)}\sum_{i=1}^{S} h_i.$$

The last inequality implies that

$$\sum_{i=1}^{S} h_i \geq (1 - \varepsilon/\delta)(n+1)^2.$$

Since $h_i \leq n + 1$, we have

$$\sum_{i=1}^{S} \frac{h_i w_i}{(n+1)^2} \log_2\left(\frac{(n+1)^2}{h_i w_i}\right) \geq \sum_{i=1}^{S} \frac{h_i w_i}{(n+1)^2} \log_2\left(\frac{(n+1)^2}{(n+1)w_i}\right)$$

$$= -\log_2(n+1) + \sum_{i=1}^{S} h_i \frac{w_i}{(n+1)^2} \log_2\left(\frac{(n+1)^2}{w_i}\right).$$

Obviously $\frac{w_i}{(n+1)^2} \geq \frac{1}{(n+1)^2}$. By Lemma 2 we get

$$\sum_{i=1}^{S} h_i \frac{w_i}{(n+1)^2} \log_2 \left(\frac{(n+1)^2}{w_i} \right) \geq \left(\sum_{i=1}^{S} h_i \right) \frac{1}{(n+1)^2} \log_2 \left((n+1)^2 \right) - 2$$

$$\geq (2 - 2\varepsilon/\delta) \log_2(n+1) - 2.$$

Thus

$$H \left(\left\{ \frac{h_i w_i}{(n+1)^2} \right\}_{i=1}^{S} \right) \geq (1 - 2\varepsilon/\delta) \log_2(n+1) - 2,$$

and the theorem is proved for deterministic protocols.

Assume now that π is a public-coin protocol with public randomness R and let r be a possible value of R. Let π_r stand for the deterministic communication protocol obtained from π by fixing $R = r$. For any protocol τ let $\|\tau\|$ denote the random variable representing communication length of τ (which may depend on the input and the randomness). Finally, set $\varepsilon_r = \Pr[X' \neq X | R = r]$

Note that π_r transmits X from Alice to Bob with error probability at most ε_r (with respect to input distribution). Since π_r is deterministic, the expected length of π_r is at least:

$$\mathrm{E}\|\pi_r\| \geq (1 - \delta - \delta/n) \log_2 \left(\frac{\delta}{\varepsilon_r + \delta/n} \right) + (\delta - 2\varepsilon_r) \log_2(n+1) - 2\delta.$$

Since $\mathrm{E}_{r \sim R}\varepsilon_r = \varepsilon$ and by concavity of log:

$$\mathrm{E}\|\pi\| = \mathrm{E}_{r \sim R}\mathrm{E}\|\pi_r\|$$

$$\geq \mathrm{E}_{r \sim R} \left[(1 - \delta - \delta/n) \log_2 \left(\frac{\delta}{\varepsilon_r + \delta/n} \right) + (\delta - 2\varepsilon_r) \log_2(n+1) - 2\delta \right]$$

$$\geq (1 - \delta - \delta/n) \log_2 \left(\frac{\delta}{\varepsilon + \delta/n} \right) + (\delta - 2\varepsilon) \log_2(n+1) - 2\delta.$$

References

1. Bauer, B., Moran, S., Yehudayoff, A.: Internal compression of protocols to entropy
2. Braverman, M., Garg, A.: Public vs private coin in bounded-round information. In: Esparza, J., Fraigniaud, P., Husfeldt, T., Koutsoupias, E. (eds.) ICALP 2014. LNCS, vol. 8572, pp. 502–513. Springer, Heidelberg (2014)
3. Braverman, M., Rao, A.: Information equals amortized communication. In: 2011 IEEE 52nd Annual Symposium on Foundations of Computer Science (FOCS), pp. 748–757. IEEE (2011)
4. Brody, J., Buhrman, H., Koucky, M., Loff, B., Speelman, F., Vereshchagin, N.: Towards a reverse newman's theorem in interactive information complexity. In: 2013 IEEE Conference on Computational Complexity (CCC), pp. 24–33. IEEE (2013)
5. Kushilevitz, E., Nisan, N.: Communication Complexity. Cambridge University Press, Cambridge (2006)

6. Newman, I.: Private vs. common random bits in communication complexity. Inf. Process. Lett. **39**(2), 67–71 (1991)
7. Orlitsky, A.: Average-case interactive communication. IEEE Trans. Inf. Theory **38**(5), 1534–1547 (1992)
8. Shannon, C.E.: A mathematical theory of communication. ACM SIGMOBILE Mob. Comput. Commun. Rev. **5**(1), 3–55 (2001)
9. Slepian, D., Wolf, J.K.: Noiseless coding of correlated information sources. IEEE Trans. Inf. Theory **19**(4), 471–480 (1973)
10. Weinstein, O.: Information complexity and the quest for interactive compression. ACM SIGACT News **46**(2), 41–64 (2015)
11. Yeung, R.W.: Information Theory and Network Coding. Springer, New York (2008)

Level Two of the Quantifier Alternation Hierarchy over Infinite Words

Manfred Kufleitner and Tobias Walter[(✉)]

FMI, Universität Stuttgart, Stuttgart, Germany
{kufleitner,walter}@fmi.uni-stuttgart.de

Abstract. The study of various decision problems for logic fragments has a long history in computer science. This paper is on the membership problem for a fragment of first-order logic over infinite words; the membership problem asks for a given language whether it is definable in some fixed fragment. The alphabetic topology was introduced as part of an effective characterization of the fragment Σ_2 over infinite words. Here, Σ_2 consists of the first-order formulas with two blocks of quantifiers, starting with an existential quantifier. Its Boolean closure is $\mathbb{B}\Sigma_2$. Our first main result is an effective characterization of the Boolean closure of the alphabetic topology, that is, given an ω-regular language L, it is decidable whether L is a Boolean combination of open sets in the alphabetic topology. This is then used for transferring Place and Zeitoun's recent decidability result for $\mathbb{B}\Sigma_2$ from finite to infinite words.

1 Introduction

Over finite words, the connection between finite monoids and regular languages is highly successful for studying logic fragments, see e.g. [2,19]. Over infinite words, the algebraic approach uses infinite repetitions. Not every logic fragment can express whether some definable property P occurs infinitely often. For instance, the usual approach for saying that P occurs infinitely often is as follows: for every position x there is a position $y > x$ satisfying $P(y)$. Similarly, P occurs only finitely often if there is a position x such that all positions $y > x$ satisfy $\neg P(y)$. Each of these formulas requires (at least) one additional change of quantifiers, which not all fragments can provide. It turns out that topology is a very useful tool for restricting the infinite behaviour of the algebraic approach accordingly, see e.g. [3,5,10,22]. In particular, the combination of algebra and topology is convenient for the study of languages in Γ^∞, the set of finite and infinite words over the alphabet Γ. In this paper, an *ω-regular language* is a regular subset of Γ^∞.

Topological ideas have a long history in the study of ω-regular languages. The Cantor topology is the most famous example in this context. We write G for the Cantor-open sets and F for the closed sets. The open sets in G are the languages of the form $W\Gamma^\infty$ for $W \subseteq \Gamma^*$. If X is a class of languages, then

---

This work was supported by the German Research Foundation (DFG) under grants DI 435/5-2 and DI 435/6-1.

© Springer International Publishing Switzerland 2016
A.S. Kulikov and G.J. Woeginger (Eds.): CSR 2016, LNCS 9691, pp. 223–236, 2016.
DOI: 10.1007/978-3-319-34171-2_16

X_δ consists of the countable intersections of languages in X and X_σ are the countable unions; moreover, we write $\mathbb{B}X$ for the Boolean closure of X. Since F contains the complements of languages in G, we have $\mathbb{B}F = \mathbb{B}G$. The Borel hierarchy is defined by iterating the operations $X \mapsto X_\delta$ and $X \mapsto X_\sigma$. The Borel hierarchy over the Cantor topology has many appearances in the context of ω-regular languages. For instance, an ω-regular language is deterministic if and only if it is in G_δ, see [8,21]. By McNaughton's Theorem [9], every ω-regular language is in $\mathbb{B}(G_\delta) = \mathbb{B}(F_\sigma)$. The inclusion $\mathbb{B}G \subset G_\delta \cap F_\sigma$ is strict, but the ω-regular languages in $\mathbb{B}G$ and $G_\delta \cap F_\sigma$ coincide [17].

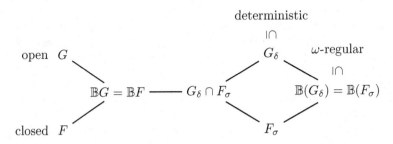

Let FO^k be the fragment of first-order logic which uses (and reuses) at most k variables. By Σ_m we denote the formulas with m quantifier blocks, starting with a block of existential quantifiers. Here, we assume that $x < y$ is the only binary predicate. Let us consider FO^1 as a toy example. With only one variable, we cannot make use of the binary predicate $x < y$. Therefore, in FO^1 we can say nothing but which letters occur, that is, a language is definable in FO^1 if and only if it is a Boolean combination of languages of the form $\Gamma^* a \Gamma^\infty$ for $a \in \Gamma$. Thus $\mathrm{FO}^1 \subseteq \mathbb{B}G$. It is an easy exercise to show that an ω-regular language is in FO^1 if and only if it is in $\mathbb{B}G$ and its syntactic monoid is both idempotent and commutative. The algebraic condition without the topology is too powerful since this would also include the language $\{a, b\}^* a^\omega$, which is not definable in FO^1. For the fragment $\mathbb{B}\Sigma_1$, the same topology $\mathbb{B}G$ with a different algebraic condition works, cf. [10, Theorems VI.3.7,VI.7.4 and VIII.4.5].

In the fragment Σ_2, we can define the language $\{a, b\}^* ab^\infty$ which is not deterministic and hence not in G_δ. Since the next level of the Borel hierarchy already contains all ω-regular languages, another topology is required. For this purpose, Diekert and the first author introduced the *alphabetic topology* [3]: the open sets in this topology are arbitrary unions of languages of the form uA^∞ for $u \in \Gamma^*$ and $A \subseteq \Gamma$. They showed that an ω-regular language is definable in Σ_2 if and only if it satisfies some particular algebraic property and if it is open in the alphabetic topology. Therefore, the canonical ingredient for an effective characterization of $\mathbb{B}\Sigma_2$ is the Boolean closure of the open sets in the alphabetic topology. Our first main result, Theorem 2, shows that, for a given ω-regular language L, it is decidable whether L is a Boolean combination of open sets in the alphabetic topology. As a by-product, we see that every ω-regular language which is a Boolean combination of arbitrary open sets in the alphabetic topology can be written as a Boolean combination of ω-regular open sets. This resembles a similar result for the Cantor topology [17].

A major breakthrough in the theory of regular languages over finite words is due to Place and Zeitoun [14]. They showed that, for a given regular language $L \subseteq \Gamma^*$, it is decidable whether L is definable in $\mathbb{B}\Sigma_2$. This solved a longstanding open problem, see e.g. [13, Section 8] for an overview. To date, no effective characterization of $\mathbb{B}\Sigma_3$ is known. Our second main result, Theorem 4, is to show that this decidability result transfers to languages in Γ^∞. If $\mathbf{V}_2$ is the algebraic counterpart of $\mathbb{B}\Sigma_2$ over finite words, then we show that $\mathbf{V}_2$ combined with the Boolean closure of the alphabetic topology yields a characterization of $\mathbb{B}\Sigma_2$ over Γ^∞. Combining the decidability of $\mathbf{V}_2$ with our first main result, the latter characterization is effective. The proof that $\mathbb{B}\Sigma_2$ satisfies both the algebraic and the topological restrictions follows a rather straightforward approach. The main difficulty is to show the converse: every language satisfying both the algebraic and the topological conditions is definable in $\mathbb{B}\Sigma_2$.

2   Preliminaries

Words

Let Γ be a finite alphabet. By Γ^* we denote the set of finite words over Γ; we write 1 for the empty word. The set of infinite words is Γ^ω and the set of finite and infinite words is $\Gamma^\infty = \Gamma^* \cup \Gamma^\omega$. By u, v, w we denote finite words and by α, β, γ we denote words in Γ^∞. In this paper a *language* is a subset of Γ^∞. Let $L \subseteq \Gamma^*$ and $K \subseteq \Gamma^\infty$. As usual L^* is the union of powers of L and $LK = \{u\alpha \mid u \in L, \alpha \in K\} \subseteq \Gamma^\infty$ is the concatenation of L and K. By L^ω we denote the set of words which are an infinite concatenation of words in L and the infinite concatenation $uu \cdots$ of the word u is written u^ω. A word $u = a_1 \ldots a_n$ is a *scattered subword* of v if $v \in \Gamma^* a_1 \Gamma^* \ldots a_n \Gamma^*$. The *alphabet* of a word is the set of all letters which appear in the word. The *imaginary alphabet* $\mathrm{im}(\alpha)$ of a word $\alpha \in \Gamma^\infty$ is the set of letters which appear infinitely often in α. Let $A^{\mathrm{im}} = \{\alpha \in \Gamma^\infty \mid \mathrm{im}(\alpha) = A\}$ be the set of words with imaginary alphabet A. In the following, we restrict ourselves to the study of ω-regular languages. A language $L \subseteq \Gamma^*$ is regular if it is recognized by a (deterministic) finite automaton. A language $K \subseteq \Gamma^\omega$ is ω-regular if it is recognized by a Büchi automaton. A language $L \subseteq \Gamma^\infty$ is ω-regular if $L \cap \Gamma^*$ is regular and $L \cap \Gamma^\omega$ is ω-regular.

First-Order logic

We consider first order logic FO over Γ^∞. Variables range over positions of the word. The atomic formulas in this logic are $\top$ for true, $x < y$ to compare two positions x and y and $\lambda(x) = a$ which is true if the word has an a at position x. One may combine those atomic formulas with the boolean connectives $\neg, \wedge$ and $\vee$ and quantifiers $\forall$ and $\exists$. A *sentence* φ is an FO formula without free variables. We write $\alpha \models \varphi$ if $\alpha \in \Gamma^\infty$ satisfies the sentence φ. The language defined by φ is $L(\varphi) = \{\alpha \in \Gamma^\infty \mid \alpha \models \varphi\}$. We classify the formulas of FO by counting the number of quantifier alternations, that is the number of alternations of $\exists$ and $\forall$. The fragment Σ_i of FO contains all FO-formulas in prenex normal form with i blocks of quantifiers $\exists$ or $\forall$, starting with a block of existential quantifiers. The fragment $\mathbb{B}\Sigma_i$ contains all Boolean combinations of formulas in Σ_i. We are particularly interested in the fragment Σ_2 and the Boolean combinations of formulas in Σ_2. A language L is definable in a fragment $\mathcal{F}$ (e.g. $\mathcal{F}$ is Σ_2 or $\mathbb{B}\Sigma_2$) if there exists a formula $\varphi \in \mathcal{F}$ such that $L = L(\varphi)$, i.e., if L is definable by some $\varphi \in \mathcal{F}$. The classes of languages defined by Σ_i and $\mathbb{B}\Sigma_i$ form a hierarchy, the quantifier alternation hierarchy. This hierarchy is strict, i.e., $\Sigma_i \subsetneq \mathbb{B}\Sigma_i \subsetneq \Sigma_{i+1}$ holds for all i, cf. [1,20].

Monomials

A *monomial* is a language of the form $A_0^* a_1 A_1^* a_2 \cdots A_{n-1}^* a_n A_n^\infty$ for $n \geq 0$, $a_i \in \Gamma$ and $A_i \subseteq \Gamma$. The number n is called the *degree*. In particular, A_0^∞ is a monomial of degree 0. A monomial is called k-monomial if it has degree at most k. In [3] it is shown that a language $L \subseteq \Gamma^\infty$ is in Σ_2 if and only if it is a finite union of monomials. We are interested in $\mathbb{B}\Sigma_2$ and thus in finite Boolean combination of monomials. For this, let $\equiv_k^\infty$ be the equivalence relation on Γ^∞ such that $\alpha \equiv_k^\infty \beta$ if α and β are contained in exactly the same k-monomials. Thus, $\equiv_k^\infty$-classes are Boolean combinations of monomials and every language in $\mathbb{B}\Sigma_2$ is a union of $\equiv_k^\infty$-classes for some k. Further, since there are only finitely many monomials of degree k, there are only finitely many $\equiv_k^\infty$-classes. The equivalence class of some word α in $\equiv_k^\infty$ is denoted by $[\alpha]_k^\infty$. Note, that such a characterization of $\mathbb{B}\Sigma_2$ in terms of monomials does not yield a decidable characterization.

Our characterization of languages $L \subseteq \Gamma^\infty$ in $\mathbb{B}\Sigma_2$ is based on the characterization of languages in $\mathbb{B}\Sigma_2$ over finite words. For this, we also introduce monomials over Γ^*. A *monomial* over Γ^* is a language of the form $A_0^* a_1 A_1^* a_2 \cdots A_{n-1}^* a_n A_n^*$ for $n \geq 1$, $a_i \in \Gamma$ and $A_i \subseteq \Gamma$. The degree is defined as above. Let $\equiv_k$ be the congruence on Γ^* which is defined by $u \equiv_k v$ if and only if u and v are contained in the same k-monomials over Γ^*. Again, a language $L \subseteq \Gamma^*$ is in $\mathbb{B}\Sigma_2$ if and only if it is a union of $\equiv_k$-classes for some k.

Algebra

In this paper all monoids are either finite or free. Finite monoids are a common way for defining regular and ω-regular languages. A monoid element e is *idempotent* if $e^2 = e$. An *ordered monoid* $(M, \leq)$ is a monoid equipped with a partial order which is compatible with the monoid multiplication, i.e., $s \leq t$ and $s' \leq t'$ implies $ss' \leq tt'$. Every monoid can be ordered using the identity as partial order. A homomorphism $h : (N, \leq) \to (M, \leq)$ between two ordered monoids must hold $s \leq t \Rightarrow h(s) \leq h(t)$ for $s, t \in N$. A *divisor* is the homomorphic image of a submonoid. A class of monoids which is closed under division and finite direct products is a *pseudovariety*. Eilenberg showed a correspondence between certain classes of languages (of finite words) and pseudovarieties [4]. A pseudovariety of ordered monoids is defined the same way as with unordered monoids, using homomorphisms of ordered monoids. The Eilenberg correspondence also holds for ordered monoids [12]. Let $\mathbf{V_{3/2}}$ be the pseudovariety of ordered monoids which corresponds to Σ_2 and $\mathbf{V_2}$ be the pseudovariety of monoids which corresponds to languages in $\mathbb{B}\Sigma_2$. Since $\Sigma_2 \subseteq \mathbb{B}\Sigma_2$, we obtain $\mathbf{V_{3/2}} \subseteq \mathbf{V_2}$ when ignoring the order. The connection between monoids and languages is given by the notion of *recognizability*. A language $L \subseteq \Gamma^*$ is *recognized* by an ordered monoid $(M, \leq)$ if there is a monoid homomorphism $h : \Gamma^* \to M$ such that $L = \cup \{h^{-1}(t) \mid s \leq t \text{ for some } s \in h(L)\}$. If M is not ordered, then this means that L is an arbitrary union of languages of the form $h^{-1}(t)$.

For ω-languages $L \subseteq \Gamma^\infty$ the notion of recognizability is slightly more technical. For simplicity, we only consider recognition by unordered monoids. Let $h : \Gamma^* \to M$ be a monoid homomorphism. If the homomorphism h is understood, we write $[s]$ for the language $h^{-1}(s)$. We call $(s, e) \in M \times M$ a *linked pair* if $e^2 = e$ and $se = s$. By Ramsey's Theorem [15] for every word $\alpha \in \Gamma^\infty$ there exists a linked pair (s, e) such that $\alpha \in [s][e]^\omega$. A language $L \subseteq \Gamma^\infty$ is *recognized* by h if

$$L = \bigcup \{[s][e]^\omega \mid (s, e) \text{ is a linked pair with } [s][e]^\omega \cap L \neq \emptyset\}.$$

Since $1^\omega = 1$, the language $[1]^\omega$ also contains finite words. We thus obtain recognizability of languages of finite words as a special case.

Next, we define syntactic homomorphisms and syntactic monoids; these are the minimal recognizers of an ω-regular language. Let $L \subseteq \Gamma^\infty$ be an ω-regular language. The *syntactic monoid* of L is defined as the quotient $\mathrm{Synt}(L) = \Gamma^* / \approx_L$ where $u \approx_L v$ holds if and only if for all $x, y, z \in \Gamma^*$ we have both $xuyz^\omega \in L \Leftrightarrow xvyz^\omega \in L$ and $x(uy)^\omega \in L \Leftrightarrow x(vy)^\omega \in L$. The syntactic monoid can be ordered by the partial order $\preceq_L$ defined by $u \preceq_L v$ if for all $x, y, z \in \Gamma^*$ we have $xuyz^\omega \in L \Rightarrow xvyz^\omega \in L$ and $x(uy)^\omega \in L \Rightarrow x(vy)^\omega \in L$. The *syntactic homomorphism* $h_L : \Gamma^* \to \mathrm{Synt}(L)$ is given by $h_L(u) = [u]_{\approx_L}$. One can effectively compute the syntactic homomorphism of L. The syntactic monoid $\mathrm{Synt}(L)$ satisfies the property that L is ω-regular if and only if $\mathrm{Synt}(L)$ is finite and the syntactic homomorphism h_L recognizes L, see e.g. [10,21]. Every pseudovariety is generated by its syntactic monoids [4], i.e., every monoid in a given pseudovariety is a divisor of a direct product of syntactic monoids. The importance of the syntactic monoid of a language $L \subseteq \Gamma^\infty$ is that it is the smallest monoid recognizing L:

Lemma 1. *Let $L \subseteq \Gamma^\infty$ be a language which is recognized by a homomorphism $h : \Gamma^* \to (M, \leq)$. Then, $(\mathrm{Synt}(L), \preceq_L)$ is a divisor of $(M, \leq)$.*

Proof. We assume that h is surjective and show that $\mathrm{Synt}(L)$ is a homomorphic image of M. If h is not surjective, we can therefore conclude that $\mathrm{Synt}(L)$ is a divisor of M. We show that $h(u) \leq h(v) \Rightarrow u \preceq_L v$. Let u, v be words with $h(u) \leq h(v)$ and denote $h^{-1}(h(w)) = [h(w)]$ for words w. Assume $xuyz^\omega \in L$, then there exists an index i such that $(h(xuyz^i), h(z)^\omega)$ is a linked pair. Thus, $[h(xuyz^i)][h(z)]^\omega \subseteq L$ and by $h(u) \leq h(v)$ also $[h(xvyz^i)][h(z)]^\omega \subseteq L$. This implies $xvyz^\omega \in L$. The proof that $x(uy)^\omega \in L \Rightarrow x(vy)^\omega \in L$ is similar. Thus, $u \preceq_L v$ holds which shows the claim. $\square$

3   Alphabetic Topology

As mentioned in the introduction, combining algebraic and topological conditions is a successful approach for characterizations of language classes over Γ^∞. A topology on a set X is given by a family of subsets of X (called open) which are closed under finite intersections and arbitrary unions. We define the *alphabetic topology* on Γ^∞ by its basis $\{uA^\infty \mid u \in \Gamma^*, A \subseteq \Gamma\}$. Hence, an open set is given by $\bigcup_A W_A A^\infty$ with $W_A \subseteq \Gamma^*$. The alphabetic topology has been introduced in [3], where it is used as a part of the characterization of Σ_2.

Theorem 1 ([3]). *Let $L \subseteq \Gamma^\infty$ be an ω-regular language. Then $L \in \Sigma_2$ if and only if $\mathrm{Synt}(L) \in \mathbf{V}_{3/2}$ and L is open in the alphabetic topology.*

The alphabetic topology has by itself been the subject of further study [16]. We are particularly interested in Boolean combinations of open sets. An effective characterization of a language L being a Boolean combination of open sets in the alphabetic topology is given in the theorem below.

Theorem 2. *Let $L \subseteq \Gamma^\infty$ be an ω-regular language which is recognized by $h : \Gamma^* \to M$. Then the following are equivalent:*

1. *L is a Boolean combination of open sets in the alphabetic topology where each open set is ω-regular.*
2. *L is a Boolean combination of open sets in the alphabetic topology.*
3. *For all linked pairs $(s, e), (t, f)$ it holds that if there exists an alphabet C and words $\hat{e}, \hat{f}$ with $h(\hat{e}) = e, h(\hat{f}) = f$, $\mathrm{alph}(\hat{e}) = \mathrm{alph}(\hat{f}) = C$ and $s \cdot h(C^*) = t \cdot h(C^*)$, then $[s][e]^\omega \subseteq L \Leftrightarrow [t][f]^\omega \subseteq L$.*

Proof. "1 $\Rightarrow$ 2": This is immediate.

"2 $\Rightarrow$ 3": Let L be a Boolean combination of open sets in the alphabetic topology. Note that for $P, Q \subseteq \Gamma^*$ and $A, B \subseteq \Gamma$ it holds $PA^\infty \cap QB^\infty = (PA^* \cap QB^*)(A \cap B)^\infty$. Therefore, we may assume

$$L = \bigcup_{i=1}^{n} \left((P_i A_i^\infty) \setminus \left(\bigcup_{j=1}^{m_i} Q_{i,j} B_{i,j}^\infty \right) \right)$$

for some $P_i, Q_{i,j} \subseteq \Gamma^*$ and alphabets $A_i, B_{i,j} \subseteq \Gamma$.

Let (s, e) and (t, f) be some linked pairs, $C \subseteq \Gamma$ be an alphabet such that $s \cdot h(C^*) = t \cdot h(C^*)$ holds and there exist words $\hat{e}, \hat{f}$ with $h(\hat{e}) = e, h(\hat{f}) = f$ and $\mathrm{alph}(\hat{e}) = \mathrm{alph}(\hat{f}) = C$. Assume $[s][e]^\omega \subseteq L$, but $[t][f]^\omega \not\subseteq L$. Since h recognizes L, it suffices to show that $[t][f]^\omega \cap L$ is nonempty to obtain a contradiction. Let $u\hat{e}^\omega \in [s][e]^\omega \subseteq L$ for some $u \in [s]$. Since $s \cdot h(C^*) = t \cdot h(C^*)$, we may choose $x, y \in C^*$ such that $s \cdot h(x) = t$ and $t \cdot h(y) = s$.

The idea is to find an increasing sequence of words $u_\ell \in [s]$ and sets $I_\ell \subseteq \{1, \ldots, n\}$ such that $u_\ell C^\infty \cap \left(P_i A_i^\infty \setminus \left(\bigcup_{j=1}^{m_i} Q_{i,j} B_{i,j}^\infty \right) \right) = \emptyset$ for all $i \in I_\ell$. We can set $u_0 = u$ and $I_0 = \emptyset$. Consider the word $u_\ell \hat{e}^\omega \in L$. There exists an index $i \in$

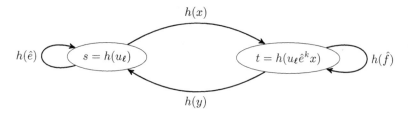

Fig. 1. Part of the right Cayley graph of M in the proof of "2 $\Rightarrow$ 3".

$\{1, \ldots, n\} \setminus I_\ell$ such that $u_\ell \hat{e}^\omega \in P_i A_i^\infty \setminus \left(\bigcup_{j=1}^{m_i} Q_{i,j} B_{i,j}^\infty \right)$. Choose a number k, such that $u_\ell \hat{e}^k \in P_i A_i^*$. Since $C = \text{alph}(\hat{e}) \subseteq A_i$, we conclude $\beta_\ell = u_\ell \hat{e}^k x \hat{f}^\omega \in P_i A_i^\infty$. By construction we have $\beta_\ell \in [t][f]^\omega$ and therefore, assuming $[t][f]^\omega \cap L = \emptyset$, there exists an index j such that $\beta_\ell \in Q_{i,j} B_{i,j}^\infty$. Analogously, there exists k' such that $u_\ell \hat{e}^k x \hat{f}^{k'} y C^\infty \subseteq Q_{i,j} B_{i,j}^\infty$. Hence we can choose $u_{\ell+1} = u_\ell \hat{e}^k x \hat{f}^{k'} y$ and $I_{\ell+1} = I_\ell \cup \{i\}$. Figure 1 gives an overview of the construction.

Since $u_\ell[e]^\omega \subseteq L \cap u_\ell C^\infty$, this construction has to fail at an index $\ell \leq n$. Therefore, the assumption is not justified and we have $[t][f]^\omega \cap L \neq \emptyset$, proving the claim.

"3 $\Rightarrow$ 1": Let $\alpha \in [s][e]^\omega \subseteq L$ for a linked pair (s, e) and let $C = \text{im}(\alpha)$ denote the imaginary alphabet of α. By $\alpha \in [s][e]^\omega$ and the definition of C, there exists an $\hat{e} \in C^*$ with $\text{alph}(\hat{e}) = C$ and $h(\hat{e}) = e$. Define

$$L(s, C) = [s]C^\infty \setminus \left(\bigcup_{D \subsetneq C} \Gamma^* D^\infty \cup \bigcup_{s \notin t \cdot h(C^*)} [t] C^\infty \right).$$

We have $\alpha \in L(s, C)$ and $L(s, C)$ is a Boolean combination of open sets in the alphabetic topology where each open set is ω-regular. There are only finitely many sets of the type $L(s, C)$. The idea is to saturate L with sets of this type, i.e., it suffices to show $L(s, C) \subseteq L$. For $C = \emptyset$, we have $L(s, C) = [s] \subseteq L$. Thus, we may assume $C \neq \emptyset$. Let $\beta \in L(s, C)$ be an arbitrary element and let (t, f) be a linked pair such that $\beta \in [t][f]^\omega$. Since β is in $L(s, C)$, there exists a prefix u of β such that $\beta \in uC^\omega$ and $u \in [s]$.

By $\beta \in [t][f]^\omega$, one gets $\beta = v\beta'$ with $v \in [t], \beta' \in [f]^\omega$. Using $tf = t$ and $C \neq \emptyset$, we may assume that u is a prefix of v, which implies $\beta' \in C^\omega$. Hence we have $t = h(v) \in h(uC^*) = s \cdot h(C^*)$. By construction $\beta \notin \bigcup_{s \notin t \cdot h(C^*)} [t] C^\infty$ and therefore $s \in t \cdot h(C^*)$. It follows $s \cdot h(C^*) = t \cdot h(C^*)$. Since $\beta \notin \bigcup_{D \subsetneq C} \Gamma^* D^\infty$, there must be a preimage of f of full alphabet C. Therefore, $\beta \in [t][f]^\omega \subseteq L$. $\square$

The alphabetic topology above is a refinement of the well-known Cantor topology. The Cantor topology is given by the basis $u\Gamma^\infty$ for $u \in \Gamma^*$. An ω-regular language L is a Boolean combination of open sets in the Cantor topology if and only if $[s][e]^\omega \subseteq L \Leftrightarrow [t][f]^\omega \subseteq L$ for all linked pairs (s, e) and (t, f) of the syntactic monoid of L with $s \mathcal{R} t$; cf. [3,10,21]. Here $s \mathcal{R} t$ denotes one of Green's relations: $s \mathcal{R} t$ if and only if $s \cdot \text{Synt}(L) = t \cdot \text{Synt}(L)$. Theorem 2 is a similar result, but one had to consider the alphabetic information of the linked pairs. Hence, one does not have $s \, \mathcal{R} \, t$ as condition, but rather $\mathcal{R}$-equivalence within a certain alphabet C.

Remark 1. The *strict alphabetic topology* on Γ^∞, which is introduced in [3], is given by the basis $\{uA^\infty \cap A^{\mathrm{im}} \mid u \in \Gamma^*, A \subseteq \Gamma\}$ and the open sets are of the form $\bigcup_A W_A A^\infty \cap A^{\mathrm{im}}$ with $W_A \subseteq \Gamma^*$. Reusing the proof of Theorem 2 it turns out, that it is equivalent to be a Boolean combination of open sets in the alphabetic topology and in the strictly alphabetic topology. Since $uA^\infty = \bigcup_{B \subseteq A} uA^* B^\infty \cap B^{\mathrm{im}}$, every open set in the alphabetic topology is also open in the strict alphabetic topology. Further, one can adapt the proof of "2 $\Rightarrow$ 3" of Theorem 2 to show that if L is a Boolean combination of open sets in the strict alphabetic topology, then item 3 of Theorem 2 holds.

4 The Fragment $\mathbb{B}\Sigma_2$

Place and Zeitoun have shown that $\mathbb{B}\Sigma_2$ is decidable over finite words. In particular, they have shown that, given the syntactic homomorphism of a language $L \subseteq \Gamma^*$, it is decidable if $L \in \mathbb{B}\Sigma_2$. Since every pseudovariety is generated by its syntactic monoids, the result of Place and Zeitoun can be stated as follows:

Theorem 3 ([14]). *The pseudovariety* $\mathbf{V_2}$ *corresponding to the* $\mathbb{B}\Sigma_2$-*definable languages in* Γ^* *is decidable.*

Our second main result charaterizes $\mathbb{B}\Sigma_2$-definable ω-regular languages. We use Theorem 3 as a black-box result.

Theorem 4. *Let* $L \subseteq \Gamma^\infty$ *be* ω-*regular. Then the following are equivalent:*

1. *L is a finite Boolean combination of monomials.*
2. *L is definable in* $\mathbb{B}\Sigma_2$.
3. *The syntactic homomorphism h of L satisfies:*
   (a) $\mathrm{Synt}(L) \in \mathbf{V_2}$ *and*
   (b) for all linked pairs $(s, e), (t, f)$ it holds that if there exists an alphabet C and words $\hat{e}, \hat{f}$ with $h(\hat{e}) = e, h(\hat{f}) = f$, $\mathrm{alph}(\hat{e}) = \mathrm{alph}(\hat{f}) = C$ and $s \cdot h(C^*) = t \cdot h(C^*)$, then $[s][e]^\omega \subseteq L \Leftrightarrow [t][f]^\omega \subseteq L$.

Note that item 3 of Theorem 4 is decidable: 3a is decidable by Theorem 3 and 3b is decibable since we can effectively compute the syntactic homomorphism h and $h(C^*)$ for all alphabets C.[1] We start with the difficult direction "3 $\Rightarrow$ 1" in the proof of Theorem 4. This is Proposition 1. The following lemma is an auxiliary result for Proposition 1.

---

[1] During the preparation of this submission, we learned that Pierron, Place and Zeitoun [11] independently found another proof for the decidability of $\mathbb{B}\Sigma_2$ over infinite words. For documenting the independency of the two proofs, we also include the technical report of our submission in the list of references [6].

Lemma 2. *For all k there exists a number ℓ such that for every set $\{M_1, \ldots, M_d\}$ of k-monomials over Γ^* and every w with $w \in M_i$ for all $i \in \{1, \ldots, d\}$, there exists an ℓ-monomial N over Γ^* with $w \in N$ and $N \subseteq \bigcap M_i$.*

Proof. Since the number of k-monomials over Γ^* is bounded, this induces a bound on d and one can iterate the statement. Therefore, it suffices to show the case $d = 2$. Consider two k-monomials $M_1 = A_0^* a_1 A_1^* a_2 \cdots A_{n-1}^* a_n A_n^*$ and $M_2 = B_0^* b_1 B_1^* b_2 \cdots B_{m-1}^* b_m B_m^*$. Since $w \in M_1$ and $w \in M_2$, it admits factorizations $w = u_0 a_1 u_1 a_2 \cdots u_{n-1} a_n u_n$ and $w = v_0 b_1 v_1 b_2 \cdots v_{m-1} b_m v_m$ such that $u_i \in A_i^*$ and $v_j \in B_j^*$. The factorizations mark the positions of the a_is and the b_js and pose an alphabetic condition for the factors in between. Thus, there exists a factorization $w = w_0 c_1 w_1 c_2 \cdots w_{\ell-1} c_\ell w_\ell$, such that the positions of c_i are exactly those, that are marked by a_i or b_j, i.e., $c_i = a_j$ or $c_i = b_j$ for some j. The words w_i are over some alphabet C_i such that $C_i = A_j \cap B_k$ for some j and k induced by the factorizations. In the case of consecutive marked positions, one can set $C_i = \emptyset$. Thus, we obtain a monomial $N = C_0^* c_1 C_1^* c_2 \cdots c_{p-1} C_{p-1}^* c_p C_p^*$ with $C_p = A_n \cap B_m$. An illustration of this construction can be found in Fig. 2. By construction $N \subseteq M_1$, $N \subseteq M_2$ and $w \in N$ holds. Since there are only finitely many monomials of degree k, the size of the number ℓ is bounded. $\square$

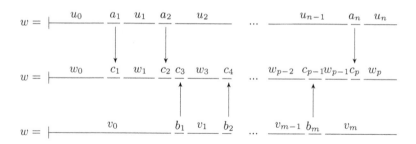

Fig. 2. Different factorizations in the proof of Lemma 2. In the situation of the figure it holds $C_0 = A_0 \cap B_0$, $C_1 = A_1 \cap B_0$, $C_2 = \emptyset$, $C_3 = A_2 \cap B_1$, $C_{p-2} = A_{n-1} \cap B_{m-1}$, $C_{p-1} = A_{n-1} \cap B_m$ and $C_p = A_n \cap B_m$.

An analysis of the proof of Lemma 2 yields the bound $\ell \leq n_k \cdot k$, where n_k is the number of distinct k-monomials over Γ^*. Next, we show that a language which is in $\mathbf{V_2}$ and is a Boolean combination of alphabetic open sets is a finite Boolean combination of monomials. One ingredient of the proof is Lemma 2: we are able to compress the information of a set of k-monomials which contain a fixed word into the information of a single ℓ-monomial that contains this fixed word.

Proposition 1. *Let $L \subseteq \Gamma^\infty$ be a Boolean combination of alphabetic open sets such that $\mathrm{Synt}(L) \in \mathbf{V_2}$. Then L is a finite Boolean combination of monomials.*

Proof. Let $h : \Gamma^* \to \mathrm{Synt}(L)$ be the syntactic homomorphism of L and consider the languages $h^{-1}(p)$ for $p \in \mathrm{Synt}(L)$. By Theorem 3 we obtain $h^{-1}(p) \in \mathbb{B}\Sigma_2$. Thus, there exists a number k such that for every $p \in M$ the language $h^{-1}(p)$ is saturated by $\equiv_k$, i.e., $u \equiv_k v \Rightarrow h(u) = h(v)$. By Lemma 2 there exists a number ℓ such that for every set $\{M_1, \ldots, M_n\}$ of k-monomials and every w with $w \in M_i$ for all $i \in \{1, \ldots, n\}$, there exists an ℓ-monomial N with $w \in N \subseteq \cap_{i=1}^n M_i$. Let $\alpha \equiv_\ell^\infty \beta$ and $\alpha \in L$. We show $\beta \in L$ which implies $L = \cup_{\alpha \in L}[\alpha]_\ell^\infty$ and thus that L is a finite Boolean combination of ℓ-monomials. Using Boolean combinations of monomials of the form $\Gamma^* a A^\infty$, one can test the imaginary alphabet of α and β. Hence we obtain $\mathrm{im}(\alpha) = \mathrm{im}(\beta)$ for the imaginary alphabets. For simplicity, we write $C = \{c_1, \ldots, c_m\}$ for the imaginary alphabet of α and β.

Let $u' \le \alpha$ and $v' \le \beta$ be prefixes such that for every ℓ-monomial $N = N' \cdot C^\infty$ with $\alpha, \beta \in N$ we have that some prefix of u', v' is in N'. Further, let $\alpha = u\alpha'$ and $\beta = v\beta'$ such that

- $u' \le u = u'u''$, $v' \le v = v'v''$,
- $(c_1 c_2 \cdots c_m)^k$ is a scattered subword of u'' and v'',
- and there exists linked pairs (s, e) and (t, f) such that $s = h(u)$, $t = h(v)$, $\alpha' \in [e]^\omega$ and $\beta' \in [f]^\omega$.

Note that, by the choice of u', v', we have $\alpha', \beta' \in C^\infty$. We show that $s \cdot h(C^*) = t \cdot h(C^*)$, which implies $\beta \in L$ by Theorem 2. By symmetry, it suffices to show $t \in s \cdot h(C^*)$. Consider the set of k-monomials $N_i = N_i' C^\infty$ which hold at u, i.e., such that $u \in N_i'$ and $\alpha' \in C^\infty$. By the choice of ℓ, there exists an ℓ-monomial N' such that $u \in N'$ and $N' \subseteq \cap_i N_i'$. Since $u \in N'$, we obtain $\alpha \in N := N'C^\infty$ and by $\alpha \equiv_\ell^\infty \beta$ the membership $\beta \in N$ holds. By construction of v, there exists a word $\hat{v}$ with $\hat{v} \le v' \le v$, $\hat{v} \in N'$ and $\hat{\beta} \in C^\infty$ with $\hat{\beta}$ being defined by $\beta = \hat{v}\hat{\beta}$. Let $v = \hat{v}x$, then $x \in C^*$. The situation is depicted in Fig. 3. We show that $ux \equiv_k v$ which then implies $t \in sh(C^*)$.

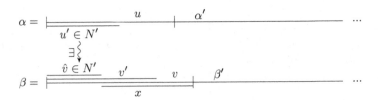

Fig. 3. Factorization of α and β in the proof of Proposition 1

Let M be a k-monomial. If $ux \in M$, then there exists a factorization $M = M_1 M_2$ where M_1, M_2 are k-monomials with $u \in M_1$ and $x \in M_2$. Since $u\beta' \in M_1 C^\infty$, we obtain $\hat{v} \in N' \subseteq M_1$ by the definition of N'. We conclude that $v = \hat{v}x \in M_1 M_2 = M$.

If $v = \hat{v}x \in M$, then there exists a factorization of the monomial $M = M_1 M_2$ where M_1, M_2 are k-monomials with $\hat{v} \in M_1$ and $x \in M_2$. Since $(c_1 c_2 \cdots c_m)^k$

is a scattered subword of x, there must be some A_i^* in the monomial M_2 such that $C \subseteq A_i$ by the pigeonhole principle. Thus, there exists a factorisation $M_2 = M_{21} M_{22}$ in k-monomials M_{21}, M_{22} such that $M_{21} \cdot C^* = M_{21}$. Let $x = x'x''$ such that $x' \in M_{21}$ and $x'' \in M_{22}$ and consider $\beta = \hat{v}x\beta' \in M_1 M_{21} \cdot C^\infty$. Since $\alpha \equiv_\ell^\infty \beta$, we obtain $\alpha \in M_1 M_{21} \cdot C^\infty$. By construction, some prefix of u is in $M_1 M_{21}$ and by $M_{21} \cdot C^* = M_{21}$ and $x' \in C^*$, we obtain $ux' \in M_1 M_{21}$. Thus, $ux = ux' \cdot x'' \in M_1 M_{21} \cdot M_{22} = M$ holds. We conclude $ux \equiv_k v$ and thus $t = h(v) = h(ux) \in s \cdot h(C^*)$. $\qquad\square$

It is well-known, that the direct product $(g \times h) : \Gamma^* \to M \times N, w \mapsto (g(w), h(w))$ of the homomorphisms $g : \Gamma^* \to M$ and $h : \Gamma^* \to N$ recognizes Boolean combinations:

Lemma 3. *Let L and K be languages such that L is recognized by $g : A^* \to M$ and K is recognized by $h : A^* \to N$. Then, any Boolean combination of L and K is recognized by $(g \times h)$.*

Proof. Since $L \cap [s][e]^\omega \neq \emptyset$ implies $[s][e]^\omega \subseteq L$ for some linked pair (s, e), we obtain $\overline{L} = \cup\{[s][e]^\omega \mid [s][e]^\omega \cap \overline{L} \neq \emptyset\}$ for the complement of L. Thus, it suffices to show that $L \cup K$ is recognized by $(g \times h)$. Obviously, L is covered by $[(s,t)][(e,f)]^\omega$, where (s, e) is a linked pair of M with $[s][e]^\omega \subseteq L$ and (t, f) is any linked pair of N. Similiarly one can cover K and thus $M \times N$ recognizes $L \cup K$. $\qquad\square$

Next, we show that the algebraic characterisation $\mathbf{V_2}$ of $\mathbb{B}\Sigma_2$ over finite words also holds over finite and infinite words simultaneously. The proof of this is based on the fact that the algebraic part of the characterisation of Σ_2 over finite words and finite and infinite words is the same [3] and on the fact that every language of Σ_2 is in $\mathbb{B}\Sigma_2$, i.e., $\mathbf{V_{3/2}} \subseteq \mathbf{V_2}$.

Lemma 4. *If $L \subseteq \Gamma^\infty$ is definable in $\mathbb{B}\Sigma_2$, then $\mathrm{Synt}(L) \in \mathbf{V_2}$.*

Proof. By definition, $L \in \mathbb{B}\Sigma_2$ implies that L is a Boolean combination of languages $L_i \in \Sigma_2$. We have $\mathrm{Synt}(L_i) \in \mathbf{V_{3/2}} \subseteq \mathbf{V_2}$ by [3]. Since L is a Boolean combination of L_i, the direct product of all $\mathrm{Synt}(L_i)$ recognizes L by Lemma 3. In particular, $\mathrm{Synt}(L)$ is a divisor of the direct product of $\mathrm{Synt}(L_i)$ by Lemma 1. Hence, we obtain $\mathrm{Synt}(L) \in \mathbf{V_2}$. $\qquad\square$

The proof that monomials are definable in Σ_2 is straightforward which yields:

Lemma 5. *Every monomial $L \subseteq \Gamma^\infty$ is definable in Σ_2.*

Proof. Let $L = A_0^* a_1 A_1^* a_2 \cdots A_{n-1}^* a_n A_n^\infty$. The Σ_2-formula

$$\exists x_1 \ldots \exists x_n \forall y : \bigwedge_{i=1}^{n} \lambda(x_i) = a_i \wedge \bigwedge_{i=1}^{n-1} x_i < y < x_{i+1} \Rightarrow \lambda(y) \in A_i \wedge$$

$$(y > x_n \Rightarrow \lambda(y) \in A_n) \wedge (y < x_1 \Rightarrow \lambda(y) \in A_0).$$

defines L. $\qquad\square$

Combining our results we are ready to prove Theorem 4.

Proof (Theorem 4). "1 ⇒ 2": Since $\mathbb{B}\Sigma_2$ is closed under Boolean combinations, it suffices to find a Σ_2-formula for a single monomial. This is provided by Lemma 5.

"2 ⇒ 3": 3a is proved by Lemma 4. Since $A_0^* a_1 A_1^* a_2 \cdots A_{n-1}^* a_n$ is a set of finite words, a monomial $A_0^* a_1 A_1^* a_2 \cdots A_{n-1}^* a_n A_n^\infty$ is open in the alphabetic topology. The languages in Σ_2 are unions of such monomials [3] and thus languages in $\mathbb{B}\Sigma_2$ are Boolean combinations of open sets. This implies 3b by Theorem 2.

"3 ⇒ 1": This is Proposition 1 and Theorem 2.   □

Example 1. In this example we show that $\mathrm{Synt}(L) \in \mathbf{V_2}$ for some language $L \subseteq \Gamma^\infty$ does not imply $L \in \mathbb{B}\Sigma_2$, i.e., the topological property 3b of Theorem 4 is necessary. For this define $L = (\{a, b\}^* aa \{a, b\}^*)^\omega$. We show that $\mathrm{Synt}(L) \in \mathbf{V_2}$, but L is not a Boolean combination of open sets of the alphabetic topology. Computing the syntactic monoid of L yields $\mathrm{Synt}(L) = \{1, a, b, aa, ab, ba\}$. The equations $b^2 = b$, $xaa = aax = aa$ and $bab = b$ hold in $\mathrm{Synt}(L)$. In particular, $(ab)^2 = ab$ and $(aa)^2 = aa$. Thus, $(s, e) = (aa, aa)$ and $(t, f) = (aa, ab)$ are linked pairs. Let h denote the syntactic homomorphism of L. Choosing aab as a preimage for $aa \in \mathrm{Synt}(L)$ yields the alphabetic condition $\mathrm{alph}(aab) = \mathrm{alph}(ab) = C$ on the idempotents. Since $s = t$, we trivially have $s \cdot h(C^*) = t \cdot h(C^*)$. However, $[aa][ab]^\omega \cap L = \emptyset$ but $[aa][aa]^\omega \subseteq L$. Thus, L does not satisfy the topological condition 3b of Theorem 4. It remains to check $\mathrm{Synt}(L) \in \mathbf{V_2}$. It is enough to show that the preimages are in $\mathbb{B}\Sigma_2$.

- $[1] = 1$
- $[a] = (ab^+)^* a$
- $[b] = (b^+ a)^* b^+$
- $[ab] = (ab^+)^+$
- $[ba] = (b^+ a)^+$
- $[aa] = \{a, b\}^* aa \{a, b\}^*$

One can find $\mathbb{B}\Sigma_2$ formulas for these languages, e.g., $[ab] = L(\varphi)$ with

$$\varphi \equiv (\exists x \forall y\colon x \leq y \wedge \lambda(x) = a) \wedge (\exists x \forall y\colon x \geq y \wedge \lambda(x) = b) \wedge$$
$$(\forall x \forall y\colon x \geq y \vee (\exists z\colon x < z < y) \vee (\lambda(x) \neq \lambda(y))$$

and thus $\mathrm{Synt}(L) \in \mathbf{V_2}$.   ◇

5   Summary and Open Problems

The alphabetic topology is an essential ingredient in the study of the fragment Σ_2. Thus, in order to study Boolean combinations of Σ_2 formulas, i.e., the fragment $\mathbb{B}\Sigma_2$ over infinite words, we looked closely at properties of Boolean combinations of its open sets. It turns out, that it is decidable whether an ω-regular language is a Boolean combination of open sets. This does not follow immediately from the decidability of the open sets. We used linked pairs of the syntactic homomorphism (which are effectively computable) to get decidability of the topological condition. Combining this result with the decidability of $\mathbf{V_2}$ we obtained an effective characterization of $\mathbb{B}\Sigma_2$ over Γ^∞, the finite and infinite words over the alphabet Γ.

In this paper we dealt with $\mathbb{B}\Sigma_2$, which is the second level of the Straubing-Thérien hierarchy. Another well-known hierarchy is the dot-depth hierarchy. On the level of logic, the difference between the Straubing-Thérien hierarchy and the dot-depth hierarchy is that formulas for the dot-depth hierarchy may also use the successor predicate. A deep result of Straubing is that over finite words each level of the Straubing-Thérien hierarchy is decidable if and only if it is decidable in the dot-depth hierarchy [18]. Thus, the decidability result for $\mathbb{B}\Sigma_2$ by Place and Zeitoun also yields a decidability result of $\mathbb{B}\Sigma_2[<, +1]$. The fragment $\Sigma_2[<, +1]$ is decidable for ω-regular languages [5]. This result also uses topological ideas, namely the factor topology. The open sets in this topology describe which factors of a certain length k may appear in the "infinite part" of the words. The study of Boolean combinations of open sets in the factor topology is an interesting line of future work, and it may yield a decidability result for $\mathbb{B}\Sigma_2[<, +1]$ over infinite words.

Another interesting class of predicates are modular predicates. In [7] the authors have studied $\Sigma_2[<, \text{MOD}]$ over finite words. The results of [7] can be generalised to infinite words by adapting the alphabetic topology to the modular setting. As for successor predicates, we believe that an appropriate effective characterization of this topology might help in deciding $\mathbb{B}\Sigma_2[<, \text{MOD}]$ over infinite words. To the best of our knowledge however, modular predicates have not yet been considered over infinite words.

References

1. Brzozowski, J.A., Knast, R.: The dot-depth hierarchy of star-free languages is infinite. J. Comput. Syst. Sci. **16**(1), 37–55 (1978)
2. Diekert, V., Gastin, P., Kufleitner, M.: A survey on small fragments of first-order logic over finite words. Int. J. Found. Comput. Sci. **19**(3), 513–548 (2008)
3. Diekert, V., Kufleitner, M.: Fragments of first-order logic over infinite words. Theor. Comput. Syst. **48**(3), 486–516 (2011)
4. Eilenberg, S.: Automata, Languages, and Machines, vol. B. Academic Press, New York (1976)
5. Kallas, J., Kufleitner, M., Lauser, A.: First-order fragments with successor over infinite words. In: STACS 2011, Proceedings, vol. 9. LIPIcs, pp. 356–367. Dagstuhl Publishing (2011)
6. Kufleitner, M., Walter, T.: Level two of the quantifier alternation hierarchy over infinite words (2015). CoRR, abs/1509.06207
7. Kufleitner, M., Walter, T.: One quantifier alternation in first-order logic with modular predicates. RAIRO-Theor. Inf. Appl. **49**(1), 1–22 (2015)
8. Landweber, L.H.: Decision problems for ω-automata. Math. Syst. Theor. **3**(4), 376–384 (1969)
9. McNaughton, R.: Testing and generating infinite sequences by a finite automaton. Inf. Control **9**, 521–530 (1966)
10. Perrin, D., Pin, J.-É.: Infinite words. Pure and Applied Mathematics. Elsevier, Amsterdam (2004)
11. Pierron, T., Place, T., Zeitoun, M.: Quantifier alternation for infinite words (2015). CoRR, abs/1511.09011

12. Pin, J.-É.: A variety theorem without complementation. Russ. Math. (Iz. VUZ) **39**, 80–90 (1995)

13. Pin, J.-É.: Syntactic Semigroups. Handbook of Formal Languages. Springer, Heidelberg (1997)

14. Place, T., Zeitoun, M.: Going higher in the first-order quantifier alternation hierarchy on words. In: Esparza, J., Fraigniaud, P., Husfeldt, T., Koutsoupias, E. (eds.) ICALP 2014, Part II. LNCS, vol. 8573, pp. 342–353. Springer, Heidelberg (2014)

15. Ramsey, F.P.: On a problem of formal logic. Proc. London Math. Soc. **30**, 264–286 (1930)

16. Schwarz, S., Staiger, L.: Topologies refining the Cantor topology on X^ω. In: Calude, C.S., Sassone, V. (eds.) Theoretical Computer Science. IFIP Advances in Information and Communication Technology, vol. 323. Springer, Heidelberg (2010)

17. Staiger, L., Wagner, K.W.: Automatentheoretische und automatenfreie Charakterisierungen topologischer Klassen regulärer Folgenmengen. Elektron. Inform.-verarb. Kybernetik **10**, 379–392 (1974)

18. Straubing, H.: Finite semigroup varieties of the form $\mathbf{V} * \mathbf{D}$. J. Pure Appl. Algebra **36**(1), 53–94 (1985)

19. Straubing, H.: Finite Automata, Formal Logic, and Circuit Complexity. Birkhäuser, Boston (1994)

20. Thomas, W.: Classifying regular events in symbolic logic. J. Comput. Syst. Sci. **25**, 360–376 (1982)

21. Thomas, W.: Automata on infinite objects. In: van Leeuwen, J. (ed.) Handbook of Theoretical Computer Science, pp. 133–191. Elsevier, Cambridge (1990). chap. 4

22. Wilke, T.: Locally threshold testable languages of infinite words. In: Enjalbert, P., Finkel, A., Wagner, K.W. (eds.) STACS 1993. LNCS, vol. 665, pp. 607–616. Springer, Heidelberg (1993)

The Word Problem for Omega-Terms over the Trotter-Weil Hierarchy
(Extended Abstract)

Manfred Kufleitner and Jan Philipp Wächter[(✉)]

Institut für Formale Methoden der Informatik,
Universität Stuttgart, Stuttgart, Germany
{kufleitner,jan-philipp.waechter}@fmi.uni-stuttgart.de

Abstract. Over finite words, there is a tight connection between the quantifier alternation hierarchy inside two-variable first-order logic FO^2 and a hierarchy of finite monoids: the Trotter-Weil Hierarchy. The various ways of climbing up this hierarchy include Mal'cev products, deterministic and codeterministic concatenation as well as identities of ω-terms. We show that the word problem for ω-terms over each level of the Trotter-Weil Hierarchy is decidable; this means, for every variety $\mathbf{V}$ of the hierarchy and every identity $u = v$ of ω-terms, one can decide whether all monoids in $\mathbf{V}$ satisfy $u = v$. More precisely, for every fixed variety $\mathbf{V}$, our approach yields nondeterministic logarithmic space (NL) and deterministic polynomial time algorithms, which are more efficient than straightforward translations of the NL-algorithms. From a language perspective, the word problem for ω-terms is the following: for every language variety $\mathcal{V}$ in the Trotter-Weil Hierarchy and every language variety $\mathcal{W}$ given by an identity of ω-terms, one can decide whether $\mathcal{V} \subseteq \mathcal{W}$. This includes the case where $\mathcal{V}$ is some level of the FO^2 quantifier alternation hierarchy. As an application of our results, we show that the separation problems for the so-called corners of the Trotter-Weil Hierarchy are decidable.

1 Introduction

For the study of many regular language classes, it turned out to be fruitful if one finds multiple characterizations for the class. For instance, one can consider the class of languages recognized by *extensive* deterministic finite automata (i. e. automata whose states can be ordered topologically). This is algebraically characterized by the variety $\mathbf{R}$ of $\mathcal{R}$-trivial monoids [3, Chap. 10]. Another example is the class of star-free languages: it is the class of languages definable by a regular expression which may use complementation instead of Kleene's star. Schützenberger's famous theorem [20] yields an algebraic characterization for this class: it coincides with the class of languages which are recognized by aperiodic monoids. A monoid

M. Kufleitner—The first author was supported by the German Research Foundation (DFG) under grants DI 435/5-2 and KU 2716/1-1.

© Springer International Publishing Switzerland 2016
A.S. Kulikov and G.J. Woeginger (Eds.): CSR 2016, LNCS 9691, pp. 237–250, 2016.
DOI: 10.1007/978-3-319-34171-2_17

M is aperiodic if $x^{|M|!} = x^{|M|!}x$ holds for all $x \in M$. In the case of star-free languages (as in many other cases), this algebraic characterization is particularly useful as it makes it possible to decide whether a given language is star-free: compute the language's syntactic monoid M (which, for a regular language, must be finite) and check it for aperiodicity. The latter can be achieved by checking the equation $x^{|M|!} = x^{|M|!}x$ for all $x \in M$. Often, this equation is also stated as $x^\omega = x^\omega x$ since this notation is independent of the monoid's size. More formally, we can see the equation as a pair of ω-terms: these are finite words built using letters, which are interpreted as variables, concatenation and an additional formal ω power. In order to check whether the equation $\alpha = \beta$ consisting of the two ω-terms α and β holds in a monoid M, one first substitutes the formal ω exponents in α and β by $|M|!$, which results in a finite word in variables. One, then, needs to substitute each variable by all element of M, which is possible if M is finite. These substitutions yield a monoid element belonging to α and one belonging to β. If and only if the respective pairs of monoid elements are equal for all variable substitutions, the equation holds in M.

Often, the question whether an equation holds is not only interesting for a single finite monoid but for a (possibly infinite) class of such monoids. For example, one may ask whether all monoids in a certain class are aperiodic. This is trivially decidable if the class is finite. But what if it is infinite? If the class forms a variety (of finite monoids, sometimes also referred to as a pseudo-variety), i. e. a class of finite monoids closed under (possibly empty) direct products, submonoids and homomorphic images, then this problem is called the variety's *word problem for ω-terms*. Usually, the study of a variety's word problem for ω-terms also gives more insight into the variety's structure, which is interesting in its own right. McCammond showed that the word problem for ω-terms of the variety **A** of aperiodic finite monoids is decidable [14]. The problem was shown to be decidable in linear time for **J** by Almeida [1] and for **R** by Almeida and Zeitoun [2]. Later Moura applied their ideas to show decidability in time $\mathcal{O}((nk)^5)$ where k is the maximal nesting depth of the ω-power (which can be linear in n) of the problem for the variety **DA** [16]. The variety **DA** is the class of finite monoids whose regular $\mathcal{D}$-classes form aperiodic semigroups. This class is interesting because of another characterization of **A** and, therefore, star-free languages: a language is star-free if and only if it can be defined by a sentence in first-order logic over words [15]. It is easy to see that any first-order sentence over words is equivalent to one which uses only three variables. Therefore, it is a natural question to ask what happens if one restricts the number of variables to two. This leads to *two-variable* first-order logic (over words). As it turns out, this class of languages is characterized by **DA** [23]; see [22] for a survey.

In this paper, we consider the word problems for ω-terms of the varieties of the *Trotter-Weil Hierarchy*. Trotter and Weil [24] used the good understanding of the band varieties (cf. [4]) for studying the lattice of sub-varieties of **DA**; bands are semigroups satisfying $x^2 = x$. An important aspect of the Trotter-Weil Hierarchy is its connection with the quantifier alternation hierarchy inside two-variable first-order logic. In addition, many characterizations of two-variable first-order logic naturally appear within this hierarchy, see [8]. The Trotter-Weil Hierarchy has a zig-zag shape, see Fig. 3. There are non-symmetric varieties, the so-called *corners*; amongst them is the variety **R** as well as its symmetric dual **L**, the variety of $\mathcal{L}$-trivial monoids. Then there are the intersections of corners, the *intersection levels*; and finally there are the joins of the corners, the *join levels*. Two-variable quantifier alternation corresponds to the intersection levels [11]; in particular, the variety **J** of $\mathcal{J}$-trivial monoids is one of them. The union of all levels is **DA** [10].

In this paper, we present the following results.

- Our main tool for studying a variety **V** of the Trotter-Weil Hierarchy is a family of finite index congruences $\equiv_{\mathbf{V},n}$ for $n \in \mathbb{N}$. These congruences have the property that a monoid M is in **V** if and only if there exists n for which M divides a quotient by $\equiv_{\mathbf{V},n}$. The congruences are not new but they differ in some minor but crucial details (and these details necessitate new proofs). In the literature, the congruences are usually introduced in terms of rankers [8, 11, 12].

- We lift the combinatorics from finite words to ω-terms using the "linear order approach" introduced by Huschenbett and the first author [6]. They showed that, over varieties of aperiodic monoids, one can use the order $\mathbb{N}+\mathbb{Z}\cdot\mathbb{Q}+(-\mathbb{N})$ for the formal ω-power. In this paper, we use the simpler order $\mathbb{N}+(-\mathbb{N})$. We show that two ω-terms α and β are equal in some variety **V** of the Trotter-Weil hierarchy if and only if $[\![\alpha]\!]_{\mathbb{N}+(-\mathbb{N})} \equiv_{\mathbf{V},n} [\![\beta]\!]_{\mathbb{N}+(-\mathbb{N})}$ for all $n \in \mathbb{N}$. Here, $[\![\alpha]\!]_{\mathbb{N}+(-\mathbb{N})}$ denotes the labeled linear order obtained from replacing every ω-power by the linear order $\mathbb{N} + (-\mathbb{N})$. Note that this order is tailor-made for the Trotter-Weil Hierarchy and does not result from simple arguments which work in any variety.

- We show that one can effectively check whether $[\![\alpha]\!]_{\mathbb{N}+(-\mathbb{N})} \equiv_{\mathbf{V},n} [\![\beta]\!]_{\mathbb{N}+(-\mathbb{N})}$ for all $n \in \mathbb{N}$. For some varieties in the Trotter-Weil Hierarchy this is rather straightforward but for the so-called intersection levels it additionally requires some kind of synchronization.

- We further improve the algorithms and show that, for every variety **V** of the Trotter-Weil Hierarchy, the word problem for ω-terms over **V** is decidable in nondeterministic logarithmic space. The main difficulty is to avoid some blow-up which (naively) is caused by the nesting depth of the ω-power. For the variety **R** of $\mathcal{R}$-trivial monoids, this result is incomparable to Almeida and Zeitoun's linear time algorithm [2].

- We also introduce polynomial time algorithms, which are more efficient than the direct translation of these NL algorithms.

- As an application, we show that the separation problem for each corner of the Trotter-Weil Hierarchy is decidable; for **J** we adapt the proof of van Rooijen and Zeitoun [25].

- With little additional effort, we also obtain all of the above results for the limit of the Trotter-Weil hierarchy, the variety **DA**. The decidability of the separation problem re-proves a result of Place, van Rooijen and Zeitoun [19]. The algorithms for the word problem for ω-terms are more efficient than Moura's results [16].

Separability of the join-levels and the intersection-levels is still open. We conjecture that these problems can be solved with similar but more technical reductions.

Proofs omitted from this paper can be found in the technical report [9].

2   The Trotter-Weil Hierarchy

Let $\mathbb{N} = \{1, 2, \dots\}$, $\mathbb{N}_0 = \{0, 1, \dots\}$ and $-\mathbb{N} = \{-1, -2, \dots\}$. For the rest of this paper, we fix a finite alphabet Σ. By Σ^*, we denote the set of all finite words over the alphabet Σ, including the empty word ε; Σ^+ denotes that excluding the empty word. Let $w = a_1 a_2 \dots a_n \in \Sigma^*$ be a word of length $n \in \mathbb{N}_0$. The set $\{a_i \mid i = 1, 2, \dots, n\}$ of letters appearing in w shall be denoted by $\mathrm{alph}(w)$. As a finite word $w \in \Sigma^*$ can be seen as a mapping $w : \{1, 2, \dots, n\} \to \Sigma$, we use $\mathrm{dom}(w)$ to denote the set of positions in w.

For a pair $(l, r) \in (\{-\infty\} \uplus \mathrm{dom}(w)) \times (\mathrm{dom}(w) \uplus \{+\infty\})$, define $w_{(l,r)}$ as the restriction of w (seen as a mapping) to the set of positions (strictly) larger than l and (strictly) smaller than r. Note that $w = w_{(-\infty, +\infty)}$ and $w_{(l,r)} = \varepsilon$ for any pair (l, r) with no position between l and r.

Monoids, Divisors, Congruences and Recognition. In this paper, the term *monoid* refers to a finite monoid (except when stated otherwise). It is well known that, for any monoid M, there is a smallest number $n \in \mathbb{N}$ such that m^n is idempotent (i.e. $m^{2n} = m^n$) for every element $m \in M$; this number is called the *exponent* of M and shall be denoted by $M! = n$.[1] A monoid N is a *divisor* of (another) monoid M, written as $N \prec M$, if N is an homomorphic image of a submonoid of M.

A *congruence* (relation) in a (not necessarily finite) monoid M is an equivalence relation $\mathcal{C} \subseteq M \times M$ such that $x_1 \, \mathcal{C} \, x_2$ and $y_1 \, \mathcal{C} \, y_2$ implies $x_1 y_1 \, \mathcal{C} \, x_2 y_2$ for all $x_1, x_2, y_1, y_2 \in M$. If M is a (possibly infinite) monoid and $\mathcal{C} \subseteq M \times M$ is a congruence, then the set of equivalence classes of $\mathcal{C}$, denoted by $M/\mathcal{C}$, is a well-defined monoid (which might still be infinite), whose size is called the *index* of $\mathcal{C}$. For any two congruences $\mathcal{C}_1$ and $\mathcal{C}_2$, one can define their *join* $\mathcal{C}_1 \vee \mathcal{C}_2$ as the smallest congruence which includes $\mathcal{C}_1$ and $\mathcal{C}_2$; its index is at most as large as the index of $\mathcal{C}_1$ and the index of $\mathcal{C}_2$.

A (possibly infinite) monoid M *recognizes* a language of finite words $L \subseteq \Sigma^*$ if there is a homomorphism $\varphi : \Sigma^* \to M$ with $L = \varphi^{-1}(\varphi(L))$. A language is *regular* if and only if it is recognized by a finite monoid. It is well known that there is a unique smallest monoid which recognizes a given regular language: the *syntactic monoid*.

Varieties, π-Terms and Equations. A *variety* (of finite monoids) – sometimes also referred to as a *pseudo-variety* – is a class of monoids which is closed under submonoids, homomorphic images and – possibly empty – finite direct products. For example, the class $\mathbf{R}$ of $\mathcal{R}$-trivial monoids and the class $\mathbf{L}$ of $\mathcal{L}$-trivial

---

[1] Note that all statements remain valid if one assumes that $M!$ is used to denote $|M|!$.

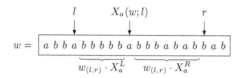

Fig. 1. Application of X_a^L and X_a^R to an example word.

monoids both form a variety, see e.g. [18]. Clearly, if $\mathbf{V}$ and $\mathbf{W}$ are varieties, then so is $\mathbf{V} \cap \mathbf{W}$. For example, the class $\mathbf{J} = \mathbf{R} \cap \mathbf{L}$ is a variety; in fact, it is the variety of all $\mathcal{J}$-trivial monoids. For two varieties $\mathbf{V}$ and $\mathbf{W}$, the smallest variety containing $\mathbf{V} \cup \mathbf{W}$, the so called *join*, is denoted by $\mathbf{V} \vee \mathbf{W}$.

Many varieties can be defined in terms of *equations* (or *identities*). Because it will be useful later, we take a more formal approach towards equations by using π-*terms*[2]. A π-term is a finite word, built using letters, concatenation and an additional formal π-power (and appropriate parentheses), whose π-exponents act as a placeholder for a substitution value.

To state equations using π-terms, one needs to substitute these placeholders by actual values resulting in an ordinary finite word. We define $[\![\gamma]\!]_n$ as the result of substituting the π-exponents in γ by $n \in \mathbb{N}_0$. An equation $\alpha = \beta$ consists of two π-terms α and β over the same alphabet Σ, which, here, can be seen as a set of *variables*. A homomorphism $\sigma : \Sigma^* \to M$ is called an *assignment of variables* in this context. An equation $\alpha = \beta$ *holds* in a monoid M if for every assignment of variables $\sigma\left([\![\alpha]\!]_{M!}\right) = \sigma\left([\![\beta]\!]_{M!}\right)$ is satisfied. It holds in a variety $\mathbf{V}$, if it holds in all monoids in $\mathbf{V}$.

Relations for the Trotter-Weil Hierarchy. In this paper, we approach the Trotter-Weil Hierarchy by using certain congruences. First, however, we give some definitions for factorizations of words at the first or last a-position (i. e. an a-labeled position). For a word w, a position $p \in \mathrm{dom}(w) \uplus \{-\infty\}$ and a letter $a \in \mathrm{alph}(w)$, let $X_a(w; p)$ denote the first a-position (strictly) larger than p (or the first a-position in w if $p = -\infty$). It is undefined if there is no such position. Define $Y_a(w; p)$ symmetrically as the first a-position from the right which is (strictly) smaller than p.

Let w be a word, define

$$w \cdot X_a^L = w_{(-\infty, X_a(w; -\infty))}, \qquad w \cdot X_a^R = w_{(X_a(w; -\infty), +\infty)},$$
$$w \cdot Y_a^L = w_{(-\infty, Y_a(w; +\infty))} \text{ and } \qquad w \cdot Y_a^R = w_{(Y_a(w; +\infty), +\infty)}$$

for all $a \in \mathrm{alph}(w)$. Additionally, define $C_{a,b}$ as a special form of applying X_a^L first and then Y_b^R which is only defined if $X_a(w; -\infty)$ is strictly larger than $Y_b(w; +\infty)$. For an example of X_a^L and X_a^R acting on a word see Fig. 1. Note that we have $w = (w \cdot X_a^L)a(w \cdot X_a^R) = (w \cdot Y_a^L)a(w \cdot Y_a^R) = (w \cdot Y_b^L)b(w \cdot C_{a,b})a(w \cdot X_a^R)$ (whenever these factors are defined).

---

[2] Usually, π-terms are referred to as ω-terms. In this paper, however, we use ω to denote the order type of the natural numbers. Therefore, we follow the approach of Perrin and Pin [17] and use π instead of ω.

With these definitions in place, we define the relations[3] $\equiv_{m,n}^{X}$, $\equiv_{m,n}^{Y}$ and $\equiv_{m,n}^{WI}$ of words for $m, n \in \mathbb{N}$. The idea is that these relations hold on two words u and v if both words allow for the same sequence of factorizations at the first or last occurrence of a letter. The parameter m is the remaining number of direction changes (which are caused by an X_a^L or Y_a^R factorizations) in such a sequence and the parameter n is the number of remaining factorization moves (independent of their direction). Thus, if m or n is zero, then all of the three relations shall be satisfied for all words. For m and n larger than zero, our first assertion is that both words have the same alphabet; otherwise, one of them would admit a factorization at a letter while the other would not, as the letter is not in its alphabet. Furthermore, for $u \equiv_{m,n}^{X} v$

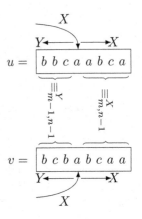

Fig. 2. $\equiv_{m,n}^{X}$ illustrated.

to hold, we require $u \cdot X_a^L \equiv_{m-1,n-1}^{Y} v \cdot X_a^L$ and $u \cdot X_a^R \equiv_{m,n-1}^{X} v \cdot X_a^R$ for all a in the common alphabet of u and v. The former states that, after an X_a factorization, the left parts of this factorization in both words have to admit the same factorization sequences where the number of moves as well as the direction changes has decreased by one. We lose one direction change because we factorize at the first a to the right of the words' beginnings but take the factors to the left. On the other hand, if we take the factors to the right, we only lose one move but no change in direction; this is stated in the latter requirement. Figure 2 gives a graphical overview of this. Additionally, we can also change the starting point of our factorization (which, normally, is the beginning of the words for $\equiv_{m,n}^{X}$); for this, we lose one move and one change in direction. Therefore, we also require $u \equiv_{m-1,n-1}^{Y} v$ for $u \equiv_{m,n}^{X} v$ to hold.

Symmetrically, we define $u \equiv_{m,n}^{Y} v$ if and only if we have $\mathrm{alph}(u) = \mathrm{alph}(v)$, $u \equiv_{m-1,n-1}^{X} v$ and $u \cdot Y_a^L \equiv_{m,n-1}^{Y} v \cdot Y_a^L$ as well as $u \cdot Y_a^R \equiv_{m-1,n-1}^{X} v \cdot Y_a^R$ for all $a \in \mathrm{alph}(u)$. Additionally, we define $\equiv_{m,n}^{R}$ as the intersection for $\equiv_{m,n}^{X}$ and $\equiv_{m,n}^{Y}$ for all $m, n \in \mathbb{N}$.

For $u \equiv_{m,n}^{WI} v$ with $m, n \in \mathbb{N}$ to hold, we require $\mathrm{alph}(u) = \mathrm{alph}(v)$ and, for all $a \in \mathrm{alph}(u)$, $u \cdot X_a^L \equiv_{m-1,n-1}^{WI} v \cdot X_a^L$, $u \cdot X_a^R \equiv_{m,n-1}^{WI} v \cdot X_a^R$, $u \cdot Y_a^L \equiv_{m,n-1}^{WI} v \cdot Y_a^L$ and $u \cdot Y_a^R \equiv_{m-1,n-1}^{WI} v \cdot Y_a^R$, as well as that $u \cdot C_{a,b}$ and $v \cdot C_{a,b}$ are either both undefined or both defined and $u \cdot C_{a,b} \equiv_{m-1,n-1}^{WI} v \cdot C_{a,b}$ holds. All of these requirements except for the last one are analogous to the cases for $\equiv_{m,n}^{X}$ and $\equiv_{m,n}^{Y}$. The last assertion states that the first a is to the right of the last b in u if

---

and only if it is so in v and that, in this case, we can continue to factorize in the middle part between b and a with one less move and one less direction change.

By simple inductions, one can see that the relations are congruences of finite index over Σ^*. Also note that $u \equiv_{m,n}^Z v$ implies $u \equiv_{m,k}^Z v$ and, if $m > 0$, also $u \equiv_{m-1,k}^Z v$ for all $k \leq n$ and $Z \in \{X, Y, R, \mathrm{WI}\}$.

The Trotter-Weil Hierarchy. Using these relations, we can define the *Trotter-Weil Hierarchy.* As the name implies, this hierarchy was first studied by Trotter and Weil [24], who obtained it by taking a different approach. For more information on the equivalence of the two definitions see also [7,12] and [5, Corollary 4.3].

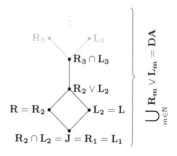

Fig. 3. Trotter-Weil Hierarchy

The Trotter-Weil Hierarchy consists of *corners, join levels* and *intersection levels*. The corners of the layer $m \in \mathbb{N}$ are the varieties $\mathbf{R_m}$ and $\mathbf{L_m}$. A monoid M is in $\mathbf{R_m}$ if and only if $M \prec \Sigma^*/\equiv_{m,n}^X$ for an $n \in \mathbb{N}_0$ and it is in $\mathbf{L_m}$ if and only if $M \prec \Sigma^*/\equiv_{m,n}^Y$ for an $n \in \mathbb{N}_0$. The corresponding join level is $\mathbf{R_m} \vee \mathbf{L_m}$ and the corresponding intersection level is $\mathbf{R_m} \cap \mathbf{L_m}$. A monoid M is in $\mathbf{R_m} \vee \mathbf{L_m}$ if and only if $M \prec \Sigma^*/\equiv_{m,n}^R$ for an $n \in \mathbb{N}$ and it is in $\mathbf{R_m} \cap \mathbf{L_m}$ if and only if $M \prec \Sigma^*/\equiv_{m,n}^{\mathrm{WI}}$ for an $n \in \mathbb{N}$.

The term "hierarchy" is justified by the following inclusions: we have $\mathbf{R_m} \cap \mathbf{L_m} \subseteq \mathbf{R_m}, \mathbf{L_m} \subseteq \mathbf{R_m} \vee \mathbf{L_m}$ and $\mathbf{R_m} \vee \mathbf{L_m} \subseteq \mathbf{R_{m+1}} \cap \mathbf{L_{m+1}}$. The Trotter-Weil Hierarchy contains some well known varieties: we have $\mathbf{R_1} = \mathbf{L_1} = \mathbf{J}$, $\mathbf{R_2} = \mathbf{R}$ and $\mathbf{L_2} = \mathbf{L}$ (for the last two, see [18]).

By taking the union of all varieties in the hierarchy, one gets the variety $\mathbf{DA}$ [10], which is usually defined as the class of monoids whose regular $\mathcal{D}$-classes form aperiodic semigroups[4]. Though we state this as a fact here, it can also be seen as the definition of $\mathbf{DA}$ for this paper. These considerations yield the graphic representation given in Fig. 3. We also note that the intersection levels corresponds to the quantifier alternation hierarchy of first-order logic with at most two variables [11]. A first-order sentence using at most two variables belongs to FO_m^2 if, on any path in its syntax tree, there is no quantifier after the first negation and there are at most m blocks of quantifiers. A language is definable by a sentence in FO_m^2 if and only if its syntactic monoid is in $\mathbf{R_{m+1}} \cap \mathbf{L_{m+1}}$.

3 Relations and Equations

Order Types. A linearly ordered set $(P, \leq_P)$ consists of a (possibly infinite) set P and a linear ordering relation $\leq_P$ of P, i.e. a reflexive, anti-symmetric, transitive

---

[4] For finite monoids, $\mathcal{D}$-classes coincide with $\mathcal{J}$-classes; a $\mathcal{D}$-class is called *regular* if it contains an idempotent. A semigroup is called *aperiodic* (or *group-free*) if it has no divisor which is a nontrivial group.

and total binary relation $\leq_P \subseteq P \times P$. To simplify notation we define two special objects $-\infty$ and $+\infty$. The former is always smaller with regard to $\leq_P$ than any element in P while the latter is always larger. We call two linearly ordered sets $(P, \leq_P)$ and $(Q, \leq_Q)$ *isomorphic* if there is an order-preserving bijection $\varphi : P \to Q$. Isomorphism between linearly order sets is an equivalence relation; its classes are called (linear) *order types*.

The *sum* of two linearly ordered sets $(P, \leq_P)$ and $(Q, \leq_Q)$ is $(P \uplus Q, \leq_{P+Q})$ where $P \uplus Q$ is the disjoint union of P and Q and $\leq_{P+Q}$ orders all elements of P to be smaller than those of Q while it behaves as $\leq_P$ and $\leq_Q$ on elements from their respective sets. Similarly, the *product* of $(P, \leq_P)$ and $(Q, \leq_Q)$ is $(P \times Q, \leq_{P*Q})$ where $(p, q) \leq_{P*Q} (\tilde{p}, \tilde{q})$ holds if and only if either $q \leq_Q \tilde{q}$ and $q \neq \tilde{q}$ or $q = \tilde{q}$ and $p \leq_P \tilde{p}$ holds. Sum and product of linearly ordered sets are compatible with taking the order type. This allows for writing $\mu + \nu$ and $\mu * \nu$ for order types μ and ν.

We re-use $n \in \mathbb{N}_0$ to denote the order type of $(\{1, 2, \ldots, n\}, \leq)$. One should note that this use of natural numbers to denote order types does not result in contradictions with sums and products: the usual calculation rules apply. Besides finite linear order types, we need ω, the order type of $(\mathbb{N}, \leq)$, and its dual ω^* the order type of $(-\mathbb{N}, \leq)$. Another important order type in the scope of this paper is $\omega + \omega^*$, whose underlying set is $\mathbb{N} \uplus (-\mathbb{N})$. Note that, here, natural numbers and the (strictly) negative numbers are ordered as $1, 2, 3, \ldots, \quad \ldots, -3, -2, -1$; therefore, in this order type, we have for example $-1 \geq_{\omega+\omega^*} 1$.

Generalized Words. As already mentioned, any finite word $w = a_1 a_2 \ldots a_n$ of length $n \in \mathbb{N}_0$ with $a_i \in \Sigma$ can be seen as a function which maps a *position* $i \in \text{dom}(w)$ to the corresponding letter a_i (or, possibly, the empty map). By relaxing the requirement of $\text{dom}(w)$ to be finite, one obtains the notion of *generalized words*: a (generalized) word w over the alphabet Σ of order type μ is a function $w : \text{dom}(w) \to \Sigma$, where $\text{dom}(w)$ is a linearly ordered set in μ. For $\text{dom}(w)$, we usually choose $(\mathbb{N}, \leq)$, $(-\mathbb{N}, \leq)$ and $(\mathbb{N} \uplus (-\mathbb{N}), \leq_{\omega+\omega^*})$ as representative of ω, ω^* and $\omega + \omega^*$, respectively. The order type of a finite word of length n is n.

Like finite words, generalized words can be concatenated, i.e. we write u to the left of v and obtain uv. In that case, the order type of uv is the sum of the order types of u and v. Beside concatenation, we can also take powers of generalized words. Let w be a generalized word of order type μ which belongs to $(P_\mu, \leq_\mu)$ and let ν be an arbitrary order type belonging to $(P_\nu, \leq_\nu)$. Then, w^ν is a generalized word of order type $\mu * \nu$ which determines the ordering of its letters; w maps $(p_1, p_2) \in P_\mu \times P_\nu$ to $w(p_1)$. If $\nu = n$ for some $n \in \mathbb{N}$, then $w^\nu = w^n$ is equal to the n-fold concatenation of w.

In this paper, the term *word* refers to a generalized word. If it is important for a word to be finite, it is referred to explicitly as a *finite word*. One may verify that all previous results still apply if a "word" is considered to be a generalized word instead of a finite word and that previous definitions extend naturally to generalized words. Especially, we can define $\text{alph}(w)$ as the image of w and apply the $\equiv_{m,n}^Z$ relations also to generalized words. We also extend the notation $[\![\gamma]\!]_\mu$ to arbitrary order types μ. The result of the π-substitution now, of course, is a generalized word. Only useful for generalized words, however, is the following congruence: for $m \in \mathbb{N}_0$ and $Z \in \{X, Y, R, WI\}$, define $u \equiv_m^Z v \Leftrightarrow \forall n \in \mathbb{N} : u \equiv_{m,n}^Z v$.

Word Problem for π-terms. The *word problem for π-terms* over a variety $\mathbf{V}$ is the problem to decide whether $\alpha = \beta$ holds in $\mathbf{V}$ for the input π-terms α and β.

In order to solve the word problem for π-terms over the varieties in the Trotter-Weil Hierarchy, one can use the following connection between the relations defined above and equations in these varieties, which is straightforward if one make the transition from finite to infinite words. Besides its use for the word problem for π-terms, this connection is also interesting in its own right as it can be used to prove or disprove equations in any of the varieties. As the class of monoids in which an equation $\alpha = \beta$ holds is a variety, one can see the assertion for the join levels as an implication of the ones for the corners.

Theorem 1. *Let α and β be two π-terms. For every $m \in \mathbb{N}$, we have:*

$$[\![\alpha]\!]_{\omega+\omega^*} \equiv_m^X [\![\beta]\!]_{\omega+\omega^*} \Leftrightarrow \alpha = \beta \text{ holds in } \mathbf{R_m}$$

$$[\![\alpha]\!]_{\omega+\omega^*} \equiv_m^Y [\![\beta]\!]_{\omega+\omega^*} \Leftrightarrow \alpha = \beta \text{ holds in } \mathbf{L_m}$$

$$[\![\alpha]\!]_{\omega+\omega^*} \equiv_m^R [\![\beta]\!]_{\omega+\omega^*} \Leftrightarrow \alpha = \beta \text{ holds in } \mathbf{R_m} \vee \mathbf{L_m}$$

$$[\![\alpha]\!]_{\omega+\omega^*} \equiv_m^{WI} [\![\beta]\!]_{\omega+\omega^*} \Leftrightarrow \alpha = \beta \text{ holds in } \mathbf{R_{m+1}} \cap \mathbf{L_{m+1}}$$

Corollary 1. $\left(\forall m \in \mathbb{N} : [\![\alpha]\!]_{\omega+\omega^*} \equiv_m^R [\![\beta]\!]_{\omega+\omega^*} \right) \Leftrightarrow \alpha = \beta$ *holds in* $\mathbf{DA}$

4   Decidability

In the previous section, we saw that checking whether $\alpha = \beta$ holds in a variety of the Trotter-Weil Hierarchy boils down to checking $[\![\alpha]\!]_{\omega+\omega^*} \equiv_m^Z [\![\beta]\!]_{\omega+\omega^*}$ (where $\equiv_m^Z$ depends on the variety in question). In this section, we give an introduction on how to do this. The presented approach works uniformly for all varieties in the Trotter-Weil Hierarchy (in particular, it also works for the intersection levels, which tend to be more complicated) and is designed to yield efficient algorithms.

The definition of the relations which need to be tested is inherently recursive. One would factorize $[\![\alpha]\!]_{\omega+\omega^*}$ and $[\![\beta]\!]_{\omega+\omega^*}$ on the first a and/or last b (for $a, b \in \Sigma$) and test the factors recursively. Therefore, the computation is based on working with factors of words of the form $[\![\gamma]\!]_{\omega+\omega^*}$ where γ is a π-term. We have already introduced the notation $w_{(l,r)}$ to denote the factor of a finite w which arises by restricting the domain of w to the open interval (l, r). This notation can easily be extended to the case of generalized words.

What happens if we consecutively factorize at a first/last a is best understood if one considers the structure of $[\![(\alpha)^\pi]\!]_{\omega+\omega^*} = \left([\![\alpha]\!]_{\omega+\omega^*} \right)^{\omega+\omega^*} = u^{\omega+\omega^*} = w$, which is schematically represented in Fig. 4.

Suppose u only contains a single a and
we start with the whole word $w_{(-\infty,+\infty)}$. If
we factorize on the first a taking the part
to the right, then we end up with the fac-
tor $w_{(X_a(w;-\infty),+\infty)}$ with $X_a(w;-\infty) = (p,1)$
where p is the single a-position in u. If we do
this again, we obtain $w_{((p,2),+\infty)}$. If we now
factorize on the next a but take the part to
the left, then we get $w_{((p,2),(p,3))}$. Notice that

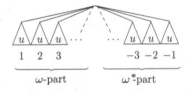

Fig. 4. Representation of $u^{\omega+\omega^*}$

the difference between 2 and 3 is 1 and that there is no way of getting a (finite)
difference larger than one by factorizing on the respective first a. On the other
hand, we can reach any number in $\mathbb{N}$ as long as the right position is not in the
ω-part.

Notice that there is also no way of reaching $(p,-2)$ as left border without
having $(q,-1)$ or $(q,-2)$ as right border for a position $q \in \mathrm{dom}(u)$. These observations (and their symmetrical duals) lead to the notion of *normalizable* pairs
of positions.

The choice of words indicates that normalizability of a pair (l,r) can be used
to define a normalization. We omit a formal – unfortunately, quite technical –
definition of this, but give a description of its idea. Let us refer back to the
schematic representation of $[\![(\alpha)^\pi]\!]_{\omega+\omega^*} = w$ as given in Fig. 4. Basically, there
are three different cases for relative positions of the left border l and the right
border r which describe the factor $w_{(l,r)}$:

1. l is in the ω-part and r is in the ω^*-part,
2. l and r are either both in the ω-part or both in the ω^*-part and have the
   same value there, or
3. l and r are either both in the ω-part or both in the ω^*-part but r has a value
   exactly larger by one than l.

This is ensured by the normalizability of (l,r). Now, in the first case, we can
safely move l to value 1 (the first position) and r to value -1 (the last position)
without changing the described factor. In the second and third case, we can move
l and r to any value – as long as we retain the difference between the values –
without changing the described factor. Here, we move them to the left-most
values (which are $1,1$ or $1,2$). Afterwards, we go on recursively.

Unfortunately, things get a bit more complicated because l might be $-\infty$
and r might be $+\infty$. In these cases, we normalize to the left-most or right-most
value without changing the factor.

For concatenations of π-terms, we have a similar situation: either l and r
belong both to the left or to the right factor, in which case we can continue
by normalization with respect to that, or l belongs to the left factor and r
belongs to the right one. In this case, we have to continue the normalization
with $(l,+\infty)$ and $(-\infty,r)$ in the respective concatenation parts, as this ensures
that the described factor remains unchanged.

One should note that if we normalize a normalizable pair (l, r), then the resulting pair is normalizable itself. Indeed, if we normalize an already normalized pair again, we do not change any values.

Another observation is crucial for the proof of the decidability: after normalizing a pair (l, r) the values belonging to the $\omega + \omega^*$ parts for the two positions are all in $\{1, 2, -2, -1\}$. But: there are only finitely many such positions in any word $w = [\![\gamma]\!]_{\omega+\omega^*}$ for a π-term γ. Because the normalization preserves the described factor, this means that there are only finitely many factors which can result from a sequence of first/last a factorizations.

Plugging all these ideas and observations together yields a proof for the next theorem (note that decidability for **DA** has already been shown by Moura [16]). Here, we only give a sketch of the proof.

Theorem 2. *The word problems for π-terms over $\mathbf{R_m}$, $\mathbf{L_m}$, $\mathbf{R_m} \vee \mathbf{L_m}$ and $\mathbf{R_m} \cap \mathbf{L_m}$ are decidable for any $m \in \mathbb{N}$. Moreover, the word problem for π-terms over **DA** is decidable.*

Proof (Sketch). The proof is structurally equivalent for all stated varieties. Though it can also be proved directly, decidability for the join levels can be seen as an implication of the decidability for the corners.

The basic idea is to construct a finite automaton for each input π-term γ. The nodes consist of the normalized position pairs and the edges are labeled by Z_a^D for variables a, $Z \in \{X, Y\}$ and $D \in \{L, R\}$. The node (l, r) has an out-going Z_a^D-edge if $w' = w_{(l,r)} \cdot Z_a^D$ is defined for $w = [\![\gamma]\!]_{\omega+\omega^*}$; its target is obtained by normalizing the pair describing w'. Except for **DA**, we additionally have to keep track of the alternations between X_a and Y_a factorizations; this can be done by taking the intersection of two automata. For the intersection levels, we also need $C_{a,b}$-edges which are defined analogously. If there is a path labeled by $Z_1 Z_2 \ldots Z_k$ in the automaton for α but not in the one for β, we know that $[\![\alpha]\!]_{\omega+\omega^*}$ is not in relation with $[\![\beta]\!]_{\omega+\omega^*}$ under the appropriate relation given by Theorem 1. Therefore, checking $\alpha = \beta$ reduces to checking the automata's symmetric difference for emptiness. □

In the presented algorithm, we have to store and compute normalized pairs of positions in words of the form $[\![\gamma]\!]_{\omega+\omega^*}$ for a π-term γ. To store a single position of such a pair, one could simply store the values for the π-exponents and a position in γ. While this would be sufficient to exactly determine the position, it is impossible to do in logarithmic space. With some additional ideas, however, it is, in fact, possible to solve the problems in nondeterministic logarithmic space, which we state in the following theorem (see the technical report [9] for more details).

Theorem 3. *The word problems for π-term over $\mathbf{R_m}$, $\mathbf{L_m}$, $\mathbf{R_m} \vee \mathbf{L_m}$, $\mathbf{R_m} \cap \mathbf{L_m}$ and **DA** can be solved by a nondeterministic Turing machine in logarithmic space (for every $m \in \mathbb{N}$).*

While NL is quite efficient from a complexity class perspective, directly translating the algorithm to polynomial time does not result in a better running time than the algorithm for **DA** given by Moura [16]. However, with some additional tweaks, the algorithm's efficiency can be improved, which yields the following theorem [9].

Theorem 4. *The word problems for π-terms over* $\mathbf{R_m}$, $\mathbf{L_m}$, $\mathbf{R_m} \vee \mathbf{L_m}$ *and* $\mathbf{R_m} \cap \mathbf{L_m}$ *can be solved by a deterministic algorithm with running time in* $\mathcal{O}(n^7 m^2)$ *where n is the length of the input π-terms. Moreover, the word problem for π-terms over* **DA** *can be solved by a deterministic algorithms in time* $\mathcal{O}(n^7)$.

5   Separability

Two languages $L_1, L_2 \subseteq \Sigma^*$ are *separable* by a variety **V** if there is a language $S \subseteq \Sigma^*$ with $L_1 \subseteq S$ and $L_2 \cap S = \emptyset$ such that S can be recognized by a monoid $M \in \mathbf{V}$. The *separation problem* of a variety **V** is the problem to decide whether two regular input languages of finite words are separable by **V**.

We are going to show the decidability of the separations problems of $\mathbf{R_m}$ for all $m \in \mathbb{N}$ as well as for **DA** using the techniques presented in this paper[5]. Note that, by symmetry, this also shows decidability for $\mathbf{L_m}$.

The general idea is as follows. If the input languages are separable, then we can find a separating language S which is recognized by a monoid in the variety in question. This, we can do by recursively enumerating all monoids and all languages in a suitable representation. For the other direction, we show that, if the input languages are inseparable, then there are π-terms α and β which witness their inseparability. Since we can also recursively enumerate these π-terms, we have decidability.

To construct suitable π-terms we need an additional combinatoric property of the $\equiv^X_{m,n}$ relations (which, in a slightly different form, can also be found in [12]). Using that, one can prove the following lemma concerning the π-term construction and plug everything together.

Lemma 1. *Let M be a monoid, $\varphi : \Sigma^* \to M$ a homomorphism and $m \in \mathbb{N}_0$. Let $(u_n, v_n)_{n \in \mathbb{N}_0}$ be an infinite sequence of word pairs $(u_n, v_n)_{n \in \mathbb{N}_0}$ with $u_n, v_n \in \Sigma^*$, $u_n \equiv^X_{m,n} v_n$, $\varphi(u_n) = m_u$ and $\varphi(v_n) = m_v$ for fixed monoid elements $m_u, m_v \in M$ and all $n \in \mathbb{N}_0$. Then, the sequence yields π-terms α and β (over Σ) such that $\varphi(\llbracket \alpha \rrbracket_{M!}) = m_u$, $\varphi(\llbracket \beta \rrbracket_{M!}) = m_v$ and $\llbracket \alpha \rrbracket_{\omega + \omega^*} \equiv^X_m \llbracket \beta \rrbracket_{\omega + \omega^*}$ hold.*

Theorem 5. *The separation problem for $\mathbf{R_m}$ and $\mathbf{L_m}$ is decidable for all $m \in \mathbb{N}$.*

Proof (idea). The idea is to recursively enumerate all separating languages and also all the π-terms which, by the last lemma, witness inseparability. $\square$

Since two languages are separable by $\mathbf{R_m}$ for some $m \in \mathbb{N}$ which depends only on the size of Σ [26] if they are separable by **DA**, we also get decidability for **DA**, which has already been shown by Place, van Rooijen and Zeitoun [19].

Corollary 2. *The separation problem for* **DA** *is decidable.*

---

[5] Decidability for **DA** is already known [19]. The proof, however, uses a fix point saturation, which is different from our approach.

References

1. Almeida, J.: Finite semigroups: an introduction to a unified theory of pseudovarieties. In: dos Gomes Moreira da Cunha, G.M., da Silva, P.V.A., Pin, J.É. (eds.) Semigroups, Algorithms, Automata and Languages, pp. 3–64. World Scientific, Singapore (2002)
2. Almeida, J., Zeitoun, M.: An automata-theoretic approach to the word problem for ω-terms over **R**. Theor. Comput. Sci. **370**(1), 131–169 (2007)
3. Eilenberg, S.: Automata, Languages, and Machines, vol. B. Academic press, New York (1976)
4. Gerhard, J., Petrich, M.: Varieties of bands revisited. Proc. Lond. Math. Soc. **58**(3), 323–350 (1989)
5. Hall, T., Weil, P.: On radical congruence systems. Semigroup Forum **59**(1), 56–73 (1999)
6. Huschenbett, M., Kufleitner, M.: Ehrenfeucht-fraïssé games on omega-terms. In: Mayr, E.W., Portier, N. (eds.) STACS 2014, Proceedings. LIPIcs, vol. 25, pp. 374–385. Dagstuhl Publishing, Dagstuhl (2014)
7. Krohn, K., Rhodes, J.L., Tilson, B.: Homomorphisms and semilocal theory. In: Arbib, M.A. (ed.) Algebraic Theory of Machines, Languages, and Semigroups, pp. 191–231. Academic Press, New York (1968). Chap. 8
8. Kufleitner, M., Lauser, A.: The join levels of the trotter-weil hierarchy are decidable. In: Rovan, B., Sassone, V., Widmayer, P. (eds.) MFCS 2012. LNCS, vol. 7464, pp. 603–614. Springer, Heidelberg (2012)
9. Kufleitner, M., Wächter, J.Ph.: The word problem for omega-terms over the Trotter-Weil hierarchy (2015). CoRR abs/1509.05364
10. Kufleitner, M., Weil, P.: On the lattice of sub-pseudovarieties of **DA**. Semigroup Forum **81**, 243–254 (2010)
11. Kufleitner, M., Weil, P.: The FO^2 alternation hierarchy is decidable. In: Cégielski, P., Durand, A. (eds.) Proceedings. LIPIcs, CSL 2012, vol. 16, pp. 426–439. Dagstuhl Publishing, Dagstuhl (2012)
12. Kufleitner, M., Weil, P.: On logical hierarchies within FO^2-definable languages. Logical Methods Comput. Sci. **8**(3), 1–30 (2012)
13. Lodaya, K., Pandya, P., Shah, S.: Marking the chops: an unambiguous temporal logic. In: Ausiello, G., Karhumäki, J., Mauri, G., Ong, L. (eds.) IFIP TCS 2008. IFIP, pp. 461–476. Springer, US (2008)
14. McCammond, J.P.: Normal forms for free aperiodic semigroups. Int. J. Algebra Comput. **11**(5), 581–625 (2001)
15. McNaughton, R., Papert, S.: Counter-Free Automata. The MIT Press, Cambridge (1971)
16. Moura, A.: The word problem for ω-terms over **DA**. Theor. Comput. Sci. **412**(46), 6556–6569 (2011)
17. Perrin, D., Pin, J.É.: Infinite words, Pure and Applied Mathematics, vol. 141. Elsevier, Amsterdam (2004)
18. Pin, J.: Varieties of Formal Languages. North Oxford Academic Publishers Ltd, London (1986)
19. Place, Th., van Rooijen, L., Zeitoun, M.: Separating regular languages by piecewise testable and unambiguous languages. In: Chatterjee, K., Sgall, J. (eds.) MFCS 2013. LNCS, vol. 8087, pp. 729–740. Springer, Heidelberg (2013)
20. Schützenberger, M.: On finite monoids having only trivial subgroups. Inf. Control **8**, 190–194 (1965)

21. Schwentick, Th., Thérien, D., Vollmer, H.: Partially-ordered two-way automata: a new characterization of **DA**. In: Kuich, W., Rozenberg, G., Salomaa, A. (eds.) DLT 2001. LNCS, vol. 2295, pp. 239–250. Springer, Heidelberg (2002)

22. Tesson, P., Thérien, D.: Diamonds are forever: the variety **DA**. In: dos Gomes Moreira da Cunha, G.M., da Silva, P.V.A., Pin, J.É. (eds.) Semigroups, Algorithms, Automata and Languages, pp. 475–500. World Scientific, Singapore (2002)

23. Thérien, D., Wilke, Th.: Over words, two variables are as powerful as one quantifier alternation. In: Proceedings of the Thirtieth Annual ACM Symposium on Theory of Computing. pp. 234–240. ACM (1998)

24. Trotter, P., Weil, P.: The lattice of pseudovarieties of idempotent semigroups and a non-regular analogue. Algebra Univers. **37**(4), 491–526 (1997)

25. van Rooijen, L., Zeitoun, M.: The separation problem for regular languages by piecewise testable languages (2013). CoRR abs/1303.2143

26. Weis, Ph., Immerman, N.: Structure theorem and strict alternation hierarchy for FO^2 on words. Logical Methods Comput. Sci. **5**(3), 1–23 (2009)

Some Complete and Intermediate Polynomials in Algebraic Complexity Theory

Meena Mahajan and Nitin Saurabh[⊠]

The Institute of Mathematical Sciences, Chennai, India
{meena,nitin}@imsc.res.in

Abstract. We provide a list of new natural VNP-Intermediate polynomial families, based on basic (combinatorial) NP-Complete problems that are complete under *parsimonious* reductions. Over finite fields, these families are in VNP, and under the plausible hypothesis $\mathsf{Mod}_p\mathsf{P} \not\subseteq \mathsf{P}/\mathsf{poly}$, are neither VNP-hard (even under oracle-circuit reductions) nor in VP. Prior to this, only the Cut Enumerator polynomial was known to be VNP-intermediate, as shown by Bürgisser in 2000.

We next show that over rationals and reals, two of our intermediate polynomials, based on satisfiability and Hamiltonian cycle, are not monotone affine polynomial-size projections of the permanent. This augments recent results along this line due to Grochow.

Finally, we describe a (somewhat natural) polynomial defined independent of a computation model, and show that it is VP-complete under polynomial-size projections. This complements a recent result of Durand et al. (2014) which established VP-completeness of a related polynomial but under constant-depth oracle circuit reductions. Both polynomials are based on graph homomorphisms. A simple restriction yields a family similarly complete for VBP.

1 Introduction

The algebraic analogue of the P versus NP problem, famously referred to as the VP versus VNP question, is one of the most significant problem in algebraic complexity theory. Valiant [27] showed that the PERMANENT polynomial is VNP-Complete (over fields of char $\neq 2$). A striking aspect of this polynomial is that the underlying decision problem, in fact the search problem, is in P. Given a graph, we can decide in polynomial time whether it has a perfect matching, and if so find a maximum matching in polynomial time [11]. Since the underlying problem is an easier problem, it helped in establishing VNP-Completeness of a host of other polynomials by a reduction from the PERMANENT polynomial (cf. [3]). Inspired from classical results in structural complexity theory, in particular [19], Bürgisser [4] proved that if Valiant's hypothesis (i.e. VP $\neq$ VNP) is true, then, over any field there is a p-family in VNP which is neither in VP nor VNP-Complete with respect to c-reductions. Let us call such polynomial families VNP-Intermediate (i.e. in VNP, not VNP-Complete, not in VP). Further, Bürgisser [4] showed that over finite fields, a *specific* family of polynomials

© Springer International Publishing Switzerland 2016
A.S. Kulikov and G.J. Woeginger (Eds.): CSR 2016, LNCS 9691, pp. 251–265, 2016.
DOI: 10.1007/978-3-319-34171-2_18

is VNP-Intermediate, provided the polynomial hierarchy PH does not collapse to the second level. On an intuitive level these polynomials enumerate *cuts* in a graph. This is a remarkable result, when compared with the classical P-NP setting or the BSS-model. Though the existence of problems with intermediate complexity has been established in the latter settings, due to the involved "diagonalization" arguments used to construct them, these problems seem highly unnatural. That is, their definitions are not motivated by an underlying combinatorial problem but guided by the needs of the proof and, hence, seem artificial. The question of whether there are other naturally-defined VNP-Intermediate polynomials was left open by Bürgisser [3]. We remark that to date the *cut enumerator* polynomial from [4] is the only known example of a natural polynomial family that is VNP-Intermediate.

The question of whether the classes VP and VNP are distinct is often phrased as whether Perm_n is *not* a quasi-polynomial-size projection of Det_n. The importance of this reformulation stems from the fact that it is a purely algebraic statement, devoid of any dependence on circuits. While we have made very little progress on this question of determinantal complexity of the permanent, the progress in restricted settings has been considerable. One of the success stories in theoretical computer science is unconditional lower bound against monotone computations [1,23,24]. In particular, Razborov [24] proved that computing the permanent over the Boolean semiring requires monotone circuits of size at least $n^{\Omega(\log n)}$. Jukna [17] observed that if the Hamilton cycle polynomial is a monotone p-projection of the permanent, then from the observation that the clique polynomial is a monotone projection of the Hamiltonian cycle [27] and that the clique requires monotone circuits of exponential size [1], one would get a lower bound of $2^{n^{\Omega(1)}}$ for monotone circuits computing the permanent, thus improving on [24]. The importance of this observation is also highlighted by the fact that such a monotone p-projection, over the reals, would give an alternate proof of the fact that computing permanent by monotone circuits over $\mathbb{R}$ requires size at least $2^{n^{\Omega(1)}}$. Jerrum and Snir [16] proved that the permanent requires monotone circuits of size $2^{\Omega(n)}$ over $\mathbb{R}$ and tropical semiring. The first progress on this question, raised in [17], was made recently by Grochow [14]. He showed that the Hamiltonian cycle polynomial is not a monotone sub-exponential-size projection of the permanent. This answered Jukna's question in its entirety, but Grochow [14] using his techniques further established that polynomials like the perfect matching polynomial, and even the VNP-Intermediate cut enumerator polynomial of Bürgisser [4], are not monotone polynomial-size projections of the permanent. This raises an intriguing question of whether there are other such non-negative polynomials which share this property.

While the Perm vs Det problem has become synonymous with the VP vs VNP question, there is a strange feeling about it. This rises from two facts: one, that the VP-hardness of the determinant is known only under the more powerful quasi-polynomial-size projections, and, second, the lack of natural VP-complete polynomials (with respect to polynomial-size projections) in the literature. (In fact, with respect to p-projections, the determinant is complete for the possibly smaller class VBP of polynomial-sized algebraic branching programs.) To remedy this situation, it seems crucial to understand the computation in VP. Bürgisser [3] showed that a generic polynomial family constructed using a topological sort of a generic VP circuit, while controlling the degree, is complete for VP. Raz [22], using the depth reduction of [28], showed that a family of "universal circuits" is VP-Complete. Thus both families directly depend on the circuit definition or characterization of VP. Last year, Durand et al. [10] made significant progress and provided a natural, first of its kind, VP-Complete polynomial. However, the natural polynomials studied by Durand et al. lacked a bit of punch because their completeness was established under polynomial-size *constant depth c-reductions* rather than projections.

In this paper, we make progress on all three fronts. First, we provide a list of new natural polynomial families, based on basic (combinatorial) NP-Complete problems [13] whose completeness is via *parsimonious* reductions [26], that are VNP-Intermediate over finite fields (Theorem 1). Then, we show that over reals, some of our intermediate polynomials are not monotone affine polynomial-size projections of the permanent (Theorem 2). As in [14], the lower bound results about monotone affine projections are unconditional. Finally, we improve upon [10] by characterizing VP and establishing a natural VP-Complete polynomial under polynomial-size projections (Theorem 5). A modification yields a family similarly complete for VBP (Theorem 6).

2   Preliminaries

Algebraic Complexity: We say that a polynomial f is a *projection* of g if f can be obtained from g by setting the variables of g to either constants in the field, or to the variables of f. A sequence (f_n) is a *p-projection* of (g_m), if f_n is a projection of g_t such that t is polynomially bounded in n. There are other notions of reductions between families of polynomials, like *c-reductions* (polynomial-size oracle circuit reductions), *constant-depth c-reductions*, and *linear p-projections*. For more on these reductions, see [3].

An arithmetic circuit is a directed acyclic graph with leaves labeled by variables or constants from an underlying field, internal nodes labeled by field operations $+$ and $\times$, and a designated output gate. Each node computes a polynomial in a natural way. The polynomial computed by a circuit is the polynomial computed at its output gate. A *parse tree* of a circuit captures monomial generation within the circuit. For a complete definition see [20]. A circuit is said to be *skew* if at every $\times$ gate, at most one incoming edge is the output of another gate.

A family of polynomials $(f_n(x_1, \ldots, x_{m(n)}))$ is called a *p-family* if both the degree $d(n)$ of f_n and the number of variables $m(n)$ are polynomially bounded. A *p-family* is in VP (resp. VBP) if a circuit family (skew circuit family, resp.) (C_n) of size polynomially bounded in n computes it. A sequence of polynomials (f_n) is in VNP if there exist a sequence (g_n) in VP, and polynomials m and t such that for all n, $f_n(\bar{x}) = \sum_{\bar{y} \in \{0,1\}^{t(\bar{x})}} g_n(x_1, \ldots, x_{m(n)}, y_1, \ldots, y_{t(n)})$. (VBP denotes the algebraic analogue of branching programs. Since these are equivalent to skew circuits, we directly use a skew circuit definition of VBP.)

Boolean Complexity: We need some basics from Boolean complexity theory. Let P/poly denote the class of languages decidable by polynomial-sized Boolean circuit families. A function $\phi : \{0,1\}^* \to \mathbb{N}$ is in #P if there exists a polynomial p and polynomial time deterministic Turing machine M such that for all $x \in \{0,1\}^*$, $f(x) = |\{y \in \{0,1\}^{p(|x|)} \mid M(x,y) = 1\}|$. For p a prime, define

$$\#_p P = \{\psi : \{0,1\}^* \to \mathbb{F}_p \mid \psi(x) = \phi(x) \bmod p \text{ for some } \phi \in \#P\},$$

$$\mathrm{Mod}_p P = \{L \subseteq \{0,1\}^* \mid \text{ for some } \phi \in \#P, x \in L \iff \phi(x) \equiv 1 \bmod p\}$$

It is easy to see that if $\phi : \{0,1\}^* \to \mathbb{N}$ is #P-complete with respect to parsimonious reductions, then the language $L = \{x \mid \phi(x) \equiv 1 \bmod p\}$ is Mod_pP-complete with respect to many-one reductions.

Graph Theory: We consider the treewidth and pathwidth parameters for an undirected graph. We will work with a "canonical" form of decompositions which is generally useful in dynamic-programming algorithms.

A *(nice) tree decomposition* of a graph G is a pair $\mathcal{T} = (T, \{X_t\}_{t \in V(T)})$, where T is a tree, rooted at X_r, whose every node t is assigned a vertex subset $X_t \subseteq V(G)$, called a bag, such that the following conditions hold:

1. $X_r = \emptyset$, $|X_\ell| = 1$ for every leaf ℓ of T, and $\cup_{t \in V(T)} X_t = V(G)$.
   That is, the root contain the empty bag, the leaves contain singleton sets, and every vertex of G is in at least one bag.
2. For every $(u,v) \in E(G)$, there exists a node t of T such that $\{u,v\} \subseteq X_t$.
3. For every $u \in V(G)$, the set $T_u = \{t \in V(T) \mid u \in X_t\}$ induces a connected subtree of T.
4. Every non-leaf node t of T is of one of the following three types:
   - **Introduce node:** t has exactly once child t', and $X_t = X_{t'} \cup \{v\}$ for some vertex $v \notin X_{t'}$. We say that v is *introduced* at t.
   - **Forget node:** t has exactly one child t', and $X_t = X_{t'} \backslash \{w\}$ for some vertex $w \in X_{t'}$. We say that w is *forgotten* at t.
   - **Join node:** t has two children t_1, t_2, and $X_t = X_{t_1} = X_{t_2}$.

The *width* of a tree decomposition $\mathcal{T}$ is one less than the size of the largest bag; that is, $\max_{t \in V(T)} |X_t| - 1$. The *tree-width* of a graph G is the minimum possible width of a tree decomposition of G.

In a similar way we can also define a *nice path decomposition* of a graph. For a complete definition we refer to [7].

A sequence (G_n) of graphs is called a *p-family* if the number of vertices in G_n is polynomially bounded in n. It is further said to have *bounded* tree(path)-width if for some absolute constant c independent of n, the tree(path)-width of each graph in the sequence is bounded by c.

A *homomorphism* from G to H is a map from $V(G)$ to $V(H)$ preserving edges. A graph is called *rigid* if it has *no* homomorphism to itself other than the identity map. Two graphs G and H are called *incomparable* if there are *no* homomorphisms from $G \to H$ as well as $H \to G$. It is known that asymptotically almost all graphs are rigid, and almost all pairs of nonisomorphic graphs are also incomparable. For the purposes of this paper, we only need a collection of three rigid and mutually incomparable graphs. For more details, we refer to [15].

3   VNP-Intermediate

In [4], Bürgisser showed that unless PH collapses to the second level, an explicit family of polynomials, called the cut enumerator polynomial, is VNP-intermediate. He raised the question, recently highlighted again in [14], of whether there are other such natural VNP-intermediate polynomials. In this section we show that in fact his proof strategy itself can be adapted to other polynomial families as well. The strategy can be described abstractly as follows: Find an explicit polynomial family $h = (h_n)$ satisfying the following properties.

M: Membership. The family is in VNP.

H: Hardness. The monomials of h encode solutions to a problem that is #P-hard via parsimonious reductions. Thus if h is in VP, then the number of solutions, modulo p, can be extracted using coefficient computation.

E: Ease. Over a field $\mathbb{F}_q$ of size q and characteristic p, h can be evaluated in P. Thus if h is VNP-hard, then we can efficiently compute #P-hard functions, modulo p.

Then, unless $\mathsf{Mod}_p\mathsf{P} \subseteq \mathsf{P/poly}$ (which in turn implies that PH collapses to the second level, [18]), h is VNP-intermediate.

We provide a list of p-families that, under the same condition $\mathsf{Mod}_p\mathsf{P} \not\subseteq \mathsf{P/poly}$, are VNP-intermediate. All these polynomials are based on basic combinatorial NP-complete problems that are complete under parsimonious reduction.

(1) The *satisfiablity* polynomial $\mathsf{Sat}^q = (\mathsf{Sat}^q{}_n)$: For each n, let Cl_n denote the set of all possible clauses of size 3 over $2n$ literals. There are n variables $\tilde{X} = \{X_i\}_{i=1}^n$, and also $8n^3$ clause-variables $\tilde{Y} = \{Y_c\}_{c\in \mathsf{Cl}_n}$, one for each 3-clause c.

$$\mathsf{Sat}^q{}_n := \sum_{a\in\{0,1\}^n} \left(\prod_{i=1}^n X_i^{a_i(q-1)}\right) \left(\prod_{\substack{c\ \in \mathsf{Cl}_n \\ a\ \text{satisfies}\ c}} Y_c^{q-1}\right).$$

(2) The *vertex cover* polynomial $\mathsf{VC}^q = (\mathsf{VC}^q{}_n)$: For a complete graph G_n on n nodes, we have the set of variables $\tilde{X} = \{X_e\}_{e\in E_n}$ and $\tilde{Y} = \{Y_v\}_{v\in V_n}$.

$$\mathsf{VC}^q{}_n := \sum_{S\subseteq V_n} \left(\prod_{e\in E_n\,:\ e\ \text{is incident on}\ S} X_e^{q-1}\right) \left(\prod_{v\in S} Y_v^{q-1}\right).$$

(3) The *clique/independent set* polynomial $\mathsf{CIS}^q = (\mathsf{CIS}^q{}_n)$:

$$\mathsf{CIS}^q{}_n := \sum_{T\subseteq E_n} \left(\prod_{e\in T} X_e^{q-1}\right) \left(\prod_{v\ \text{incident on}\ T} Y_v^{q-1}\right).$$

(4) The *3D-matching* polynomial $3DM^q = (3DM^q{}_n)$: Consider the complete tri-partite hyper-graph, where each part in the partition contain n nodes, and each hyperedge has exactly one node from each part. There are variables X_e for hyperedge e and Y_v for node v.

$$3DM^q{}_n := \sum_{M \subseteq A_n \times B_n \times C_n} \left(\prod_{e \in M} X_e^{q-1} \right) \left(\prod_{\substack{v \in M \\ \text{(counted only once)}}} Y_v^{q-1} \right).$$

(5) The *clow* polynomial $Clow^q = (Clow^q{}_n)$: A clow in an n-vertex graph is a closed walk of length exactly n, in which the minimum numbered vertex (called the head) appears exactly once.

$$Clow^q{}_n := \sum_{w:\ \text{clow of length } n} \left(\prod_{e:\ \text{edges in } w} X_e^{q-1} \right) \left(\prod_{\substack{v:\ \text{vertices in } w \\ \text{(counted only once)}}} Y_v^{q-1} \right).$$

(If an edge e is used k times in a clow, it contributes $X_e^{k(q-1)}$ to the monomial.)

We show that if $\mathsf{Mod}_p\mathsf{P} \not\subseteq \mathsf{P/poly}$, then all five polynomials defined above are VNP-intermediate.

Theorem 1. *Over a finite field $\mathbb{F}_q$ of characteristic p, the polynomial families Sat^q, VC^q, CIS^q, $3DM^q$, and $Clow^q$, are in VNP. Further, if $\mathsf{Mod}_p\mathsf{P} \not\subseteq \mathsf{P/poly}$, then they are all VNP-intermediate; that is, neither in VP nor VNP-hard with respect to c-reductions.*

Proof (Sketch). (M) An easy way to see membership in VNP is to use Valiant's criterion ([27]; see also Proposition 2.20 in [3]); the coefficient of any monomial can be computed efficiently, hence the polynomial is in VNP.

We illustrate the rest of the proof by showing that the polynomial Sat^q satisfies the properties (H), (E).

(H): Assume $(Sat^q{}_n)$ is in VP, via polynomial-sized circuit family $\{C_n\}_{n \geq 1}$. We will use C_n to give a P/poly upper bound for computing the number of satisfying assignments of a 3-CNF formula, modulo p. Since this question is complete for $\mathsf{Mod}_p\mathsf{P}$, the upper bound implies $\mathsf{Mod}_p\mathsf{P}$ is in P/poly.

Given an instance ϕ of 3SAT, with n variables and m clauses, consider the projection of $Sat^q{}_n$ obtained by setting all Y_c for $c \in \phi$ to t, and all other variables to 1. This gives the polynomial $Sat^q\phi(t) = \sum_{j=1}^{m} d_j t^{j(q-1)}$ where d_j is the number of assignments (modulo p) that satisfy exactly j clauses in ϕ. Our goal is to compute d_m.

We convert the circuit C into a circuit D that compute elements of $\mathbb{F}_q[t]$ by explicitly giving their coefficient vectors, so that we can pull out the desired coefficient. (Note that after the projection described above, C works over the polynomial ring $\mathbb{F}_q[t]$). Since the polynomial computed by C is of degree $m(q-1)$, we need to compute the coefficients of all intermediate polynomials too only upto degree $m(q-1)$. Replacing $+$ by gates performing coordinate-wise addition, $\times$ by a sub-circuit performing (truncated) convolution, and supplying appropriate coefficient vectors at the leaves gives the desired circuit. Since the number of clauses, m, is polynomial in n, the circuit D is also of polynomial size. Given the description of C as advice, the circuit D can be evaluated in P, giving a P/poly algorithm for computing #3-SAT(ϕ) mod p. Hence Mod$_p$P $\subseteq$ P/poly.

(E) Consider an assignment to $\tilde{X}$ and $\tilde{Y}$ variables in $\mathbb{F}_q$. Since all exponents are multiples of $(q-1)$, it suffices to consider $0/1$ assignments to $\tilde{X}$ and $\tilde{Y}$. Each assignment a contributes 0 or 1 to the final value; call it a contributing assignment if it contributes 1. So we just need to count the number of contributing assignments. An assignment a is contributing exactly when $\forall i \in [n]$, $X_i = 0 \implies a_i = 0$, and $\forall c \in \text{Cl}_n$, $Y_c = 0 \implies a$ does not satisfy c. These two conditions, together with the values of the X and Y variables, constrain many bits of a contributing assignment; an inspection reveals how many (and which) bits are so constrained. If any bit is constrained in conflicting ways (for example, $X_i = 0$, and $Y_c = 0$ for some clause c containing the literal $\bar{x}_i$), then no assignment is contributing (either $a_i = 1$ and the X part becomes zero due to $X_i^{a_i}$, or $a_i = 0$ and the Y part becomes zero due to Y_c). Otherwise, some bits of a potentially contributing assignment are constrained by X and Y, and the remaining bits can be set in any way. Hence the total sum is precisely $2^{(\#\text{ unconstrained bits})}$ mod p.

Now assume Sat^q is VNP-hard. Let L be any language in Mod$_p$P, witnessed via #P-function f. (That is, $x \in L \iff f(x) \equiv 1$ mod p.) By the results of [3,5], there exists a p-family $r = (r_n) \in \text{VNP}_{\mathbb{F}_p}$ such that $\forall n$, $\forall x \in \{0,1\}^n$, $r_n(x) = f(x)$ mod p. By assumption, there is a c-reduction from r to Sat^q. We use the oracle circuits from this reduction to decide instances of L. On input x, the advice is the circuit C of appropriate size reducing r to Sat^q. We evaluate this circuit bottom-up. At the leaves, the values are known. At $+$ and $\times$ gates, we perform these operations in $\mathbb{F}_q$. At an oracle gate, the paragraph above tells us how to evaluate the gate. So the circuit can be evaluated in polynomial time, showing that L is in P/poly. Thus Mod$_p$P $\subseteq$ P/poly.   □

It is worth noting that the cut enumerator polynomial Cut^q, showed by Bürgisser to be VNP-intermediate over field $\mathbb{F}_q$, is in fact VNP-complete over the rationals if $q = 2$, [8]. Thus the above technique is specific to finite fields.

4 Monotone Projection Lower Bounds

We now show that some of our intermediate polynomials are not *monotone p-projections* of the PERMANENT polynomial. The results here are motivated by the recent results of Grochow [14]. Recall that a polynomial $f(x_1, \ldots, x_n)$ is a *projection* of a polynomial $g(y_1, \ldots, y_m)$ if $f(x_1, \ldots, x_n) = g(a_1, \ldots, a_m)$, where a_i's are either constants or x_j for some j. The polynomial f is an *affine projection* of g if f can be obtained from g by replacing each y_i with an affine linear function

$\ell_i(\tilde{x})$. Over any subring of $\mathbb{R}$, or more generally any totally ordered semi-ring, a *monotone projection* is a projection in which all constants appearing in the projection are non-negative. We say that the family (f_n) is a (monotone affine) projection of the family (g_n) with *blow-up* $t(n)$ if for all sufficiently large n, f_n is a (monotone affine) projection of $g_{t(n)}$.

Theorem 2. *Over the reals (or any totally ordered semi-ring), for any q, the families Sat^q and Clow^q are not monotone affine p-projections of the* PERMANENT *family. Any monotone affine projection from* PERMANENT *to* Sat^q *must have a blow-up of at least $2^{\Omega(\sqrt{n})}$. Any monotone affine projection from* PERMANENT *to* Clow^q *must have a blow-up of at least $2^{\Omega(n)}$.*

Before giving the proof, we set up some notation. For more details, see [2,14,25]. For any polynomial p in n variables, let $\mathsf{Newt}(p)$ denote the polytope in $\mathbb{R}^n$ that is the convex hull of the vectors of exponents of monomials of p. For any Boolean formula ϕ on n variables, let $\mathsf{p\text{-}SAT}(\phi)$ denote the polytope in $\mathbb{R}^n$ that is the convex hull of all satisfying assignments of ϕ. Let $K_n = (V_n, E_n)$ denote the n-vertex complete graph. The travelling salesperson (TSP) polytope is defined as the convex hull of the characteristic vectors of all subsets of E_n that define a Hamiltonian cycle in K_n.

For a polytope P, let $\mathsf{c}(P)$ denote the minimal number of linear inequalities needed to define P. A polytope $Q \subseteq \mathbb{R}^m$ is an *extension* of $P \subseteq \mathbb{R}^n$ if there is an affine linear map $\pi \colon \mathbb{R}^m \to \mathbb{R}^n$ such that $\pi(Q) = P$. The *extension complexity* of P, denoted $\mathsf{xc}(P)$, is the minimum size $\mathsf{c}(Q)$ of any extension Q (of any dimension) of P. The following are straightforward, see for instance [12,14].

Fact 3. *1.* $\mathsf{c}(\mathsf{Newt}(\mathsf{Perm}_n)) \leqslant 2n$.
2. If polytope Q is an extension of polytope P, then $\mathsf{xc}(P) \leqslant \mathsf{xc}(Q)$.

We use the following recent results.

Proposition 1. *1. Let $f(x_1, \ldots, x_n)$ and $g(y_1, \ldots, y_m)$ be polynomials over a totally ordered semi-ring R, with non-negative coefficients. If f is a monotone projection of g, then the intersection of $\mathsf{Newt}(g)$ with some linear subspace is an extension of $\mathsf{Newt}(f)$. In particular, $\mathsf{xc}(\mathsf{Newt}(f)) \leqslant m + \mathsf{c}(\mathsf{Newt}(g))$. [14]*
2. For every n there exists a 3SAT formula ϕ with $O(n)$ variables and $O(n)$ clauses such that $\mathsf{xc}(\mathsf{p\text{-}SAT}(\phi)) \geqslant 2^{\Omega(\sqrt{n})}$. [2]
3. The extension complexity of the TSP polytope is $2^{\Omega(n)}$. [25]

Proof (of Theorem 2). Let ϕ be a 3SAT formula with n variables and m clauses as given by Proposition 1(2). For the polytope $P = \mathsf{p\text{-}SAT}(\phi)$, $\mathsf{xc}(P)$ is high.

Let Q be the Newton polytope of $\mathsf{Sat}^q{}_n$. It resides in N dimensions, where $N = n + |\mathsf{Cl}_n| = n + 8n^3$, and is the convex hull of vectors of the form $(q-1)\langle \tilde{a}\tilde{b}\rangle$ where $\tilde{a} \in \{0,1\}^n$, $\tilde{b} \in \{0,1\}^{N-n}$, and for all $c \in \mathsf{Cl}_n$, $\tilde{a}$ satisfies c iff $b_c = 1$. For each $\tilde{a} \in \{0,1\}^n$, there is a unique $\tilde{b} \in \{0,1\}^{N-n}$ such that $(q-1)\langle \tilde{a}\tilde{b}\rangle$ is in Q.

Define the polytope R, also in N dimensions, to be the convex hull of vectors that are vertices of Q and also satisfy the constraint $\sum_{c \in \phi} b_c \geq m$. This constraint discards vertices of Q where $\tilde{a}$ does not satisfy ϕ. Thus R is an extension of P (projecting the first n coordinates of points in R gives a $(q-1)$-scaled version of P), so by Fact 3(2), $\mathsf{xc}(P) \leq \mathsf{xc}(R)$. Further, we can obtain an extension of R from any extension of Q by adding just one inequality; hence $\mathsf{xc}(R) \leq 1 + \mathsf{xc}(Q)$.

Suppose Sat^q is a monotone affine projection of Perm_n with blow-up $t(n)$. By Fact 3(1) and Proposition 1(1), $\mathsf{xc}(\mathsf{Newt}(\mathsf{Sat}^q)) = \mathsf{xc}(Q) \leq t(n) + c(\mathsf{Perm}_{t(n)}) \leq O(t(n))$. From the preceding discussion and By Proposition 1(2), we get $2^{\Omega(\sqrt{n})} \leq \mathsf{xc}(P) \leq \mathsf{xc}(R) \leq 1 + \mathsf{xc}(Q) \leq O(t(n))$. It follows that $t(n)$ is at least $2^{\Omega(\sqrt{n})}$.

For the Clow^q polynomial, let P be the TSP polytope and Q be $\mathsf{Newt}(\mathsf{Clow}^q)$. The vertices of Q are of the form $(q-1)\tilde{a}\tilde{b}$ where $\tilde{a} \in \{0,1\}^{\binom{n}{2}}$ picks a subset of edges, $\tilde{b} \in \{0,1\}^n$ picks a subset of vertices, and the picked edges form a length-n clow touching exactly the picked vertices. Define polytope R by discarding vertices of Q where $\sum_{i \in [n]} b_i < n$. Now the same argument as above works, using Proposition 1(3) instead of (2).   □

5   Complete Families for VP and VBP

The quest for a natural VP-Complete polynomial has generated a significant amount of research [3,6,10,21,22]. The first success story came from [10]. They studied naturally defined homomorphism polynomials and showed that a host of them are complete for the class VP. But the results came with minor caveats. When the completeness was established under projections, there were non-trivial restrictions on the set of homomorphisms $\mathcal{H}$, and sometimes even on the target graph H. On the other hand, when all homomorphisms were allowed, completeness could only be shown under seemingly more powerful reductions, namely, constant-depth c-reductions. Furthermore, the graphs were either directed or had weights on them. It is worth noting that the reductions in [10] actually do not use the full power of generic constant-depth c-reductions; a closer analysis reveals that they are in fact *linear p-projection*. That is, the reductions are linear combinations of polynomially many p-projections (see Chap. 3, [3]). Still, this falls short of p-projections.

In this work, we remove all such restrictions and show that there is a simple explicit homomorphism polynomial that is complete for VP under p-projections. More formally, there is a p-family (G_m) of *bounded* tree-width graphs such that the family of homomorphism polynomial $(f_{G_m, H_m, \mathbf{Hom}}(\bar{Y}))_m$ is complete for VP with respect to projections.

We start the discussion with an *upper bound* for computing a generic homomorphism polynomial defined over bounded tree-width graphs. Let $G = (V(G), E(G))$ and $H = (V(H), E(H))$ be two graphs. Consider the set of variables $\bar{Z} \cup \bar{Y}$ where $\bar{Z} := \{Z_{u,a} \mid u \in V(G) \text{ and } a \in V(H)\}$ and $\bar{Y} := \{Y_{(u,v)} \mid (u,v) \in E(H)\}$. Let $\mathcal{H}$ be a set of homomorphisms from G to H. The generalised homomorphism polynomial $f_{G,H,\mathcal{H}}$ is defined as follows:

$$f_{G,H,\mathcal{H}} = \sum_{\phi \in \mathcal{H}} \left(\prod_{u \in V(G)} Z_{u,\phi(u)} \right) \left(\prod_{(u,v) \in E(G)} Y_{(\phi(u),\phi(v))} \right).$$

In [10], it was shown that the homomorphism polynomial $f_{T_m,K_n,\mathbf{Hom}}$ where T_m is a binary tree on m leaves, K_n is a complete graph on n nodes, and **Hom** is the set of all homomorphisms, is computable by an arithmetic circuit of size $O(m^3 n^3)$. Their proof idea is based on recursion: group the homomorphisms based on where they map the root of T_m and its children, and recursively compute the sub-polynomials within each group. The sub-polynomials of a specific group have a special set of variables in their monomials. Hence, the homomorphism polynomial can be computed by suitably combining partial derivatives of the sub-polynomials.

Generalizing the above idea to polynomials where T_m is not a binary tree but a bounded tree-width graph seems hard. The very first obstacle we encounter is to generalize the concept of partial derivative to monomial extension. Combining sub-polynomials to obtain the original polynomial also gets hairy and unnecessarily complicated, if possible.

We present a simple algorithm (based on dynamic programming [9]) showing that the homomorphism polynomial $f_{G,H,\mathbf{Hom}}$ is computable by an arithmetic circuit of size at most $2|V(G)|\cdot|V(H)|^{tw(G)+1}\cdot(2|V(H)|+2|E(H)|)$, where $tw(G)$ is the tree-width of G. From this algorithm we obtain the following theorem.

Theorem 4. *Consider the family of homomorphism polynomials* (f_m), *where* $f_m = f_{G_m,H_m,\mathbf{Hom}}(\bar{Z},\bar{Y})$ *and,* (H_m) *is a p-family of complete graphs.*

- *If* (G_m) *is a p-family of graphs of bounded tree-width,* $(f_m) \in \mathsf{VP}$.
- *If* (G_m) *is a p-family of graphs of bounded path-width,* $(f_m) \in \mathsf{VBP}$.

Proof (Sketch). We use a nice tree decomposition $\mathcal{T} = (T, \{X_t\}_{t\in V(T)})$ of G. For each $t \in V(T)$, let M_t be the set of all mappings from X_t to $V(H)$. Let T_t be the subtree of T rooted at node t, $V_t := \bigcup_{t'\in V(T_t)} X_{t'}$, and $G_t := G[V_t]$ be the subgraph of G induced on V_t. Let r be the root of T; note that $G_r = G$.

The circuit is built inductively. For each $t \in V(T)$ and $\phi \in M_t$, we have gates $\langle t,\phi \rangle$ and $\langle t,\phi \rangle'$ in the circuit. The gate $\langle t,\phi \rangle$ computes the sum of all monomials corresponding to those homomorphisms from G_t to H that map X_t according to ϕ. The gate $\langle t,\phi \rangle'$, computes the "partial derivative" (or, quotient) of the polynomial computed at $\langle t,\phi \rangle$ with respect to the monomial given by ϕ. Due to the normal form of the decomposition, a reasonably straightforward decomposition works. The output gate of the circuit is $\langle r,\emptyset \rangle$.

A path decomposition has no *join* nodes and yields a *skew* circuits.   □

(Note that the circuit constructed above is a constant-free circuit. This was the case with [10] too. Furthermore, the same construction specialises from treewidth to pathwidth and gives skew circuits. The algorithm from [10] does not give skew circuits when T_m is a path. It seems the obstacle there lies in computing partial-derivatives using skew circuits.)

We now turn towards establishing VP-*hardness* of the homomorphism polynomials. We show that there exist a p-family (G_m) of bounded tree-width graphs such that $(f_{G_m,H_m,\mathbf{Hom}}(\bar{Z},\bar{Y}))$ is hard for VP under projections. For our hardness proof, the $\bar{Z}$ variables are in fact redundant and can be set to 1.

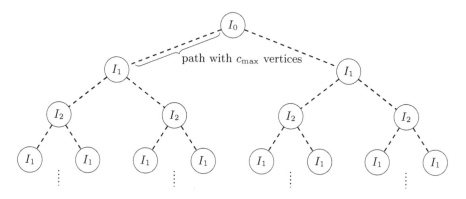

Fig. 1. The graph G_m.

We use *rigid* and mutually *incomparable* graphs in the construction of G_m. Let $I := \{I_0, I_1, I_2\}$ be a fixed set of three connected, rigid and mutually incomparable graphs. Note that they are necessarily *non-bipartite*. Let $c_{I_i} = |V(I_i)|$. Choose an integer $c_{\max} > \max\{c_{I_0}, c_{I_1}, c_{I_2}\}$. Identify two distinct vertices $\{v_\ell^0, v_r^0\}$ in I_0, three distinct vertices $\{v_\ell^1, v_r^1, v_p^1\}$ in I_1, and three distinct vertices $\{v_\ell^2, v_r^2, v_p^2\}$ in I_2.

For every m a power of 2, we denote a complete (perfect) binary tree with m leaves by T_m. We construct a sequence of graphs G_m (Fig. 1) from T_m as follows: first replace the root by the graph I_0, then all the nodes on a particular level are replaced by either I_1 or I_2 alternately (cf. Fig. 1). Now we add edges; suppose we are at a 'node' which is labeled I_i and the left child and right child are labeled I_j, we add an edge between v_ℓ^i and v_p^j in the left child, and an edge between v_r^i and v_p^j in the right child. Finally, to obtain G_m we expand each added edge into a simple path with $c_{\max}$ vertices on it (cf. Fig. 1). That is, a left-edge connection between two incomparable graphs in the tree looks like, $I_i(v_\ell^i) - (\text{path with } c_{\max} \text{vertices}) - (v_p^j)I_j$.

Theorem 5. *Over any field, the family of homomorphism polynomials (f_m), with $f_m(\bar{Y}) = f_{G_m, H_m, \mathbf{Hom}}(\bar{Y})$, where*

- *G_m is defined as above (see Fig. 1), and*
- *H_m is an undirected complete graph on $\mathsf{poly}(m)$, say m^6, vertices,*

is complete for VP *under p-projections.*

Proof (Sketch). *Membership* in VP follows from Theorem 4.

We sketch the *hardness* proof here. The idea is to obtain the VP-complete universal polynomial from [22] as a projection of f_m. Let C_n denote the universal circuit in a nice normal form as described in [10]. Our starting point is the related graph J'_n in [10]. In [10], it was observed that there is a one-to-one correspondence between parse trees of C_n and subgraphs of J'_n that are rooted at $root_L$ and isomorphic to $\mathsf{T}_{2^{k(n)}}$.

We now transform J'_n using the set $I = \{I_0, I_1, I_2\}$. This is similar to the transformation we did to the balanced binary tree T_m. We replace each node by a graph in I; $root_L$ gets I_0 and the rest of the layers get I_1 or I_2 alternately (as in Fig. 1). Edge connections are made so that a left/right child is connected to its parent via the edge $(v_p^j, v_\ell^i)/(v_p^j, v_r^i)$. Finally we replace each edge connection by a path with c_{max} vertices on it (as in Fig. 1), to obtain the graph J_n. All edges of J_n are labeled 1, except for the following edges. Every input node contains the same rigid graph I_i. It has a vertex v_p^i. Each path connection to other nodes has this vertex as its end point. Label such path edges that are incident on v_p^i by the label of the input gate.

Let $m := 2^{k(n)}$. The choice of $\mathsf{poly}(m)$ is such that $4s_n \leqslant \mathsf{poly}(m)$, where s_n is the size of J_n. The $\bar{Y}$ variables are set to $\{0, 1, \bar{x}\}$ such that the non-zero variables pick out the graph J_n. From the observations of [10] it follows that for each parse tree $p\text{-}T$ of C_n, there exists a homomorphism $\phi : G_{2^{k(n)}} \to J_n$ such that $mon(\phi)$ is exactly equal to $mon(p\text{-}T)$. By $mon(\cdot)$ we mean the monomial associated with an object. We claim that these are the only valid homomorphisms from $G_{2^{k(n)}} \to J_n$. We observe the following properties of homomorphisms from $G_{2^{k(n)}} \to J_n$, from which the claim follows.

(i) Any homomorphic image of a rigid-node-subgraph of $G_{2^{k(n)}}$ in J_n, cannot split across two mutually incomparable rigid-node-subgraphs in J_n. That is, there cannot be two vertices in a rigid subgraph of $G_{2^{k(n)}}$ such that one of them is mapped into a rigid subgraph say n_1, and the other one is mapped into another rigid subgraph say n_2. This follows because homomorphisms do not increase distance.

(ii) Because of (i), with each homomorphic image of a rigid node $g_i \in G_{2^{k(n)}}$, we can associate at most one rigid node of J_n, say n_i, such that the homomorphic image of g_i is a subgraph of n_i and the paths (corresponding to incident edges) emanating from it. But such a subgraph has a homomorphism to n_i itself: fold each hanging path into an edge and then map this edge into an edge within n_i. (For instance, let ρ be a path hanging off n_i and attached to n_i at u, and let v be any neighbour of u within n_i. Mapping vertices of ρ to u and v alternately preserves all edges and hence is a homomorphism.) Therefore, we note that in such a case we have a homomorphism from $g_i \to n_i$. By rigidity, g_i must be the same as n_i. The other scenario, where we cannot associate any n_i because g_i is mapped entirely within a path, is not possible since it contradicts *non-bipartiteness* of mutually-incomparable graphs.

Root Must be Mapped to the Root: The rigidity of I_0 and Property (ii) implies that $I_0 \in G_{2^{k(n)}}$ is mapped identically to I_0 in J_n.

Every Level Must be Mapped Within the Same Level: The children of I_0 in $G_{2^{k(n)}}$ are mapped to the children of the root while respecting left-right behaviour. Firstly, the left child cannot be mapped to the root because of incomparability of the graphs I_1 and I_0. Secondly, the left child cannot be mapped to the right child (or vice versa) even though they are the same graphs, because the minimum distance between the vertex in I_0 where the left path emanates and the right child is $c_{\max} + 1$ whereas the distance between the vertex in I_0 where the left path emanates and the left child is $c_{\max}$. So some vertex from the left child must be mapped into the path leading to the right child and hence the rest of the left child must be mapped into a proper subgraph of right child. But this contradicts rigidity of I_1. Continuing like this, we can show that every level must map within the same level and that the mapping within a level is correct.   □

Finally, we show that homomorphism polynomials are rich enough to characterize computation by algebraic branching programs. Here we establish that there exist a p-family (G_k) of undirected *bounded path-width* graphs such that the family $(f_{G_k,H_k,\mathbf{Hom}}(\bar{Y}))$ is VBP-complete with respect to p-projections.

We note that for VBP-completeness under projections, the construction in [10] required directions on their graph, whereas in the undirected setting they could establish hardness under *linear p-projection* using 0-1 valued weights.

As before, we use rigid and mutually incomparable graphs in the construction of G_k. Let $I := \{I_1, I_2\}$ be a set of connected, non-bipartite, rigid and mutually incomparable graphs. Let $c_{I_i} = |V(I_i)|$, and $c_{max} = \max\{c_{I_1}, c_{I_2}\}$. Consider the sequence of graphs G_k (Fig. 2); for every k, there is a simple path with $(k-1) + 2c_{max}$ edges between a copy of I_1 and I_2. The path is between the vertices $u \in V(I_1)$ and $v \in V(I_2)$. These vertices are arbitrarily chosen. The path between vertices a and b in G_k contains $(k-1)$ edges.

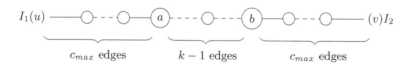

Fig. 2. The graph G_k.

Theorem 6. *Over any field, the family of homomorphism polynomials (f_k), f_k $(\bar{Y}) = f_{G_k,H_k,\mathbf{Hom}}(\bar{Y})$, where*

- *G_k is defined as above (see Fig. 2),*
- *H_k is undirected complete graph on $O(k^2)$ vertices,*

is complete for VBP *with respect to p-projections.*

If the field characteristic is not 2, then completeness also holds when G_k is the simple undirected cycle of length $2k + 1$.

6   Conclusion

In this paper, we have shown that over finite fields, five families of polynomials are intermediate in complexity between VP and VNP, assuming the PH does not collapse. Over rationals and reals, we have established that two of these families are provably not monotone p-projections of the permanent polynomials. Finally, we have obtained a natural family of polynomials, defined via graph homomorphisms, that is complete for VP with respect to projections; this is the first family defined independent of circuits and with such hardness. An analogous family is also shown to be complete for VBP.

Several interesting questions remain.

The definitions of our intermediate polynomials use the size q of the field $\mathbb{F}_q$, not just the characteristic p. Can we find families of polynomials with integer coefficients, that are VNP-intermediate (under some natural complexity assumption of course) over all fields of characteristic p? Even more ambitiously, can we find families of polynomials with integer coefficients, that are VNP-intermediate over all fields with non-zero characteristic? at least over all finite fields? over fields $\mathbb{F}_p$ for all (or even for infinitely many) primes p?

Equally interestingly, can we find an explicit family of polynomials that is VNP-intermediate in characteristic zero?

A related question is whether there are any polynomials defined over the integers, that are VNP-intermediate over $\mathbb{F}_q$ (for some fixed q) but that are monotone p-projections of the permanent.

Can we show that the remaining intermediate polynomials are also not polynomial-sized monotone projections of the permanent? Do such results have any interesting consequences, say, improved circuit lower bounds?

References

1. Alon, N., Boppana, R.B.: The monotone circuit complexity of Boolean functions. Combinatorica **7**(1), 1–22 (1987)
2. Avis, D., Tiwary, H.R.: On the extension complexity of combinatorial polytopes. In: Fomin, F.V., Freivalds, R., Kwiatkowska, M., Peleg, D. (eds.) ICALP 2013, Part I. LNCS, vol. 7965, pp. 57–68. Springer, Heidelberg (2013)
3. Bürgisser, P.: Completeness and Reduction in Algebraic Complexity Theory. Algorithms and Computation in Mathematics, vol. 7. Springer, Heidelberg (2000)
4. Bürgisser, P.: On the structure of Valiant's complexity classes. Discrete Math. Theor. Comput. Sci. **3**(3), 73–94 (1999)
5. Bürgisser, P.: Cook's versus Valiant's hypothesis. Theor. Comput. Sci. **235**(1), 71–88 (2000)
6. Capelli, F., Durand, A., Mengel, S.: The arithmetic complexity of tensor contractions. In: Symposium on Theoretical Aspects of Computer Science STACS. LIPIcs, vol. 20, pp. 365–376 (2013)
7. Cygan, M., Fomin, F.V., Kowalik, Ł., Lokshtanov, D., Marx, D., Pilipczuk, M., Pilipczuk, M., Saurabh, S.: Parameterized Algorithms. Springer, Switzerland (2015)
8. de Rugy-Altherre, N.: A dichotomy theorem for homomorphism polynomials. In: Rovan, B., Sassone, V., Widmayer, P. (eds.) MFCS 2012. LNCS, vol. 7464, pp. 308–322. Springer, Heidelberg (2012)
9. Díaz, J., Serna, M.J., Thilikos, D.M.: Counting h-colorings of partial k-trees. Theor. Comput. Sci. **281**(1–2), 291–309 (2002)
10. Durand, A., Mahajan, M., Malod, G., de Rugy-Altherre, N., Saurabh, N.: Homomorphism polynomials complete for VP. In: 34th Foundation of Software Technology and Theoretical Computer Science Conference, FSTTCS, pp. 493–504 (2014)

11. Edmonds, J.: Paths, trees, and flowers. Can. J. Math. **17**(3), 449–467 (1965)
12. Fiorini, S., Massar, S., Pokutta, S., Tiwary, H.R., Wolf, R.D.: Exponential lower bounds for polytopes in combinatorial optimization. J. ACM **62**(2), 17 (2015)
13. Garey, M.R., Johnson, D.S.: Computers and Intractability: A Guide to the Theory of NP-Completeness. W.H. Freeman, New York (1979)
14. Grochow, J.A.: Monotone projection lower bounds from extended formulation lower bounds. [cs.CC] (2015). arXiv:1510.08417
15. Hell, P., Nešetřil, J.: Graphs And Homomorphisms. Oxford Lecture Series in Mathematics and its Applications. Oxford University Press, Oxford (2004)
16. Jerrum, M., Snir, M.: Some exact complexity results for straight-line computations over semirings. J. ACM **29**(3), 874–897 (1982)
17. Jukna, S.: Why is Hamilton cycle so different from permanent? (2014). http://cs-theory.stackexchange.com/questions/27496/why-is-hamiltonian-cycle-so-different-from-permanent
18. Karp, R.M., Lipton, R.: Turing machines that take advice. L'enseignement mathématique **28**(2), 191–209 (1982)
19. Ladner, R.E.: On the structure of polynomial time reducibility. J. ACM **22**(1), 155–171 (1975)
20. Malod, G., Portier, N.: Characterizing Valiant's algebraic complexity classes. J. Complex. **24**(1), 16–38 (2008)
21. Mengel, S.: Characterizing arithmetic circuit classes by constraint satisfaction problems. In: Aceto, L., Henzinger, M., Sgall, J. (eds.) ICALP 2011, Part I. LNCS, vol. 6755, pp. 700–711. Springer, Heidelberg (2011)
22. Raz, R.: Elusive functions and lower bounds for arithmetic circuits. Theor. Comput. **6**, 135–177 (2010)
23. Razborov, A.A.: Lower bounds on the monotone complexity of some Boolean functions. Dokl. Akad. Nauk SSSR **281**(4), 798–801 (1985)
24. Razborov, A.A.: Lower bounds on monotone complexity of the logical permanent. Math. Notes Acad. Sci. USSR **37**(6), 485–493 (1985)
25. Rothvoß, T.: The matching polytope has exponential extension complexity. In: Symposium on Theory of Computing, STOC, pp. 263–272. New York, 31 May–03 June 2014
26. Simon, J.: On the difference between one and many. In: Salomaa, A., Steinby, M. (eds.) Automata, Languages and Programming. LNCS, vol. 52, pp. 480–491. Springer, Heidelberg (1977)
27. Valiant, L.G.: Completeness classes in algebra. In: Symposium on Theory of Computing STOC, pp. 249–261 (1979)
28. Valiant, L.G., Skyum, S., Berkowitz, S., Rackoff, C.: Fast parallel computation of polynomials using few processors. SIAM J. Comput. **12**(4), 641–644 (1983)

Sums of Read-Once Formulas: How Many Summands Suffice?

Meena Mahajan and Anuj Tawari[✉]

The Institute of Mathematical Sciences, Chennai, India
{meena,anujvt}@imsc.res.in

Abstract. An arithmetic read-once formula (ROF) is a formula (circuit of fan-out 1) over $+$, $\times$ where each variable labels at most one leaf. Every multilinear polynomial can be expressed as the sum of ROFs. In this work, we prove, for certain multilinear polynomials, a tight lower bound on the number of summands in such an expression.

1 Introduction

Read-once formulas (ROF) are formulas (circuits of fan-out 1) in which each variable appears at most once. A formula computing a polynomial that depends on all its variables must read each variable at least once. Therefore, ROFs compute some of the simplest possible functions that depend on all of their variables. The polynomials computed by such formulas are known as read-once polynomials (ROPs). Since every variable is read at most once, ROPs are multilinear[1]. But not every multilinear polynomial is a ROP. For example, $x_1x_2 + x_2x_3 + x_1x_3$.

We investigate the following question: Given an n-variate multilinear polynomial, can it be expressed as a sum of at most k ROPs? It is easy to see that every bivariate multilinear polynomial is a ROP. Any tri-variate multilinear polynomial can be expressed as a sum of 2 ROPs. With a little thought, we can obtain a sum-of-3-ROPs expression for any 4-variate multilinear polynomial. An easy induction on n then shows that any n-variate multilinear polynomial, for $n \geq 4$, can be written as a sum of at most $3 \times 2^{n-4}$ ROPs. Also, the sum of two multilinear monomials is a ROP, so any n-variate multilinear polynomial with M monomials can be written as the sum of $\lceil M/2 \rceil$ ROPs. We ask the following question: Does there exist a strict hierarchy among k-sums of ROPs? We answer this affirmatively for $k \leq \lceil n/2 \rceil$. In particular, for $k = \lceil n/2 \rceil$, we describe an explicit n-variate multilinear polynomial which cannot be written as a sum of less than k ROPs but it admits a sum-of-k-ROPs representation.

Note that n-variate ROPs are computed by linear sized formulas. Thus if an n-variate polynomial p is in $\sum^k \cdot \text{ROP}$, then p is computed by a formula of size $O(kn)$ where every intermediate node computes a multilinear polynomial. Since superpolynomial lower bounds are already known for the model of multilinear formulas [8], we know that for those polynomials (including the determinant and

---

[1] A polynomial is multilinear if the individual degree of each variable is at most one.

© Springer International Publishing Switzerland 2016
A.S. Kulikov and G.J. Woeginger (Eds.): CSR 2016, LNCS 9691, pp. 266–279, 2016.
DOI: 10.1007/978-3-319-34171-2_19

the permanent), a $\sum^k \cdot$ROP expression must have k at least quasi-polynomial in n. However the best upper bound on k for these polynomials is only exponential in n, leaving a big gap between the lower and upper bound. On the other hand, our lower bound is provably tight.

A counting argument shows that a random multilinear polynomial requires exponentially many ROPs; there are multilinear polynomials requiring $k = \Omega(2^n/n^2)$. Our general upper bound on k is $O(2^n)$, leaving a gap between the lower and upper bound. One challenge is to close this gap. A perhaps more interesting challenge is to find explicit polynomials that require exponentially large k in any $\sum^k \cdot$ROP expression.

A natural question to ask is whether stronger lower bounds than the above result can be proven. In particular, to separate $\sum^{k-1} \cdot$ROP from $\sum^k \cdot$ROP, how many variables are needed? The above hierarchy result says that $2k-1$ variables suffice, but there may be simpler polynomials (with fewer variables) witnessing this separation. We demonstrate another technique which improves upon the previous result for $k = 3$, showing that 4 variables suffice. In particular, we show that over the field of reals, there exists an explicit multilinear 4-variate multilinear polynomial which cannot be written as a sum of 2 ROPs. This lower bound is again tight, as there is a sum of 3 ROPs representation for every 4-variate multilinear polynomial.

Our Results and Techniques: We now formally state our results.

Theorem 1. *For each $n \geq 1$, the n-variate degree $n-1$ symmetric polynomial S_n^{n-1} cannot be written as a sum of less than $\lceil n/2 \rceil$ ROPs, but it can be written as a sum of $\lceil n/2 \rceil$ ROPs.*

The idea behind the lower bound is that if g can be expressed as a sum of less than $\lceil n/2 \rceil$ ROFs, then one of the ROFs can be eliminated by taking partial derivative with respect to one variable and substituting another by a field constant. We then use the inductive hypothesis to arrive at a contradiction. This approach necessitates a stronger hypothesis than the statement of the theorem, and we prove this stronger statement in Lemma 3 as part of Theorem 7.

Theorem 2. *There is an explicit 4-variate multilinear polynomial f which cannot be written as the sum of 2 ROPs over $\mathbb{R}$.*

The proof of this theorem mainly relies on a structural lemma (Lemma 6) for sum of 2 read-once formulas. In particular, we show that if f can be written as a sum of 2 ROPs then one of the following must be true: 1. Some 2-variate restriction is a linear polynomial. 2. There exist variables $x_i, x_j \in \text{Var}(f)$ such that the polynomials $x_i, x_j, \partial_{x_i}(f), \partial_{x_j}(f), 1$ are linearly dependent. 3. We can represent f as $f = l_1 \cdot l_2 + l_3 \cdot l_4$ where (l_1, l_2) and (l_3, l_4) are variable-disjoint linear forms. Checking the first two conditions is easy. For the third condition we use the commutator of f, introduced in [9], to find one of the l_i's. The knowledge of one of the l_i's suffices to determine all the linear forms. Finally, we construct a 4-variate polynomial which does not satisfy any of the above mentioned conditions. This construction does not work over algebraically closed fields. We do not yet know how to construct an explicit 4-variate multilinear polynomial not expressible as the sum of 2 ROPs over such fields, or even whether such polynomials exist.

Related Work: Despite their simplicity, ROFs have received a lot of attention both in the arithmetic as well as in the Boolean world [2–5,9,10]. The most fundamental question that can be asked about polynomials is polynomial identity testing (PIT): Given an arithmetic circuit $\mathcal{C}$, is the polynomial computed by $\mathcal{C}$ identically zero? PIT has a randomized polynomial time algorithm: Evaluate the polynomial at random points. It is not known whether PIT has a deterministic polynomial time algorithm. In 2004, Kabanets and Impagliazzo established a connection between PIT algorithms and proving general circuit lower bounds [6]. However, for restricted arithmetic circuits, no such result is known. For instance, consider the case of multilinear formulas. Even though strong lower bounds are known for this model, there is no efficient deterministic PIT algorithm. For this reason, PIT was studied for the weaker model of sum of read-once formulas. Notice that multilinear depth 3 circuits are a special case of this model.

Shpilka and Volkovich gave a deterministic PIT algorithm for the sum of a small number of ROPs [10]. Interestingly, their proof uses a lower bound for a weaker model, that of 0-justified ROFs (setting some variables to zero does not kill any other variables). In particular, they show that the polynomial $\mathcal{M}_n = x_1 x_2 \cdots x_n$, consisting of just a single monomial, cannot be represented as a sum of less than $n/3$ weakly justified ROPs. More recently, Kayal showed that if $\mathcal{M}_n$ is represented as a sum of powers of low degree (at most d) polynomials, then the number of summands is at most $\exp(\Omega(n/d))$ [7]. He used this lower bound to give a PIT algorithm. Our lower bound from Theorem 1 is orthogonal to both these results and is provably tight. An interesting question is whether it can be used to give a PIT algorithm.

Similar to ROPs, one may also study read-restricted formulas. For any number k, RkFs are formulas that read every variable at most k times. For $k > 1$, RkFs for $k \geq 2$ need not be multilinear, and thus are strictly more powerful than ROPs. However, even when restricted to multilinear polynomials, they are more powerful; in [1], Anderson, Melkebeek and Volkovich show that there is a multilinear n-variate polynomial in R2F requiring $\Omega(n)$ summands when written as a sum of ROPs.

Organization: The paper is organized as follows. In Sect. 2 we give the basic definitions and notations. In Sect. 3, we establish Theorem 1. showing that the hierarchy of k-sums of ROPs is proper. In Sect. 4 we establish Theorem 2, showing an explicit 4-variate multilinear polynomial that is not expressible as the sum of two ROPs. We conclude in Sect. 5 with some further open questions.

2 Preliminaries

For a positive integer n, we denote $[n] = \{1, 2, \ldots, n\}$. For a polynomial f, $\mathrm{Var}(f)$ denotes the set of variables occurring in f. Further, for a variable x_i and a field element α, we denote by $f \mid_{x_i = \alpha}$ the polynomial resulting from setting $x_i = \alpha$. Let f be an n-variate polynomial. We say that g is a k-variate restriction of f if g is obtained by setting some variables in f to field constants and $|\mathrm{Var}(g)| \le k$. A set of polynomials $f_1, f_2, \ldots, f_k$ over the field $\mathbb{F}$ is said to be linearly dependent if there exist constants $\alpha_1, \alpha_2, \ldots, \alpha_k$ such that $\sum_{i \in [k]} \alpha_i f_i = 0$.

The n-variate degree k elementary symmetric polynomial, denoted S_n^k, is defined as follows: $S_n^k(x_1, \ldots, x_n) = \sum_{A \subseteq [n], |A| = k} \prod_{i \in A} x_i$.

A circuit is a directed acyclic graph with variables and field constants labeling the leaves, field operations $+, \times$ labeling internal nodes, and a designated output node. Each node naturally computes a polynomial; the polynomial at the output node is the polynomial computed by the circuit. If the underlying undirected graph is a tree, then the circuit is called a formula. A formula is said to be read-k if each variable appears as a leaf label at most k times. For read-once formulas, it is more convenient to use the following "normal form" from [10].

Definition 1 (Read-once formulas [10]). *A read-once arithmetic formula (ROF) over a field $\mathbb{F}$ in the variables $\{x_1, x_2, \ldots, x_n\}$ is a binary tree as follows. The leaves are labeled by variables and internal nodes by $\{+, \times\}$. In addition, every node is labeled by a pair of field elements $(\alpha, \beta) \in \mathbb{F}^2$. Each input variable labels at most once leaf. The computation is performed as follows. A leaf labeled by x_i and (α, β) computes $\alpha x_i + \beta$. If a node v is labeled by $\star \in \{+, \times\}$ and (α, β) and its children compute the polynomials f_1 and f_2, then v computes $\alpha(f_1 \star f_2) + \beta$.*

We say that f is a read-once polynomial (ROP) if it can be computed by a ROF, and is in $\sum^k \cdot$ROP if it can be expressed as the sum of at most k ROPs.

Proposition 1. *For every n, every n-variate multilinear polynomial can be written as the sum of at most $\lceil 3 \times 2^{n-4} \rceil$ ROPs.*

Proposition 2. *For every n, every n-variate multilinear polynomial with M monomials can be written as the sum of at most $\lceil \frac{M}{2} \rceil$ ROPs.*

The partial derivative of a polynomial is defined naturally over continuous domains. The definition can be extended in more than one way over finite fields. However, for multilinear polynomials, these definitions coincide. We consider only multilinear polynomials in this paper, and the following formulation is most useful for us: The partial derivative of a polynomial $p \in \mathbb{F}[x_1, x_2, \ldots, x_n]$ with respect to a variable x_i, for $i \in [n]$, is given by $\partial_{x_i}(p) \triangleq p \mid_{x_i = 1} - p \mid_{x_i = 0}$. For multilinear polynomials, the sum, product, and chain rules continue to hold.

Fact 3 ([10]). *The partial derivatives of ROPs are also ROPs.*

Proposition 3 (3-variate ROPs). *Let $f \in \mathbb{F}[x_1, x_2, x_3]$ be a 3-variate ROP. Then there exists $i \in [3]$ and $a \in \mathbb{F}$ such that $\deg(f \mid_{x_i=a}) \leq 1$.*

A special case of ROFs, multiplicative ROFs defined below, will be relevant.

Definition 2 (Multiplicative Read-once formulas). *A ROF is said to be a multiplicative ROF if it does not contain any addition gates. We say that f is a multiplicative ROP if it can be computed by a multiplicative ROF.*

Fact 4 ([10] (Lemma 3.10)). *A ROP p is a multiplicative ROP if and only if for any two variables $x_i, x_j \in \text{Var}(p)$, $\partial_{x_i}\partial_{x_j}(p) \neq 0$.*

Multiplicative ROPs have the following useful property, observed in [10]. (See Lemma 3.13 in [10]. For completeness, and since we refer to the proof later, we include a proof sketch here.)

Lemma 1 ([10]). *Let g be a multiplicative ROP with $|\text{Var}(g)| \geq 2$. For every $x_i \in \text{Var}(g)$, there exists $x_j \in \text{Var}(g) \setminus \{x_i\}$ and $\gamma \in \mathbb{F}$ such that $\partial_{x_j}(g) \mid_{x_i=\gamma} = 0$.*

Proof. Let φ be a multiplicative ROF computing g. Pick any $x_i \in \text{Var}(g)$. As $|\text{Var}(\varphi)| = |\text{Var}(g)| \geq 2$, φ has at least one gate. Let v be the unique neighbour (parent) of the leaf labeled by x_i, and let w be the other child of v. We denote by $P_v(\bar{x})$ and $P_w(\bar{x})$ the ROPs computed by v and w. Since v is a $\times$ gate and we use the normal form from Definition 1, P_v is of the form $(\alpha x_i + \beta) \times P_w$ for some $\alpha \neq 0$.

Replacing the output from v by a new variable y, we obtain from φ another multiplicative ROF ψ in the variables $\{y\} \cup \text{Var}(g) \setminus \text{Var}(P_v)$. Let ψ compute the polynomial Q; then $g = Q \mid_{y=P_v}$.

Note that the sets $\text{Var}(Q), \{x_i\}, \text{Var}(P_w)$ are non-empty and disjoint, and form a partition of $\{y\} \cup \{\text{Var}(g)\}$.

By the chain rule, for every variable $x_j \in \text{Var}(P_w)$ we have:

$$\partial_{x_j}(g) = \partial_y(Q) \cdot \partial_{x_j}(P_v) = \partial_y(Q) \cdot (\alpha x_i + \beta) \cdot \partial_{x_j}(P_w)$$

It follows that for $\gamma = -\beta/\alpha$, $\partial_{x_j}(g) \mid_{x_i=\gamma} = 0$.   □

Along with partial derivatives, another operator that we will find useful is the commutator of a polynomial. The commutator of a polynomial has previously been used for polynomial factorization and in reconstruction algorithms for read-once formulas, see [9].

Definition 3 (Commutator [9]). *Let $P \in \mathbb{F}[x_1, x_2, \ldots, x_n]$ be a multilinear polynomial and let $i, j \in [n]$. The commutator between x_i and x_j, denoted $\triangle_{ij}P$, is defined as follows.*

$$\triangle_{ij}P = \left(P \mid_{x_i=0,x_j=0} \right) \cdot \left(P \mid_{x_i=1,x_j=1} \right) - \left(P \mid_{x_i=0,x_j=1} \right) \cdot \left(P \mid_{x_i=1,x_j=0} \right)$$

The following property of the commutator will be useful to us.

Lemma 2. *Let $f = l_1(x_1, x_2) \cdot l_2(x_3, x_4) + l_3(x_1, x_3) \cdot l_4(x_2, x_4)$ where the l_i's are linear polynomials. Then l_2 divides $\triangle_{12}(f)$.*

3   A Proper Hierarchy in $\sum^k \cdot$ROP

This section is devoted to proving Theorem 1.

We prove the lower bound for S_n^{n-1} by induction. This necessitates a stronger induction hypothesis, so we will actually prove the lower bound for a larger class of polynomials. The upper bound will also hold for this larger class. For any $\alpha, \beta \in \mathbb{F}$, we define the polynomial $\mathcal{M}_n^{\alpha,\beta} = \alpha S_n^n + \beta S_n^{n-1}$. We note the following recursive structure of $\mathcal{M}_n^{\alpha,\beta}$:

$$(\mathcal{M}_n^{\alpha,\beta})\,|_{x_n=\gamma} = \mathcal{M}_{n-1}^{\alpha\gamma+\beta,\beta\gamma} \quad ; \quad \partial_{x_n}(\mathcal{M}_n^{\alpha,\beta}) = \mathcal{M}_{n-1}^{\alpha,\beta} \quad .$$

We show below that each $\mathcal{M}_n^{\alpha,\beta}$ is expressible as the sum of $\lceil n/2 \rceil$ ROPs (Lemma 4); however, for any non-zero $\beta \neq 0$, $\mathcal{M}_n^{\alpha,\beta}$ cannot be written as the sum of fewer than $\lceil n/2 \rceil$ ROPs (Lemma 3). At $\alpha = 0$, $\beta = 1$, we get S_n^{n-1}, the simplest such polynomials, establishing Theorem 1.

Lemma 3. *Let $\mathbb{F}$ be a field. For every $\alpha \in \mathbb{F}$ and $\beta \in \mathbb{F} \setminus \{0\}$, the polynomial $\mathcal{M}_n^{\alpha,\beta} = \alpha S_n^n + \beta S_n^{n-1}$ cannot be written as a sum of $k < n/2$ ROPs.*

Proof. The proof is by induction on n. The cases $n = 1, 2$ are easy to see. We now assume that $k \geq 1$ and $n > 2k$. Assume to the contrary that there are ROPs $f_1, f_2, \ldots, f_k$ over $\mathbb{F}[x_1, x_2, \ldots, x_n]$ such that $f \triangleq \sum_{m \in [k]} f_m = \mathcal{M}_n^{\alpha,\beta}$. The main steps in the proof are as follows:

1. Show using the inductive hypothesis that for all $m \in [k]$ and $a, b \in [n]$, $\partial_{x_a} \partial_{x_b}(f_m) \neq 0$.
2. Conclude that for all $m \in [k]$, f_m must be a multiplicative ROP. That is, the ROF computing f_m does not contain any addition gate.
3. Use the multiplicative property of f_k to show that f_k can be eliminated by taking partial derivative with respect to one variable and substituting another by a field constant. If this constant is non-zero, we contradict the inductive hypothesis.
4. Otherwise, use the sum of (multiplicative) ROPs representation of $\mathcal{M}_n^{\alpha,\beta}$ to show that the degree of f can be made at most $(n-2)$ by setting one of the variables to zero. This contradicts our choice of f since $\beta \neq 0$.

We now proceed with the proof.

Claim 5. *For all $m \in [k]$ and $a, b \in [n]$, $\partial_{x_a} \partial_{x_b}(f_m) \neq 0$.*

Proof. Suppose to the contrary that $\partial_{x_a} \partial_{x_b}(f_m) = 0$. Assume without loss of generality that $a = n$, $b = n - 1$, $m = k$, so $\partial_{x_n} \partial_{x_{n-1}}(f_k) = 0$. Then,

$$\mathcal{M}_n^{\alpha,\beta} = f = \sum_{m=0}^{k} f_m \qquad \text{(by assumption)}$$

$$\partial_{x_n}\partial_{x_{n-1}}(\mathcal{M}_n^{\alpha,\beta}) = \sum_{m=0}^{k} \partial_{x_n}\partial_{x_{n-1}}(f_m) \quad \text{(by additivity of partial derivative)}$$

$$\mathcal{M}_{n-2}^{\alpha,\beta} = \sum_{m=0}^{k-1} \partial_{x_n}\partial_{x_{n-1}}(f_m) \quad \text{(by recursive structure of } \mathcal{M}_n,$$

$$\text{and since } \partial_{x_n}\partial_{x_{n-1}}(f_k) = 0)$$

Thus $\mathcal{M}_{n-2}^{\alpha,\beta}$ can be written as the sum of $k-1$ polynomials, each of which is a ROP (by Fact 3). By the inductive hypothesis, $2(k-1) \geq (n-2)$. Therefore, $k \geq n/2$ contradicting our assumption. $\qquad\square$

From Claim 5 and Fact 4, we can conclude:

Observation 6. *For all $m \in [k]$, f_m is a multiplicative ROP.*

Observation 6 and Lemma 1 imply that for each $m \in [k]$ and $a \in [n]$, there exist $b \neq a \in [n]$ and $\gamma \in \mathbb{F}$ such that $\partial_{x_b}(f_m)\,|_{x_a=\gamma}= 0$. There are two cases.

First, consider the case when for some m, a and the corresponding b, γ, it turns out that $\gamma \neq 0$. Assume without loss of generality that $m = k$, $a = n-1$, $b = n$, so that $\partial_{x_n}(f_k)\,|_{x_{n-1}=\gamma}= 0$. (For other indices the argument is symmetric.) Then

$$\mathcal{M}_n^{\alpha,\beta} = \sum_{i\in[k]} f_i \qquad \text{(by assumption)}$$

$$\partial_{x_n}(\mathcal{M}_n^{\alpha,\beta})\,|_{x_{n-1}=\gamma}= \sum_{i\in[k]} \partial_{x_n}(f_i)\,|_{x_{n-1}=\gamma} \qquad \text{(additivity of partial derivative)}$$

$$\mathcal{M}_{n-1}^{\alpha,\beta}\,|_{x_{n-1}=\gamma}= \sum_{i\in[k-1]} \partial_{x_n}(f_i)\,|_{x_{n-1}=\gamma} \quad \text{(since } \gamma \text{ is chosen from Lemma 1)}$$

$$\mathcal{M}_{n-2}^{\alpha\gamma+\beta,\beta\gamma}= \sum_{i\in[k-1]} \partial_{x_n}(f_i)\,|_{x_{n-1}=\gamma} \quad \text{(recursive structure of } \mathcal{M}_n)$$

Therefore, $\mathcal{M}_{n-2}^{\alpha\gamma+\beta,\beta\gamma}$ can be written as a sum of at most $k-1$ polynomials, each of which is a ROP (Fact 3). By the inductive hypothesis, $2(k-1) \geq n-2$ implying that $k \geq n/2$ contradicting our assumption.

(Note: the term $\mathcal{M}_{n-2}^{\alpha\gamma+\beta,\beta\gamma}$ is what necessitates a stronger induction hypothesis than working with just $\alpha = 0, \beta = 1$.)

It remains to handle the case when for all $m \in [k]$ and $a \in [n]$, the corresponding value of γ to some x_b (as guaranteed by Lemma 1) is 0. Examining the proof of Lemma 1, this implies that each leaf node in any of the ROFs can be made zero only by setting the corresponding variable to zero. That is, the linear forms at all leaves are of the form $a_i x_i$.

Since each φ_m is a multiplicative ROP, setting $x_n = 0$ makes the variables in the polynomial computed at the sibling of the leaf node $a_n x_n$ redundant. Hence setting $x_n = 0$ reduces the degree of each f_m by at least 2. That is, $\deg(f \mid_{x_n=0}) \leq n-2$. But $\mathcal{M}_n^{\alpha,\beta} \mid_{x_n=0}$ equals $\mathcal{M}_{n-1}^{\beta,0} = \beta S_{n-1}^{n-1}$, which has degree $n-1$, contradicting the assumption that $f = \mathcal{M}_n^{\alpha,\beta}$. $\qquad\square$

The following lemma shows that the above lower bound is indeed optimal.

Lemma 4. *For any field* $\mathbb{F}$ *and* $\alpha, \beta \in \mathbb{F}$, *the polynomial* $f = \alpha S_n^n + \beta S_n^{n-1}$ *can be written as a sum of at most* $\lceil n/2 \rceil$ *ROPs.*

Proof. (Sketch) For n odd, this follows immediately from Proposition 2. For even n, a small tweak works: combine αS_n^n with any one pair of monomials from βS_n^{n-1} to get a single ROP. $\qquad\square$

Combining the results of Lemmas 3 and 4, we obtain the following theorem. At $\alpha = 0, \beta = 1$, it yields Theorem 1.

Theorem 7. *For each* $n \geq 1$, *any* $\alpha \in \mathbb{F}$ *and any* $\beta \in \mathbb{F} \setminus \{0\}$, *the polynomial* $\alpha S_n^n + \beta S_n^{n-1}$ *is in* $\sum^k \cdot ROP$ *but not in* $\sum^{k-1} \cdot ROP$, *where* $k = \lceil n/2 \rceil$.

4   A 4-Variate Multilinear Polynomial Not in $\sum^2 \cdot$ROP

This section is devoted to proving Theorem 2. We want to find an explicit 4-variate multilinear polynomial that is not expressible as the sum of 2 ROPs.

Note that the proof of Theorem 1 does not help here, since the polynomials separating $\sum^2 \cdot$ROP from $\sum^3 \cdot$ROP have 5 or 6 variables. One obvious approach is to consider other combinations of the symmetric polynomials. This fails too; we can show that all such combinations are in $\sum^2 \cdot$ROP.

Proposition 4. *For every choice of field constants* a_i *for each* $i \in \{0, 1, 2, 3, 4\}$, *the polynomial* $\sum_{i=0}^4 a_i S_4^i$ *can be expressed as the sum of two ROPs.*

Instead, we define a polynomial that gives carefully chosen weights to the monomials of S_4^2. Let $f^{\alpha,\beta,\gamma}$ denote the following polynomial:

$$f^{\alpha,\beta,\gamma} = \alpha \cdot (x_1 x_2 + x_3 x_4) + \beta \cdot (x_1 x_3 + x_2 x_4) + \gamma \cdot (x_1 x_4 + x_2 x_3).$$

To keep notation simple, we will omit the superscript when it is clear from the context. In the theorem below, we obtain necessary and sufficient conditions on α, β, γ under which f can be expressed as a sum of two ROPs.

Theorem 8 (Hardness of representation for sum of 2 ROPs). *Let* f *be the polynomial* $f^{\alpha,\beta,\gamma} = \alpha \cdot (x_1 x_2 + x_3 x_4) + \beta \cdot (x_1 x_3 + x_2 x_4) + \gamma \cdot (x_1 x_4 + x_2 x_3)$. *The following are equivalent:*

1. *f is not expressible as the sum of two ROPs.*
2. *α, β, γ satisfy all the three conditions C1, C2, C3 listed below.*

C1: $\alpha\beta\gamma \neq 0$.
C2: $(\alpha^2 - \beta^2)(\beta^2 - \gamma^2)(\gamma^2 - \alpha^2) \neq 0$.
C3: *None of the equations* $X^2 - d_i = 0$, $i \in [3]$, *has a root in* $\mathbb{F}$, *where*

$$d_1 = (+\alpha^2 - \beta^2 - \gamma^2)^2 - (2\beta\gamma)^2$$
$$d_2 = (-\alpha^2 + \beta^2 - \gamma^2)^2 - (2\alpha\gamma)^2$$
$$d_3 = (-\alpha^2 - \beta^2 + \gamma^2)^2 - (2\alpha\beta)^2$$

Remark 1. 1. It follows that $2(x_1x_2 + x_3x_4) + 4(x_1x_3 + x_2x_4) + 5(x_1x_4 + x_2x_3)$ cannot be written as a sum of 2 ROPs over reals, yielding Theorem 2.
2. If $\mathbb{F}$ is an algebraically closed field, then for every α, β, γ, condition C3 fails, and so every $f^{\alpha,\beta,\gamma}$ can be written as a sum of 2 ROPs. However we do not know if there are other examples, or whether all multilinear 4-variate polynomials are expressible as the sum of two ROPs.
3. Even if $\mathbb{F}$ is not algebraically closed, condition C3 fails if for each $a \in \mathbb{F}$, the equation $X^2 = a$ has a root.

Our strategy for proving Theorem 8 is a generalization of an idea used in [11]. While Volkovich showed that 3-variate ROPs have a nice structural property in terms of their partial derivatives and commutators, we show that the sums of two 4-variate ROPs have at least one nice structural property in terms of their bivariate restrictions, partial derivatives, and commutators. Then we show that provided α, β, γ are chosen carefully, the polynomial $f^{\alpha,\beta,\gamma}$ will not satisfy any of these properties and hence cannot be a sum of two ROPs.

To prove Theorem 8, we first consider the easier direction, $1 \Rightarrow 2$, and prove the contrapositive.

Lemma 5. *If* α, β, γ *do not satisfy all of C1, C2, C3, then the polynomial* f *can be written as a sum of 2 ROPs.*

Proof. **C1 false:** If any of α, β, γ is zero, then by definition f is the the sum of at most two ROPs.
C2 false: Without loss of generality, assume $\alpha^2 = \beta^2$, so $\alpha = \pm\beta$. Then f is computed by $f = \alpha \cdot (x_1 \pm x_4)(x_2 \pm x_3) + \gamma \cdot (x_1x_4 + x_2x_3)$.
C1 true; C3 false: Without loss of generality, the equation $X^2 - d_1 = 0$ has a root τ. We try to express f as

$$\alpha(x_1 - ax_3)(x_2 - bx_4) + \beta(x_1 - cx_2)(x_3 - dx_4).$$

The coefficients for x_3x_4 and x_2x_4 force $ab = 1$, $cd = 1$, giving the form

$$\alpha(x_1 - ax_3)(x_2 - \frac{1}{a}x_4) + \beta(x_1 - cx_2)(x_3 - \frac{1}{c}x_4).$$

Comparing the coefficients for x_1x_4 and x_2x_3, we obtain the constraints

$$-\frac{\alpha}{a} - \frac{\beta}{c} = \gamma; \qquad -\alpha a - \beta c = \gamma$$

Expressing a as $\frac{-\gamma-\beta c}{\alpha}$, we get a quadratic constraint on c; it must be a root of the equation

$$Z^2 + \frac{-\alpha^2 + \beta^2 + \gamma^2}{\beta\gamma} Z + 1 = 0.$$

Using the fact that $\tau^2 = d_1 = (-\alpha^2 + \beta^2 + \gamma^2)^2 - (2\beta\gamma)^2$, we see that indeed this equation does have roots. The left-hand size splits into linear factors, giving

$$(Z - \delta)(Z - \frac{1}{\delta}) = 0 \text{ where } \delta = \frac{\alpha^2 - \beta^2 - \gamma^2 + \tau}{2\beta\gamma}.$$

It is easy to verify that $\delta \neq 0$ and $\delta \neq -\frac{\gamma}{\beta}$ (since $\alpha \neq 0$). Further, define $\mu = \frac{-(\gamma+\beta\delta)}{\alpha}$. Then μ is well-defined (because $\alpha \neq 0$) and is also non-zero. Now setting $c = \delta$ and $a = \mu$, we satisfy all the constraints, and we can write f as the sum of 2 ROPs as $f = \alpha(x_1 - \mu x_3)(x_2 - \frac{1}{\mu}x_4) + \beta(x_1 - \delta x_2)(x_3 - \frac{1}{\delta}x_4)$. □

Now we consider the harder direction: $2 \Rightarrow 1$. Again, we consider the contrapositive. We first show (Lemma 6) a structural property satisfied by every polynomial in $\sum^2 \cdot$ROP: it must satisfy at least one of the three properties $C1', C2', C3'$ described in the lemma. We then show (Lemma 7) that under the conditions $C1, C2, C3$ from the theorem statement, f does not satisfy any of $C1', C2', C3'$; it follows that f is not expressible as the sum of 2 ROPs.

Lemma 6. *Let g be a 4-variate multilinear polynomial over the field $\mathbb{F}$ which can be expressed as a sum of 2 ROPs. Then at least one of the following conditions is true:*

C1': *There exist $i, j \in [4]$ and $a, b \in \mathbb{F}$ such that $g \mid_{x_i=a, x_j=b}$ is linear.*
C2': *There exist $i, j \in [4]$ such that $x_i, x_j, \partial_{x_i}(g), \partial_{x_j}(g), 1$ are linearly dependent.*
C3': *$g = l_1 \cdot l_2 + l_3 \cdot l_4$ where l_is are linear forms, l_1 and l_2 are variable-disjoint, and l_3 and l_4 are variable-disjoint.*

Proof. Let φ be a sum of 2 ROFs computing g. Let v_1 and v_2 be the children of the topmost + gate. The proof is in two steps. First, we reduce to the case when $|\text{Var}(v_1)| = |\text{Var}(v_2)| = 4$. Then we use a case analysis to show that at least one of the aforementioned conditions hold true. In both steps, we will repeatedly use Proposition 3, which showed that any 3-variate ROP can be reduced to a linear polynomial by substituting a single variable with a field constant. We now proceed with the proof.

Suppose $|\text{Var}(v_1)| \leq 3$. Applying Proposition 3 first to v_1 and then to the resulting restriction of v_2, one can see that there exist $i, j \in [4]$ and $a, b \in \mathbb{F}$ such that $g \mid_{x_i=a, x_j=b}$ is a linear polynomial. So condition $C1'$ is satisfied.

Now assume that $|\text{Var}(v_1)| = |\text{Var}(v_2)| = 4$. Depending on the type of gates of v_1 and v_2, we consider 3 cases.

Case 1: Both v_1 and v_2 are $\times$ gates. Then g can be represented as $M_1 \cdot M_2 + M_3 \cdot M_4$ where (M_1, M_2) and (M_3, M_4) are variable-disjoint ROPs.

Suppose that for some i, $|\text{Var}(M_i)| = 1$. Then, $g\,|_{M_i \to 0}$ is a 3-variate restriction of f and is clearly an ROP. Applying Proposition 3 to this restriction, we see that condition $C1'$ holds.

Otherwise each M_i has $|\text{Var}(M_i)| = 2$.

Suppose (M_1, M_2) and (M_3, M_4) define distinct partitions of the variable set. Assume without loss of generality that $g = M_1(x_1, x_2) \cdot M_2(x_3, x_4) + M_3(x_1, x_3) \cdot M_4(x_2, x_4)$. If all M_is are linear forms, it is clear that condition $C3'$ holds. If not, assume that M_1 is of the form $l_1(x_1) \cdot m_1(x_2) + c_1$ where l_1, m_1 are linear forms and $c_1 \in \mathbb{F}$. Now $g\,|_{l_1 \to 0} = c_1 \cdot M_2(x_3, x_4) + M_3'(x_3) \cdot M_4(x_2, x_4)$. Either set x_3 to make M_3' zero, or, if that is not possible because M_3' is a non-zero field constant, then set $x_4 \to b$ where $b \in \mathbb{F}$. In both cases, by setting at most 2 variables, we obtain a linear polynomial, so $C1'$ holds.

Otherwise, (M_1, M_2) and (M_3, M_4) define the same partition of the variable set. Assume without loss of generality that $g = M_1(x_1, x_2) \cdot M_2(x_3, x_4) + M_3(x_1, x_2) \cdot M_4(x_3, x_4)$. If one of the M_is is linear, say without loss of generality that M_1 is a linear form, then $g\,|_{M_4 \to 0}$ is a 2-variate restriction which is also a linear form, so $C1'$ holds. Otherwise, none of the M_is is a linear form. Then each M_i can be represented as $l_i \cdot m_i + c_i$ where l_i, m_i are univariate linear forms and $c_i \in \mathbb{F}$. We consider a 2-variate restriction which sets l_1 and m_4 to 0. (Note that $\text{Var}(l_1) \cap \text{Var}(m_4) = \emptyset$.) Then the resulting polynomial is a linear form, so $C1'$ holds.

Case 2: Both v_1 and v_2 are $+$ gates. Then g can be written as $f = M_1 + M_2 + M_3 + M_4$ where (M_1, M_2) and (M_3, M_4) are variable-disjoint ROPs.

Suppose (M_1, M_2) and (M_3, M_4) define distinct partitions of the variable set.

Suppose further that there exists M_i such that $|\text{Var}(M_i)| = 1$. Without loss Of generality, $\text{Var}(M_1) = \{x_1\}$, $\{x_1, x_2\} \subseteq \text{Var}(M_3)$, and $x_3 \in \text{Var}(M_4)$. Any setting to x_2 and x_4 results in a linear polynomial, so $C1'$ holds.

So assume without loss of generality that $g = M_1(x_1, x_2) + M_2(x_3, x_4) + M_3(x_1, x_3) + M_4(x_2, x_4)$. Then for $a, b \in \mathbb{F}$, $g\,|_{x_1 = a, x_4 = b}$ is a linear polynomial, so $C1'$ holds.

Otherwise, (M_1, M_2) and (M_3, M_4) define the same partition of the variable set. Again, if say $|\text{Var}(M_1)| = 1$, then setting two variables from M_2 shows that $C1'$ holds. So assume without loss of generality that $g = M_1(x_1, x_2) + M_2(x_3, x_4) + M_3(x_1, x_2) + M_4(x_3, x_4)$. Then for $a, b \in \mathbb{F}$, $g\,|_{x_1 = a, x_3 = b}$ is a linear polynomial, so again $C1'$ holds.

Case 3: One of v_1, v_2 is a $+$ gate and the other is a $\times$ gate. Then g can be written as $g = M_1 + M_2 + M_3 \cdot M_4$ where (M_1, M_2) and (M_3, M_4) are variable-disjoint ROPs. Suppose that $|\text{Var}(M_3)| = 1$. Then $g\,|_{M_3 \to 0}$ is a 3-variate restriction which is a ROP. Using Proposition 3, we get a 2-variate restriction of g which is also linear, so $C1'$ holds. The same argument works when $|\text{Var}(M_4)| = 1$. So assume that M_3 and M_4 are bivariate polynomials.

Suppose that (M_1, M_2) and (M_3, M_4) define distinct partitions of the variable set. Assume without loss of generality that $g = M_1 + M_2 + M_3(x_1, x_2) \cdot M_4(x_3, x_4)$, and x_3, x_4 are separated by M_1, M_2. Then $g \mid_{M_3 \to 0}$ is a 2-variate restriction which is also linear, so $C1'$ holds.

Otherwise (M_1, M_2) and (M_3, M_4) define the same partition of the variable set. Assume without loss of generality that $g = M_1(x_1, x_2) + M_2(x_3, x_4) + M_3(x_1, x_2) \cdot M_4(x_3, x_4)$. If M_1 (or M_2) is a linear form, then consider a 2-variate restriction of g which sets M_4 (or M_3) to 0. The resulting polynomial is a linear form. Similarly if M_3 (or M_4) is of the form $l \cdot m + c$ where l, m are univariate linear forms, then we consider a 2-variate restriction which sets l to 0 and some $x_i \in \mathrm{Var}(M_4)$ to a field constant. The resulting polynomial again is a linear form. In all these cases, $C1'$ holds.

The only case that remains is that M_3 and M_4 are linear forms while M_1 and M_2 are not. Assume that $M_1 = (a_1 x_1 + b_1)(a_2 x_2 + b_2) + c$ and $M_3 = a_3 x_1 + b_3 x_2 + c_3$. Then $\partial_{x_1}(g) = a_1(a_2 x_2 + b_2) + a_3 M_4$ and $\partial_{x_2}(g) = (a_1 x_1 + b_1)a_2 + b_3 M_4$. It follows that $b_3 \cdot \partial_{x_1}(g) - a_3 \cdot \partial_{x_2}(g) + a_1 a_2 a_3 x_1 - a_1 a_2 b_3 x_2 = a_1 b_2 b_3 - b_1 a_2 a_3 \in \mathbb{F}$, and hence the polynomials $x_1, x_2, \partial_{x_1}(g), \partial_{x_2}(g)$ and 1 are linearly dependent. Therefore, condition $C2'$ of the lemma is satisfied. $\qquad\square$

Lemma 7. *If α, β, γ satisfy conditions $C1, C2, C3$ from the statement of Theorem 8, then the polynomial $f^{\alpha, \beta, \gamma}$ does not satisfy any of the properties $C1'$, $C2'$, $C3'$ from Lemma 6.*

Proof. **C1 $\Rightarrow \neg$C1':** Since $\alpha\beta\gamma \neq 0$, f contains all possible degree 2 monomials. Hence after setting $x_i = a$ and $x_j = b$, the monomial $x_k x_l$ where $k, l \in [4]\backslash\{i, j\}$ still survives.

C2 $\Rightarrow \neg$C2': The proof is by contradiction. Assume to the contrary that for some i, j, without loss of generality say for $i = 1$ and $j = 2$, the polynomials $x_1, x_2, \partial_{x_1}(f), \partial_{x_2}(f), 1$ are linearly dependent. Note that $\partial_{x_1}(f) = \alpha x_2 + \beta x_3 + \gamma x_4$ and $\partial_{x_2}(f) = \alpha x_1 + \gamma x_3 + \beta x_4$. This implies that the vectors $(1, 0, 0, 0, 0)$, $(0, 1, 0, 0, 0)$, $(0, \alpha, \beta, \gamma, 0)$, $(\alpha, 0, \gamma, \beta, 0)$ and $(0, 0, 0, 0, 1)$ are linearly dependent. This further implies that the vectors (β, γ) and (γ, β) are linearly dependent. Therefore, $\beta = \pm\gamma$, contradicting C2.

C1 $\wedge$ C2 $\wedge$ C3 $\Rightarrow \neg$C3': Suppose, to the contrary, that $C3'$ holds. That is, f can be written as $f = l_1 \cdot l_2 + l_3 \cdot l_4$ where (l_1, l_2) and (l_3, l_4) are variable-disjoint linear forms. By the preceding arguments, we know that f does not satisfy $C1'$ or $C2'$.

First consider the case when (l_1, l_2) and (l_3, l_4) define the same partition of the variable set. Assume without loss of generality that $\mathrm{Var}(l_1) = \mathrm{Var}(l_3)$, $\mathrm{Var}(l_2) = \mathrm{Var}(l_4)$, and $|\mathrm{Var}(l_1)| \leq 2$. Setting the variables in l_1 to any field constants yields a linear form, so f satisfies $C1'$, a contradiction.

Hence it must be the case that (l_1, l_2) and (l_3, l_4) define different partitions of the variable set. Since all degree-2 monomials are present in f, each pair x_i, x_j must be separated by at least one of the two partitions. This implies that both partitions have exactly 2 variables in each part. Assume without loss of generality that $f = l_1(x_1, x_2) \cdot l_2(x_3, x_4) + l_3(x_1, x_3) \cdot l_4(x_2, x_4)$.

At this point, we use properties of the commutator of f; recall Definition 3. By Lemma 2, we know that l_2 divides $\triangle_{12}f$. We compute $\triangle_{12}f$ explicitly for our candidate polynomial:

$$\begin{aligned}
\triangle_{12}f &= (\alpha x_3 x_4)(\alpha + (\beta + \gamma)(x_3 + x_4) + \alpha x_3 x_4) \\
&\quad - (\beta x_4 + \gamma x_3 + \alpha x_3 x_4)(\beta x_3 + \gamma x_4 + \alpha x_3 x_4) \\
&= -\beta\gamma(x_3^2 + x_4^2) + (\alpha^2 - \beta^2 - \gamma^2)x_3 x_4
\end{aligned}$$

Since l_2 divides $\triangle_{12}f$, $\triangle_{12}f$ is not irreducible but is the product of two linear factors. Since $\triangle_{12}f(0,0) = 0$, at least one of the linear factors of $\triangle_{12}f$ must vanish at $(0,0)$. Let $x_3 - \delta x_4$ be such a factor. Then $\triangle_{12}(f)$ vanishes not only at $(0,0)$, but whenever $x_3 = \delta x_4$. Substituting $x_3 = \delta x_4$ in $\triangle_{12}f$, we get

$$-\delta^2\beta\gamma - \beta\gamma + \delta(\alpha^2 - \beta^2 - \gamma^2) = 0$$

Hence δ is of the form

$$\delta = \frac{-(\alpha^2 - \beta^2 - \gamma^2) \pm \sqrt{(\alpha^2 - \beta^2 - \gamma^2)^2 - 4\beta^2\gamma^2}}{-2\beta\gamma}$$

Hence $2\beta\gamma\delta - (\alpha^2 - \beta^2 - \gamma^2)$ is a root of the equation $X^2 - D_1 = 0$, contradicting the assumption that C3 holds.

Hence it must be the case that $C3'$ does not hold.   □

With this, the proof of Theorem 8 is complete.

The conditions imposed on α, β, γ in Theorem 8 are tight and irredundant. Below we give some explicit examples over the field of reals.

1. $f = 2(x_1 x_2 + x_3 x_4) + 2(x_1 x_3 + x_2 x_4) + 3(x_1 x_4 + x_2 x_3)$ satisfies conditions C1 and C3 from the Theorem but not C2; $\alpha = \beta$. A $\sum^2 \cdot$ROP representation for f is $f = 2(x_1 + x_4)(x_2 + x_3) + 3(x_1 x_4 + x_2 x_3)$.
2. $f = 2(x_1 x_2 + x_3 x_4) - 2(x_1 x_3 + x_2 x_4) + 3(x_1 x_4 + x_2 x_3)$ satisfies conditions C1 and C3 but not C2; $\alpha = -\beta$. A $\sum^2 \cdot$ROP representation for f is $f = 2(x_1 - x_4)(x_2 - x_3) + 3(x_1 x_4 + x_2 x_3)$.
3. $f = (x_1 x_2 + x_3 x_4) + 2(x_1 x_3 + x_2 x_4) + 3(x_1 x_4 + x_2 x_3)$ satisfies conditions C1 and C2 but not C3. A $\sum^2 \cdot$ROP representation for f is $f = (x_1 + x_3)(x_2 + x_4) + 2(x_1 + x_2)(x_3 + x_4)$.

5   Conclusions

1. We have seen in Proposition 1 that every n-variate multilinear polynomial $(n \geq 4)$ can be written as the sum of $3 \times 2^{n-4}$ ROPs. A counting argument shows that there exist multilinear polynomials f requiring exponentially many ROPs summands; if $f \in \sum^k \cdot$ROP then $k = \Omega(2^n/n^2)$. Our general upper bound on k is $O(2^n)$, leaving a small gap between the lower and upper bound. What is the true tight bound? Can we find explicit polynomials that require exponentially large k in any $\sum^k \cdot$ROP expression?

2. We have shown in Theorem 1 that for each k, $\sum^k \cdot$ROP can be separated from $\sum^{k-1} \cdot$ROP by a polynomial on $2k-1$ variables. Can we separate these classes with fewer variables? Note that any separating polynomial must have $\Omega(\log k)$ variables.

3. In particular, can 4-variate multilinear polynomials separate sums of 3 ROPs from sums of 2 ROPs over every field? If not, what is an explicit example?

References

1. Anderson, M., van Melkebeek, D., Volkovich, I.: Deterministic polynomial identity tests for multilinear bounded-read formulae. Comput. Complex. **24**(4), 695–776 (2015)

2. Bshouty, D., Bshouty, N.H.: On interpolating arithmetic read-once formulas with exponentiation. J. Comput. Syst. Sci. **56**(1), 112–124 (1998)

3. Bshouty, N.H., Cleve, R.: Interpolating arithmetic read-once formulas in parallel. SIAM J. Comput. **27**(2), 401–413 (1998)

4. Bshouty, N.H., Hancock, T.R., Hellerstein, L.: Learning boolean read-once formulas over generalized bases. J. Comput. Syst. Sci. **50**(3), 521–542 (1995)

5. Hancock, T.R., Hellerstein, L.: Learning read-once formulas over fields and extended bases. In: Warmuth, M.K., Valiant, L.G. (eds.) Proceedings of the Fourth Annual Workshop on Computational Learning Theory, COLT 1991, Santa Cruz, California, USA, 5–7 August 1991, pp. 326–336. Morgan Kaufmann (1991)

6. Kabanets, V., Impagliazzo, R.: Derandomizing polynomial identity tests means proving circuit lower bounds. Comput. Complex. **13**(1–2), 1–46 (2004)

7. Kayal, N., Koiran, P., Pecatte, T., Saha, C.: Lower bounds for sums of powers of low degree univariates. In: Halldórsson, M.M., Iwama, K., Kobayashi, N., Speckmann, B. (eds.) ICALP 2015, Part I. LNCS, vol. 9134, pp. 810–821. Springer, Heidelberg (2015)

8. Raz, R.: Multi-linear formulas for permanent and determinant are of super-polynomial size. J. ACM **56**(2), 1–17 (2009)

9. Shpilka, A., Volkovich, I.: On reconstruction and testing of read-once formulas. Theor. Comput. **10**, 465–514 (2014)

10. Shpilka, A., Volkovich, I.: Read-once polynomial identity testing. Comput. Complex. **24**(3), 477–532 (2015). (combines results from papers in RANDOM 2009 and STOC 2008)

11. Volkovich, I.: Characterizing arithmetic read-once formulae. ACM Trans. Comput. Theor. **8**(1), 2:1–2:19 (2016)

Algorithmic Statistics: Normal Objects and Universal Models

Alexey Milovanov[1,2(✉)]

[1] Moscow State University, Moscow, Russian Federation
almas239@gmail.com
[2] National Research University Higher School of Economics,
Moscow, Russian Federation
https://www.hse.ru/en/org/persons/176000050

Abstract. In algorithmic statistics quality of a statistical hypothesis (a model) P for a data x is measured by two parameters: Kolmogorov complexity of the hypothesis and the probability $P(x)$. A class of models S_{ij} that are the best at this point of view, were discovered. However these models are too abstract.

To restrict the class of hypotheses for a data, Vereshchaginintroduced a notion of a strong model for it. An object is called normal if it can be explained by using strong models not worse than without this restriction. In this paper we show that there are "many types" of normal strings. Our second result states that there is a normal object x such that all models S_{ij} are not strong for x. Our last result states that every best fit strong model for a normal object is again a normal object.

Keywords: Algorithmic statistics · Minimum description length · Stochastic strings · Total conditional complexity · Sufficient statistic · Denoising

1 Introduction

Let us recall the basic notion of algorithmic information theory and algorithmic statistics (see [4,6,8] for more details).

We consider strings over the binary alphabet $\{0,1\}$. The set of all strings is denoted by $\{0,1\}^*$ and the length of a string x is denoted by $l(x)$. The empty string is denoted by Λ.

1.1 Algorithmic Information Theory

Let D be a partial computable function mapping pairs of strings to strings. *Conditional Kolmogorov complexity* with respect to D is defined as

$$C_D(x|y) = \min\{l(p) \mid D(p, y) = x\}.$$

In this context the function D is called a *description mode* or a *decompressor*. If $D(p, y) = x$ then p is called a *description of x conditional to y* or a *program mapping y to x*.

© Springer International Publishing Switzerland 2016
A.S. Kulikov and G.J. Woeginger (Eds.): CSR 2016, LNCS 9691, pp. 280–293, 2016.
DOI: 10.1007/978-3-319-34171-2_20

A decompressor D is called *universal* if for every other decompressor D' there is a string c such that $D'(p, y) = D(cp, y)$ for all p, y. By Solomonoff—Kolmogorov theorem [2] universal decompressors exist. We pick arbitrary universal decompressor D and call $C_D(x|y)$ *the Kolmogorov complexity* of x conditional to y, and denote it by $C(x|y)$. Then we define the unconditional Kolmogorov complexity $C(x)$ of x as $C(x|\Lambda)$.

Kolmogorov complexity can be naturally extended to other finite objects (pairs of strings, finite sets of strings, etc.). We fix some computable bijection ("encoding") between these objects are binary strings and define the complexity of an object as the complexity of the corresponding binary string. It is easy to see that this definition is invariant (change of the encoding changes the complexity only by $O(1)$ additive term).

In particular, we fix some computable bijection between strings and finite subsets of $\{0, 1\}^*$; the string that corresponds to a finite $A \subset \{0, 1\}^*$ is denoted by $[A]$. Then we understand $C(A)$ as $C([A])$. Similarly, $C(x|A)$ and $C(A|x)$ are understood as $C(x|[A])$ and $C([A]|x)$, etc.

1.2 Algorithmic Statistics: Basic Notions

Algorithmic statistics studies explanations of observed data that are suitable in the algorithmic sense: an explanation should be simple and capture all the algorithmically discoverable regularities in the data. The data is encoded, say, by a binary string x. In this paper we consider explanations (statistical hypotheses) of the form "x was drawn at random from a finite set A with uniform distribution".

Kolmogorov suggested in a talk [3] in 1974 to measure the quality of an explanation $A \ni x$ by two parameters: Kolmogorov complexity $C(A)$ of A and the log-cardinality $\log |A|$[1] of A. The smaller $C(A)$ is the simpler the explanation is. The log-cardinality measures the *fit* of A—the lower is $|A|$ the more A fits as an explanation for any of its elements. For each complexity level m any model A for x with smallest $\log |A|$ among models of complexity at most m for x is called a *best fit hypothesis for x*. The trade off between $C(A)$ and $\log |A|$ is represented by the *profile* of x.

Definition 1. *The* profile *of a string x is the set P_x consisting of all pairs (m, l) of natural numbers such that there exists a finite set $A \ni x$ with $C(A) \leq m$ and $\log_2 |A| \leq l$.*

Both parameters $C(A)$ and $\log |A|$ cannot be very small simultaneously unless the string x has very small Kolmogorov complexity. Indeed, $C(A) + \log |A| \gtrsim C(x)$, since x can be specified by A and its index in A. A model (we also use the word "statistic") $A \ni x$ is called *sufficient* if $C(A) + \log |A| \approx C(x)$. The value

$$\delta(x|A) = C(A) + \log |A| - C(x)$$

is called the *optimality deficiency* of A as a model for x. On Fig. 1 parameters of sufficient statistics lie on the segment BD. A sufficient statistic that has the

---

[1] by log we denote $\log_2$.

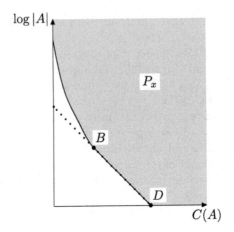

Fig. 1. The profile P_x of a string x.

minimal complexity is called *minimal* (MSS), its parameters are represented by the point B on Fig. 1.

Example 1. Consider a string $x \in \{0,1\}^{2n}$ such that leading n bits of x are zeros, and the remaining bits are random, i.e. $C(x) \approx n$. Consider the model A for x that consists of all strings from $\{0,1\}^{2n}$ that have n leading zeros. Then $C(A) + \log|A| = \log n + O(1) + n \approx C(x)$, hence A is a sufficient statistic for x. As the complexity of A is negligible, A is a minimal sufficient statistic for x.

The string from this example has a sufficient statistic of negligible complexity. Such strings are called *stochastic*. Are there strings that have no sufficient statistics of negligible complexity? The positive to this question was obtained in [7]. Such strings are called *non-stochastic*. Moreover, under some natural constraints for every set P there is a string whose profile is close to P. The constraints are listed in the following theorem:

Theorem 1. *Let x be a string of length n and complexity k. Then P_x has the following properties:*

(1) $(k + O(\log n), 0) \in P_x$.
(2) $(O(\log n), n) \in P_x$.
(3) if $(a, b + c) \in P_x$ then $(a + b + O(\log n), c) \in P_x$.
(4) if $(a, b) \in P_x$ then $a + b > k - O(\log n)$.

In other words, with logarithmic accuracy, the boundary of P_x contains a point $(0, a)$ with $a \le l(x)$, contains the point $(C(x), 0)$, decreases with the slope at least -1 and lies above the line $C(A) + \log|A| = C(x)$. Conversely, given a curve with these property that has low complexity one can find a string x of length n and complexity about k such that the boundary of P_x is close to that curve:

Theorem 2 [10]. *Assume that we are given k, n and an upward closed set P of pairs of natural numbers such that $(0, n), (k, 0) \in P$, $(a, b+c) \in P \Rightarrow (a+c, b) \in P$ and $(a, b) \in P \Rightarrow a + b \geq k$. Then there is a string x of length n and complexity $k + O(\log n)$ whose profile is $C(P) + O(\log n)$-close to P. (We call subsets of N^2 ϵ-close if each of them is in the ϵ-neighborhood of the other.) By $C(P)$ we denote the Kolmogorov complexity of the boundary of P, which is a finite object.*

1.3   Universal Models

Assume that A is a sufficient statistic for x. Then A provides a two-part code $y =$ (the shortest description of A, the index of x in A) for x whose total length is close to the complexity of x. The symmetry of information implies that $C(y|x) \approx C(y) + C(x|y) - C(x)$. Obviously, the term $C(x|y)$ here is negligible and $C(y)$ is at most its total length, which by assumption is close to $C(x)$. Thus $C(y|x) \approx 0$, that is, x and y have almost the same information. That is, the two-part code y for x splits the information from x in two parts: the shortest description of A, the index of x in A. The second part of this two-part code is incompressible (random) conditional to the first part (as otherwise, the complexity of x would be smaller than the total length of y). Thus the second part of this two-part code can be considered as accidental information (noise) in the data x. In a sense every sufficient statistic A identifies about $C(x) - C(A)$ bits of accidental information in x. And thus any minimal sufficient statistic for x extracts almost all useful information from x.

However, it turns out that this viewpoint is inconsistent with the existence of universal models, discovered in [1]. Let L_m denote the list of strings of complexity at most m. Let p be an algorithm that enumerates all strings of L_m in some order. Notice that there is such algorithm of complexity $O(\log m)$. Denote by Ω_m the cardinality of L_m. Consider its binary representation, i.e., the sum:

$$\Omega_m = 2^{s_1} + 2^{s_2} + \dots + 2^{s_t}, \text{ where } s_1 > s_2 > \dots > s_t.$$

According to this decomposition and p, we split L_m into groups: first 2^{s_1} elements, next 2^{s_2} elements, etc. Let us denote by $S_{m,s}^p$ the group of size 2^s from the partition. Notice that $S_{m,s}^p$ is defined only for s that correspond to ones in the binary representation of Ω_m, so $m \geq s$.

If x is a string of complexity at most m, it belongs to some group $S_{m,s}^p$ and this group can be considered as a model for x. We may consider different values of m (starting from $C(x)$). In this way we get different models $S_{m,s}^p$ for the same x. The complexity of $S_{m,s}^p$ is $m - s + O(\log m + C(p))$. Indeed, chop L_m into portions of size 2^s each, then $S_{m,s}^p$ is the last full portion and can be identified by m, s and the number of full portions, which is less than $\Omega_m/2^s < 2^{m-s+1}$. Thus if m is close to $C(x)$ and $C(p)$ is small then $S_{m,s}^p$ is a sufficient statistic for x. More specifically $C(S_{m,s}^p) + \log |S_{m,s}^p| = C(S_{m,s}^p) + s = m + O(\log m + C(p))$.

For every m there is an algorithm p of complexity $O(\log m)$ that enumerates all strings of complexity at most m. We will fix for every m any such algorithm p_m and denote $S_{m,s}^{p_m}$ by $S_{m,s}$.

The models $S_{m,s}$ were introduced in [1]. The models $S_{m,s}^p$ are universal in the following sense:

Theorem 3 [10]. [2] *Let A be any finite set of strings containing a string x of length n. Then for every p there are $s \leq m \leq n + O(1)$ such that*

(1) $x \in S_{m,s}^p$,
(2) $C(S_{m,s}^p|A) = O(\log n + C(p))$ (and hence $C(S_{m,s}^p) \leq C(A) + O(\log n + C(p))$),
(3) $\delta(x|S_{m,s}^p) \leq \delta(x|A) + O(\log n + C(p))$.

It turns out that the model $S_{m,s}^p$ has the same information as the the number Ω_{m-s}:

Lemma 1 [10]. *For every $a \leq b$ and for every $s \leq m$:*

(1) $C(\Omega_a|\Omega_b) = O(\log b)$.
(2) $C(\Omega_{m-s}|S_{m,s}^p) = O(\log m + C(p))$ and $C(S_{m,s}^p|\Omega_{m-s}) = O(\log m + C(p))$.
(3) $C(\Omega_a) = a + O(\log a)$.

By Theorem 3 for every data x there is a minimal sufficient statistic for x of the form $S_{m,s}$. Indeed, let A be any minimal sufficient statistic for x and let $S_{m,s}$ be any model for x that exists by Theorem 3 for this A. Then by item 3 the statistic $S_{m,s}$ is sufficient as well and by item 2 its complexity is also close to minimum. Moreover, since $C(S_{m,s}|A)$ is negligible and $C(S_{m,s}) \approx C(A)$, by symmetry of information $C(A|S_{m,s})$ is negligible as well. Thus A has the same information as $S_{m,s}$, which has the same information as Ω_{m-s} (Lemma 1(2)). Thus if we agree that every minimal sufficient statistic extracts all useful information from the data, we must agree also that information is the same as the information in the number of strings of complexity at most i for some i.

1.4   Total Conditional Complexity and Strong Models

The paper [9] suggests the following explanation to this situation. Although conditional complexities $C(S_{m,s}|A)$ and $C(S_{m,s}|x)$ are small, the short programs that map A and x, respectively, to $S_{m,s}$ work in a huge time. A priori their work time is not bounded by any total computable function of their input. Thus it may happen that practically we are not able to find $S_{m,s}$ (and also Ω_{m-s}) from a MSS A for x or from x itself.

Let us consider now programs whose work time is bounded by a total computable function for the input. We get the notion of *total conditional complexity* $CT(y|x)$, which is the length of the shortest *total* program that maps x to y. Total conditional complexity can be much greater than plain one, see for example [5]. Intuitively, good sufficient statistics A for x must have not only negligible

---

[2] This theorem was proved in [10, Theorem VIII.4] with accuracy $O(\max\{\log C(y) \mid y \in A\} + C(p))$ instead of $O(\log n)$. Applying [10, Theorem VIII.4] to $A' = \{y \in A \mid l(y) = n\}$ we obtain the theorem in the present form.

conditional complexity $C(A|x)$ (which follows from definition of a sufficient statistic) but also negligible *total* conditional complexity $CT(A|x)$. The paper [9] calls such models A *strong models for* x.

Is it true that for some x there is no **strong** MSS $S_{m,s}$ for x? The positive answer to this question was obtained in [9]: there are strings x for which all minimal sufficient statistics are not strong for x. Such strings are called *strange*. In particular, if $S_{m,s}$ is a MSS for a strange string x then $CT(S_{m,s}|x)$ is large. However, a strange string has no strong MSS at all. An interesting question is whether there are strings x that do have strong MSS but have no strong MSS of the form $S_{m,s}$? This question was left open in [9]. In this paper we answer this question in positive. Moreover, we show that there is a "normal" string x that has no strong MSS of the form $S_{m,s}$ (Theorem 7). A string x is called *normal* if for every complexity level i there is a best fitting model A for x of complexity at most i (whose parameters thus lie on the border of the set P_x) that is strong. In particular, every normal string has a strong MSS.

Our second result answers yet another question asked in [9]. Assume that A is a strong MSS for a normal string x. Is it true that the code $[A]$ of A is a normal string itself? Our Theorem 10 states that this is indeed the case.

Our last result (which comes first in the following exposition) states that there are normal strings with any given profile, under the same restrictions as in Theorem 1 (Theorem 4 in Sect. 2).

2 Normal Strings with a Given Profile

In this section we prove an analogue of Theorem 2 for normal strings. We start with a rigorous definition of strong models and normal strings.

Definition 2. *A set $A \ni x$ is called ϵ-strong statistic (model) for a string x if $CT(A|x) < \epsilon$.*

To represent the trade off between size and complexity of ϵ-strong models for x consider the ϵ-*strong profile of* x:

$$P_x^\epsilon = \{(a, b) \mid \exists A \ni x : CT(A|x) \le \epsilon, C(A) \le a, \log|A| \le b\}.$$

It is not hard to see that the set P_x^ϵ satisfies the item (3) from Theorem 1:

for all $x \in \{0, 1\}^n$ if $(a, b+c) \in P_x^\epsilon$ then $(a+b+O(\log n), c) \in P_x^{\epsilon+O(\log n)}$.

It follows from the definition that $P_x^\epsilon \subset P_x$ for all x, ϵ. Informally a string is called normal if for a negligible ϵ we have $P_x \approx P_x^\epsilon$.

Definition 3. *a string x is called (ϵ, δ)-normal if $(a, b) \in P_x$ implies $(a + \delta, b + \delta) \in P_x^\epsilon$ for all a, b.*

The smaller ϵ, δ are the stronger is the property of (ϵ, δ)-normality. The main result of this section shows that for some $\epsilon, \delta = o(n)$ for every set P satisfying the assumptions of Theorem 1 there is an ϵ, δ-normal string of length n with $P_x \approx P$:

Theorem 4. *Assume that we are given an upward closed set P of pairs of natural numbers satisfying assumptions of Theorem 2. Then there is an $(O(\log n), O(\sqrt{n \log n}))$-normal string x of length n and complexity $k + O(\log n)$ whose profile P_x is $C(P) + O(\sqrt{n \log n})$-close to P.*

To prove this theorem we do an excursus to Algorithmic statistics with models of restricted type.

Models of Restricted Type. It turns out that Theorems 1 and 2 remain valid (with smaller accuracy) even if we restrict (following [11]) the class of models under consideration to models from a class $\mathcal{A}$ provided the class $\mathcal{A}$ has the following properties.

(1) The family $\mathcal{A}$ is enumerable. This means that there exists an algorithm that prints elements of $\mathcal{A}$ as lists of strings, with some separators (saying where one element of $\mathcal{A}$ ends and another one begins).

(2) For every n the class $\mathcal{A}$ contains the set $\{0,1\}^n$.

(3) There exists some polynomial p with the following property: for every $A \in \mathcal{A}$, for every natural n and for every natural $c < |A|$ the set of all n-bit strings in A can be covered by at most $p(n) \cdot |A|/c$ sets of cardinality at most c from $\mathcal{A}$.

Any family of finite sets sets of strings that satisfies these three conditions is called *acceptable*.

Let us define the *profile of x with respect to $\mathcal{A}$*:

$$P_x^{\mathcal{A}} = \{(a,b) \mid \exists A \ni x : A \in \mathcal{A}, C(A) \le a, \log |A| \le b\}.$$

Obviously $P_x^{\mathcal{A}} \subseteq P_x$. Let us fix any acceptable class $\mathcal{A}$ of models.

Theorem 5 [11]. *Let x be a string of length n and complexity k. Then $P_x^{\mathcal{A}}$ has the following properties:*

(1) $(k + O(\log n), 0) \in P_x^{\mathcal{A}}$.
(2) $(O(\log n), n) \in P_x^{\mathcal{A}}$.
(3) if $(a, b+c) \in P_x^{\mathcal{A}}$ then $(a + b + O(\log n), c) \in P_x^{\mathcal{A}}$.
(4) if $(a,b) \in P_x^{\mathcal{A}}$ then $a + b > k - O(\log n)$.

Theorem 6 [11]. *Assume that we are given k, n and an upward closed set P of pairs of natural numbers such that $(0, n), (k, 0) \in P$, $(a, b+c) \in P \Rightarrow (a+c, b) \in P$ and $(a, b) \in P \Rightarrow a + b \ge k$. Then there is a string x of length n and complexity $k + O(\log n)$ such that both sets $P_x^{\mathcal{A}}$ and P_x are $C(P) + O(\sqrt{n \log n})$-close to P.*

Remark 1. Originally, the conclusion of Theorem 6 stated only that the set $P_x^{\mathcal{A}}$ is close to the given set P. However, as observed in [8], the proof from [11] shows also that P_x is close to P.

Proof (Proof of Theorem 4). We will derive this theorem from Theorem 6. To this end consider the following family $\mathcal{B}$ of sets. A set B is in this family if it has the form

$$B = \{uv \mid v \in \{0,1\}^m\},$$

where u is an arbitrary binary string and m is an arbitrary natural number. Obviously, the family $\mathcal{B}$ is acceptable, that is, it satisfies the properties (1)–(3) above.

Note that for every x and for every $A \ni x$ from $\mathcal{B}$ the total complexity of A given x is $O(\log n)$. So $P_x^{\mathcal{B}} \subseteq P_x^{O(\log n)}$. By Theorem 6 there is a string x such that P_x and $P_x^{\mathcal{B}}$ are $C(P) + O(\sqrt{n \log n})$-close to P. Since $P_x^{\mathcal{B}} \subseteq P_x^{O(\log n)} \subseteq P_x$ we conclude that x is $(O(\log n), O(\sqrt{n \log n}))$-normal.

Instead of using Theorem 4 one can, in special cases, show this result directly even within a better accuracy range.

For instance, this happens for the smallest set P, satisfying the assumptions of Theorem 6, namely for the set

$$P = \{(m, l) \mid m \geq k, \text{ or } m + l \geq n\}.$$

Strings with such profile are called "antistochastic".

Definition 4. *A string x of length n and complexity k is called ϵ-antistochastic if for all $(m, l) \in P_x$ either $m > k - \epsilon$, or $m + l > n - \epsilon$ (Fig. 2).*

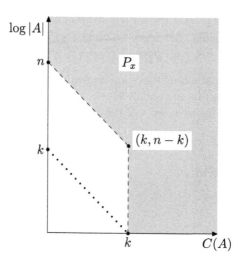

Fig. 2. The profile of an ϵ-antistochastic string x for a small ϵ.

We will need later the fact that for every n there is an $O(\log n)$-antistochastic string x of length n and that such strings are normal:

Lemma 2. *For all n and all $k \leq n$ there is an $O(\log n)$-antistochastic string x of length n and complexity $k + O(\log n)$. Any such string x is $(O(\log n), O(\log n))$-normal.*

Proof. Let x be the lexicographic first string of length n that is not covered by any set A of cardinality 2^{n-k} and complexity less than k. By a direct counting such a string exists. The string x can be computed from k, n and the number of halting programs of length less than k hence $C(x) \le k + O(\log n)$. To prove that x is normal it is enough to show that for every $i \le k$ there is a $O(\log n)$-strong statistics A_i for x with $C(A_i) \le i + O(\log n)$ and $\log |A_i| = n - i$.

Let $A_k = \{x\}$ and for $i < k$ let A_i be the set of all strings of length n whose the first i bits are the same as those of x. By the construction $C(A_i) \le i + O(\log n)$ and $\log |A_i| = n - i$.

3   Normal Strings Without Universal MSS

Our main result of this section is Theorem 7 which states that there is a normal string x such that no set $S_{m,l}$ is a strong MSS for x.

Theorem 7. *For all large enough k there exist an $(O(\log k), O(\log k))$-normal string x of complexity $3k + O(\log k)$ and length $4k$ such that:*

(1) The profile P_x of x is $O(\log k)$-close to the gray set on Fig. 3.
(2) The string x has a strong MSS. More specifically, there is an $O(\log k)$-strong model A for x with complexity $k + O(\log k)$ and log-cardinality $2k$.
(3) For all simple q and all m, l the set $S^q_{m,l}$ cannot be a strong sufficient statistic for x. More specifically, for every ϵ-strong ϵ-sufficient model $S^q_{m,l}$ for x of complexity at most $k + \delta$ we have $O(\epsilon + \delta + C(q)) \ge k - O(\log k)$

(The third condition means that there are constants r and t such that $r(\epsilon + \delta + C(q)) \ge k - t \log k$ for all large enough k).

In the proof of this theorem we will need a rigorous definition of MSS and a related result from [9].

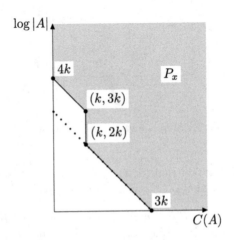

Fig. 3. The profile P_x of a string x from Theorem 7.

Definition 5. *A set A is called a (δ, ϵ, D)-minimal sufficient statistic (MSS) for x if A is an ϵ-sufficient statistic for x and there is no model B for x with $C(B) < C(A) - \delta$ and $C(B) + \log|B| - C(x) < \epsilon + D \log C(x)$.*

The next theorem states that for every strong MSS B and for every sufficient statistic A for x the total conditional complexity $CT(B|A)$ is negligible.

Theorem 8 ([9], **Theorem 13**). *For some constant D if B is ϵ-strong (δ, ϵ, D)-minimal sufficient statistic for x and A is an ϵ-sufficient statistic for x then $CT(B|A) = O(\epsilon + \delta + \log C(x))$.*

Let us fix a constant D satisfying Theorem 8 and call a model (δ, ϵ)-MSS if it is (δ, ϵ, D)-MSS. Such models have the following property.

Theorem 9 ([9], **Theorem 14**). *Let x be a string of length n and A be an ϵ-strong ϵ-sufficient statistic for x. Then for all $b \geq \log|A|$ we have*

$$(a, b) \in P_x \Leftrightarrow (a + O(\epsilon + \log n), b - \log|A| + O(\epsilon + \log n)) \in P_{[A]}$$

and for $b \leq \log|A|$ we have $(a, b) \in P_x \Leftrightarrow a + b \geq C(x) - O(\log n)$.

Proof (The proof of Theorem 7). Define x as the concatenation of strings y and z, where y is an $O(\log k)$-antistochastic string of complexity k and length $2k$ (existing by Lemma 2) and z is a string of length $2k$ such that $C(z|y) = 2k - O(\log k)$ (and hence $C(x) = 3k + O(\log k)$). Consider the following set $A = \{yz' \mid l(z') = 2k\}$. From the shape of P_x it is clear that A is an $(O(\log k), O(\log k))$-MSS for x. Also it is clear that A is an $O(\log k)$-strong model for x. So, by Theorem 9 the profile of x is $O(\log k)$-close to the gray set on Fig. 3. From normality of y (Lemma 2) it is not difficult to see that x is $(O(\log k), O(\log k))$-normal.

Let $S^q_{m,l}$ be an ϵ-strong ϵ-sufficient model for x of complexity at most $k + \delta$. We claim that $S^q_{m,l}$ is an $(\epsilon, \delta + O(\log k))$-MSS for x.

By Theorem 8 we get $CT(S^q_{m,l}|A) = O(\epsilon + \delta + \log k)$ and thus $CT(s_0|y) = O(\epsilon + \delta + \log k)$, where s_0 is the lexicographic least element in $S^q_{m,l}$. Denote by p a total program of length $O(\epsilon + \delta + \log k)$ that transforms y to s_0. Consider the following set $B := \{p(y') \mid l(y') = 2k\}$. We claim that if ϵ and δ are not very big, then the complexity of any element from B is not greater than m. Indeed, if $\epsilon + \delta \leq dk$ for a small constant d, then $l(p) < k - O(\log k)$ and hence every element from B has complexity at most $C(B) + \log|B| + O(\log k) \leq 3k - O(\log k) \leq m$. The last inequality holds because $S^q_{m,l}$ is a model for x and hence $m \geq C(x) = 3k + O(\log k)$. (Otherwise, if $\epsilon + \delta > dk$ then the conclusion of the theorem is straightforward.)

Let us run the program q until it prints all elements from B. Since $s_0 \in B$, there are at most 2^l elements of complexity m that we have been printed yet. So, we can find the list of all strings of complexity at most m from B, q and some extra l bits. Since this list has complexity at least $m - O(\log m)$ (as from this list and m we can compute a string of complexity more than m), we get $O(C(B) + C(q)) + l \geq m - O(\log m)$.

Recall that the $C(S_{m,l}^q) + \log |S_{m,l}^q|$ is equal to $m + O(\log m + C(q))$ and is at most $C(x) + \epsilon$ (since $S_{m,l}^q$ is the strong statistic for x). Hence $m \leq 4k$ unless $\epsilon > k + O(\log k + C(q))$. Therefore the term $O(\log m)$ in the last inequality can be re-written as $O(\log k)$.

Recall that the complexity of $S_{m,l}^q$ is $m - l + O(\log m + C(q))$. From the shape of P_x it follows that $C(S_{m,l}^q) \geq k - O(\log k)$ or $C(S_{m,l}^q) + \log |S_{m,l}^q| \geq C(x) + k - O(\log k)$. In the latter case $\epsilon \geq k - O(\log k)$ and we are done. In the former case $m - l \geq k - O(\log k + C(q))$ hence $O(C(B) + C(q)) \geq k - O(\log k + C(q))$ and so $O(\epsilon + \delta + C(q)) \geq k - O(\log k)$.

4   Hereditary of Normality

In this section we prove that every strong MSS for a normal string is itself normal. Recall that a string x is called (ϵ, δ)-normal if for every model B for x there is a model A for x with $CT(A|x) \leq \epsilon$ and $C(A) \leq C(B) + \delta$, $\log |A| \leq \log |B| + \delta$.

Theorem 10. *There is a constant D such that the following holds. Assume that A is an ϵ-strong (δ, ϵ, D)-MSS for an (ϵ, ϵ)-normal string x of length n. Assume that $\epsilon \leq \sqrt{n}/2$. Then the code $[A]$ of A is $O((\epsilon + \delta + \log n) \cdot \sqrt{n})$-normal.*

The rest of this section is the proof of this theorem. We start with the following lemma, which is a simple corollary of Theorem 3 and Lemma 1.

Lemma 3. *For all large enough D the following holds: if A is a (δ, ϵ, D)-MSS for $x \in \{0,1\}^n$ then $C(\Omega_{C(A)}|A) = O(\delta + \log n)$.*

We fix a constant D satisfying Lemma 3 and call a model (δ, ϵ)-MSS if it (δ, ϵ, D)-MSS. This D is the constant satisfying Theorem 10

A family of sets $\mathcal{A}$ is called *partition* if for every $A_1, A_2 \in \mathcal{A}$ we have $A_1 \cap A_2 \neq \varnothing \Rightarrow A_1 = A_2$. Note that for a finite partition we can define its complexity. The next lemma states that every strong statistic A can be transformed into a strong statistic A_1 such that A_1 belongs to some partition of similar complexity.

Lemma 4. *Let A be an ϵ-strong statistic for $x \in \{0,1\}^n$. Then there is a set A_1 and a partition $\mathcal{A}$ of complexity at most $\epsilon + O(\log n)$ such that:*

(1) A_1 is $\epsilon + O(\log n)$-strong statistic for x.
(2) $CT(A|A_1) < \epsilon + O(\log n)$ and $CT(A_1|A) < \epsilon + O(\log n)$.
(3) $|A_1| \leq |A|$.
(4) $A_1 \in \mathcal{A}$.

Proof. Assume that A is an ϵ-strong statistic for x. Then there is a total program p such that $p(x) = A$ and $l(p) \leq \epsilon$.

We will use the same construction as in Remark 1 in [9]. For every set B denote by B' the following set: $\{x' \in B \mid p(x') = B, \ x' \in \{0,1\}^n\}$. Notice that $CT(A'|A)$, $CT(A|A')$ and $CT(A'|x)$ are less than $l(p) + O(\log n) = \epsilon + O(\log n)$ and $|A'| \leq |A|$.

For any $x_1, x_2 \in \{0,1\}^n$ with $p(x_1) \neq p(x_2)$ we have $p(x_1)' \cap p(x_2)' = \varnothing$. Hence $\mathcal{A} := \{p(x)'|x \in \{0,1\}^n\}$ is a partition of complexity at most $\epsilon + O(\log n)$.

By Theorem 3 and Lemma 1 for every $A \ni x$ there is a $B \ni x$ such that B is informational equivalent to $\Omega_{C(B)}$ and parameters of B are not worse than those of A. We will need a similar result for normal strings and for strong models.

Lemma 5. *Let x be an (ϵ, α)-normal string with length n such that $\epsilon \leq n$, $\alpha < \sqrt{n}/2$. Let A be an ϵ-strong statistic for x. Then there is a set H such that:*

(1) H is an ϵ-strong statistic for x.
(2) $\delta(x|H) \leq \delta(x|A) + O((\alpha + \log n) \cdot \sqrt{n})$ and $C(H) \leq C(A)$.
(3) $C(H|\Omega_{C(H)}) = O(\sqrt{n})$.

Proof (Sketch of proof). Consider the sequence $A_1, B_1, A_2, B_2, \ldots$ of statistics for x defined as follows. Let $A_1 := A$ and let B_i be an improvement of A_i such that B_i is informational equivalent to $\Omega_{C(B_i)}$, which exists by Theorem 3. Let A_{i+1} be a strong statistic for x that has a similar parameters as B_i, which exists because x is normal. (See Fig. 4.)

Denote by N the minimal integer such that $C(A_N) - C(B_N) \leq \sqrt{n}$. For $i < N$ the complexity of B_i is more than $\sqrt{n}$ less that of A_i. On the other hand, the complexity of A_{i+1} is at most $\alpha < \sqrt{n}/2$ larger than that of B_i. Hence $N = O(\sqrt{n})$. Let $H := A_N$. By definition A_N (and H) is strong. From $N = O(\sqrt{n})$ it follows that the second condition is satisfied. From $C(A_N) - C(B_N) \leq \sqrt{n}$ and definition of B_N it is follows that the third condition is satisfied too (use symmetry of information).

Proof (Sketch of proof of Theorem 10). Assume that A is a ϵ-strong (δ, ϵ, D)-minimal statistic for x, where D satisfies Lemma 3. By Lemma 3 A is informational equivalent to $\Omega_{C(A)}$. We need to prove that the profile of $[A]$ is close to the strong profile of $[A]$.

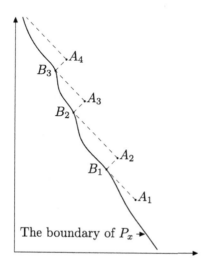

Fig. 4. Parameters of statistics A_i and B_i

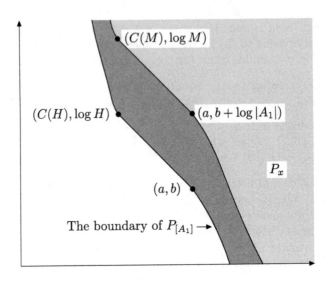

Fig. 5. P_x is located $\log|A_1|$ higher than $P_{[A_1]}$

Let $\mathcal{A}$ be a simple partition and A_1 a model from $\mathcal{A}$ which exists by Lemma 4 applied to A, x. As the total conditional complexities $CT(A_1|A)$ and $CT(A|A_1)$ are small, the profiles of A and A_1 are close to each other. This also applies to strong profiles. Therefore it suffices to show that (the code of) A_1 is normal.

Let $(a, b) \in P_{[A_1]}$. The parameters (complexity and log-cardinality) of A_1 are not larger than those of A and hence A_1 is a sufficient statistic for x. By Theorem 9 we have $(a, b + \log|A_1|) \in P_x$ (see Fig. 5).

As x is normal, the pair $(a, b + \log|A_1|)$ belongs to the strong profile of x as well. By Lemma 5 there is a **strong** model M for x that has low complexity conditional to $\Omega_{C(M)}$ and whose parameters (complexity, optimality deficiency) are not worse than those of A_1.

We claim that $C(M|A_1)$ is small. As A is informational equivalent to $\Omega_{C(A)}$, so is A_1. From $\Omega_{C(A)}$ we can compute $\Omega_{C(M)}$ (Lemma 1) and then compute M (as $C(M|\Omega_{C(M)}) \approx 0$). This implies that $C(M|A_1) \approx 0$.

However we will need a stronger inequality $CT(M|A_1) \approx 0$. To find such M, we apply Lemma 4 to M, x and change it to a model M_1 with the same parameters that belongs to a simple partition $\mathcal{M}$. Item (2) of Lemma 4 guarantees that M_1 is also simple given A_1 and that M_1 is a strong model for x. Since $C(M|A_1) \approx 0$, we have $C(M_1|A_1) \approx 0$ as well.

As A_1 lies on the border line of P_x and $C(M_1|A_1) \approx 0$, the intersection $A_1 \cap M_1$ cannot be much less than A_1, that is, $\log|A_1 \cap M_1| \approx \log|A_1|$ (otherwise the model $A_1 \cap M_1$ for x would have much smaller cardinality and almost the same complexity as A_1). The model M_1 can be computed by a total program from A_1 and its index among all $M' \in \mathcal{M}$ with $\log|A_1 \cap M'| \approx \log|A_1|$. As $\mathcal{M}$ is a partition, there are few such sets M'. Hence $CT(M_1|A_1) \approx 0$.

Finally, let $H = \{A' \in \mathcal{A} \mid \log |A' \cap M_1| = \log |A_1 \cap M_1|\}$. The model H for A_1 is strong because the partition $\mathcal{A}$ is simple and $CT(M_1|A_1) \approx 0$. The model H can be computed from M_1, $\mathcal{A}$ and $\log |A_1 \cap M_1|$. As $\mathcal{A}$ is simple, we conclude that $C(H) \lesssim C(M_1)$. Finally $\log |H| \leq \log |M_1| - \log |A_1|$, because $\mathcal{A}$ is a partition and thus it has few sets that have $\log |A_1 \cap M_1| \approx \log |A_1|$ common elements with M_1.

Thus the complexity of H is not larger than that of M_1 and the sum of complexity and cardinality of H is at most $a + b - \log |A_1|$. As the strong profile of x has the third property from Theorem 1, we can conclude that it includes the point (a, b).

Acknowledgments. The author is grateful to professor N. K. Vereshchagin for statements of questions, remarks and useful discussions.

This work is supported by RFBR grant 16-01-00362 and partially supported by RaCAF ANR-15-CE40-0016-01 grant. The study has been funded by the Russian Academic Excellence Project '5-100'.

References

1. Gács, P., Tromp, J., Vitányi, P.M.B.: Algorithmic statistics. IEEE Trans. Inform. Theory **47**(6), 2443–2463 (2001)
2. Kolmogorov, A.N.: Three approaches to the quantitative definition of information. Probl. Inf. Trans. **1**(1), 1–7 (1965)
3. Kolmogorov, A.N.: The complexity of algorithms, the objective definition of randomness. Usp. Matematicheskich Nauk **29**(4(178)), 155 (1974). Summary of the talk presented April 16, at Moscow Mathematical Society
4. Li, M., Vitányi, P.: An Introduction to Kolmogorov complexity and its applications, 3rd edn., xxiii+790 p. Springer, New York (2008). (1st edn. 1993; 2nd edn. 1997), ISBN 978-0-387-49820-1
5. Shen, A.: Game arguments in computability theory and algorithmic information theory. In: Cooper, S.B., Dawar, A., Löwe, B. (eds.) CiE 2012. LNCS, vol. 7318, pp. 655–666. Springer, Heidelberg (2012)
6. Shen, A.: Around kolmogorov complexity: basic notions and results. In: Vovk, V., Papadoupoulos, H., Gammerman, A. (eds.) Measures of Complexity: Festschrift for Alexey Chervonenkis. Springer, Heidelberg (2015)
7. Shen, A.: The concept of (α, β)-stochasticity in the Kolmogorov sense, and its properties. Sov. Math. Dokl. **271**(1), 295–299 (1983)
8. Shen, A., Uspensky, V., Vereshchagin, N.: Kolmogorov complexity and algorithmic randomness. MCCME (2013) (Russian). English translation: http://www.lirmm.fr/~ashen/kolmbook-eng.pdf
9. Vereshchagin, N.: Algorithmic minimal sufficient statistics: A new approach. Theory Comput. Syst. **56**(2), 291–436 (2015)
10. Vereshchagin, N., Vitányi, P.: Kolmogorov's structure functions with an application to the foundations of model selection. IEEE Trans. Inf. Theory **50**(12), 3265–3290 (2004). Preliminary version: Proceedings of the 47th IEEE Symposium on Foundations of Computer Science, pp. 751–760 (2002)
11. Vitányi, P., Vereshchagin, N.: On algorithmic rate-distortion function. In: Proceedings of 2006 IEEE International Symposium on Information Theory, Seattle, Washington, 9–14 July 2006

Subquadratic Algorithms for Succinct Stable Matching

Daniel Moeller, Ramamohan Paturi, and Stefan Schneider[(✉)]

University of California, San Diego, La Jolla, USA
{dmoeller,paturi,stschnei}@cs.ucsd.edu

Abstract. We consider the stable matching problem when the preference lists are not given explicitly but are represented in a succinct way and ask whether the problem becomes computationally easier. We give subquadratic algorithms for finding a stable matching in special cases of two very natural succinct representations of the problem, the d-attribute and d-list models. We also give algorithms for verifying a stable matching in the same models. We further show that for $d = \omega(\log n)$ both finding and verifying a stable matching in the d-attribute model requires quadratic time assuming the Strong Exponential Time Hypothesis. The d-attribute model is therefore as hard as the general case for large enough values of d.

Keywords: Stable matching · Attribute model · Subquadratic algorithms · Conditional lower bounds · SETH

1 Introduction

The stable matching problem has applications that vary from coordinating buyers and sellers to assigning students to public schools and residents to hospitals [16,21,27]. Gale and Shapley [14] proposed a quadratic time *deferred acceptance* algorithm for this problem which has helped clear matching markets in many real-world settings. For arbitrary preferences, the deferred acceptance algorithm is optimal and even verifying that a given matching is stable requires quadratic time [15,24,28]. This is reasonable since representing all participants' preferences requires quadratic space. However, in many applications the preferences are not arbitrary and can have more structure. For example, top doctors are likely to be universally desired by residency programs and students typically seek highly

This research was partially supported by the Army Research Office grant number W911NF-15-1-0253. This research is supported by NSF grant CCF-1213151 from the Division of Computing and Communication Foundations. Any opinions, findings and conclusions or recommendations expressed in this material are those of the authors and do not necessarily reect the views of the National Science Foundation. This work was done in part while the author was visiting the Simons Institute for the Theory of Computing.

A.S. Kulikov and G.J. Woeginger (Eds.): CSR 2016, LNCS 9691, pp. 294–308, 2016.
DOI: 10.1007/978-3-319-34171-2_21

ranked schools. In these cases participants can represent their preferences succinctly. It is natural to ask whether the same quadratic time bounds apply with compact and structured preference models that have subquadratic representations. This will provide a more nuanced understanding of where the complexity lies: Is stable matching inherently complex, or is the complexity merely a result of the large variety of possible preferences? To this end, we examine two restricted preference models originally proposed by Bhatnagar et al. [7], the d-attribute and d-list models. Using a wide range of techniques we provide algorithms and conditional hardness results for several settings of these models.

In the d-*attribute* model, we assume that there are d different attributes (e.g. income, height, sense of humor, etc.) with a fixed, possibly objective, ranking of the men for each attribute. Each woman's preference list is based on a linear combination of the attributes of the men, where each woman can have different weights for each attribute. Some women may care more about, say, height whereas others care more about sense of humor. Men's preferences are defined analogously. This model is applicable in large settings, such as online dating systems, where participants lack the resources to form an opinion of every other participant. Instead the system can rank the members of each gender according to the d attributes and each participant simply needs to provide personalized weights for the attributes. The combination of attribute values and weights implicitly represents the entire preference matrix. Bogomolnaia and Laslier [8] show that representing all possible $n \times n$ preference matrices requires $n - 1$ attributes. Therefore it is reasonable to expect that when $d \ll n - 1$, we could beat the worst case quadratic lower bounds for the general stable matching problem.

In the d-*list* model, we assume that there are d different rankings of the men. Each women selects one of the d lists as her preference list. Similarly, each man chooses one of d lists of women as his preference list. This model captures the setting where members of one group (i.e. student athletes, sorority members, engineering majors) may all have identical preference lists. Mathematically, this model is actually a special case of the d-attribute model where each participant places a positive weight on exactly one attribute. However, its motivation is distinct and we can achieve improved results for this model.

Chebolu et al. prove that approximately counting stable matchings in the d-attribute model for $d \geq 3$ is as hard as the general case [11]. Bhatnagar et al. showed that sampling stable matchings using random walks can take exponential time even for a small number of attributes or lists but left it as an open question whether subquadratic algorithms exist for these models [7].

We show that faster algorithms exist for finding a stable matching in some special cases of these models. In particular, we provide subquadratic algorithms for the d-attribute model, where all values and weights are from a small set, and the one-sided d-attribute model, where one side of the market has only one attribute. These results show we can achieve meaningful improvement over the general setting for some restricted preferences.

While we only provide subquadratic algorithms to find stable matchings in special cases of the attribute model, we have stronger results concerning verification of stable matchings. We demonstrate optimal subquadratic stability testing algorithms for the d-list and boolean d-attribute settings as well as a subquadratic algorithm for the general d-attribute model with constant d. These algorithms provide a clear distinction between the attribute model and the general setting. Moreover, these results raise the question of whether verifying and finding a stable matching are equally hard problems for these restricted models, as both require quadratic time in the general case.

Finally, we show that the stable matching problem in the d-attribute model for $d = \omega(\log n)$ cannot be solved in subquadratic time under the Strong Exponential Time Hypothesis (SETH) [18,20]. We show SETH-hardness for both finding and verifying a stable matching, even if the weights and attributes are boolean. This adds the stable matching problem to a growing list of SETH-hard problems, including Fréchet distance [9], edit distance [5], string matching [1], k-dominating set [25], orthogonal vectors [30], model checking on sparse graphs [10], and vector domination [19]. Thus the quadratic time hardness of the stable matching problem in the general case extends to the more restricted and succinct d-attribute model. This limits the space of models where we can hope to find subquadratic algorithms.

Dabney and Dean [12] study an alternative succinct preference representation where there is a canonical preference list for each side and individual deviations from this list are specified separately. They provide an adaptive $O(n + k)$ time algorithm for the special one-sided case, where k is the number of deviations. Bartholdi and Trick [6] present a subquadratic time algorithm for stable roommates with narcissistic, single-peaked preferences. Arkin et al. [4] derive a subquadratic algorithm for stable roommates with constant dimensional geometric preferences.

2   Summary of Results

Section 4.1 gives an $O(C^{2d}n(d+\log n))$ time algorithm for finding a stable matching in the d-attribute model if both the attributes and weights are from a set of size at most C. This gives a strongly subquadratic algorithm (i.e. $O(n^{2-\varepsilon})$ for $\varepsilon > 0$) if $d < \frac{1}{2\log C}\log n$.

Section 4.2 considers an asymmetric case, where one side of the matching market has d attributes, while the other side has a single attribute. We allow both the weights and attributes to be arbitrary real values. Our algorithm for finding a stable matching in this model has time complexity $\tilde{O}(n^{2-1/\lfloor d/2 \rfloor})$, which is strongly subquadratic for constant d.

In Sect. 5.1 we consider the problem of verifying that a given matching is stable in the d-attribute model with real attributes and weights. The time complexity of our algorithm is $\tilde{O}(n^{2-1/2d})$, which is again strongly subquadratic for constant d.

Section 5.2 gives an $O(dn)$ time algorithm for verifying a stable matching in the d-list model. This is linear in its input size and is therefore optimal.

In Sect. 5.3 we give a randomized $\tilde{O}(n^{2-1/O(c\log^2(c))})$ time algorithm for $d = c\log n$ for verifying a stable matching in the d-attribute model when both the weights and attributes are boolean. This algorithm is strongly subquadratic for $d = O(\log n)$.

Finally, in Sect. 6 we give a conditional lower bound for both finding and verifying a stable matching in the d-attribute model. We show that there is no strongly subquadratic algorithm when $d = \omega(\log n)$ assuming the Strong Exponential Time Hypothesis.

3 Preliminaries

A *matching market* consists of a set of men M and a set of women W with $|M| = |W| = n$. We further have a permutation of W for every $m \in M$, and a permutation of M for every $w \in W$, called *preference lists*. Note that representing a general matching market requires size $\Omega(n^2)$.

For a perfect bipartite matching μ, a *blocking pair* with respect to μ is a pair $(m, w) \notin \mu$ where $m \in M$ and $w \in W$, such that w appears before $\mu(m)$ in m's preference list and m appears before $\mu(w)$ in w's preference list. A perfect bipartite matching is called *stable* if there are no blocking pairs. In settings where ties in the preference lists are possible, we consider weakly stable matchings where (m, w) is a blocking pair if and only if both strictly prefer each other to their partner.

Gale and Shapley's deferred acceptance algorithm [14] works as follows. While there is an unmatched man m, have m *propose* to his most preferred woman who has not already rejected him. A woman accepts a proposal if she is unmatched or if she prefers the proposing man to her current partner, leaving her current partner unmatched. Otherwise, she rejects the proposal. This process finds a stable matching in time $O(n^2)$.

A matching market in the d-attribute model consists of n men and n women as before. A participant p has attributes $A_i(p)$ for $1 \leq i \leq d$ and weights $\alpha_i(p)$ for $1 \leq i \leq d$. For a man m and woman w, m's *value* of w is given by $\mathrm{val}_m(w) = \langle \alpha(m), A(w) \rangle = \sum_{i=1}^{d} \alpha_i(m)A_i(w)$. m ranks the women in decreasing order of value. Symmetrically, w's value of m is $\mathrm{val}_w(m) = \sum_{i=1}^{d} \alpha_i(w)A_i(m)$. Note that representing a matching market in the d-attribute model requires size $O(dn)$. Unless otherwise specified, both attributes and weights can be negative.

A matching market in the d-list model is a matching market where both sides have at most d distinct preference lists. Describing a matching market in this model requires $O(dn)$ numbers.

Throughout the paper, we use $\tilde{O}$ to suppress polylogarithmic factors in the time complexity.

4   Finding Stable Matchings

4.1   Small Set of Attributes and Weights

We first present a stable matching algorithm for the d-attribute model when the attribute and weight values are limited to a set of constant size. In particular, we assume that the number of possible values for each attribute and weight for all participants is bounded by a constant C.

Theorem 1. *There is an algorithm to find a stable matching in the d-attribute model with at most a constant C distinct attribute and weight values in time $O(C^{2d}n(d + \log n))$.*

Proof. First group the women into sets S_i with a set for each of the $O(C^{2d})$ types of women. ($O(C^d)$ possible attribute values and $O(C^d)$ possible weight vectors.) Observe that each man is indifferent between the women in a given set S_i because each woman has identical attribute values. Thus we can enumerate the full preference list over the sets S_i for each man. Moreover, the women in a set S_i share the same ranking of the men, since they have identical weight vectors. Therefore we can treat each set of women S_i as an individual entity in a many to one stable matching problem where the capacity for each S_i is the number of women it contains. With these observations, we can run the standard deferred acceptance algorithm for many-one stable matching and the stability follows directly from the stability of this algorithm.

Grouping the women requires $O(C^{2d} + dn)$ time to initialize the groups and place each woman in the appropriate group. Creating the men's preference lists requires $O(dC^{2d}n)$ time to evaluate and sort the groups of women for every man. The deferred acceptance algorithm requires $O(C^{2d}n(d + \log n))$ time since each man will propose to at most C^{2d} sets of women and each proposal requires $O(d + \log n)$ time since we must evaluate the proposer and keep track of which tentative partner is least preferred (using a min-heap). This results in an overall running time of $O(C^{2d}n(d + \log n))$.

As long as $d < \frac{1}{2\log C} \log n$, the time complexity in Theorem 1 will be sub-quadratic. It is worth noting that the algorithm and proof actually do not rely on any restriction of the men's attribute and weight values. Thus, this result holds whenever one side's attributes and weight values come from a set of constant size. Pseudocode for the algorithm can be found in the full version of the paper.

4.2   One-Sided Real Attributes

In this section we consider a one-sided attribute model with real attributes and weights. In this model, women have d attributes and men have d weights, and the preference list of a man is given by the weighted sum of the women's attributes as in the two-sided attribute model. On the other hand there is only one attribute for the men. The women's preferences are thus determined by whether they have a positive or negative weight on this attribute. For simplicity, we first assume that all women have a positive weight on the men's attribute and show a subquadratic algorithm for this case. Then we extend it to allow for negative weights.

To find a stable matching when the women have a global preference list over the men, we use a greedy approach: process the men from the most preferred to the least preferred and match each man with the highest unmatched woman in his preference list. This general technique is not specific to the attribute model but actually works for any market where one side has a single global preference list. (e.g. [12] uses a similar approach for their algorithm.) The complexity lies in repeatedly finding which of the available women is most preferred by the current top man.

This leads us to the following algorithm: for every woman w consider a point with $A(w)$ as its coordinates and organize the set of points into a data structure. Then, for the men in order of preference, query the set of points against a direction vector consisting of the man's weight and find the point with the largest distance along this direction. Remove that point and repeat.

The problem of finding a maximal point along a direction is typically considered in its dual setting, where it is called the *ray shooting problem*. In the ray shooting problem we are given n hyperplanes and must maintain a data structure to answer queries. Each query consists of a vertical ray and the data structure returns the first hyperplane hit by that ray.

The relevant results are in Lemma 1 which follows from several papers for different values of d. For an overview of the ray shooting problem and related range query problems, see [2].

Lemma 1 [13,17,23]. *Given an n point set in $\mathbb{R}^d$ for $d \geq 2$, there is a data structure for ray shooting queries with preprocessing time $\tilde{O}(n)$ and query time $\tilde{O}(n^{1-1/\lfloor d/2 \rfloor})$. The structure supports deletions with amortized update time $\tilde{O}(1)$.*

For $d = 1$, queries can trivially be answered in constant time. We use this data structure to provide an algorithm when there is a global list for one side of the market.

Lemma 2. *For $d \geq 2$ there is an algorithm to find a stable matching in the one-sided d-attribute model with real-valued attributes and weights in $\tilde{O}(n^{2-1/\lfloor d/2 \rfloor})$ time when there is a single preference list for the other side of the market.*

Proof. For a man m, let $\dim(m)$ denote the index of the last non-zero weight, i.e. $\alpha_{\dim(m)+1}(m) = \cdots = \alpha_d(m) = 0$. We assume $\dim(m) > 0$, as otherwise m is indifferent among all women and we can pick any woman as $\mu(m)$. We assume without loss of generality $\alpha_{\dim(m)}(m) \in \{-1, 1\}$. For each d' such that $1 \leq d' \leq d$ we build a data structure consisting of n hyperplanes in $\mathbb{R}^{d'}$. For each woman w, consider the hyperplanes

$$H_{d'}(w) = \left\{ x_{d'} = \sum_{i=1}^{d'-1} A_i(w)x_i - A_{d'}(w) \right\} \tag{1}$$

and for each d' preprocess the set of all hyperplanes according to Lemma 1. Note that $H_{d'}(w)$ is the dual of the point $(A_1(w), \ldots, A_{d'}(w))$.

For a man m we can find his most preferred partner by querying the $\dim(m)$-dimensional data structure. Let $s = \alpha_{\dim(m)}(m)$. Consider a ray $r(m) \in \mathbb{R}^{\dim(m)}$ originating at

$$(-\frac{\alpha_1(m)}{s}, \ldots, -\frac{\alpha_{\dim(m)-1}(m)}{s}, -s \cdot \infty) \tag{2}$$

in the direction $(0, \ldots, 0, s)$. If $\alpha_{\dim(m)} = 1$ we find the lowest hyperplane intersecting the ray, and if $\alpha_{\dim(m)} = -1$ we find the highest hyperplane. We claim that the first hyperplane $r(m)$ hits corresponds to m's most preferred woman. Let woman w be preferred over woman w', i.e. $\mathrm{val}_m(w) = \sum_{i=1}^{\dim(m)} A_i(w)\alpha_i(m) \geq \sum_{i=1}^{\dim(m)} A_i(w')\alpha_i(m) = \mathrm{val}_m(w')$. Since the ray $r(m)$ is vertical in coordinate $x_{d'}$, it is sufficient to evaluate the right-hand side of the definition in Eq. 1. Indeed we have $\mathrm{val}_m(w) \geq \mathrm{val}_m(w')$ if and only if

$$\sum_{i=1}^{\dim(m)-1} -A_i(w)\frac{\alpha_i(m)}{s} - A_{\dim(m)}(w) \leq \sum_{i=1}^{\dim(m)-1} -A_i(w')\frac{\alpha_i(m)}{s} - A_{\dim(m)}(w') \tag{3}$$

when $s = 1$. The case $s = -1$ is symmetrical.

Note that the query ray is the standard dual of the set of hyperplanes with normal vector $(\alpha_1(m), \ldots, \alpha_d(m))$.

Now we pick the highest man m in the (global) preference list, consider the ray as above and find the first hyperplane $H_{\dim(m)}(w)$ hit by the ray. We then match the pair (m, w), remove $H(w)$ from all data structures and repeat. Correctness follows from the correctness of the greedy approach when all women share the same preference list and the properties of the halfspaces proved above.

The algorithm preprocesses d data structures, then makes n queries and dn deletions. The time is dominated by the n ray queries each requiring time $\tilde{O}(n^{1-1/\lfloor d/2 \rfloor})$. Thus the total time complexity is bounded by $\tilde{O}(n^{2-1/\lfloor d/2 \rfloor})$, as claimed.

We use the following lemma to extend the above algorithm to account for positive and negative weights for the women. It deals with settings where the women choose one of two lists (σ_1, σ_2) as their preference lists over the men while the men's preferences can be arbitrary.

Lemma 3. *Suppose there are k women who use σ_1. If the top k men in σ_1 are in the bottom k places in σ_2, then the women using σ_1 will only match with those men and the $n - k$ women using σ_2 will only match with the other $n - k$ men in the woman-optimal stable matching.*

Proof. Consider the operation of the woman-proposing deferred acceptance algorithm for finding the woman-optimal stable matching. Suppose the lemma is false so that at some point a woman using σ_1 proposed to one of the last $n - k$ men in σ_1. Let w be the first such woman. w must have been rejected by all of the top k, so at least one of those men received a proposal from a woman, w', using σ_2.

However, since the top k men in σ_1 are the bottom k men in σ_2, w' must have been rejected by all of the top $n - k$ men in σ_2. But there are only $n - k$ women using σ_2, so one of the top $n - k$ men in σ_2 must have already received a proposal from a woman using σ_1. This is a contradiction because w was the first woman using σ_1 to propose to one of the bottom $n - k$ men in σ_1 (which are the top $n - k$ men in σ_2).

We can now prove the following theorem where negative values are allowed for the women's weights.

Theorem 2. *For $d \geq 2$ there is an algorithm to find a stable matching in the one-sided d-attribute model with real-valued attributes and weights in time $\tilde{O}(n^{2-1/\lfloor d/2 \rfloor})$.*

Proof. Suppose there are k women who have a positive weight on the men's attribute. Since the remaining $n - k$ women's preference list is the reverse, we can use Lemma 3 to split the problem into two subproblems. Namely, in the woman-optimal stable matching the k women with a positive weight will match with the top k men, and the $n - k$ women with a negative weight will match with the bottom $n - k$ men. Now the women in each of these subproblems all have the same list. Therefore we can use Lemma 2 to solve each subproblem. Splitting the problem into subproblems can be done in time $O(n)$ so the running time follows immediately from Lemma 2.

Table 1. Preference lists where a greedy approach will not work

σ_1	σ_2	π_1	π_2	Man	List	Woman	List
m_1	m_3	w_1	w_3	m_1	π_1	w_1	σ_2
m_2	m_5	w_2	w_5	m_2	π_1	w_2	σ_2
m_3	m_1	w_3	w_1	m_3	π_2	w_3	σ_1
m_4	m_4	w_4	w_4	m_4	π_1	w_4	σ_2
m_5	m_2	w_5	w_2	m_5	π_2	w_5	σ_1

As a remark, this "greedy" approach where we select a man, find his most preferred available woman, and permanently match him to her will not work in general. Table 1 describes a simple 2-list example where the unique stable matching is $(m_1w_2, m_2w_3, m_3w_5, m_4w_4, m_5w_1)$. In this instance, no participant is matched with their top choice. Therefore, the above approach cannot work for this instance. This illustrates to some extent why the general case seems more difficult than the one-sided case.

5   Verification

We now turn to the problem of verifying whether a given matching is stable. While this is as hard as finding a stable matching in the general setting, the verification algorithms we present here are more efficient than our algorithms for finding stable matchings in the attribute model.

5.1   Real Attributes and Weights

In this section we adapt the geometric approach for finding a stable matching in the one-sided d-attribute model to the problem of verifying a stable matching in the (two-sided) d-attribute model. We express the verification problem as a *simplex range searching problem* in $\mathbb{R}^{2d}$, which is the dual of the ray shooting problem. In simplex range searching we are given n points and answer queries that ask for the number of points inside a simplex. In our case we only need degenerate simplices consisting of the intersection of two halfspaces. Simplex range searching queries can be done in sublinear time for constant d.

Lemma 4 [22]. *Given a set of n points in $\mathbb{R}^d$, one can process it for simplex range searching in time $O(n \log n)$, and then answer queries in time $\tilde{O}(n^{1-\frac{1}{d}})$.*

For $1 \leq d' \leq d$ we use the notation $(x_1, \ldots, x_d, y_1, \ldots, y_{d'-1}, z)$ for points in $\mathbb{R}^{d+d'}$. We again let $\dim(w)$ be the index of w's last non-zero weight, assume without loss of generality $\alpha_{\dim(w)} \in \{-1, 1\}$, and let $\text{sgn}(w) = \text{sgn}(\alpha_{\dim(w)})$. We partition the set of women into $2d$ sets $W_{d',s}$ for $1 \leq d' \leq d$ and $s \in \{-1, 1\}$ based on $\dim(w)$ and $\text{sgn}(w)$. Note that if $\dim(w) = 0$, then w is indifferent among all men and can therefore not be part of a blocking pair. We can ignore such women.

For a woman w, consider the point

$$P(w) = (A_1(w), \ldots, A_d(w), \alpha_1(w), \ldots, \alpha_{\dim(w)-1}(w), \text{val}_w(m)) \qquad (4)$$

where $m = \mu(w)$ is the partner of w in the input matching μ. For a set $W_{d',s}$ we let $P_{d',s}$ be the set of points $P(w)$ for $w \in W_{d',s}$. The basic idea is to construct a simplex for every man and query it against all sets $P_{d',s}$.

Given d',s, and a man m, let $H_1(m)$ be the halfspace

$$\left\{ \sum_{i=1}^{d} \alpha_i(m)x_i > \text{val}_m(w) \right\} \qquad (5)$$

where $w = \mu(m)$. For $w' \in W_{d',s}$ we have $P(w') \in H_1(m)$ if and only if m strictly prefers w' to w. Further let $H_2(m)$ be the halfspace

$$\left\{ \sum_{i=1}^{d'-1} A_i(m)y_i + A_{d'}(m)s > z \right\} \qquad (6)$$

For $w' \in W_{d',s}$ we have $P(w') \in H_2(m)$ if and only if w' strictly prefers m to $\mu(w')$. Hence (m, w') is a blocking pair if and only if $P(w') \in H_1(m) \cap H_2(m)$.

Using Lemma 4 we immediately have an algorithm to verify a stable matching.

Theorem 3. *There is an algorithm to verify a stable matching in the d-attribute model with real-valued attributes and weights in time $\tilde{O}(n^{2-1/2d})$*

Proof. Partition the set of women into sets $W_{d',s}$ for $1 \le d' \le d$ and $s \in \{-1, 1\}$ and for $w \in W_{d',s}$ construct $P(w) \in \mathbb{R}^{d+d'}$ as above. Then preprocess the sets according to Lemma 4. For each man m query $H_1(m) \cap H_2(m)$ against the points in all sets. By the definitions of $H_1(m)$ and $H_2(m)$, there is a blocking pair if and only if for some man m there is a point $P(w) \in H_1(m) \cap H_2(m)$ in one of the sets $P_{d',s}$. The time to preprocess is $O(n \log n)$. There are $2dn$ queries of time $\tilde{O}(n^{1-1/2d})$. Hence the whole process requires time $\tilde{O}(n^{2-1/2d})$ as claimed.

5.2   Lists

Algorithm 1. Verify d-List Stable Matching

---

for $i = 1$ *to* d **do**

   for $j = 1$ *to* d **do**

      $w \leftarrow \mathrm{head}(\pi_i, j)$.

      $m \leftarrow \mathrm{head}(\sigma_j, i)$.

      while $m \ne \perp$ *and* $w \ne \perp$ **do**

         if $\mathrm{rank}(w, i) > \mathrm{rank}(\mu(m), i)$ **then**

            $m \leftarrow \mathrm{next}(m, j)$.

         else

            if $\mathrm{rank}(m, j) > \mathrm{rank}(\mu(w), j)$ **then**

               $w \leftarrow \mathrm{next}(w, i)$.

            else

               return (m, w) *is a blocking pair.*

return μ *is stable.*

When there are d preference orders for each side, and each participant uses one of the d lists, we provide a more efficient algorithm. Here, assume μ is the given matching between M and W. Let $\{\pi_i\}_{i=1}^d$ be the set of d permutations on the women and $\{\sigma_i\}_{i=1}^d$ be the set of d permutations on the men. Define $\mathrm{rank}(w, i)$ to be the position of w in permutation π_i. This can be determined in constant time after $O(dn)$ preprocessing of the permutations. Let $\mathrm{head}(\pi_i, j)$ be the first woman in π_i who uses permutation σ_j and $\mathrm{next}(w, i)$ be the next highest ranked woman after w in permutation π_i who uses the same permutation as w or $\perp$ if no such woman exists. These can also be determined in constant time after $O(dn)$ preprocessing by splitting the lists into sublists, with one sublist for the women using each permutation of men. The functions rank, head, and next are defined analogously for the men.

Theorem 4. *There is an algorithm to verify a stable matching in the d-list model in $O(dn)$ time.*

Proof. We claim that Algorithm 1 satisfies the theorem. Indeed, if the algorithm returns a pair (m, w) where m uses π_i and w uses σ_j, then (m, w) is a blocking pair because w appears earlier in π_i than $\mu(m)$ and m appears earlier in σ_j than $\mu(w)$.

On the other hand, suppose the algorithm returns that μ is stable but there is a blocking pair, (m, w), where m uses π_i and w uses σ_j. The algorithm considers permutations π_i and σ_j since it does not terminate early. Clearly if the algorithm evaluates m and w simultaneously when considering permutations π_i and σ_j, it will detect that (m, w) is a blocking pair. Therefore, the algorithm either moves from m to $\text{next}(m, j)$ before considering w or it moves from w to $\text{next}(w, i)$ before considering m. In the former case, $\text{rank}(\mu(m), i) < \text{rank}(w', i)$ for some w' that comes before w in π_i. Therefore m prefers $\mu(m)$ to w. Similarly, in the latter case, $\text{rank}(\mu(w), j) < \text{rank}(m', i)$ for some m' that comes before m in σ_j so w prefers $\mu(w)$ to m. Thus (m, w) is not a blocking pair and we have a contradiction.

The **for** and **while** loops proceed through all men and women once for each of the d lists in which they appear. Since at each step we are either proceeding to the next man or the next woman unless we find a blocking pair, the algorithm requires time $O(dn)$. This is optimal since the input size is dn.

5.3   Boolean Attributes and Weights

In this section we consider the problem of verifying a stable matching when the d attributes and weights are restricted to boolean values and $d = c \log n$. The algorithm closely follows an algorithm for the maximum inner product problem by Alman and Williams [3]. The idea is to express the existence of a blocking pair as a probabilistic polynomial with a bounded number of monomials and use fast rectangular matrix multiplication to evaluate it. A probabilistic polynomial for a function f is a polynomial p such that for every input x, $\Pr[f(x) \neq p(x)] \leq \frac{1}{3}$.

We construct a probabilistic polynomial that outputs 1 if there is a blocking pair. To minimize the degree of the polynomial, we pick a parameter s and divide the men and women into sets of size at most s. The polynomial takes the description of s men $m_1, \ldots, m_s$ and s women $w_1, \ldots, w_s$ along with their respective partners as input, and outputs 1 if and only if there is a blocking pair (m_i, w_j) among the s^2 pairs of nodes with high probability.

Proofs for this section, using lemmas from [26,29,31] are omitted here and can be found in the full version.

Lemma 5. *Let u be a large constant and $s = n^{1/uc \log^2 c}$. There is a probabilistic polynomial with the following inputs:*

- *The attributes and weights of s men, $A(m_1), \ldots, A(m_s), \alpha(m_1), \ldots, \alpha(m_s)$*
- *The attributes of women matched with these men, $A(\mu(m_1)), \ldots, A(\mu(m_s))$*
- *The attributes and weights of s women, $A(w_1), \ldots, A(w_s), \alpha(w_1), \ldots, \alpha(w_s)$*
- *The attributes of men matched with these women, $A(\mu(w_1)), \ldots, A(\mu(w_s))$*

The output of the polynomial is 1 if and only if there is a blocking pair with respect to the matching μ among the s^2 pairs in the input. The number of monomials is at most $n^{0.17}$ and the polynomial can be constructed efficiently.

Given such a probabilistic polynomial, we can decide if a given matching is stable by evaluating the polynomial on $\left(\frac{n}{s}\right)^2$ inputs efficiently.

Theorem 5. *In the d-attribute model with n men and women, and $d = c \log n$ boolean attributes and weights, there is a randomized algorithm to decide if a given matching is stable in time $\tilde{O}(n^{2-1/O(c \log^2(c))})$ with error probability at most $1/3$.*

6   Conditional Hardness

6.1   Background

The Strong Exponential Time Hypothesis has proved useful in arguing conditional hardness for a large number of problems. We show SETH-hardness for both verifying and finding a stable matching in the d-attribute model, even if the weights and attributes are boolean.

Definition 1 [18,20]. *The Strong Exponential Time Hypothesis (SETH) stipulates that for each $\varepsilon > 0$ there is a k such that k-SAT requires $\Omega(2^{(1-\varepsilon)n})$ time.*

The main step of the proof is a reduction from the maximum inner product problem, which is known to be SETH-hard. For any d and input l, the *maximum inner product problem* is to decide if two input sets $U, V \subseteq \mathbb{R}^d$ with $|U| = |V| = n$ have a pair $u \in U$, $v \in V$ such that $\langle u, v \rangle \geq l$. The *boolean maximum inner product problem* is the variant where $U, V \subseteq \{0,1\}^d$.

Lemma 6 [3,20,30]. *Assuming SETH, for any $\varepsilon > 0$, there is a c such that solving the boolean maximum inner product problem on $d = c \log n$ dimensions requires time $\Omega(n^{2-\varepsilon})$.*

6.2   Finding Stable Matchings

In this subsection we give a fine-grained reduction from the maximum inner product problem to the problem of finding a stable matching in the boolean d-attribute model. This shows that the stable matching problem in the d-attribute model is SETH-hard, even if we restrict the attributes and weights to booleans.

Theorem 6. *Assuming SETH, for any $\varepsilon > 0$, there is a c such that finding a stable matching in the boolean d-attribute model with $d = c \log n$ dimensions requires time $\Omega(n^{2-\varepsilon})$.*

Proof. The proof is a reduction from maximum inner product to finding a stable matching. Given an instance of the maximum inner product problem with sets $U, V \subseteq \{0,1\}^d$ where $|U| = |V| = n$ and threshold l, we construct a matching market with n men and n women. For every $u \in U$ we have a man m_u with $A(m_u) = u$ and $\alpha(m_u) = u$. Similarly, for vectors $v \in V$ we have women w_v with $A(w_v) = v$ and $\alpha(w_v) = v$. This matching market is symmetric in the sense that for m_u and w_v, $\text{val}_{m_u}(w_v) = \text{val}_{w_v}(m_u) = \langle u, v \rangle$.

We claim that any stable matching contains a pair (m_u, w_v) such that the inner product $\langle u, v \rangle$ is maximized. Indeed, suppose there are vectors $u \in U$, $v \in V$ with $\langle u, v \rangle \geq l$ but there exists a stable matching μ with $\langle u', v' \rangle < l$ for all pairs $(m_{u'}, w_{v'}) \in \mu$. Then (m_u, w_v) is clearly a blocking pair for μ which is a contradiction.

6.3   Verifying Stable Matchings

In this section we give a reduction from the maximum inner product problem to the problem of verifying a stable matching, showing that this problem is also SETH-hard.

Theorem 7. *Assuming* SETH, *for any $\varepsilon > 0$, there is a c such that verifying a stable matching in the boolean d-attribute model with $d = c \log n$ dimensions requires time $\Omega(n^{2-\varepsilon})$.*

Proof. We give a reduction from maximum inner product with sets $U, V \subseteq \{0,1\}^d$ where $|U| = |V| = n$ and threshold l. We construct a matching market with $2n$ men and women in the d'-attribute model with $d' = d + 2(l - 1)$. Since $d' < 3d$ the theorem then follows immediately from the SETH-hardness of maximum inner product.

For $u \in U$, let m_u be a man in the matching market with attributes and weights $A(m_u) = \alpha(m_u) = u \circ 1^{l-1} \circ 0^{l-1}$ where we use $\circ$ for concatenation. Similarly, for $v \in V$ we have women w_v with $A(w_v) = \alpha(w_v) = v \circ 0^{l-1} \circ 1^{l-1}$. We further introduce *dummy women* w'_u for $u \in U$ with $A(w'_u) = \alpha(w'_u) = 0^d \circ 1^{l-1} \circ 0^{l-1}$ and *dummy men* m'_v for $v \in V$ with $A(m'_v) = \alpha(m'_v) = 0^d \circ 0^{l-1} \circ 1^{l-1}$.

We claim that the matching consisting of pairs (m_u, w'_u) for all $u \in U$ and (m'_v, w_v) for all $v \in V$ is stable if and only if there is no pair $u \in U, v \in V$ with $\langle u, v \rangle \geq l$. For $u, u' \in U$ we have $\mathrm{val}_{m_u}(w'_{u'}) = \mathrm{val}_{w'_{u'}}(m_u) = l - 1$, and for $v, v' \in V$ we have $\mathrm{val}_{w_v}(m'_{v'}) = \mathrm{val}_{m'_{v'}}(w_v) = l - 1$. In particular, any pair in μ has (symmetric) value $l - 1$. Hence there is a blocking pair with respect to μ if and only if there is a pair with value at least l. For $u \neq u'$ and $v \neq v'$ the pairs $(m_u, w'_{u'})$ and $(w_v, m'_{v'})$ can never be blocking pairs as their value is $l - 1$. Furthermore for any pair of dummy nodes w'_u and m'_v we have $\mathrm{val}_{m'_v}(w'_u) = \mathrm{val}_{w'_u}(m'_v) = 0$, thus no such pair can be a blocking pair either. This leaves pairs of real nodes as the only candidates for blocking pairs. For non-dummy nodes m_u and w_v we have $\mathrm{val}_{m_u}(w_v) = \mathrm{val}_{w_v}(m_u) = \langle u, v \rangle$ so (m_u, w_v) is a blocking pair if and only if $\langle u, v \rangle \geq l$.

7   Conclusion and Open Problems

We give subquadratic algorithms for finding and verifying stable matchings in the d-attribute model and d-list model. We also show that, assuming SETH, one can only hope to find such algorithms if the number of attributes d is bounded by $O(\log n)$.

For a number of cases there is a gap between the conditional lower bound and the upper bound. Our algorithms with real attributes and weights are only subquadratic if the dimension is constant. It would be interesting to either close or explain this gap. Even for small constants our algorithm to find a stable matching is not tight, as it is not subquadratic for any $d = O(\log n)$. The techniques we use when the attributes and weights are small constants do not readily apply to the more general case.

We also lack a subquadratic time algorithm for the problem of finding a stable matching with general real attributes and weights. Even for the arbitrary 2-list case we do not currently have a subquadratic algorithm. This 2-list case seems to be a good starting place for further research.

Acknowledgment. We would like to thank Russell Impagliazzo, Vijay Vazirani, and the anonymous reviewers for helpful discussions and comments.

References

1. Abboud, A., Backurs, A., Williams, V.V.: Quadratic-time hardness of LCS and other sequence similarity measures. In: 2015 IEEE 56th Annual Symposium on Foundations of Computer Science (FOCS). IEEE (2015)
2. Agarwal, P.K., Erickson, J., et al.: Geometric range searching and its relatives. Contemp. Math. **223**, 1–56 (1999)
3. Alman, J., Williams, R.: Probabilistic polynomials and hamming nearest neighbors. In: 2015 IEEE 56th Annual Symposium on Foundations of Computer Science (FOCS). IEEE (2015)
4. Arkin, E.M., Bae, S.W., Efrat, A., Okamoto, K., Mitchell, J.S., Polishchuk, V.: Geometric stable roommates. Inf. Process. Lett. **109**(4), 219–224 (2009)
5. Backurs, A., Indyk, P.: Edit distance cannot be computed in strongly subquadratic time (unless SETH is false). In: Proceedings of the Forty-Seventh Annual ACM on Symposium on Theory of Computing, STOC 2015, Portland, OR, USA, 14–17 June 2015, pp. 51–58 (2015)
6. Bartholdi, J., Trick, M.A.: Stable matching with preferences derived from a psychological model. Oper. Res. Lett. **5**(4), 165–169 (1986)
7. Bhatnagar, N., Greenberg, S., Randall, D.: Sampling stable marriages: why spouse-swapping won't work. In: Proceedings of the Nineteenth Annual ACM-SIAM Symposium on Discrete Algorithms. Society for Industrial and Applied Mathematics, pp. 1223–1232 (2008)
8. Bogomolnaia, A., Laslier, J.F.: Euclidean preferences. J. Math. Econ. **43**(2), 87–98 (2007)
9. Bringmann, K.: Why walking the dog takes time: Fréchet distance has no strongly subquadratic algorithms unless seth fails. In: 2014 IEEE 55th Annual Symposium on Foundations of Computer Science (FOCS), pp. 661–670. IEEE (2014)
10. Carmosino, M.L., Gao, J., Impagliazzo, R., Mihajlin, I., Paturi, R., Schneider, S.: Nondeterministic extensions of the strong exponential time hypothesis and consequences for non-reducibility. In: Proceedings of the 2016 ACM Conference on Innovations in Theoretical Computer Science, pp. 261–270. ACM (2016)
11. Chebolu, P., Goldberg, L.A., Martin, R.: The complexity of approximately counting stable matchings. In: Serna, M., Shaltiel, R., Jansen, K., Rolim, J. (eds.) APPROX and RANDOM 2010. LNCS, vol. 6302, pp. 81–94. Springer, Heidelberg (2010)
12. Dabney, J., Dean, B.C.: Adaptive stable marriage algorithms. In: Proceedings of the 48th Annual Southeast Regional Conference, p. 35. ACM (2010)
13. Dobkin, D.P., Kirkpatrick, D.G.: A linear algorithm for determining the separation of convex polyhedra. J. Algorithms **6**(3), 381–392 (1985)
14. Gale, D., Shapley, L.S.: College admissions and the stability of marriage. Am. Math. Mon. **69**(1), 9–15 (1962)

15. Gonczarowski, Y.A., Nisan, N., Ostrovsky, R., Rosenbaum, W.: A stable marriage requires communication. In: Proceedings of the Twenty-Sixth Annual ACM-SIAM Symposium on Discrete Algorithms, pp. 1003–1017. SIAM (2015)
16. Gusfield, D., Irving, R.W.: The Stable Marriage Problem: Structure and Algorithms. Foundations of Computing Series. MIT Press, Cambridge (1989)
17. Hershberger, J., Suri, S.: A pedestrian approach to ray shooting: shoot a ray, take a walk. J. Algorithms 18(3), 403–431 (1995)
18. Impagliazzo, R., Paturi, R.: On the complexity of k-SAT. J. Comput. Syst. Sci. 62(2), 367–375 (2001)
19. Impagliazzo, R., Paturi, R., Schneider, S.: A satisfiability algorithm for sparse depth two threshold circuits. In: 2013 IEEE 54th Annual Symposium on Foundations of Computer Science (FOCS), pp. 479–488. IEEE (2013)
20. Impagliazzo, R., Paturi, R., Zane, F.: Which problems have strongly exponential complexity? J. Comput. Syst. Sci. 63, 512–530 (2001)
21. Knuth, D.E.: Stable Marriage and Its Relation to Other Combinatorial Problems: An Introduction to the Mathematical Analysis of Algorithms, vol. 10. American Mathematical Society, Providence (1997)
22. Matoušek, J.: Efficient partition trees. Discrete Comput. Geom. 8(1), 315–334 (1992)
23. Matoušek, J., Schwarzkopf, O.: Linear optimization queries. In: Proceedings of the Eighth Annual Symposium on Computational Geometry, pp. 16–25. ACM (1992)
24. Ng, C., Hirschberg, D.S.: Lower bounds for the stable marriage problem and its variants. SIAM J. Comput. 19(1), 71–77 (1990)
25. Patrascu, M., Williams, R.: On the possibility of faster SAT algorithms. In: Proceedings of the Twenty-First Annual ACM-SIAM Symposium on Discrete Algorithms, SODA 2010, Austin, Texas, USA, 17–19 January 2010, pp. 1065–1075 (2010)
26. Razborov, A.A.: Lower bounds on the size of bounded depth circuits over a complete basis with logical addition. Math. Notes 41(4), 333–338 (1987)
27. Roth, A.E., Sotomayor, M.A.O.: Two-sided Matching: A Study in Game - Theoretic Modeling and Analysis. Econometric Society Monographs. Cambridge University, Cambridge (1990)
28. Segal, I.: The communication requirements of social choice rules and supporting budget sets. J. Econ. Theory 136(1), 341–378 (2007)
29. Smolensky, R.: Algebraic methods in the theory of lower bounds for boolean circuit complexity. In: Proceedings of the Nineteenth Annual ACM Symposium on Theory of Computing, pp. 77–82. ACM (1987)
30. Williams, R.: A new algorithm for optimal constraint satisfaction and its implications. In: Díaz, J., Karhumäki, J., Lepistö, A., Sannella, D. (eds.) ICALP 2004. LNCS, vol. 3142, pp. 1227–1237. Springer, Heidelberg (2004)
31. Williams, R.: Faster all-pairs shortest paths via circuit complexity. In: Proceedings of the 46th Annual ACM Symposium on Theory of Computing, pp. 664–673. ACM (2014)

Depth-4 Identity Testing and Noether's Normalization Lemma

Partha Mukhopadhyay$^{(\boxtimes)}$

Chennai Mathematical Institute, Chennai, India
partham@cmi.ac.in

Abstract. We consider the *black-box* polynomial identity testing (PIT) problem for a sub-class of depth-4 $\Sigma\Pi\Sigma\Pi(k,r)$ circuits. Such circuits compute polynomials of the following type: $C(X) = \sum_{i=1}^{k}\prod_{j=1}^{d_i} Q_{i,j}$, where k is the fan-in of the top Σ gate and r is the maximum degree of the polynomials $\{Q_{i,j}\}_{i\in[k],j\in[d_i]}$, and $k,r = O(1)$. We consider a subclass of such circuits satisfying a *generic* algebraic-geometric restriction, and we give a deterministic polynomial-time black-box PIT algorithm for such circuits.

Our study is motivated by two recent results of Mulmuley (FOCS 2012, [Mul12]), and Gupta (ECCC 2014, [Gup14]). In particular, we obtain the derandomization by solving a particular instance of derandomization problem of Noether's Normalization Lemma (NNL). Our result can also be considered as a unified way of viewing the depth-4 PIT problems closely related to the work of Gupta [Gup14], and the approach suggested by Mulmuley [Mul12].

1 Introduction

Polynomial Identity Testing (PIT) is the following problem: Given an arithmetic circuit C computing a polynomial in $\mathbb{F}[x_1,\ldots,x_n]$, decide whether $C(X) \equiv 0$ or not. The problem can be presented either in *white-box* model or in *black-box* model. In the white-box model, the arithmetic circuit is given explicitly as the input. In the black-box model, the arithmetic circuit is given as a black-box, and the circuit can be evaluated over any point in the field (or in a suitable extension field). Over the years, the problem has played pivotal role in many important results in complexity theory and algorithms: Primality Testing [AKS04], the PCP Theorem [ALM+98], IP = PSPACE [Sha90], graph matching algorithms [Lov79,MVV87]. The problem PIT admits a co-RP algorithm via the Schwartz-Zippel Lemma [Sch80,Zip79], but an efficient derandomized algorithm is not known.

An important result of Impagliazzo and Kabanets [KI04] (also, see [HS80]) showed a connection between the derandomization of PIT and arithmetic circuit lower bound. In particular, it is now known that if PIT can be derandomized using a certain type of pseudo-random generator, then the Permanent polynomial can not be computed by a polynomial-size arithmetic circuit [Agr05,KI04]. As a result, it will prove the algebraic analogue of P vs NP problem: VP $\neq$ VNP.

© Springer International Publishing Switzerland 2016
A.S. Kulikov and G.J. Woeginger (Eds.): CSR 2016, LNCS 9691, pp. 309–323, 2016.
DOI: 10.1007/978-3-319-34171-2_22

We refer the reader to the survey of Shpilka and Yehudayoff [SY10] for the exposition to many important results in arithmetic circuit complexity, and polynomial identity testing problem.

In a surprising result, Agrawal and Vinay [AV08] showed that the derandomization of PIT only for depth-4 $\Sigma\Pi\Sigma\Pi$ circuits is sufficient to derandomize the PIT for the general arithmetic circuits. The main technical ingredient in their proof is an ingenious depth-reduction technique. As a result, it is now known that a sufficiently strong lower bound for only $\Sigma\Pi\Sigma\Pi$ circuits (even for depth-3 $\Sigma\Pi\Sigma$ circuits over large fields [GKKS13b]) will separate VP from VNP. Currently, there are many impressive partial results in this direction showing depth four lower bounds for explicit polynomials in VP and in VNP. All these papers use the shifted partial derivative technique (for example, see [GKKS13a]).

Motivated by the results of [KI04, Agr05, AV08], a large body of works consider the polynomial identity testing problem for restricted classes of depth-3 and depth-4 circuits. A particularly popular model in depth three arithmetic circuits is $\Sigma\Pi\Sigma(k)$ circuit, where the fan-in of the top Σ gate is bounded. Dvir-Shpilka showed a *white-box* quasi-polynomial time deterministic PIT algorithm for $\Sigma\Pi\Sigma(k)$ circuits [DS07]. Kayal-Saxena gave a polynomial-time white-box algorithm for the same problem [KS07]. Following the result of [KS07], Arvind-Mukhopadhyay gave a somewhat simpler algorithm of same running time [AM10]. Karnin and Shpilka gave the first *black-box* quasi-polynomial time algorithm for $\Sigma\Pi\Sigma(k)$ circuits [KS11]. Later, Kayal and Saraf [KS09] gave polynomial-time deterministic black-box PIT algorithm for the same class of circuits over $\mathbb{Q}$ or $\mathbb{R}$. Finally, Saxena and Sheshadhri settled the situation completely by giving a deterministic polynomial-time *black-box* algorithm for $\Sigma\Pi\Sigma(k)$ circuits [SS12] over any field.

For $\Sigma\Pi\Sigma\Pi$ circuits, relatively a fewer deterministic algorithms are known. Just like in depth three, in depth four also the model $\Sigma\Pi\Sigma\Pi(k)$ is of considerable interest (where the top Σ gate is of bounded fan-in). Karnin et al. showed a quasi-polynomial time black-box identity testing algorithm for *multilinear* $\Sigma\Pi\Sigma\Pi(k)$ circuits [KMSV13]. Later, Saraf and Volkovich improved it to a deterministic polynomial time algorithm [SV11]. In 2013, Beecken et al. [BMS13] considered an algebraic restriction on $\Sigma\Pi\Sigma\Pi(k)$ circuits: bounded *transcendence degree*, and they showed an efficient deterministic black-box algorithm for such a class of circuits. Finally, Agrawal et al. showed that all these results can be proved under a unified framework using *Jacobian Criterion* [ASSS12].

Now we briefly discuss the recent results of Mulmuley [Mul12] and Gupta [Gup14]. Noether's Normalization Lemma (NNL) is a fundamental result in algebraic geometry. Recently, Mulmuley observed a close connection between a certain formulation of *derandomization of* NNL, and the problem of showing explicit circuit lower bounds in arithmetic complexity [Mul12]. His main result is that these seemingly different looking problems are computationally equivalent. We explain the setting briefly.

Let $V \subseteq \mathbb{P}(\mathbb{C}^n)$ be any projective variety and $\dim V = m$. Then any homogeneous and random (generic) linear map $\Psi : \mathbb{P}^n \to \mathbb{P}^m$ restricts to a finite-to-one surjective closed map: $\Psi : V \to \mathbb{P}^m$. By derandomization of NNL, we mean an explicit construction of the map Ψ for all *explicit* variety. Mulmuley showed that this problem is equivalent to the problem of black-box derandomization of polynomial identity testing (PIT) [Mul12]. Here we note that efficient explicit derandomization of NNL is known for particular explicit varieties [FS13a]. This result is closely related to the breakthrough result of the same authors where they first showed a quasi-polynomial time *black-box* derandomization of non-commutative ABPs [FS13b].

In a recent work, Gupta [Gup14] takes a fresh approach to the black-box identity testing of depth-4 circuits. He considers a class of depth-4 circuits denoted by $\Sigma\Pi\Sigma\Pi(k,r)$. Such a circuit C computes a polynomial (over a field $\mathbb{F}$) of the following form: $C(X) = \sum_{i=1}^{k} Q_i = \sum_{i=1}^{k} \prod_{j=1}^{d_i} Q_{i,j}(x_1, \ldots, x_n)$, where $Q_{i,j}$s are polynomials over $\mathbb{F}$ of maximum degree r and $\{x_1, x_2, \ldots, x_n\}$ are the variables appearing in the polynomial, and $k, r = O(1)$. It is an open problem to find an efficient deterministic black-box algorithm to identity test the circuit class $\Sigma\Pi\Sigma\Pi(k,r)$. Gupta considers an interesting sub-class of $\Sigma\Pi\Sigma\Pi(k,r)$ circuits by applying an algebraic-geometric restriction which he defines as *Sylvester-Gallai* property. By standard reasoning it can be assumed that Q_{ij}s are homogeneous polynomials over the variables $x_0, \ldots, x_n$. For that reason we can just work in a projective space $\mathbb{P}^n$ over $\mathbb{C}$. The circuit C is not Sylvester-Gallai (SG) if the following property is true:

$$\exists i_k \in [k] : V(Q_{i_1}, \ldots, Q_{i_{k-1}}) \not\subseteq V(Q_{i_k}). \tag{1}$$

It is easy to observe that such a circuit class is *generic* in the sense that, when the polynomials $Q_{i,j}$s are selected uniformly and independently at random, the circuit is not SG with high probability. The indices $i_1, \ldots, i_{k-1}$ are the indices in $[k] \setminus \{i_k\}$. Gupta gives an efficient deterministic polynomial-time algorithm for polynomial identity testing for such a class of depth-4 circuits. He further conjectures that if C is SG, then the transcendence degree of the polynomials Q_{ij}s is $O(1)$. Then one can use the result of [BMS13], to solve the problem completely. His algorithm is interesting for several reasons. Firstly, his approach gives a clean and systematic algebraic-geometric approach to an interesting sub-class of depth-4 identity testing. Secondly, the algorithm connects the classical algebraic-geometric results such as Bertini's Theorem, and Ideal Membership Testing to the PIT problem for depth four circuits. We note that for $\Sigma\Pi\Sigma(k)$ circuits, Arvind-Mukhopadhyay [AM10] used ideal membership testing to give a simplified and alternative proof of Kayal-Saxena's algorithm [KS07].

Our Results

The main motivation of our study comes from the work of Mulmuley [Mul12], and from the work of Gupta [Gup14]. In this paper, we try to connect their approaches from a conceptual perspective. More precisely, we try to answer the following question: is there an interesting sub-class of $\Sigma\Pi\Sigma\Pi(k,r)$ circuits for which we can find a black-box polynomial-time deterministic PIT algorithm by derandomizing a special instance of NNL? We give an affirmative answer. One of our key ideas is to start from a slightly different assumption (than Gupta's assumption) on the algebraic structure of the circuit, which is still generic.

The Class of Circuits $(\mathcal{C})$: Let $\max\{d_i\} \le D$. The family of $\Sigma\Pi\Sigma\Pi(k,r)$ circuits that we consider has the following property $\mathcal{P}$. There exists distinct indices $i_1, i_2, \ldots, i_{k-1} \in [k]$, and $j_1, j_2, \ldots, j_{k-1} \in [D]$ such that $\forall S \subseteq [D]$ of size at most r^k, the following is true: In $\mathbb{P}^{n1}$,

$$\dim(V(Q_{i_1,j_1}, \ldots, Q_{i_{k-1},j_{k-1}}, \prod_{j_k \in S} Q_{i_k,j_k})) < \dim(V(Q_{i_1,j_1}, \ldots, Q_{i_{k-1},j_{k-1}})).$$
(2)

It is shown in Lemma 3 that the family $\mathcal{C}$ does not contain the identically zero polynomial. Our main result is the following.

Theorem 1. *Suppose that we are given an arithmetic circuit C by a black-box with the promise that either $C \in \mathcal{C}$ or $C \equiv 0$. Then there is a deterministic polynomial-time algorithm that always decides correctly whether $C \equiv 0$ or not and runs in time $(D \cdot n)^{r^{O(k^2)}}$. Moreover, the derandomization is achieved by solving a special instance of NNL deterministically.*

It is straightforward to observe that the class of circuits that we consider (for identity testing) is subsumed by the result of Gupta [Gup14]. By Observation 1 and Hilbert's Nullstellensatz, for each $S \subseteq [D]$ of size at most r^k, we have that $\prod_{j_k \in S} Q_{i_k,j_k} \notin \sqrt{\langle Q_{i_1,j_1}, \ldots, Q_{i_{k-1},j_{k-1}} \rangle}$. By the Lemma 2, it follows that $Q_{i_k} \notin \sqrt{\langle Q_{i_1}, \ldots, Q_{i_{k-1}} \rangle}$. Again, applying Hilbert's Nullstellensatz, we observe that $V(Q_{i_1}, \ldots, Q_{i_{k-1}}) \not\subseteq V(Q_{i_k})$. In this work, our main motivation is to find deterministic identity testing algorithm for a generic class of depth-4 circuits, which can be obtained by efficient derandomization of some special instance of Noether's Normalization Lemma. In this context, our result should be seen as a very tiny step where Mulmuley's formulation to attack general PIT is implemented for a subclass of depth four circuits. Another difference of our work with [Gup14] is that we avoid the use of deep results from algebraic geometry like Bertini's second theorem and sharp nullstellensatz result of Dubé.

Some Comments About the Circuit Class $\mathcal{C}$: Why the circuit class $\mathcal{C}$ is interesting? The fact that the top Σ gate fan-in k is $O(1)$ is not a serious complain. Even for depth-3 results such restriction is there. If $r = 1$ the situation is already complicated for identity testing and it is finally resolved after a series of serious works as we have already mentioned [DS07,KS07,KS09,KS11,SS12]. To the best of our knowledge, the identity testing for $\Sigma\Pi\Sigma\Pi(k,2)$ is still open. Given the situation, a natural way to progress is to identify some structural weakness in some special subclass of $\Sigma\Pi\Sigma\Pi(k,r)$ circuits and try to exploit that for algorithm design. This is precisely the reason behind considering such

---

[1] We work in the projective spaces $\mathbb{P}^n$, since by standard reason the polynomials $Q_{i,j}$s can be assumed to be homogeneous polynomials over $n+1$ variables.

a model. We hope that when the circuit does not satisfy the property $\mathcal{P}$, there must be some intrinsic reason for that which can be exploited to tackle the situation completely. The fact that the property $\mathcal{P}$ is a generic property follows easily from the fact that when the polynomials $Q_{i,j}$ are selected uniformly and independently at random, the circuit will have the property $\mathcal{P}$ with high probability. The genericness of our circuit class should not be confused with the following algorithmic idea: Fix any point $\bar{a} \in \mathbb{P}^n$. Sample a circuit C uniformly from the class of circuits $\mathcal{C}$. Since it is a generic circuit class, it will be almost always the case that $C(\bar{a}) \neq 0$. So we can essentially ignore the input circuit. This situation has nothing to do with our problem. Notice that our algorithm never makes mistake. Thus, although the average case solution of the PIT problem for the circuit class $\mathcal{C}$ is trivial (which is the case for general PIT too), we do not see any completely elementary way to solve it in the worst case. Our algorithm still uses at least one classical concept from basic algebraic geometry: Noether's Normalization Lemma. Now we present the overview of our algorithm.

Variable Reduction. In this stage, our idea is to construct an explicit linear transformation T such that $C(\mathrm{X}) \equiv 0$ if and only if $C(T(\mathrm{X})) \equiv 0$, and that T transforms the polynomial computed by C over a fewer number of variables. In particular, $\forall i, T : x_i \mapsto L_i(y_0, \ldots, y_{2k-1})$ where L_i's are linear forms. We use Proposition 2 to argue that such a transformation always exists. The idea of this section is inspired by the work of Gupta in [Gup14] (in particular, Theorem 13, and Lemma 14), but with a key difference. Since our starting assumption on the circuit is different than Gupta's assumption, we do not need to use the classical result of Bertini directly (Theorem 13, [Gup14]).

Explicit Subspace Construction. In this stage, we find the sufficient algebraic conditions that the coefficients of the linear forms $L_i : 0 \leq i \leq n$ should satisfy. This section contains the main technical idea of our work, where we connect the problem to the derandomization of a particular instance of NNL. The idea is inspired by the work of [Mul12]. The main derandomization tool is the *multivariate resultant*. Using multivariate resultant, we reduce our problem to the problem of finding a hitting point of a product of a small number of *sparse* polynomials.

Hitting Set Construction. In this stage, we complete the algorithm by constructing a hitting set by applying the result of Klivans and Spielman [KS01]. More precisely, using the theory of multivariate resultant, and the hitting set construction of Klivans-Spielman, we argue that there is a small collection of points for the coefficients of L_is, such that at least for one such point $C(T(\mathrm{X})) \not\equiv 0$ if $C(\mathrm{X}) \not\equiv 0$. Moreover, $C(T(\mathrm{X}))$ is a polynomial over $O(1)$ variables and the individual degree of each variable is small. Then we can use the combinatorial nullstellensatz [Alo99] to find a small size hitting set for such a circuit.

The paper is organized as follows. In Sect. 1.1, we state the necessary results from algebraic geometry. In Sect. 1.2, we collect the necessary background from arithmetic complexity. We define our problem precisely in Sect. 2. The main algorithm and the proof of Theorem 1 are given in Sects. 3, 4, 4.1 and 5.

1.1 Algebraic Geometry

We recall the necessary background briefly. In this work, we focus in the setting of projective spaces. We recall the standard definition from Chap. 8 of [CLO07]. Complex projective n-space $\mathbb{P}^n$ is the set of $(n+1)$-tuples $a_0, \ldots, a_n$ of complex numbers, not all zero, modulo the equivalence relation: $(a_0, \ldots, a_n) \sim (\lambda a_0, \ldots, \lambda a_n)$, where $\lambda \in \mathbb{C} \setminus \{0\}$. A projective variety is the set of common zeros of a system of homogeneous polynomial equations in $\mathbb{P}^n$.

There are many ways to define the dimension of a variety. We use the following definition of dimension of a projective variety (Definition 10, page 453, [CLO07]).

Definition 1. *Dimension of $V \subseteq \mathbb{P}^n$, denoted by $\dim(V)$ is the degree of the Hilbert polynomial of the corresponding homogeneous ideal $I(V)$.*

Intuitively, r generic polynomials define a variety of dimension $n - r$. Co-dimension of a variety $V \subseteq \mathbb{P}^n$ is denoted by $\mathrm{codim}(V)$ and it is defined as $n - \dim(V)$. The following basic facts are useful for us. Those can be found in the standard text [CLO07].

Proposition 1. *The following facts extend our intuition from standard linear algebra:*

1. *Let $f \in \mathbb{C}[x_0, \ldots, x_n]$ be a non-zero homogenous polynomial. Then $\dim(V(f)) = n - 1$ in $\mathbb{P}^n$.*
2. *Let $V, W \subseteq \mathbb{P}^n$. If $V \subseteq W$ then $\dim(V) \leq \dim(W)$.*
3. *$V \cap W$ has co-dimension $\leq \mathrm{codim}(V) + \mathrm{codim}(W)$ in $\mathbb{P}^n$.*

We state the following version of Hilbert's Nullstellnesatz from Theorem 2 of Chap. 4 [CLO07]. Let $f, f_1, \ldots, f_s$ are polynomials in $\mathbb{C}[x_1 \ldots, x_n]$. Then, $V(f_1, f_2, \ldots, f_s) \subseteq V(f)$ if and only if $f \in \sqrt{\langle f_1, f_2, \ldots, f_s \rangle}$, where $\sqrt{\langle f_1, f_2, \ldots, f_s \rangle}$ is the radical generated by $f_1, \ldots, f_s$.

Noether's Normalization. We recall the following version of Noether's Normalization Lemma from the book by Mumford [Mum76]. Consider the following class of maps: $Y_i = \sum_{j=0}^n a_{i,j} X_j, 0 \leq i \leq r'$, be $r' + 1$ independent linear forms. Let $L \subset \mathbb{P}^n$ be the $(n - r' - 1)$-dimensional linear space $V(Y_0, \ldots, Y_{r'})$. Define the projection $p_L : \mathbb{P}^n - L \longrightarrow \mathbb{P}^{r'}$ by $(b_0, \ldots, b_n) \longrightarrow (\sum a_{0,j} b_j, \ldots, \sum a_{r',j} b_j)$.

Theorem 2 (Corollary 2.29, [Mum76]). *Let V be an r'-dimensional variety in $\mathbb{P}^n$. Then there is a linear subspace L of dimension $n - r' - 1$ such that $L \cap V$ is empty. For all such L, the projection p_L restricts to a finite-to-one surjective closed map: $p_L : V \longrightarrow \mathbb{P}^{r'}$, and the homogeneous coordinate ring $\mathbb{C}[X_0, \ldots, X_n]/I(V)$ of V is a finitely generated module over $\mathbb{C}[Y_0, \ldots, Y_{r'}]$.*

Notice that from the definition of dimension it follows easily that every $n - r'$ dimensional subspace intersects V non trivially. For example, it can be formally derived using the Proposition 7 in page 461 of [CLO07]. But the fact that there exists a $n - r' - 1$ dimensional subspace that *does not intersect* V is non-trivial[2].

One can derive the following algorithmically useful consequence of the above theorem (See Lemma 2.14, [Mul12]).

Lemma 1. *Let $V \subseteq \mathbb{P}^n$ be the variety defined by a set of homogeneous polynomials $f_1, \ldots, f_k \in \mathbb{C}[x_0, \ldots, x_n]$, and $\dim(V)$ be the dimension of V. Consider the random linear forms, $L_j(x) = \sum_{\ell=0}^{n} b_{\ell,j} x_\ell; 0 \le j \le s$. Let $H_j \subseteq \mathbb{P}^n$ be the hyperplane defined by $L_j(x) = 0$. If $s < \dim(V)$, then $V \cap \bigcap_j H_j \ne \phi$. If $s = \dim(V)$, then with high probability, $V \cap \bigcap_j H_j$ is empty.*

If $s = \dim(V)$, then Theorem 2 implies the existence of a linear subspace $L = \bigcap_j H_j$ of dimension $n - s - 1$ such that $V \cap L = \phi$. Lemma 1 implies that such a linear subspace L can be randomly picked with high probability.

Next, we recall the concept of multivariate resultant from Chap. 3 of [CLO05]. Suppose we have $n + 1$ homogeneous polynomials $F_0, F_1, \ldots, F_n$ in the variables $x_0, \ldots, x_n$, and assume that each F_i has positive total degree. We get $n + 1$ equations in $n + 1$ unknowns: $F_0(x_0, \ldots, x_n) = \ldots = F_n(x_0, \ldots, x_n) = 0$.

The multivariate resultant answers precisely the following question: what conditions must the coefficients of $F_0, \ldots, F_n$ satisfy in order that the system in equation above has a nontrivial solution. Suppose d_i be the total degree of F_i. Then F_i can be written as $F_i = \sum_{\alpha:|\alpha|=d_i} c_{i,\alpha} x^\alpha$. For each pair i, α, we introduce a variable $u_{i,\alpha}$. Now we are ready to state the following important Theorem.

Theorem 3 *(Theorem 2.3, [CLO05]). There is a unique irreducible polynomial $\text{Res}[u_{i,\alpha}] \in \mathbb{Z}[u_{i,\alpha}]$ such that the above system of polynomial equations has a nontrivial solution if and only if $\text{Res}[c_{i,\alpha}] = 0$ (i.e. we substitute $u_{i,\alpha}$ by $c_{i,\alpha}$).*

For our application, we need an upper bound on the degree of the polynomial $\text{Res}[u_{i,\alpha}]$. If $d_i \le d$ for $i \in [0; n]$, $\deg(\text{Res}) \le (n+1) \cdot d^n$ (Theorem 3.1, [CLO05]).

1.2 Arithmetic Complexity

An arithmetic circuit over a field $\mathbb{F}$ with the set of variables $x_1, x_2, \ldots, x_n$ is a directed acyclic graph such that the internal nodes are labelled by addition or multiplication gates and the leaf nodes are labelled by the variables or the field elements. The node with fan-out zero is the output gate. An arithmetic circuit computes a polynomial in the polynomial ring $\mathbb{F}[x_1, x_2, \ldots, x_n]$. Size of an arithmetic circuit is the number of nodes and the depth is the length of a longest path from the root to a leaf node.

Usually a depth-4 circuit over a field $\mathbb{F}$ is denoted by $\Sigma\Pi\Sigma\Pi$. The circuit has an addition gate at the top, then a layer of multiplication gates, followed by a layer of addition gates, and a bottom layer of multiplication gates. In this

---

[2] Notice that a zero dimensional variety contains a finite number of points.

work we focus on a class of $\Sigma\Pi\Sigma\Pi$ circuits that we denote by $\Sigma\Pi\Sigma\Pi(k,r)$, where k is the fan-in of the top Σ gate and r is the upper bound on the fan-in of the bottom Π gate. A $\Sigma\Pi\Sigma\Pi(k,r)$ circuit C computes a polynomial of the following form: $C(\mathrm{X}) = \sum_{i=1}^{k} Q_i = \sum_{i=1}^{k} \prod_{j=1}^{d_i} Q_{i,j}(x_1,\ldots,x_n)$, where $Q_{i,j}$s are polynomials over $\mathbb{F}$ and $\{x_1, x_2, \ldots, x_n\}$ are the variables appearing in the polynomial. In this work, we will consider depth four circuits with $k, r = O(1)$. We will also assume that $\forall i : d_i \leq D$. Also, we always assume that the circuit is given as a black-box.

We can homogenize the circuit w.r.t a new variable x_0 by obtaining the black-box for $C' = x_0^d C(\frac{x_1}{x_0}, \ldots, \frac{x_n}{x_0})$, where d is the degree of the polynomial computed by C. Clearly, $C' \equiv 0 \iff C \equiv 0$. We can also factorize the polynomials Q_{ij}s to their irreducible factors[3]. Since the degrees of the polynomials Q_{ij} are bounded by r, each Q_{ij} can be factored in at most r irreducible factors, increasing the fan-in of the Π-gate in the second layer by a factor r. We continue to use the notation C to represent the homogeneous circuit, and use D for the fan-in upper bound of the Π gates in the second layer.

We use the following version of Combinatorial Nullstellensatz from [Alo99].

Theorem 4. *Let $f(x_1, x_2, \ldots, x_n)$ be a polynomial in n variables over an arbitrary field $\mathbb{F}$. Suppose that the degree of f as a polynomial in x_i is at most t_i, for $1 \leq i \leq n$ and let $S_i \subseteq \mathbb{F}$ such that $|S_i| \geq t_i + 1$. If $f(a_1, a_2, \ldots, a_n) = 0$ for all n-tuples in $S_1 \times S_2 \times \cdots \times S_n$, then $f \equiv 0$.*

Let $\mathcal{C}$ be a family of arithmetic circuits computing n-variate polynomials over a field $\mathbb{F}$. A hitting set for $\mathcal{C}$ is a subset $\mathcal{H}$ of $\mathbb{F}^n$, such that for any non-zero circuit $C \in \mathcal{C}$, there exists $\boldsymbol{b} \in \mathcal{H}$ such that $C(\boldsymbol{b}) \neq 0$. If $\mathcal{H}$ can be constructed in deterministic polynomial time (in the input size), then we say that $\mathcal{H}$ is an efficiently computable explicit hitting set. The problem of black-box derandomization and efficient explicit hitting set construction are equivalent. A multivariate polynomial $f \in \mathbb{F}[x_1, \ldots, x_n]$ is t-sparse if it has at most t non-zero monomials.

For integers $a, b \geq 0$, the notation $[a; b] = \{x \in \mathbb{Z} : a \leq x \leq b\}$. We use the notation $C(\mathrm{X})$ to denote the multivariate polynomial output of a circuit C. Otherwise, we use the notation $Q(x)$ to denote a multivariate polynomial $Q(x_1, \ldots, x_n)$.

2   The Problem

The circuit C (computing a polynomial in $\mathbb{C}[x_0, \ldots, x_n]$) is given by a black-box and we need to test whether $C \equiv 0$ or not. Here we consider an assumption that either $C \equiv 0$ or C satisfies a *generic* property that we call the property $\mathcal{P}$. We recall it here from the introduction for the sake of reading.

---

[3] We do not explicitly use the fact that $Q_{i,j}$s are irreducible in the analysis. This fact is useful if we would like to formulate a conjecture in the similar spirit of Conjecture 1 in [Gup14].

We say that the circuit C satisfies the property $\mathcal{P}$, if there exist distinct indices $i_1, i_2, \ldots, i_{k-1} \in [k]$, and $j_1, j_2, \ldots, j_{k-1} \in [D]$ such that $\forall S \subseteq [D]$ of size at most r^k, the following is true.

In the projective space $\mathbb{P}^n$,

$$\dim(V(Q_{i_1,j_1}, \ldots, Q_{i_{k-1},j_{k-1}}, \prod_{j_k \in S} Q_{i_k,j_k})) < \dim(V(Q_{i_1,j_1}, \ldots, Q_{i_{k-1},j_{k-1}})).$$

$$(3)$$

The following observation is obvious from the above assumption.

Observation 1. *For all $S \subseteq [D]$ of size at most r^k, $V(Q_{i_1,j_1}, \ldots, Q_{i_{k-1},j_{k-1}}) \not\subseteq V(\prod_{j_k \in S} Q_{i_k,j_k})$. This is true for any $i_1, j_1 \ldots, i_{k-1}, j_{k-1}$, and i_k satisfying the condition in 3.*

For the simplicity, we will assume that (w.l.o.g) $i_1 = 1, \ldots, i_{k-1} = k-1, i_k = k$, and $j_1 = j_2 = \ldots = j_{k-1} = 1$. Using Bézout's theorem Gupta made the following simple but very useful observation.

Lemma 2 ([Gup14], **Claim 11**). *Let $P_1, \ldots, P_d, Q_1, \ldots, Q_k \in \mathbb{C}[x_0, \ldots, x_n]$ be homogeneous polynomials and degree of each Q_i is at most r. Then, $P_1 \ldots P_d \in \sqrt{\langle Q_1, \ldots, Q_k \rangle} \iff \exists \{i_1, \ldots, i_{r^k}\} \subseteq [d] : P_{i_1} \ldots P_{i_{r^k}} \in \sqrt{\langle Q_1, \ldots, Q_k \rangle}$.*

We use the above lemma to observe that if C satisfies property $\mathcal{P}$, then $C \not\equiv 0$.

Lemma 3. *If C is a circuit computing a polynomial that satisfies the property $\mathcal{P}$, then C can not compute an identically zero polynomial.*

Proof. By Hilbert Nullstellensatz and from Observation 1, $\prod_{j \in S} Q_{k,j} \notin \sqrt{\langle Q_{1,1}, Q_{2,1}, \ldots, Q_{k-1,1} \rangle}$ for any S of size at most r^k. Now using Lemma 2, we get that $Q_k \notin \sqrt{\langle Q_{1,1}, Q_{2,1}, \ldots, Q_{k-1,1} \rangle}$, which is not possible if $C \equiv 0$.

3 Variable Reduction Phase

The goal of this section is to find an efficiently computable explicit linear transformation T such that $C(\mathbf{X}) \equiv 0$ if and only if $C(T(\mathbf{X})) \equiv 0$, and $C(T(\mathbf{X}))$ is a polynomial over a fewer number of variables.

Let $Q_S = \prod_{j \in S} Q_{k,j}$. Recall that the subset S is of size at most r^k, and for each such S, $V(Q_{1,1}, Q_{2,1}, \ldots, Q_{k-1,1}) \not\subseteq V(Q_S)$. The total number of such sets are only $\leq D^{r^k}$ which is polynomially bounded for $r, k = O(1)$. Notice that $\forall S : \text{codim}(V(Q_{1,1}, Q_{2,1}, \ldots, Q_{k-1,1}, Q_S)) \leq k$ (Proposition 1). Now, we mention the following simple fact. It can be easily seen using the Exercise 8 in page 464 of [CLO07].

Proposition 2. *For a variety $V \subseteq \mathbb{P}^n$ of co-dimension c and a generic(random) linear subspace Λ of co-dimension $\leq n - c - 1$, $\text{codim}(V \cap \Lambda) = \text{codim}(V) + \text{codim}(\Lambda)$.*

It was first observed and used by Gupta [Gup14].

From Proposition 2, we know that for each S, $\exists$ a subspace Λ_S such that

$$\text{codim}(V(Q_{1,1},\ldots,Q_{k-1,1},Q_S,\Lambda_S)) = \text{codim}(V(Q_{1,1},\ldots,Q_{k-1,1},Q_S)) + \text{codim}(\Lambda_S),$$

and the dimension of $\Lambda_S = 2k - 1$[4].

Since the number of possible sets S is *small* (polynomially bounded), by an union bound, one can observe that $\exists \Lambda$ of dimension $2k - 1$ that satisfies the above property for all S simultaneously[5]. The following fact is also immediately clear.

Lemma 4. *There exists a subspace Λ of co-dimension $n - (2k - 1)$ in $\mathbb{P}^n$ such that, $\forall S \subseteq [D]$ of size at most r^k the following is true :* $\dim(V(Q_{1,1},\ldots,Q_{k-1,1},Q_S,\Lambda)) = \dim(V(Q_{1,1},\ldots,Q_{k-1,1},Q_S)) - (n - 2k + 1)$, *and* $\dim(V(Q_{1,1},\ldots,Q_{k-1,1},Q_S,\Lambda)) < \dim(V(Q_{1,1},\ldots,Q_{k-1,1},\Lambda))$.

The following is an easy observation.

Observation 2. *From Proposition 1, we get that in $\mathbb{P}^n$,* $\dim(V(Q_{1,1},\ldots,Q_{k-1,1})) \geq n - (k-1)$ *and* $\dim(V(Q_{1,1},\ldots,Q_{k-1,1},Q_S)) = \dim(V(Q_{1,1},\ldots,Q_{k-1,1})) - 1$ *for all subsets $S \subseteq [D]$ of size at most r^k.*

Now our goal is to explicitly construct the subspace Λ (of dimension $2k - 1$) such that $\forall S \subseteq [D]$ of size at most r^k, $\dim(V(Q_{1,1},\ldots,Q_{k-1,1},Q_S,\Lambda)) = \dim(V(Q_{1,1},\ldots,Q_{k-1,1},Q_S)) - (n - (2k - 1))$ in $\mathbb{P}^n$.

We fix a subspace Λ of co-dimension $n - (2k - 1)$ in $\mathbb{P}^n$ as follows. For each $i \in \{0, 1, \ldots, n\}$, set $x_i = \sum_{j=0}^{2k-1} a_{ij}y_j$, where $a_{ij} \in \mathbb{C}$ are the constants to be specialized later in the analysis. We define the matrix $A = (a_{i,j})_{0 \leq i \leq 2k-1, 0 \leq j \leq 2k-1}$. To ensure that the dimension of the subspace Λ is $2k - 1$, we will choose the constants in such a way that the symbolic determinant $\det(A)$ is non-zero.

After the above substitution, we identify the polynomials $Q_{1,1},\ldots,Q_{k-1,1},Q_S$ over the variables $y_0, y_1, \ldots, y_{2k-1}$ with coefficients as polynomials in $\mathbb{C}[\{a_{i,j}\}_{0 \leq i \leq n, 0 \leq j \leq 2k-1}]$. Notice that the degree of coefficient polynomials $\leq r^{k+1}$[6], and also the coefficient polynomials are $(2k(n + 1))^{r^{k+1}}$-sparse. In the next section, we fix the coefficients $\{a_{i,j}\}_{0 \leq i \leq n, 0 \leq j \leq 2k-1}$ explicitly, using an application of Noether's Normalization Lemma. To summarize, the transformation T does the following: $0 \leq i \leq n : T : x_i \rightarrow \sum_{j=0}^{2k-1} a_{ij}y_j$. After the substitution by the map T, we identify the variety $V(Q_{1,1},\ldots,Q_{k-1,1},\Lambda)$ by $V(Q_{1,1}(y),\ldots,Q_{k-1,1}(y))$. Also, for any subset S, we identify the variety $V(Q_{1,1},\ldots,Q_{k-1,1},Q_S,\Lambda)$ by $V(Q_{1,1}(y),\ldots,Q_{k-1,1}(y),Q_S(y))$.

---

[4] Notice that $\text{codim}(\Lambda_S) = n - (2k - 1) \leq n - k - 1$ for $k \geq 2$.

[5] In the next section, we show how to construct such a subspace deterministically.

[6] Recall that $\deg(Q_S) \leq r^{k+1}$ for any S.

4   An Explicit Subspace Construction

In $\mathbb{P}^n$, for each subset S, $n - k \leq \dim(V(Q_{1,1}, \ldots, Q_{k-1,1}, Q_S)) \leq n - 1$. Let $s_0 = \dim(V(Q_{1,1}, \ldots, Q_{k-1,1}, Q_S)) - (n - 2k + 1)$. Clearly $k - 1 \leq s_0 \leq 2k - 2$, but notice that we do not know the exact value of s_0. For each $s \in [k - 1; 2k - 2]$, we apply Lemma 1 to construct linear subspaces $\bigcap_{0 \leq j \leq s} H_j$ given by: $H_j(y)$: $L_j(y) = \sum_{\ell=0}^{2k-1} w_{\ell,j} y_\ell; 0 \leq j \leq s$, where $w_{\ell,j} \in \mathbb{C}$ are constants to be fixed. To ensure that the dimension of the varieties $V(Q_{1,1}(y), \ldots, Q_{k-1,1}(y), Q_S(y))$ are exactly s, we consider the multivariate resultant of the system of polynomial equations (for each fixed S): $Q_{1,1}(y) = \ldots = Q_{k-1,1}(y) = Q_S(y) = 0$, and $H_j(y) : L_j(y) = \sum_{\ell=0}^{2k-1} w_{\ell,j} y_\ell = 0; 0 \leq j \leq s$.

We use the following ideas from (Lemma 2.14, [Mul12]). It can be derived by applying the formulation of NNL in Lemma 1. For each S, we construct the following system of polynomials. $F_j^S(y) = \sum_{i=1}^{k-1} z_{i,j} Q_{i,1}(y) + z_j Q_S(y)$, $0 \leq j \leq (2k - 1) - (s + 1)$ and, $L_j(y) = \sum_{\ell=0}^{2k-1} w_{\ell,j} y_\ell$, $0 \leq j \leq s$. Notice that for each j and S, the polynomial F_j^S is a generic linear combination of the polynomials $Q_{1,1}, \ldots, Q_{k-1,1}, Q_S$.

Remark 1. From Lemma 2.11, and Lemma 2.14 of [Mul12], one observes that to implement the above idea, the polynomials $Q_{1,1}(y), \ldots, Q_{k-1,1}(y), Q_S(y)$ should be of same degree . This can be ensured by raising each $Q_{i,1}$ to the power $\left(\prod_{j=1}^{k-1} \deg(Q_{j,1}) \cdot \deg(Q_S) \right) / \deg(Q_{i,1})$. For the polynomial Q_S, we raise it to the power $\left(\prod_{j=1}^{k-1} \deg(Q_{j,1}) \right)$. So the final degree of each polynomial is at most $r^{O(k)}$.

For any subset S, the coefficients of the above system of polynomials (in the variables $y_0, \ldots, y_{2k-1}$) are polynomials in the ring $\mathbb{C}[\{a_{i,j}\}_{0 \leq i \leq n, 0 \leq j \leq 2k-1}, \{z_{ij}, z_j\}_{1 \leq i \leq k-1, 0 \leq j \leq (2k-1)-(s+1)}, \{w_{\ell,j}\}_{0 \leq \ell \leq 2k-1, 0 \leq j \leq s}]$.

So, the coefficients of the polynomials $Q_{1,1}(y), \ldots, Q_{k-1,1}(y), Q_S(y)$ can be viewed as polynomials with at most $N = ((2k(n + 1) + 2k(s + 1) + k(2k - 1 - s))$ variables and degree bounded by $r^{O(k)}$.

Now we use the estimate on the degree of the multivariate resultant polynomial given in Sect. 1.1.

Observation 3. *For each fixed S, the multivariate resultant polynomial corresponding to the above system of polynomials $\{F_j^S(y)\}_{0 \leq j \leq (2k-1)-(s+1)}$, and $\{L_j(y)\}_{0 \leq j \leq s}$ is a $\leq (2k) \cdot (r^{O(k)})^{2k} \cdot (r^{O(k)})$-degree polynomial in at most N variables.*

So the polynomials are $N^{r^{O(k^2)}}$-sparse. Also, the number of such resultant polynomials are bounded by D^{r^k}. Suppose we choose the subspace Λ (by fixing the indeterminates $\{a_{i,j}\}_{0 \leq i \leq n, 0 \leq j \leq 2k-1}$ over $\mathbb{C}$) in such a way that $s_0 = \dim(V(Q_{1,1}(y), \ldots, Q_{k-1,1}(y), Q_S(y)) = \dim(V(Q_{1,1}, \ldots, Q_{k-1,1}, Q_S) - (n - 2k + 1)$. Then if $s = s_0$, for each S, the resultant polynomial R_S is a non-identically zero polynomial. It follows as a consequence of NNL that the above system of polynomial equations has only trivial solution for some rational values of $z_{i,j}$'s, z_j's and $w_{\ell,j}$'s, when $s = s_0$. This point is discussed explicitly in Lemma 2.14 of [Mul12]. This also implies that the resultant polynomials $R_S(z_{i,j}, z_j, w_{\ell,j})$ are non identically zero. So, in particular, when $s = s_0$ and the subspace Λ is not fixed, the resultant polynomials $R_S(a_{i,j}, z_{i,j}, z_j, w_{\ell,j})$ are non identically zero polynomials.

Next, our idea is to specialize the values of the N indeterminates using the hitting set construction of Kilvans-Spielman [KS01] for the product of $N^{r^{O(k^2)}}$-sparse polynomials, so that all the resultant polynomials and the polynomial $\det(A)$, evaluate to nonzero at some point in the hitting set. In the next section, we explain that such an idea is sufficient for our problem.

4.1   The Correctness Proof

In Sect. 4, we repeat the construction for all possible values for the parameter $s \in [k-1; 2k-2]$. From the discussion in the last section, we know that if $s = s_0$, the resultant polynomials R_S are non identically zero polynomials.

Using the hitting set construction of [KS01], we can specialize the indeterminates so that for each S, the polynomial R_S and $\det(A)$ evaluate to nonzero. Once we fix the values for $\{a_{i,j}\}_{0\le i\le n, 0\le j\le 2k-1}$ to $\{a^*_{i,j}\}_{0\le i\le n, 0\le j\le 2k-1}$, we also define the subspace Λ. Moreover, since $\det(A) \ne 0$, we ensure that $\operatorname{codim}(\Lambda) = n - (2k-1)$.

Lemma 5. Let $(\{a^*_{i,j}\}, \{z^*_{ij}, z^*_j\}, \{w^*_{\ell,j}\})$ be a hitting point for $\prod_{S \in \binom{[D]}{r^k}} R_S \cdot \det(A)$. Then on such a point $\dim(V(Q_{1,1}(y), \ldots, Q_{k-1,1}(y), Q_S(y))) = s_0$.

Proof. If $\dim(V(Q_{1,1}(y), \ldots, Q_{k-1,1}(y), Q_S(y)))$ is more than s_0, then $\dim(V(\{F^S_j\}_{0\le\ell\le(2k-1)-(s_0+1)}))$ is also more than s_0. The dimension of the linear space defined by $\{L_j(y)\}_{0\le j\le s_0}$ is $\ge (2k-1)-(s_0+1)$. Then using the Definition 1, and the Proposition 7 in page 461 of [CLO07], we can easily see that the system of polynomials $\{F^S_j\}_{0\le\ell\le(2k-1)-(s_0+1)}, \{L_j(y)\}_{0\le j\le s_0}$ has a nontrivial solution and $R_S = 0$. Since the $\operatorname{codim}(\Lambda) = n - (2k-1)$ on the point $\{a^*_{i,j}\}_{0\le i\le n, 0\le j\le 2k-1}$, it is not possible that $\dim(V(Q_{1,1}(y), \ldots, Q_{k-1,1}(y), Q_S(y))) < s_0$.

From Lemma 5, it is obvious that for $s = s_0$ and on a hitting point, the following is true: $\forall S \subseteq [D]; |S| \le r^k : \dim(Q_{1,1}(y), \ldots, Q_{k-1,1}(y), Q_S(y)) < \dim(Q_{1,1}(y), \ldots, Q_{k-1,1}(y))$, which implies that, $\forall S \subseteq [D]; |S| \le r^k : V(Q_{1,1}(y), \ldots, Q_{k-1,1}(y)) \not\subseteq V(Q_S)$ (Observation 1). Finally we conclude that, $\forall S \subseteq [D]; |S| \le r^k : Q_S(y) \notin \sqrt{\langle Q_{1,1}(y), \ldots, Q_{k-1,1}(y)\rangle}$.

The last relation follows from the Hilbert Nullstellensatz. Now we apply Lemma 2, to deduce that $Q_k(y) \notin \sqrt{\langle Q_{1,1}(y), \ldots, Q_{k-1,1}(y)\rangle} \Rightarrow C(\mathrm{Y}) \not\equiv 0$. Recall that in the Remark 1, the degrees of $Q_{1,1}, \ldots, Q_{k-1,1}, Q_S$ are pretended to be increased by appropriate powering, is only for the analysis purpose.

The degree of each variable in $C(\mathrm{Y})$ is bounded by $D \cdot r$, and also $C(\mathrm{Y})$ is a $2k$-variate polynomial. We use Combinatorial Nullstellensatz (Theorem 4) to construct a hitting set for $C(\mathrm{Y})$. In the next section, we formally explain the construction of the final hitting set.

5   The Hitting Set Construction

Let $\mathcal{H}_{t,m,N,d} \subset \mathbb{C}^N$ be a hitting set for the product of $\leq m$ polynomials in N variables such that each polynomial is t-sparse, and of degree $\leq d$. One can construct $\mathcal{H}_{t,m,N,d} \subset \mathbb{C}^N$ efficiently following the result of Klivans and Spielman [KS01].

For each $s \in [k-1; 2k-2]$, we do the following. We use the hitting set $\mathcal{H}_s = \mathcal{H}_{N(s)^{r^{O(k^2)}}, D^{r^k}+1, N(s), r^{O(k^2)}}$ to substitute values to the indeterminates,

$$\{a_{i,j}\}_{0\leq i\leq n, 0\leq j\leq 2k-1}, \{z_{ij}, z_j\}_{1\leq i\leq k-1, 0\leq j\leq (2k-1)-(s+1)}, \{w_{\ell,j}\}_{0\leq \ell\leq 2k-1, 0\leq j\leq s}.$$

For each such substitution, we construct the subspace Λ by setting $\forall i \in [0; n]$: $x_i = \sum_{j=0}^{2k-1} a_{i,j} y_j$. Next we fix a set $\mathcal{S} \subset \mathbb{Q}$ such that $\mathcal{S} = D \cdot r + 1$, and test whether $C(Y)|_{y \in S^{2k}} = 0$. From the correctness proof (Sect. 4.1), we know that if $C(X) \not\equiv 0$, then for one of the subspaces Λ that we have constructed, $\exists b \in S^{2k}$ such that $C(b) \neq 0$.

The Final Algorithm

We state our final algorithm formally.

1. For each $s \in [k-1; 2k-2]$, we do the following.
   (a) For each point in the hitting set $\mathcal{H}_s$, specialize the values for $\{a_{ij}\}_{0\leq i\leq n, 0\leq j\leq 2k-1}$.
   (b) For each such specialization for $\{a_{ij}\}_{0\leq i\leq n, 0\leq j\leq 2k-1}$, construct the subspace Λ by substituting $0 \leq i \leq n : x_i = \sum_{j=0}^{2k-1} a_{ij} y_j$.
   (c) Check whether $C(Y)|_{y \in S^{2k}} = 0$, where $S = \{1, \ldots, D \cdot r + 1\}$. If anytime C evaluates to a non-zero value, we stop the procedure and announce that $C \not\equiv 0$.
2. Otherwise, output that $C \equiv 0$.

The cost of our algorithm is bounded by $(2k-2) \cdot \max_s |\mathcal{H}_s| \cdot (D \cdot r + 1)^{2k}$. Using the known estimate of $\max_s |\mathcal{H}_s|$, one can easily upper bound the cost by $(D \cdot n)^{r^{O(k^2)}}$. This also completes the proof of Theorem 1. Finally, what remains to tackle is the case when the circuit does not have the generic property $\mathcal{P}$. Can this structural information be exploited?

Acknowledgement. I thank K.V. Subrahmanyam for many helpful discussions.

References

[Agr05] Agrawal, M.: Proving lower bounds via pseudo-random generators. In: Sarukkai, S., Sen, S. (eds.) FSTTCS 2005. LNCS, vol. 3821, pp. 92–105. Springer, Heidelberg (2005)

[AKS04] Agrawal, M., Kayal, N., Saxena, N.: PRIMES is in P. Ann. Math. **160**(2), 781–793 (2004)

[ALM+98] Arora, S., Lund, C., Motwani, R., Sudan, M., Szegedy, M.: Proof verification and the hardness of approximation problems. J. ACM **45**(3), 501–555 (1998)

[Alo99] Alon, N.: Combinatorial Nullstellensatz. Comb. Probab. Comput. **8**, 7–30 (1999)

[AM10] Arvind, V., Mukhopadhyay, P.: The ideal membership problem and polynomial identity testing. Inf. Comput. **208**(4), 351–363 (2010)

[ASSS12] Agrawal, M., Saha, C., Saptharishi, R., Saxena, N.: Jacobian hits circuits: hitting-sets, lower bounds for depth-D occur-k formulas & depth-3 transcendence degree-k circuits. In: Proceedings of the 44th Symposium on Theory of Computing Conference, STOC 2012, pp. 599–614 (2012)

[AV08] Agrawal, M., Vinay, V.: Arithmetic circuits: a chasm at depth four. In: Proceedings-Annual Symposium on Foundations of Computer Science, pp. 67–75. IEEE (2008)

[BMS13] Beecken, M., Mittmann, J., Saxena, N.: Algebraic independence and blackbox identity testing. Inf. Comput. **222**, 2–19 (2013)

[CLO05] Cox, D.A., Little, J., O'Shea, D.: Using Algebraic Geometry. Springer, New York (2005)

[CLO07] Cox, D.A., Little, J., O'Shea, D.: Ideals, Varieties, and Algorithms: An Introduction to Computational Algebraic Geometry and Commutative Algebra. 3/e (Undergraduate Texts in Mathematics). Springer, New York (2007)

[DS07] Dvir, Z., Shpilka, A.: Locally decodable codes with two queries and polynomial identity testing for depth 3 circuits. SIAM J. Comput. **36**(5), 1404–1434 (2007)

[FS13a] Forbes, M.A., Shpilka, A.: Explicit Noether normalization for simultaneous conjugation via polynomial identity testing. In: Raghavendra, P., Raskhodnikova, S., Jansen, K., Rolim, J.D.P. (eds.) RANDOM 2013 and APPROX 2013. LNCS, vol. 8096, pp. 527–542. Springer, Heidelberg (2013)

[FS13b] Forbes, M.A., Shpilka, A.: Quasipolynomial-time identity testing of noncommutative and read-once oblivious algebraic branching programs. In: 54th Annual IEEE Symposium on Foundations of Computer Science, FOCS 2013, pp. 243–252 (2013)

[GKKS13a] Gupta, A., Kamath, P., Kayal, N., Saptharishi, R.: Approaching the chasm at depth four. In: IEEE Conference on Computational Complexity, pp. 65–73 (2013)

[GKKS13b] Gupta, A., Kamath, P., Kayal, N., Saptharishi, R.: Arithmetic circuits: a chasm at depth three. In: FOCS, pp. 578–587 (2013)

[Gup14] Gupta, A.: Algebraic geometric techniques for depth-4 PIT & Sylvester-Gallai conjectures for varieties. In: Electronic Colloquium on Computational Complexity (ECCC), vol. 21, p. 130 (2014)

[HS80] Heintz, J., Schnorr, C.-P.: Testing polynomials which are easy to compute (extended abstract). In: Proceedings of the 12th Annual ACM Symposium on Theory of Computing, pp. 262–272 (1980)

[KI04] Kabanets, V., Impagliazzo, R.: Derandomizing polynomial identity tests means proving circuit lower bounds. Comput. Complex. **13**(1–2), 1–46 (2004)

[KMSV13] Karnin, Z.S., Mukhopadhyay, P., Shpilka, A., Volkovich, I.: Deterministic identity testing of depth-4 multilinear circuits with bounded top fan-in. SIAM J. Comput. **42**(6), 2114–2131 (2013)

[KS01] Klivans, A., Spielman, D.A.: Randomness efficient identity testing of multivariate polynomials. In: Proceedings on 33rd Annual ACM Symposium on Theory of Computing, 6–8 July 2001, pp. 216–223 (2001)

[KS07] Kayal, N., Saxena, N.: Polynomial identity testing for depth 3 circuits. Comput. Complex. **16**(2), 115–138 (2007)

[KS09] Kayal, N., Saraf, S.: Blackbox polynomial identity testing for depth 3 circuits. In: 50th Annual IEEE Symposium on Foundations of Computer Science, FOCS 2009, pp. 198–207 (2009)

[KS11] Karnin, Z.S., Shpilka, A.: Black box polynomial identity testing of generalized depth-3 arithmetic circuits with bounded top fan-in. Combinatorica **31**(3), 333–364 (2011)

[Lov79] Lovász, L.: On determinants, matchings, and random algorithms. In: FCT, pp. 565–574 (1979)

[Mul12] Mulmuley, K.: Geometric complexity theory V: equivalence between blackbox derandomization of polynomial identity testing and derandomization of Noether's Normalization Lemma. In: CoRR (also, in FOCS 2012) (2012). abs/1209.5993

[Mum76] Mumford, D.: Algebraic Geometry I : Complex Projective Varieties. Springer, Heidelberg (1976). Grundlehren der mathematischen Wissenschaften

[MVV87] Mulmuley, K., Vazirani, U.V., Vazirani, V.V.: Matching is as easy as matrix inversion. Combinatorica **7**(1), 105–113 (1987)

[Sch80] Schwartz, J.T.: Fast probabilistic algorithms for verification of polynomial identities. J. ACM **27**(4), 701–717 (1980)

[Sha90] Shamir, A.: IP=PSPACE. In: 31st Annual Symposium on Foundations of Computer Science, St. Louis, Missouri, USA, 22–24 October 1990, vol. 1, pp. 11–15 (1990)

[SS12] Saxena, N., Seshadhri, C.: Blackbox identity testing for bounded top-fanin depth-3 circuits: the field doesn't matter. SIAM J. Comput. **41**(5), 1285–1298 (2012)

[SV11] Saraf, S., Volkovich, I.: Black-box identity testing of depth-4 multilinear circuits. In: Proceedings of the 43rd ACM Symposium on Theory of Computing, STOC 2011, pp. 421–430 (2011)

[SY10] Shpilka, A., Yehudayoff, A.: Arithmetic circuits: a survey of recent results and open questions. Found. Trends Theoret. Comput. Sci. **5**(3–4), 207–388 (2010)

[Zip79] Zippel, R.: Probabilistic algorithms for sparse polynomials. In: Symbolic and Algebraic Computation, EUROSAM 1979, An International Symposiumon Symbolic and Algebraic Computation, pp. 216–226 (1979)

Improved Approximation Algorithms for Min-Cost Connectivity Augmentation Problems

Zeev Nutov[✉]

The Open University of Israel, Ra'anana, Israel
nutov@openu.ac.il

Abstract. A graph G is k-connected if it has k internally-disjoint st-paths for every pair s, t of nodes. Given a root s and a set T of terminals is k-(s, T)-connected if it has k internally-disjoint st-paths for every $t \in T$. We consider two well studied min-cost connectivity augmentation problems, where we are given an integer $k \geq 0$, a graph $G = (V, E)$, and and an edge set F on V with costs. The goal is to compute a minimum cost edge set $J \subseteq F$ such that $G + J$ has connectivity $k + 1$. In the k-Connectivity Augmentation problem G is k-connected and $G + J$ should be $(k + 1)$-connected. In the k-(s, T)-Connectivity Augmentation problem G is k-(s, T)-connected and $G + J$ should be $(k + 1)$-(s, T)-connected.

For the k-Connectivity Augmentation problem we obtain the following results. For $n \geq 3k - 5$, we obtain approximation ratios 3 for directed graphs and 4 for undirected graphs, improving the previous ratio 5 of [26]. For directed graphs and $k = 1$, or $k = 2$ and n odd, we further improve to 2.5 the previous ratios 3 and 4, respectively.

For the undirected 2-(s, T)-Connectivity Augmentation problem we achieve ratio $4\frac{2}{3}$, improving the previous best ratio 12 of [24]. For the special case when all the edges in F are incident to s, we give a polynomial time algorithm, improving the ratio $4\frac{17}{30}$ of [21, 25] for this variant.

1 Introduction

1.1 Problems and Results

A graph is k-**connected** if it has k internally-disjoint paths from any of its nodes to any other node. A graph with a **root** s and a set T of **terminals** is k-(s, T)-**connected** if it has k internally-disjoint st-paths for every $t \in T$. We consider two extensively studied min-cost connectivity augmentation problems. In both problems we are given an integer $k \geq 0$, a graph $G = (V, E)$, and an edge set F on V with costs. The goal is to compute a minimum cost edge set $J \subseteq F$ such that $G + J$ has connectivity $k + 1$.

k-Connectivity Augmentation
Here G is k-connected and $G + J$ should be $(k + 1)$-connected.

© Springer International Publishing Switzerland 2016
A.S. Kulikov and G.J. Woeginger (Eds.): CSR 2016, LNCS 9691, pp. 324–339, 2016.
DOI: 10.1007/978-3-319-34171-2_23

> k-(s,T)-Connectivity Augmentation
> Here we are also given a root node s and a set $T \subseteq V$ of terminals, G is k-(s,T)-connected, and $G + J$ should be $(k+1)$-(s,T)-connected.

One important particular case of k-(s,T)-Connectivity Augmentation is when all edges of positive cost are incident to s. This variant is closely related to Source Location problems, see [12,21].

Both problems were studied extensively, see [1,3,4,7,10,12,17,19–21,24–26, 29] for only a small sample of papers in the area. For $k = 0$ and undirected graphs our problems include the Minimum Spanning Tree problem and the Steiner Tree problem; for directed graphs we get the Minimum Cost Strongly Connected Subgraph problem (that admits ratio 2 by taking a union of minimum cost in- and out-arborescences), and the Directed Steiner Tree problem.

We now state our results for the k-Connectivity Augmentation problem. Let $n = |V|$. In general, for both directed and undirected graphs, the problem admits ratio $O\left(\log \frac{n}{n-k}\right)$ (which is a constant unless $k = n - o(n)$), and also ratio $O(\log(n - k))$ [26]. Specifically, for $n \geq 3k - 5$, the previous best ratio was 5, for both directed and undirected graphs. For small values of k better ratios are known: $k + 2$ for $k \leq 2$ in the case of directed graphs [20], and $\lceil k/2 \rceil + 1$ for $k \leq 6$ in the case of undirected graphs [2,6,20]. We prove the following.

Theorem 1. k-Connectivity Augmentation *with* $n \geq 3k - 5$ *admits approximation ratio 3 for directed graphs and 4 for undirected graphs. Furthermore, for directed graphs the problem admits ratio 2.5 if* $k = 1$*, or if* $k = 2$ *and* n *is odd.*

For directed graphs our ratio improves over the previous ratios for any $k \geq 1$. For undirected graphs our ratio matches the best known ratio 4 for $k = 6, 7$, and it improves over the previous ratios for any $k \geq 8$.

Let $H(k)$ denote the kth harmonic number. The best known ratio for k-(s,T)-Connectivity Augmentation is $O(k \log k)$, and it was 12 for $k = 2$ [24]. For the version when all edges are incident to s the best ratio was $2H(2k+1)$ [21], which for $k = 2$ is $2H(5) = 4\frac{17}{30} > 4.5$. We consider the case $k = 2$, and significantly improve the previous best known ratios. Specifically, we prove the following.

Theorem 2. *Undirected* 2-(s,T)-Connectivity Augmentation *admits ratio* $4\frac{2}{3}$*; if all edges in* F *are incident to* s*, then the problem admits a polynomial time algorithm.*

In fact, our result for k-Connectivity Augmentation is more general that the one stated in Theorem 1. To state our generic result, we need some definitions. Let q be the largest integer such that $2q - 1 \leq n - k$, namely, $q = \lfloor \frac{n-k+1}{2} \rfloor$. Let

$$\mu = \left\lfloor \frac{n}{q+1} \right\rfloor = \left\lfloor \frac{n}{\lfloor (n-k+3)/2 \rfloor} \right\rfloor = \begin{cases} \left\lfloor \frac{2n}{n-k+3} \right\rfloor & \text{if } n - k \text{ is odd} \\[2ex] \left\lfloor \frac{2n}{n-k+2} \right\rfloor & \text{if } n - k \text{ is even} \end{cases}.$$

It is not hard to see that:

- $\mu = 1$ if and only if $k = 0$, or $k = 1$, or $k = 2$ and n is odd.
- $\mu = 2$ if and only if one of the following holds: $k = 2$ and n is even, or $k \geq 3$ and one of the following holds: $n \geq 3k - 8$ and n, k have distinct parities, or $n \geq 3k - 5$ and n, k have the same parity.
- $\mu \leq 3$ if and only if one of the following holds: $n \geq 2k - 5$ and n, k have distinct parities, or $n \geq 2k - 3$ and n, k have the same parity.

The previous best known approximation ratio for k-Connectivity Augmentation was $2H(\mu) + 2$ for both directed and undirected graphs [26], except the better ratios for small values of k listed above. We prove the following theorem, that implies Theorem 1; for comparison with previous ratios see Table 1.

Table 1. Previous and our ratios for k-Connectivity Augmentation; for $k = 2$ our ratio 2.5 for directed graphs is valid when n is odd.

Range	μ	$H(\mu)$	Directed		Undirected	
			Previous	This paper	Previous	This paper
$k = 0$	1	1	2		in P	
$k = 1, 2$	1	1	3, 4 [20]	2.5	2 [2,18]	
$3 \leq k \leq 6$	2	1.5	5 [26]	3	$\lceil k/2 \rceil + 1$ [6,20]	
$n \geq 3k - 5$	2	1.5	5 [26]	3	5 [26]	4
$n \geq 2k - 3$	3	$1\frac{5}{6}$	$5\frac{2}{3}$ [26]	$3\frac{1}{3}$	$5\frac{2}{3}$ [26]	$4\frac{2}{3}$
$n < 2k - 3$			$2H(\mu) + 2$ [26]	$H(\mu) + 1.5$	$2H(\mu) + 2$ [26]	$2H(\mu) + 1$

Theorem 3 (Implies Theorem 1). k-Connectivity Augmentation *admits the following approximation ratios:*

(i) *For directed graphs, ratio* $H(\mu) + \frac{3}{2}$. *In particular:*
- *For $k = 1$, and for $k = 2$ and n odd, $\mu = 1$, $H(\mu) = 1$, so the ratio is 2.5.*
- *For $n \geq 3k - 5$, $\mu \leq 2$, $H(\mu) \leq 3/2$, so the ratio is 3.*
- *For $n \geq 2k - 3$, $\mu \leq 3$, $H(\mu) \leq 11/6$, so the ratio is $3\frac{1}{3}$.*

(ii) *For undirected graphs, ratio* $2H(\mu) + 1$. *In particular, for $n \geq 3k - 5$, $\mu \leq 2$, $H(\mu) \leq 3/2$, so the ratio is 4.*

1.2 Techniques

Let us briefly describe how we achieve these improvements, starting with the k-Connectivity Augmentation problem. Given $J \subseteq F$ let us say that $A \subseteq V$ is a **tight set** if $|\Gamma(A)| = k$ and $A^* = V \setminus (A \cup \Gamma(A)) \neq \emptyset$, where $\Gamma(A)$ is the set of neighbors of A in $G + J$; A is a **small tight set** if $|A| \leq \lfloor \frac{n-k+1}{2} \rfloor$ and a **large tight set** otherwise. Inclusionwise minimal tight sets are called **cores**. By Menger's Theorem, the graph $G + J$ is $(k + 1)$-connected if and only if it has no cores. In the case of undirected graphs, if A is a large tight set then A^* is a small tight set, and thus just the absence of small cores already implies $(k+1)$-connectivity. In what follows, let τ denote the optimal value of the standard LP-relaxation for the problem, see Sect. 2.

Let us discuss the case of directed graphs. In this case, even if $G + J$ has no small cores, it may not be $(k+1)$-connected, since large cores may exist. We will show, by a novel proof, that instances with one small core admit ratio $3/2$; the previous ratio for this subproblem was $H(\mu)+2$. Obtaining an instance with one core is done in two steps. In the first step, we use an algorithm from [26] that for directed graphs computes an edge-set J_1 of cost $\leq \tau$ such that $G + J_1$ has at most μ small cores. In the second step, we use an algorithm of [7,26] to compute an edge set J_2 that reduces the number of small cores from μ to 1 by cost $\tau(H(\mu)-1)$. The overall ratio is therefore $1 + (H(\mu)-1) + 3/2 = H(\mu)+3/2$.

In the case of undirected graphs, $G+J$ is $(k+1)$-connected if and only if it has no small cores. We will show that instances with two small cores admit ratio 2; the previous ratio for this subproblem was 3. The other parts are similar to those of the directed case, but the bounds are $c(J_1) \leq 2\tau$ and $c(J_2) \leq 2\tau(H(\mu)-3/2)$. Thus the overall ratio we get is $2 + 2(H(\mu)-3/2) + 2 = 2H(\mu)+1$.

For k-(s,T)-Connectivity Augmentation our approach follows [24], where the problem is decomposed into a small number of "good" subproblems. Specifically, [24] decomposes 2-(s,T)-Connectivity Augmentation into 6 subproblems of covering an uncrossable biset family; the latter admits ratio 2. The number of subproblems is determined by the chromatic number of a certain auxiliary graph, and in [24] it was shown that in the case $k=2$ this graph is 5-colorable. We will show that in the case $k=2$ this graph is a forest, and thus is 2-colorable. This gives just 3 subproblems. Furthermore, we observe that 2 of the 3 uncrossable families needed to be covered are of a special type, for which we can get ratio $4/3$ using a result of Fukunaga [13]. Overall, we get ratio $2 \cdot 4/3 + 2 = 4\frac{2}{3}$.

1.3 Some Previous and Related Work

We consider *node-connectivity* problems for which classic techniques like the primal dual method [15] and iterative rounding [16] do not seem to be applicable directly. Ravi and Williamson [28] gave an example of a k-Connectivity Augmentation instance when the primal dual method has ratio $\Omega(k)$. Aazami et al. [1] presented a related instance for which the basic optimal solution to the cut-LP relaxation has all variables of value $O(1/\sqrt{k})$, ruling out the iterative rounding method. On the other hand, several works showed that node-connectivity problems can be decomposed into a small number p of "good" problems. The bound on p was subsequently improved, culminating in the currently best known bounds $O(\log \frac{n}{n-k})$ for directed/undirected k-Connectivity Augmentation [26], and $O(k)$ for undirected k-(s,T)-Connectivity Augmentation [24]. In fact, [23] shows that for $k = \Omega(n)$ the approximability of the directed and undirected variants of these problems is the same, up to a factor of 2. We refer the reader to [22] for various hardness results on k-(s,T)-Connectivity Augmentation. We note that the version of k-Connectivity Augmentation when any edge can be added by a cost of 1 can be solved in polynomial time for both directed [10] and undirected [29] graphs. But for general costs, determining whether k-Connectivity Augmentation admits a constant ratio for $k = n - o(n)$ is one of the most challenging problems in connectivity network design.

We mention some related work on the more general k-Connected Subgraph problem, where we seek a min-cost k-connected spanning subgraph. k-Connectivity Augmentation is a particular case, when the target connectivity is $k + 1$ and the edges of cost zero of the input graph form a k-connected spanning subgraph. Many papers that considered k-Connected Subgraph built on the Frank Tardos [11] algorithm for a related problem of finding a min-cost k-outconnected subgraph [2–4,6,7,18,19], but most papers that considered high values of k in fact designed algorithms for k-Connectivity Augmentation [4,7,19,26]. These papers use the fact that approximation ratio ρ w.r.t. the biset LP-relaxation for k-Connectivity Augmentation implies approximation ratio $\rho H(k) = \rho \cdot O(\log k)$ for k-Connected Subgraph [27]. Recently, Cheriyan and Végh [3] showed that for undirected graphs with $n = \Omega(k^4)$ this $O(\log k)$ factor can be saved and ratio 6 can be achieved; the bound $n = \Omega(k^4)$ of [3] was improved to $n = \Omega(k^3)$ in [14]. This algorithm easily extends to arbitrary crossing supermodular functions.

In the more general Survivable Network problem, we are given connectivity requirements $\{r_{uv} : u, v \in V\}$. The goal is to compute a min-cost subgraph that has r_{uv} internally-disjoint uv-paths for all $u, v \in V$. For undirected graphs the problem admits ratio $O(k^3 \log n)$ due to Chuzhoy and Khanna [5]. For directed graphs, no non-trivial ratio is known even for 2-(s, T)-Connectivity Augmentation.

2   Preliminaries on Biset Families

We cast our problems in terms of bisets (called also "setpairs"). Most concepts related to bisets that we need are summarized in this section.

Definition 1. *An ordered pair* $\mathbb{A} = (A, A^+)$ *of subsets of a groundset* V *is called a* **biset** *if* $A \subseteq A^+$; A *is the* **inner part** *and* A^+ *is the* **outer part** *of* $\mathbb{A}$, *and* $\partial \mathbb{A} = A^+ \setminus A$ *is the* **boundary** *of* $\mathbb{A}$. *The* **co-set** *of a biset* $\mathbb{A} = (A, A^+)$ *is* $A^* = V \setminus A^+$; *the* **co-biset** *of* $\mathbb{A}$ *is* $\mathbb{A}^* = (A^*, V \setminus A)$.

Definition 2. *A* **biset family** *is a family of bisets. The* **co-family** *of a biset family* $\mathcal{F}$ *is* $\mathcal{F}^* = \{\mathbb{A}^* : \mathbb{A} \in \mathcal{F}\}$. $\mathcal{F}$ *is* **symmetric** *if* $\mathcal{F} = \mathcal{F}^*$.

Definition 3. *An* **edge covers** *a biset* $\mathbb{A}$ *if it goes from* A *to* A^*. *Let* $\delta_E(\mathbb{A})$ *denote the set of edges in* E *that cover* $\mathbb{A}$. *The* **residual family** *of a biset family* $\mathcal{F}$ *w.r.t. an edge-set /graph* J *is denoted* $\mathcal{F}^J$ *and it consists of the members in* $\mathcal{F}$ *not covered by any* $e \in J$, *namely,* $\mathcal{F}^J = \{\mathbb{A} \in \mathcal{F} : \delta_J(\mathbb{A}) = \emptyset\}$. *We say that an* **edge set/graph** J **covers** $\mathcal{F}$ *or that* J *is an* $\mathcal{F}$**-edge-cover** *if every* $\mathbb{A} \in \mathcal{F}$ *is covered by some* $e \in J$, *namely, if* $\mathcal{F}^J = \emptyset$.

Given an instance of k-(s, T)-Connectivity Augmentation we will assume that G has no edge between s and T, by subdividing by a new node every edge $ts \in E$

with $t \in T$. The biset families $\mathcal{F}_k$ and $\mathcal{F}_{k\text{-}(s,T)}$ we need to cover in k-Connectivity Augmentation and k-(s,T)-Connectivity Augmentation, respectively, are:

$$\mathcal{F}_k = \{\mathbb{A} : |\partial\mathbb{A}| = k, \delta_E(\mathbb{A}) = \emptyset, A \neq \emptyset, A^* \neq \emptyset\} \tag{1}$$

$$\mathcal{F}_{k\text{-}(s,T)} = \{\mathbb{A} : |\partial\mathbb{A}| = k, \delta_E(\mathbb{A}) = \emptyset, A \cap T \neq \emptyset, s \in A^*\} \tag{2}$$

By the node-connectivity version of Menger's Theorem we have the following.

Fact 1

(i) *Let G be k-connected graph. Then $|\partial\mathbb{A}| \geq k$ for any biset $\mathbb{A}$ on V with $\delta_E(\mathbb{A}) = \emptyset$ and $A, A^* \neq \emptyset$. Furthermore, $G + J$ is $(k+1)$-connected if and only if J covers the family $\mathcal{F}_k$ in (1).*

(ii) *Let G be a k-(s,T)-connected graph (without edges between s and T). Then $|\partial\mathbb{A}| \geq k$ for any biset $\mathbb{A}$ on $V \setminus \{s\}$ with $A \cap T \neq \emptyset$ and $\delta_E(\mathbb{A}) = \emptyset$. Furthermore, $G + J$ is $(k+1)$-(s,T)-connected if and only if J covers the family $\mathcal{F}_{k\text{-}(s,T)}$ in (2).*

We thus consider the following generic algorithmic problem.

Biset-Family Edge-Cover

Instance: A graph (V, F) with edge-costs $\{c_e : e \in F\}$ and a biset family $\mathcal{F}$.
Objective: Find a minimum cost $\mathcal{F}$-edge-cover $J \subseteq F$.

Here the biset family $\mathcal{F}$ may not be given explicitly, and a polynomial time implementation in $n = |V|$ of our algorithms requires that the following query can be answered in time polynomial in n: Given an edge set/graph J on V and $s, t \in V$, find the inclusionwise minimal and the inclusionwise maximal members of the family $\{\mathbb{A} \in \mathcal{F}^J : s \in A, t \in V \setminus A^+\}$, if non-empty. For biset families arising from our problems, this query can be answered in polynomial time using max-flow min-cut computation (we omit the standard implementation details).

Definition 4. *The **intersection** and the **union** of two bisets $\mathbb{A}, \mathbb{B}$ are defined by $\mathbb{A} \cap \mathbb{B} = (A \cap B, A^+ \cap B^+)$ and $\mathbb{A} \cup \mathbb{B} = (A \cup B, A^+ \cup B^+)$. The biset $\mathbb{A} \setminus \mathbb{B}$ is defined by $\mathbb{A} \setminus \mathbb{B} = (A \setminus B^+, A^+ \setminus B)$. We say that $\mathbb{B}$ **contains** $\mathbb{A}$ and write $\mathbb{A} \subseteq \mathbb{B}$ if $A \subseteq B$ and $A^+ \subseteq B^+$.*

Definition 5. *Let $\mathbb{A}, \mathbb{B}$ be bisets and T be a set of terminals. We say that $\mathbb{A}, \mathbb{B}$:*

- *T-**intersect** if $A \cap B \cap T \neq \emptyset$.*
- *T-**cross** if both $A \cap B \cap T$ and $A^* \cap B^* \cap T$ are nonempty.*
- *T-**co-cross** if both $A \cap B^* \cap T$ and $B \cap A^* \cap T$ are nonempty*

*In the case $T = V$ we omit the prefix "T-", and say that $\mathbb{A}, \mathbb{B}$: **intersect**, **cross**, or **co-cross**, respectively.*

The following definition gives two fundamental types of biset families for which good approximation ratios are known.

Definition 6. *A biset family $\mathcal{F}$ is:*

- **intersecting** *if* $\mathbb{A} \cap \mathbb{B}, \mathbb{A} \cup \mathbb{B} \in \mathcal{F}$ *for any* $\mathbb{A}, \mathbb{B} \in \mathcal{F}$ *that intersect.*
- **uncrossable** *if* $\mathbb{A} \cap \mathbb{B}, \mathbb{A} \cup \mathbb{B} \in \mathcal{F}$ *or* $\mathbb{A} \setminus \mathbb{B}, \mathbb{B} \setminus \mathbb{A} \in \mathcal{F}$ *for any* $\mathbb{A}, \mathbb{B} \in \mathcal{F}$.

Let $\tau(\mathcal{F})$ denote the optimal value of a standard **biset LP-relaxation** for the problem of edge-covering a biset family $\mathcal{F}$, namely:

$$\tau(\mathcal{F}) = \min \left\{ \sum_{e \in F} c_e x_e : \sum_{e \in \delta_F(\mathbb{A})} x_e \geq 1 \ \forall \mathbb{A} \in \mathcal{F}, x \geq 0 \right\} .$$

Directed Biset-Family Edge-Cover with intersecting $\mathcal{F}$ admits a polynomial time algorithm that computes an $\mathcal{F}$-edge-cover of cost $\tau(\mathcal{F})$ [9]; for undirected graphs the cost is $2\tau(\mathcal{F})$ for intersecting $\mathcal{F}$ (by a standard "bidirection" reduction to the directed case) and for uncrossable $\mathcal{F}$ [8]. But the biset families arising from our problems have weaker "uncrossing" properties.

Definition 7. *We say that a biset family $\mathcal{F}$ is:*

- **crossing** *if* $\mathbb{A} \cap \mathbb{B}, \mathbb{A} \cup \mathbb{B} \in \mathcal{F}$ *for any* $\mathbb{A}, \mathbb{B} \in \mathcal{F}$ *that cross.*
- **k-regular** *if for any intersecting* $\mathbb{A}, \mathbb{B} \in \mathcal{F}$ *the following holds:* $\mathbb{A} \cap \mathbb{B} \in \mathcal{F}$ *if* $|A \cup B| \leq n - k$, *and* $\mathbb{A} \cup \mathbb{B} \in \mathcal{F}$ *if* $|A \cup B| \leq n - k - 1$.
- **T-uncrossable** *if for any* $\mathbb{A}, \mathbb{B} \in \mathcal{F}$ *the following holds:* $\mathbb{A} \cap \mathbb{B}, \mathbb{A} \cup \mathbb{B} \in \mathcal{F}$ *if* $\mathbb{A}, \mathbb{B}$ T-*intersect, and* $\mathbb{A} \setminus \mathbb{B}, \mathbb{B} \setminus \mathbb{A} \in \mathcal{F}$ *if* $\mathbb{A}, \mathbb{B}$ T-*co-cross.*

The following properties of bisets are known and easy to verify.

Fact 2 *For any bisets* $\mathbb{A}, \mathbb{B}$ *the following holds. If a directed/undirected edge e covers one of* $\mathbb{A} \cap \mathbb{B}, \mathbb{A} \cup \mathbb{B}$ *then e covers one of* $\mathbb{A}, \mathbb{B}$; *if e is an undirected edge, then if e covers one of* $\mathbb{A} \setminus \mathbb{B}, \mathbb{B} \setminus \mathbb{A}$, *then e covers one of* $\mathbb{A}, \mathbb{B}$. *Furthermore*

$$|\partial \mathbb{A}| + |\partial \mathbb{B}| = |\partial(\mathbb{A} \cap \mathbb{B})| + |\partial(\mathbb{A} \cup \mathbb{B})| = |\partial(\mathbb{A} \setminus \mathbb{B})| + |\partial(\mathbb{B} \setminus \mathbb{A})| .$$

The following known lemma (c.f. [17,24]) can be easily deduced from Fact 2.

Lemma 3.

(i) *If G is k-connected then $\mathcal{F}_k$ and $\mathcal{F}_k^*$ are crossing and k-regular.*
(ii) *If G is undirected and k-(s,T)-connected then $\mathcal{F}_{k\text{-}(s,T)}$ is T-uncrossable.*

Note also that Fact 2 implies that if $\mathcal{F}$ is intersecting, crossing, or k-regular, then so is the residual family $\mathcal{F}^J$ of $\mathcal{F}$, for any J. For an undirected edge set J this is also so if $\mathcal{F}$ is uncrossable or T-uncrossable.

In terms of bisets, we prove the following theorem that implies Theorem 3.

Theorem 4 (Implies Theorem 3). *Let $\mathcal{F}$ be a crossing biset family such that both $\mathcal{F}$ and $\mathcal{F}^*$ are k-regular and such that $|\partial \mathbb{A}| = k$ for all $\mathbb{A} \in \mathcal{F}$. If $n \geq k+3$ then Biset-Family Edge-Cover with such $\mathcal{F}$ admits the following approximation ratios w.r.t. the biset LP-relaxation:*

(i) *For directed graphs, ratio* $H(\mu) + \frac{3}{2}$. *In particular:*

- *For* $k = 1$, *and for* $k = 2$ *and* n *odd,* $\mu = 1, H(\mu) = 1$, *so the ratio is* 2.5.
- *For* $n \geq 3k - 5$, $\mu \leq 2$, $H(\mu) \leq 3/2$, *so the ratio is 3.*
- *for* $n \geq 2k - 3$, $\mu \leq 3$, $H(\mu) \leq 11/6$, *so the ratio is* $3\frac{1}{3}$.

(ii) *For undirected graphs, ratio* $2H(\mu) + 1$. *In particular, for* $n \geq 3k - 5, \mu \leq 2, H(\mu) \leq 3/2$, *so the ratio is 4.*

For formulating Theorem 2 purely in terms of bisets, the T-uncrossability property is not enough. In Sect. 4 we state additional properties of the family $\mathcal{F}_{2-(s,T)}$ that we need, see Lemma 10.

The following definition plays a key role in our algorithms.

Definition 8. *The inclusionwise minimal members of a biset family $\mathcal{F}$ are called $\mathcal{F}$-cores, or simply **cores**, if $\mathcal{F}$ is clear from the context. Let $\mathcal{C}(\mathcal{F})$ denote the family of $\mathcal{F}$-cores, and let $\nu(\mathcal{F}) = |\mathcal{C}(\mathcal{F})|$ denote the number of $\mathcal{F}$-cores. For $\mathbb{C} \in \mathcal{C}(\mathcal{F})$, the **halo-family** $\mathcal{F}(\mathbb{C})$ of $\mathbb{C}$ is the family of those members of $\mathcal{F}$ that contain $\mathbb{C}$ and contain no $\mathcal{F}$-core distinct from $\mathbb{C}$.*

3   Algorithm for Crossing k-Regular Families (Theorem 4)

We mention some results from [7,26] needed for the proof of Theorem 4.

Definition 9. *A biset family $\mathcal{F}$ is **intersection-closed** if $\mathbb{A} \cap \mathbb{B} \in \mathcal{F}$ for any intersecting $\mathbb{A}, \mathbb{B} \in \mathcal{F}$. An intersection-closed $\mathcal{F}$ is q-**semi-intersecting** if $|A| \leq q$ for every $\mathbb{A} \in \mathcal{F}$ and if $\mathbb{A} \cup \mathbb{B} \in \mathcal{F}$ for any intersecting $\mathbb{A}, \mathbb{B} \in \mathcal{F}$ with $|A \cup B| \leq q$.*

We obtain a q-semi-intersecting family from a k-regular family as follows.

Definition 10. *The q-**truncated family** of $\mathcal{F}$ is $\mathcal{F}_{\leq q} := \{\mathbb{A} \in \mathcal{F} : |A| \leq q\}$.*

Lemma 4. *Let $\mathcal{F}$ be a k-regular biset family. If $2q - 1 \leq n - k$ and $q \leq n - k - 1$ (in particular if $q \leq \lfloor \frac{n-k+1}{2} \rfloor$ and $n \geq k + 3$) then $\mathcal{F}_{\leq q}$ is q-semi-intersecting.*

Proof. Let $\mathbb{A}, \mathbb{B} \in \mathcal{F}_{\leq q}$ intersect. Then $|A \cup B| \leq |A| + |B| - 1 \leq 2q - 1 \leq n - k$. Thus $\mathbb{A} \cap \mathbb{B} \in \mathcal{F}_{\leq q}$. If $|A \cup B| \leq q \leq n - k - 1$ then $\mathbb{A} \cup \mathbb{B} \in \mathcal{F}_{\leq q}$. Hence if both $2q - 1 \leq n - k$ and $q \leq n - k - 1$, then $\mathcal{F}_{\leq q}$ is q-semi-intersecting. □

The following theorem is the main result of [26].

Theorem 5 [26]. *Directed Biset-Family Edge-Cover with q-semi-intersecting $\mathcal{F}$ admits a polynomial time algorithm that computes an edge-set $J \subseteq E$ such that $\nu(\mathcal{F}^J) \leq \lfloor n/(q + 1) \rfloor$ and $c(J) \leq \tau(\mathcal{F})$.*

From Theorem 5 and Lemma 4 we have the following.

Corollary 1. *Directed Biset-Family Edge-Cover with k-regular $\mathcal{F}$ and $n \geq k + 3$ admits a polynomial time algorithm that for any $q \leq \lfloor \frac{n-k+1}{2} \rfloor$ computes $J \subseteq E$ such that $\nu(\mathcal{F}^J_{\leq q}) \leq \lfloor n/(q + 1) \rfloor$ and $c(J) \leq \tau(\mathcal{F}_{\leq q})$.*

Definition 11. *Two biset families* $\mathcal{A}, \mathcal{B}$ *are* **cross/intersect-independent** *if no* $\mathbb{A} \in \mathcal{A}$ *and* $\mathbb{B} \in \mathcal{B}$ *cross/intersect.*

It is easy to see that if $\mathcal{F}$ is intersection-closed then halo families of distinct $\mathcal{F}$-cores are intersect-independent. Thus by Lemma 4, if $\mathcal{F}$ is k-regular and $n \geq k + 3$, then for any $q \leq \lfloor \frac{n-k+1}{2} \rfloor$, $\mathcal{F}_{\leq q}(\mathbb{C}_i)$ and $\mathcal{F}_{\leq q}(\mathbb{C}_j)$ are intersect independent for distinct $\mathcal{F}_{\leq q}$-cores $\mathbb{C}_i$ and $\mathbb{C}_j$. Furthermore, note that if $\mathcal{F}$ is intersection-closed and J covers some halo family $\mathcal{F}(\mathbb{C})$, then any $\mathbb{A} \in \mathcal{F}^J$ that contains $\mathbb{C}$ contains an $\mathcal{F}$-core distinct from $\mathbb{C}$, and thus $\nu(\mathcal{F}^J) \leq \nu(\mathcal{F}) - 1$. The following statement summarizes several similar relevant properties of crossing biset families, c.f. [7, 20, 26].

Lemma 5. *For any crossing biset family* $\mathcal{F}$ *the following holds.*

(i) *For any* $\mathcal{F}$-*core* $\mathbb{C}$, *the co-family of* $\mathcal{F}(\mathbb{C})$ *is an intersecting family.*

(ii) *Halo families of distinct* $\mathcal{F}$-*cores are cross-independent.*

(iii) *For every* $\mathcal{F}$-*core* $\mathbb{C}$, *if* J *is a directed edge set that covers* $\mathcal{F}(\mathbb{C})$, *then any* $\mathbb{A} \in \mathcal{F}^J$ *that contains* $\mathbb{C}$ *contains an* $\mathcal{F}$-*core distinct from* $\mathbb{C}$. *Furthermore, if* J *is an inclusionwise minimal edge set that covers* $\mathcal{F}(\mathbb{C})$, *then* J *covers no biset in other halo families, and thus* $\mathcal{C}(\mathcal{F}^J) = \mathcal{C}(\mathcal{F}) \setminus \{\mathbb{C}\}$.

In what follows, note that if $\mathcal{A}, \mathcal{B}$ are two cross-independent subfamilies of $\mathcal{F}$, then no directed edge can cover $\mathbb{A} \in \mathcal{A}$ and $\mathbb{B} \in \mathcal{B}$, and thus $\tau(\mathcal{A}) + \tau(\mathcal{B}) \leq \tau(\mathcal{F})$. Now let $\mathcal{F}$ be a crossing family. By part (i) of Lemma 5, for any core $\mathbb{C}$, an optimal $\mathcal{F}(\mathbb{C})$-cover $J_{\mathbb{C}}$ of cost $\tau(\mathcal{F}(\mathbb{C}))$ can be computed in polynomial time; this is since an edge set $J_{\mathbb{C}}$ covers $\mathcal{F}(\mathbb{C})$ if and only if the reverse edge set of $J_{\mathbb{C}}$ covers the co-family of $\mathcal{F}_{\mathbb{C}}$. By part (ii), $\sum_{\mathbb{C} \in \mathcal{C}(\mathcal{F})} c(J_{\mathbb{C}}) \leq \tau(\mathcal{F})$. Thus for any $\mathcal{C} \subseteq \mathcal{C}(\mathcal{F})$, there exists $\mathbb{C} \in \mathcal{C}$ with $c(J_{\mathbb{C}}) \leq \tau(\mathcal{F})/|\mathcal{C}|$. Based on Lemma 5, consider the following algorithm that given $\mathcal{C} \subseteq \mathcal{C}(\mathcal{F})$ and $0 \leq t \leq |\mathcal{C}|$, computes $J \subseteq E$ such that $\mathcal{C}(\mathcal{F}^J) = \mathcal{C}(\mathcal{F}) \setminus \mathcal{C}'$ for some $\mathcal{C}' \subseteq \mathcal{C}$ with $|\mathcal{C}'| = |\mathcal{C}| - t$. Start with a partial solution $J = \emptyset$, and while $|\mathcal{C}(\mathcal{F}^J) \cap \mathcal{C}| \geq t + 1$, add to J a minimum cost cover $J_{\mathbb{C}}$ of the halo family in $\mathcal{F}^J$ of a core $\mathbb{C} \in \mathcal{C}(\mathcal{F}^J) \cap \mathcal{C}$ with $c(J_{\mathbb{C}})$ minimal. By part (iii) of the lemma, at iteration i we have $|\mathcal{C}(\mathcal{F}^J) \cap \mathcal{C}| \geq |\mathcal{C}| - i + 1$, and thus $c(J_{\mathbb{C}}) \leq \tau(\mathcal{F})/(|\mathcal{C}| - i + 1)$ at iteration i. Consequently, we have:

Theorem 6 [7, 26]. *Directed* Biset-Family Edge-Cover *with crossing* $\mathcal{F}$ *admits a polynomial time algorithm that given* $\mathcal{C} \subseteq \mathcal{C}(\mathcal{F})$ *and an integer* $0 \leq t \leq |\mathcal{C}|$ *computes an edge set* $J \subseteq E$ *such that the following holds:*

- $\mathcal{C}(\mathcal{F}^J) = \mathcal{C}(\mathcal{F}) \setminus \mathcal{C}'$ *for some* $\mathcal{C}' \subseteq \mathcal{C}$ *with* $|\mathcal{C}'| = |\mathcal{C}| - t$.
- $c(J) \leq (H(|\mathcal{C}|) - H(t)) \cdot \tau(\mathcal{F}')$, *where* $\mathcal{F}'$ *is the family of those members of* $\mathcal{F}$ *that contain no core in* $\mathcal{C}(\mathcal{F}) \setminus \mathcal{C}$.

In particular, for $t = 0$, J *covers the family of those members of* $\mathcal{F}$ *that contain no core in* $\mathcal{C}(\mathcal{F}) \setminus \mathcal{C}$ *and* $c(J) \leq H(|\mathcal{C}|)\tau(\mathcal{F}')$; *if also* $\mathcal{C} = \mathcal{C}(\mathcal{F})$ *then* J *covers* $\mathcal{F}$ *and has cost* $c(J) \leq H(\nu(\mathcal{F})) \cdot \tau(\mathcal{F})$.

We note that each of the statements in Theorem 5, Corollary 1, and Theorem 6, applies also for undirected graphs and symmetric $\mathcal{F}$, but with an additional factor of 2 in the cost. In this case we have $c(J) \leq 2\tau(\mathcal{F})$ in Theorem 5 and Corollary 1, and $c(J) \leq 2H(|\mathcal{C}|) \cdot \tau(\mathcal{F}')$ in Theorem 6. This is achieved by the following standard reduction. In each of the cases, we bidirect the edges of G (namely, replace every undirected edge e with endnodes u, v by two opposite directed edges uv, vu of cost c_e each), compute a set of directed edges for the obtained directed problem, and return the corresponding set of undirected edges.

In what follows assume that $n \geq k + 3$, $q = \lfloor \frac{n-k+1}{2} \rfloor$, and $|\partial \mathbb{A}| = k$ for all $\mathbb{A} \in \mathcal{F}$. We say that $\mathbb{A}$ is a **small biset/core** if $|A| \leq q$, and $\mathbb{A}$ is a **large biset/core** otherwise. At this point we will split the proof of Theorem 4 into two cases: the case of directed graphs and the case of undirected graphs.

3.1   Directed Graphs

To prove the directed part of Theorem 4 we prove the following lemma.

Lemma 6. *Suppose that $\mathcal{F}$ is crossing, $\mathcal{F}^*$ is k-regular, and $|\partial \mathbb{A}| = k$ for all $\mathbb{A} \in \mathcal{F}$, and that $n \geq k + 3$ and $q = \lfloor \frac{n-k+1}{2} \rfloor$. Then directed* Biset-Family Edge-Cover *admits a polynomial time algorithm if $\nu(\mathcal{F}_{\leq q}) = 0$ and approximation ratio $3/2$ if $\nu(\mathcal{F}_{\leq q}) = 1$.*

Lemma 6 together with Corollary 1 and Theorem 6 implies the directed part of Theorem 4. Note that the following algorithm uses all the assumptions on $\mathcal{F}$ in Theorem 4: $\mathcal{F}$ is k-regular in Corollary 1, crossing in Theorem 6, and in Lemma 6 $\mathcal{F}$ is crossing, $\mathcal{F}^*$ is k-regular, and $|\partial \mathbb{A}| = k$ for all $\mathbb{A} \in \mathcal{F}$. In the algorithm, we sequentially compute three edge sets:

1. J_1 reduces the number of small cores to μ by cost τ (Corollary 1).
2. J_2 further reduces the number of small cores to 1 by cost $(H(\mu) - H(1))\tau$ (Theorem 6).
3. J_3 covers the remaining members of $\mathcal{F}$ by cost $\frac{3}{2}\tau$ (Lemma 6).

Algorithm 1: DIRECTED-COVER($\mathcal{F}, G, c$)

1 Using the algorithm from **Corollary 1** compute $J_1 \subseteq E$ such that $\nu(\mathcal{F}_{\leq q}^{J_1}) \leq \mu$ and $c(J_1) \leq \tau(\mathcal{F}_{\leq q})$.

2 Using the algorithm from **Theorem 6** with $\mathcal{C} = \mathcal{C}(\mathcal{F}_{\leq q}^{J_1})$ and $t = 1$, compute $J_2 \subseteq E \setminus J_1$ such that $\nu\left(\mathcal{F}_{\leq q}^{J_1 \cup J_2}\right) \leq 1$ and $c(J_2) \leq (H(\mu) - 1)\tau\left(\mathcal{F}^{J_1}\right)$.

3 Using the algorithm from **Lemma 6** compute an $\mathcal{F}^{J_1 \cup J_2}$-cover $J_3 \subseteq E \setminus (J_1 \cup J_2)$ such that $c(J_3) \leq \frac{3}{2}\tau(\mathcal{F}^{J_1 \cup J_2})$.

4 **return** $J = J_1 \cup J_2 \cup J_3$

Clearly, the algorithm computes a feasible solution. The approximation ratio is bounded by $1 + (H(\mu) - 1) + 3/2 = H(\mu) + 3/2$.

The proof of Lemma 6 follows. For a biset family $\mathcal{A}$ let us say that $\mathcal{A}$-**cover** admits LP-ratioρ if there exists a polynomial time algorithm that computes an $\mathcal{A}$-cover of cost $\rho \cdot \tau(\mathcal{A})$. The proof of Lemma 6 relies on the following lemma, in which cross-independence is exploited in a novel way.

Lemma 7. *Let $\mathcal{A}, \mathcal{B}$ be subfamilies of a crossing family $\mathcal{F}$ such that $\mathcal{A} \cup \mathcal{B} = \mathcal{F}$ and the families $\mathcal{A} \setminus \mathcal{B}$ and $\mathcal{B} \setminus \mathcal{A}$ are cross-independent. If $\mathcal{A}^J$-cover admits LP-ratio α and $\mathcal{B}^J$-cover admits LP-ratio β for any $J \subseteq E$, then $\mathcal{F}$-cover admits LP-ratio $\alpha + \beta - \frac{\alpha\beta}{\alpha+\beta}$.*

Proof. We claim that the following algorithm achieves LP-ratio $\alpha + \beta - \frac{\alpha\beta}{\alpha+\beta}$:

Algorithm 2: CROSS-INDEPENDENT-COVER$(\mathcal{A}, \mathcal{B}, G, c)$

---

1 $J_{\mathcal{A}} \leftarrow \alpha$-approximate $\mathcal{A}$-cover $J'_{\mathcal{B}} \leftarrow \beta$-approximate $\mathcal{B}^{J_{\mathcal{A}}}$-cover
2 $J_{\mathcal{B}} \leftarrow \beta$-approximate $\mathcal{B}$-cover $J'_{\mathcal{A}} \leftarrow \alpha$-approximate $\mathcal{A}^{J_{\mathcal{B}}}$-cover
3 **return** the cheaper edge set J among $J_{\mathcal{A}} \cup J'_{\mathcal{B}}, J_{\mathcal{B}} \cup J'_{\mathcal{A}}$.

Note that since $\mathcal{A} \setminus \mathcal{B}$ and $\mathcal{B} \setminus \mathcal{A}$ are cross-independent, so are $\mathcal{B}^{J_{\mathcal{A}}}$ and $\mathcal{A}^{J_{\mathcal{B}}}$. Thus no $\mathbb{B} \in \mathcal{B}^{J_{\mathcal{A}}}$ and $\mathbb{A} \in \mathcal{A}^{J_{\mathcal{B}}}$ cross, so no directed edge can cover both $\mathbb{A}$ and $\mathbb{B}$. Therefore

$$\tau\left(\mathcal{B}^{J_{\mathcal{A}}}\right) + \tau\left(\mathcal{A}^{J_{\mathcal{B}}}\right) \le \tau(\mathcal{F}).$$

Denoting $\tau = \tau(\mathcal{F})$ and $\tau' = \tau\left(\mathcal{B}^{J_{\mathcal{A}}}\right)$, we have $\tau\left(\mathcal{A}^{J_{\mathcal{B}}}\right) \le \tau - \tau'$. We also have:

$$c(J_{\mathcal{A}}) \le \alpha\tau(\mathcal{A}) \le \alpha\tau \qquad c(J'_{\mathcal{B}}) \le \beta\tau\left(\mathcal{B}^{J_{\mathcal{A}}}\right) = \beta\tau'$$
$$c(J_{\mathcal{B}}) \le \beta\tau(\mathcal{B}) \le \beta\tau \qquad c(J'_{\mathcal{A}}) \le \alpha\tau\left(\mathcal{A}^{J_{\mathcal{B}}}\right) \le \alpha(\tau - \tau')$$

Thus the cost of the edge set produced by the algorithm is bounded by

$$c(J) = \min\{c(J_{\mathcal{A}}) + c(J'_{\mathcal{B}}), c(J_{\mathcal{B}}) + c(J'_{\mathcal{A}})\} \le \min\left\{\alpha\tau + \beta\tau', \beta\tau + \alpha(\tau - \tau')\right\}.$$

The worst case is when $\alpha\tau + \beta\tau' = \beta\tau + \alpha(\tau - \tau')$, namely $\tau' = \frac{\beta}{\alpha+\beta}\tau$. Then

$$c(J) = \alpha\tau + \beta\tau' = \tau\left(\alpha + \frac{\beta^2}{\alpha + \beta}\right) = \tau\frac{\alpha^2 + \alpha\beta + \beta^2}{\alpha + \beta} = \tau\left(\alpha + \beta - \frac{\alpha\beta}{\alpha + \beta}\right).$$

This concludes the proof of the lemma. $\square$

We show that the following two subfamilies $\mathcal{A}, \mathcal{B}$ of $\mathcal{F}$ satisfy the assumptions of Lemma 7 with $\alpha = \beta = 1$; note that then $\alpha + \beta - \frac{\alpha\beta}{\alpha+\beta} = 3/2$.

 – $\mathcal{A}$ is the family of bisets in $\mathcal{F}$ that contain some small $\mathcal{F}$-core;
 – $\mathcal{B}$ is the family of bisets in $\mathcal{F}$ that contain some large $\mathcal{F}$-core.

Lemma 8. *The families $\mathcal{A}, \mathcal{B}$ above satisfy the assumption properties of Lemma 7 with $\alpha = \nu(\mathcal{F}_{\le q})$ (so $\alpha = 1$ if $\nu(\mathcal{F}_{\le q}) = 1$) and $\beta = 1$.*

Proof. Clearly, $\mathcal{A} \cup \mathcal{B} = \mathcal{F}$. We prove that $\mathcal{A} \setminus \mathcal{B}$ and $\mathcal{B} \setminus \mathcal{A}$ are cross-independent. Let $\mathbb{A} \in \mathcal{A}$ and $\mathbb{B} \in \mathcal{B}$ cross. Then $\mathbb{A} \cap \mathbb{B} \in \mathcal{F}$, since $\mathcal{F}$ is a crossing family. Thus $\mathbb{A} \cup \mathbb{B}$ contains an $\mathcal{F}$-core $\mathbb{C}$. If $\mathbb{C}$ is small then $\mathbb{A}, \mathbb{B} \in \mathcal{A}$ and thus $\mathbb{B} \notin \mathcal{B} \setminus \mathcal{A}$. If $\mathbb{C}$ is large then $\mathbb{A}, \mathbb{B} \in \mathcal{B}$ and thus $\mathbb{A} \notin \mathcal{A} \setminus \mathcal{B}$. In both cases we cannot have $\mathbb{A} \in \mathcal{A} \setminus \mathcal{B}$ and $\mathbb{B} \in \mathcal{B} \setminus \mathcal{A}$, hence $\mathcal{A} \setminus \mathcal{B}$ and $\mathcal{B} \setminus \mathcal{A}$ are cross-independent.

To prove the claimed approximability of covering $\mathcal{A}$ and $\mathcal{B}$, we show that $\mathcal{A}^*$ is a union of $\nu(\mathcal{F}_{\leq q})$ intersecting biset families, and that $\mathcal{B}^*$ is an intersecting biset family. For a core $\mathbb{C}$ denote $\mathcal{F}_{\mathbb{C}} = \{\mathbb{A} \in \mathcal{F} : \mathbb{C} \subseteq \mathbb{A}\}$. Note that $\mathcal{A} = \bigcup \mathcal{F}_{\mathbb{C}}$ and $\mathcal{A}^* = \bigcup \mathcal{F}_{\mathbb{C}}^*$, where the union is taken over all small cores $\mathbb{C}$ of $\mathcal{F}$. It is easy to see that since $\mathcal{F}$ is crossing, then each family $\mathcal{F}_{\mathbb{C}}^*$ is an intersecting family. Hence $\mathcal{A}^*$ is a union of $\nu(\mathcal{F}_{\leq q})$ intersecting families.

We prove that $\mathcal{B}^*$ is an intersecting family. Consider the inclusionwise maximal members of $\mathcal{B}^*$; each maximal member of $\mathcal{B}^*$ is the co-biset $\mathbb{C}^*$ of some large $\mathcal{F}$-core $\mathbb{C}$. We claim that if $\mathbb{C}_i, \mathbb{C}_j$ are distinct large $\mathcal{F}$-cores then $C_i^* \cap C_j^* = \emptyset$. Note that $|C_i|, |C_j| \geq q + 1$, hence $|C_i^*|, |C_j^*| \leq n - k - q - 1$. If $C_i^* \cap C_j^* \neq \emptyset$ then for $q \geq \frac{n-k-2}{2}$, and in particular for $q = \lfloor \frac{n-k+1}{2} \rfloor$, we have

$$|C_i^* \cup C_j^*| \leq |C_i^*| + |C_j^*| - 1 \leq 2n - 2k - 2q - 3 \leq n - k - 1$$

Since $\mathcal{F}^*$ is k-regular, we get that $\mathbb{C}_i^* \cup \mathbb{C}_j^* \in \mathcal{F}^*$, contradicting the maximality of $\mathbb{C}_i^*, \mathbb{C}_j^*$. This implies that if $\mathbb{A}, \mathbb{B} \in \mathcal{B}^*$ intersect, then $\mathbb{A}, \mathbb{B}$ are contained in the same inclusionwise maximal member of $\mathcal{B}^*$, namely, $\mathbb{A}, \mathbb{B} \subseteq \mathbb{C}^*$ for some large $\mathcal{F}$-core $\mathbb{C}^*$. Note that $\mathbb{C} \subseteq \mathbb{A}^* \cap \mathbb{B}^*$. Thus if $\mathbb{A}, \mathbb{B}$ cross, and since $\mathcal{F}^*$ is a crossing family, $\mathbb{A} \cap \mathbb{B}, \mathbb{A} \cup \mathbb{B} \in \mathcal{F}^*$. Moreover, $\mathbb{A} \cap \mathbb{B}, \mathbb{A} \cup \mathbb{B} \subseteq \mathbb{C}^*$, which implies $\mathbb{A} \cap \mathbb{B}, \mathbb{A} \cup \mathbb{B} \in \mathcal{B}^*$. Consequently, $\mathcal{B}^*$ is an intersecting family. $\square$

Lemma 6 easily follows from Lemmas 7 and 8, and thus the proof of the directed part of Theorem 4 is complete.

3.2   Undirected Graphs

To prove the undirected part of Theorem 4 we prove the following lemma.

Lemma 9. *Suppose that $\mathcal{F}$ is symmetric k-regular and $|\partial \mathbb{A}| = k$ for all $\mathbb{A} \in \mathcal{F}$, and that $n \geq k + 3$ and $q = \lfloor \frac{n-k+1}{2} \rfloor$. Then undirected* Biset-Family Edge-Cover *admits a polynomial time algorithm if $\nu(\mathcal{F}_{\leq q}) = 1$, and ratio 2 if $\nu(\mathcal{F}_{\leq q}) = 2$.*

Proof. We claim that if $\nu(\mathcal{F}_{\leq q}) \leq 2$ then there exist a pair $s, t \in V$ such that

$$\nu\left(\mathcal{F}_{\leq q}^{\{st\}}\right) \leq \nu(\mathcal{F}_{\leq q}) - 1. \tag{3}$$

Namely, adding the edge st reduces the number of small cores by at least 1. Note that such a pair s, t can be found in polynomial time by computing $\nu(\mathcal{F}_{\leq q})$ and $\nu\left(\mathcal{F}_{\leq q}^{\{st\}}\right)$ for every $s, t \in V$. Once such pair s, t is found, we compute a minimum cost edge cover J_{st} of the biset family $\{\mathcal{F}_{st} = \mathbb{A} \in \mathcal{F} : s \in A, t \in A^*\}$. This family is intersecting and has a unique core; such a family is sometimes called a **ring family**. Thus we get that in the case $\nu(\mathcal{F}_{\leq q}) \leq 2$, the problem of edge covering $\mathcal{F}$ is reduced to edge covering $\nu(\mathcal{F}_{\leq q})$ ring families. It is known that Biset-Family Edge-Cover with a ring family admits a polynomial time algorithm that computes a solution of cost $\tau(\mathcal{F})$. Consequently, we get a polynomial time algorithm if $\nu(\mathcal{F}_{\leq q}) = 1$ and ratio 2 if $\nu(\mathcal{F}_{\leq q}) = 2$.

We now prove existence of a pair s, t as above. Let $\mathbb{C} \in \mathcal{C}(\mathcal{F}_{\leq q})$ and let $\mathcal{M}_{\mathbb{C}}$ be the family of inclusionwise maximal bisets in $\mathcal{F}_{\leq q}$ that contain $\mathbb{C}$. If $\mathcal{M}_{\mathbb{C}}$ has a unique biset $\mathbb{M}$, then (3) holds for any $s \in C$ and $t \in M^*$. Suppose that $|\mathcal{M}_{\mathbb{C}}| \geq 2$. Note that by Lemma 4 and by the symmetry of $\mathcal{F}$, if $\mathbb{A}, \mathbb{B} \in \mathcal{F}_{\leq q}$ intersect, then $\mathbb{A} \cup \mathbb{B} \in \mathcal{F}_{\leq q}$ or $(\mathbb{A} \cup \mathbb{B})^* \in \mathcal{F}_{\leq q}$. Thus for any distinct $\mathbb{A}, \mathbb{B} \in \mathcal{M}_{\mathbb{C}}$, $(\mathbb{A} \cup \mathbb{B})^* \in \mathcal{F}_{\leq q}$ holds, by the maximality of the bisets in $\mathcal{M}_{\mathbb{C}}$. Consequently, since $\nu(\mathcal{F}_{\leq q}) \leq 2$, there is a unique $\mathcal{F}_{\leq q}$-core $\mathbb{C}'$ distinct from $\mathbb{C}$, such that $\mathbb{C}' \subseteq (\mathbb{A} \cup \mathbb{B})^*$ for any distinct $\mathbb{A}, \mathbb{B} \in \mathcal{M}_{\mathbb{C}}$. This implies that (3) holds for any $s \in C$ and $t \in C'$. □

Let us now show that Lemma 9 implies the undirected part of Theorem 4. The algorithm is similar to the one for the directed case; it returns a solution $J = J_1 \cup J_2 \cup J_3$ where:

1. J_1 reduces the number of small cores to μ by cost 2τ (Corollary 1).
2. If $\mu \geq 3$ then J_2 further reduces the number of small cores to 2 by cost $2(H(\mu) - H(2))\tau$ (Theorem 6).
3. J_3 covers the remaining members of $\mathcal{F}$ by cost τ if $\mu = 1$ and by cost 2τ otherwise (Lemma 9).

Clearly, the algorithm computes a feasible solution. In the case $\mu \geq 2$ the approximation ratio is $2 + 2(H(\mu) - H(2)) + 2 = 2H(\mu) + 1$. This is so also in the case $\mu = 1$, since then the ratio is $2 + 1 = 3 = 2H(1) + 1$.

This concludes the proof of the undirected part of Theorem 4.

4   Proof-Sketch of Theorem 2

Let $G = (V, E)$ be a 2-(s, T)-connected graph and let $\mathcal{F} = \mathcal{F}_{2\text{-}(s,T)}$ be as in (2). One can prove the following "uncrossing" properties of the bisets in $\mathcal{F}$.

Lemma 10. *Let* $\mathbb{A}, \mathbb{B} \in \mathcal{F}$ *such that* $A \cap B \cap T = \emptyset$. *Then either* $\partial \mathbb{A} \cap B, \partial \mathbb{B} \cap A$ *are both empty, or the following holds:*

(i) *Each one of the sets* $\partial \mathbb{A} \cap B, \partial \mathbb{A} \cap B^*, \partial \mathbb{B} \cap A, \partial \mathbb{B} \cap A^*$ *is a singleton.*
(ii) *If* $B \cap A^* \cap T \neq \emptyset$ *then* $\mathbb{B} \setminus \mathbb{A} \in \mathcal{F}$; *if* $A \cap B^* \cap T \neq \emptyset$ *then* $\mathbb{A} \setminus \mathbb{B} \in \mathcal{F}$.
(iii) *If* $|A \cap T| \geq 2$ *and* $|B \cap T| \geq 2$ *then* $\mathbb{A}, \mathbb{B}$ *T-co-cross.*

Corollary 2. *Let* $\mathbb{A}, \mathbb{B} \in \mathcal{C}(\mathcal{F})$. *Then either* $\mathbb{A} \subseteq \mathbb{B}^*$ *and* $\mathbb{B} \subseteq \mathbb{A}^*$, *or each one of the sets* $A \cap T, B \cap T$ *is a singleton, and* $\partial \mathbb{B} \cap A = A \cap T$ *and* $\partial \mathbb{A} \cap B = B \cap T$.

For $\mathcal{A} \subseteq \mathcal{F}$ the U-**mesh graph** $\mathcal{G} = \mathcal{G}(\mathcal{A}, U)$ of $\mathcal{A}$ has node set $\mathcal{A}$ and edge set $\{\mathbb{A}_i \mathbb{A}_j : \partial \mathbb{A}_i \cap A_j \cap U \neq \emptyset \text{ or } \partial \mathbb{A}_j \cap A_i \cap U \neq \emptyset\}$. Using Lemma 10, one can also prove the following key statement.

Lemma 11. *Let* $\mathcal{A} \subseteq \mathcal{F}$. *If* $A_i \cap A_j \cap T = \emptyset$ *for any distinct* $\mathbb{A}_i, \mathbb{A}_j \in \mathcal{A}$ *then the* V*-mesh graph* $\mathcal{G}$ *of* $\mathcal{A}$ *is a forest.*

Corollary 3. *Let* $\mathcal{A}$ *be obtained by picking for each core* $\mathbb{C}_i \in \mathcal{C}(\mathcal{F})$ *a biset* $\mathbb{A}_i$ *in the halo-family* $\mathcal{F}(\mathbb{C}_i)$ *of* $\mathbb{C}_i$. *Then the* V*-mesh graph of* $\mathcal{A}$ *is a forest.*

Proof. Since $\mathcal{F}$ is T-uncrossable, bisets from distinct halo families cannot T-intersect. Thus $A_i \cap A_j \cap T = \emptyset$ for distinct $\mathbb{A}_i, \mathbb{A}_j \in \mathcal{A}$, and the V-mesh graph of $\mathcal{A}$ is a forest by Lemma 11. We prove that if $\mathbb{A}_i = \mathbb{C}_i$ for each i then $\mathcal{G}$ has no node of degree ≥ 3. Otherwise, $\mathcal{G}$ has a node $\mathbb{C}_0$ with 3 distinct neighbors $\mathbb{C}_1, \mathbb{C}_2, \mathbb{C}_3$. Then $C_i \cap C_j \cap T = \emptyset$ for distinct $0 \leq i, j \leq 3$. By Corollary 2 $C_i \cap T \subseteq \partial \mathbb{C}_0$ for $i = 1, 2, 3$, and we get the contradiction $|\partial \mathbb{C}_0| \geq 3$. □

Corollary 4. *Let* $\mathcal{C}$ *be the set family of the inner parts of the bisets in* $\mathcal{C}(\mathcal{F})$. *Then the maximum degree of a node in the hypergraph* $(V, \mathcal{C})$ *is at most* 2.

Proof. Let $v \in V$ and let $\mathcal{C}_v = \{\mathbb{C} \in \mathcal{C}(\mathcal{F}) : v \in C\}$ be the family of cores whose inner part contains v. Consider the the V-mesh graph $\mathcal{G}_v$ of $\mathcal{C}_v$. By Corollary 2 $\mathcal{G}_v$ is a clique, while by Corollary 3 $\mathcal{G}_v$ is a path. Thus $\mathcal{G}_v$ has at most 2 nodes. □

A **simple biset family** $\mathcal{F}$ has no biset that contains 2 distinct cores, namely, $\mathcal{F}$ is the union of its halo families. The best known ratio for edge-covering uncrossable $\mathcal{F}$ is 2, even for set families. Fukunaga [13] showed that for simple uncrossable biset families one can achieve ratio 4/3. We prove the following.

Theorem 7. *If* Biset-Family Edge-Cover *admits approximation ratio* α *for simple uncrossable families and approximation ratio* β *for uncrossable families, then* 2-(s, T)-Connectivity Augmentation *admits approximation ratio* $2\alpha + \beta$.

The currently best known values of α and β are $\alpha = 4/3$ [13] and $\beta = 2$ [8], so we get ratio $2 \cdot 4/3 + 2 = 4\frac{2}{3}$.

We now prove Theorem 7. The following lemma from [24] is easy to verify.

Lemma 12 [24]. *Let* $\mathcal{F}$ *be an arbitrary* T*-uncrossable biset family and let* $\mathbb{A}_i \in \mathcal{F}(\mathbb{C}_i)$ *and* $\mathbb{A}_j \in \mathcal{F}(\mathbb{C}_j)$, *where* $\mathbb{C}_i, \mathbb{C}_j \in \mathcal{C}(\mathcal{F})$ *(possibly* $i = j$*).*

(i) *If* $i = j$ *(so* $\mathbb{A}_i, \mathbb{A}_j$ *contain the same* $\mathcal{F}$*-core) then* $\mathbb{A}_i \cap \mathbb{A}_j, \mathbb{A}_i \cup \mathbb{A}_j \in \mathcal{F}(\mathbb{C}_i)$.
(ii) *If* $i \neq j$ *and* $\mathbb{A}_i, \mathbb{A}_j$ T*-co-cross then* $\mathbb{A}_i \setminus \mathbb{A}_j \in \mathcal{F}(\mathbb{C}_i)$ *and* $\mathbb{A}_j \setminus \mathbb{A}_i \in \mathcal{F}(\mathbb{C}_j)$.

Lemma 12(i) implies that if $\mathcal{F}$ is T-uncrossable, then for every $\mathbb{C}_i \in \mathcal{C}(\mathcal{F})$, the halo family of $\mathbb{C}_i$ has a unique maximal member (the union of the bisets in $\mathcal{F}(\mathbb{C}_i)$). The following statement easily follows from Lemma 12.

Corollary 5 [24]. *Let* $\mathcal{F}$ *be an arbitrary* T*-uncrossable biset family and let* $\mathcal{A}$ *be the family of the maximal members of the halo families of the* $\mathcal{F}$*-cores. Let* $\mathcal{A}'$ *be an independent set in the* T*-mesh graph of* $\mathcal{A}$. *Then the union of the halo families of the bisets in* $\mathcal{A}'$ *is a simple uncrossable biset family.*

Let now $\mathcal{F} = \mathcal{F}_{2\text{-}(s,T)}$. Let $\mathcal{G}$ be the T-mesh graph of $\mathcal{A}$ as in Corollary 5. By Lemma 11 $\mathcal{G}$ is a forest. Thus $\mathcal{G}$ is 2-colorable, so its nodes can be partitioned into 2 independent sets $\mathcal{A}'$ and $\mathcal{A}''$. The rest of the analysis coincides with [24]. Let $\mathcal{C}'$ and $\mathcal{C}''$ the set of $\mathcal{F}$-cores that correspond to $\mathcal{A}'$ and $\mathcal{A}''$, respectively. By Corollary 5, each one of the families $\mathcal{F}' = \bigcup_{\mathbb{C} \in \mathcal{C}'} \mathcal{F}(\mathbb{C})$ and $\mathcal{F}'' = \bigcup_{\mathbb{C} \in \mathcal{C}''} \mathcal{F}(\mathbb{C})$ is uncrossable and simple. Thus the problem of covering $\mathcal{F}' \cup \mathcal{F}''$ admits ratio 2β. After the family $\mathcal{F}' \cup \mathcal{F}''$ is covered, the inner part of every core of the residual family contains at least 2 terminals. Hence by Lemma 10(iii), the residual family is uncrossable, and thus the problem of covering it admits ratio β. Consequently, the overall ratio is $2\alpha + \beta$, as claimed in Theorem 7.

In the case when all edges in F are incident to s, $J \subseteq F$ is a feasible solution for the problem if and only if the set $\{v \in V : sv \in J\}$ is a hitting set of the hypergraph formed by the inner parts of the $\mathcal{F}$-cores. By Corollary 4, the maximum degree in this hypergraph is ≤ 2, and thus its minimum-weight hitting set can be found in polynomial time. This concludes the proof of Theorem 2.

References

1. Aazami, A., Cheriyan, J., Laekhanukit, B.: A bad example for the iterative rounding method for mincost k-connected spanning subgraphs. Discrete Optim. **10**(1), 25–41 (2013)
2. Auletta, V., Dinitz, Y., Nutov, Z., Parente, D.: A 2-approximation algorithm for finding an optimum 3-vertex-connected spanning subgraph. J. Algorithms **32**(1), 21–30 (1999)
3. Cheriyan, J., Végh, L.: Approximating minimum-cost k-node connected subgraphs via independence-free graphs. SIAM J. Comput. **43**(4), 1342–1362 (2014)
4. Cheriyan, J., Vempala, S., Vetta, A.: An approximation algorithm for the min-cost k-vertex connected subgraph. SIAM J. Comput. **32**(4), 1050–1055 (2003)
5. Chuzhoy, J., Khanna, S.: An $O(k^3 \log n)$-approximation algorithm for vertex-connectivity survivable network design. Theor. Comput. **8**(1), 401–413 (2012)
6. Dinitz, Y., Nutov, Z.: A 3-approximation algorithm for finding optimum 4, 5-vertex-connected spanning subgraphs. J. Algorithms **32**(1), 31–40 (1999)
7. Fackharoenphol, J., Laekhanukit, B.: An $O(\log^2 k)$-approximation algorithm for the k-vertex connected subgraph problem. SIAM J. Comput. **41**, 1095–1109 (2012)
8. Fleischer, L., Jain, K., Williamson, D.: Iterative rounding 2-approximation algorithms for minimum-cost vertex connectivity problems. J. Comput. Syst. Sci. **72**(5), 838–867 (2006)
9. Frank, A.: Rooted k-connections in digraphs. Discrete Appl. Math. **157**(6), 1242–1254 (2009)
10. Frank, A., Jordán, T.: Minimal edge-coverings of pairs of sets. J. Comb. Theor. B **65**, 73–110 (1995)
11. Frank, A., Tardos, E.: An application of submodular flows. Linear Algebra Appl. **114**(115), 329–348 (1989)
12. Fukunaga, T.: Approximating minimum cost source location problems with local vertex-connectivity demands. J. Discrete Algorithms **19**, 30–38 (2013)
13. Fukunaga, T.: Approximating the generalized terminal backup problem via half-integral multiflow relaxation. In: STACS, pp. 316–328 (2015)

14. Fukunaga, T., Nutov, Z., Ravi, R.: Iterative rounding approximation algorithms for degree-bounded node-connectivity network design. SIAM J. Comput. **44**(5), 1202–1229 (2015)

15. Goemans, M., Goldberg, A., Plotkin, S., Shmoys, D., Tardos, E., Williamson, D.: Improved approximation algorithms for network design problems. In: SODA, pp. 223–232 (1994)

16. Jain, K.: A factor 2 approximation algorithm for the generalized Steiner network problem. Combinatorica **21**(1), 39–60 (2001)

17. Jordán, T.: On the optimal vertex-connectivity augmentation. J. Comb. Theor. B **63**, 8–20 (1995)

18. Khuller, S., Raghavachari, B.: Improved approximation algorithms for uniform connectivity problems. J. Algorithms **21**, 434–450 (1996)

19. Kortsarz, G., Nutov, Z.: Approximating node connectivity problems via set covers. Algorithmica **37**, 75–92 (2003)

20. Kortsarz, G., Nutov, Z.: Approximating k-node connected subgraphs via critical graphs. SIAM J. Comput. **35**(1), 247–257 (2005)

21. Kortsarza, G., Nutov, Z.: Approximating source location and star survivable network problems. Manuscript (2015)

22. Laekhanukit, B.: Parameters of two-prover-one-round game and the hardness of connectivity problems. In: SODA, pp. 1626–1643 (2014)

23. Lando, Y., Nutov, Z.: Inapproximability of survivable networks. Theort. Comput. Sci. **410**(21–23), 2122–2125 (2009)

24. Nutov, Z.: Approximating minimum cost connectivity problems via uncrossable bifamilies. ACM Trans. Algorithms **9**(1), 1 (2012)

25. Nutov, Z.: Approximating node-connectivity augmentation problems. Algorithmica **63**(1–2), 398–410 (2012)

26. Nutov, Z.: Approximating minimum-cost edge-covers of crossing biset families. Combinatorica **34**(1), 95–114 (2014)

27. Ravi, R., Williamson, D.P.: An approximation algorithm for minimum-cost vertex-connectivity problems. Algorithmica **18**, 21–43 (1997)

28. Ravi, R., Williamson, D.P.: Erratum: an approximation algorithm for minimum-cost vertex-connectivity problems. Algorithmica **34**(1), 98–107 (2002)

29. Végh, L.: Augmenting undirected node-connectivity by one. SIAM J. Discrete Math. **25**(2), 695–718 (2011)

The Hardest Language for Conjunctive Grammars

Alexander Okhotin$^{(\boxtimes)}$

Department of Mathematics and Statistics,
University of Turku, FI-20014 Turku, Finland
alexander.okhotin@utu.fi

Abstract. A famous theorem by Greibach ("The hardest context-free language", *SIAM J. Comp.*, 1973) states that there exists such a context-free language L_0 that every context-free language over any alphabet is reducible to L_0 by a homomorphic reduction—in other words, is representable as an inverse homomorphic image $h^{-1}(L_0)$, for a suitable homomorphism h. This paper establishes similar characterizations for conjunctive grammars, that is, for grammars extended with a conjunction operator.

1 Introduction

One of the central notions in the theory of computation is that of a *hard set for a class*, to which all other sets from that class can be *reduced*: that is to say, the membership problem for each set in this class can be solved by mapping an element to be tested to an instance of the membership problem for the hardest set, using a relatively easily computable *reduction function*. Most computational complexity classes have their hardest sets, and their existence forms the fundamental knowledge in the area. For example, the set of Turing machines recognizing co-finite languages is complete for Σ_0^2 under recursive reductions, whereas Boolean formula satisfiability is complete for NP under polynomial-time reductions.

In the world of formal grammars, there is a result similar in spirit, established for a much more restricted notion of reducibility (which indicates a stronger result). In 1973, Greibach [8] proved that among the languages defined by the formal grammars of the ordinary kind ("context-free languages"), there is a hardest set under reductions by *homomorphisms*—that is, by functions mapping each symbol of the target language's alphabet to a string over the alphabet of the hardest language. To be precise, Greibach's hardest language theorem presents a particular context-free language L_0 over an alphabet Σ_0, with the following property: for every context-free language L over any alphabet Σ, there exists such a homomorphism $h \colon \Sigma \to \Sigma_0^+$, that a string $w \in \Sigma^*$ is in L if and only if its image $h(w)$ belongs to L_0. This language L_0 is then called *the hardest context-free language*, for the reason that every context-free language is reducible to it.

Among formal language theorists, this result is regarded as a *representation theorem*, in the sense that every language in the family is representable by

A.S. Kulikov and G.J. Woeginger (Eds.): CSR 2016, LNCS 9691, pp. 340–351, 2016.
DOI: 10.1007/978-3-319-34171-2_24

applying a certain operation to base languages of a restricted form. In case of Greibach's theorem, there is just one base language L_0, and every context-free language L is represented as an *inverse homomorphic image* $L = \{ w \mid h(w) \in L_0 \} = h^{-1}(L_0)$. For context-free languages, it is additionally known that they are closed under taking inverse homomorphisms, and thus Greibach's theorem [8] gives a necessary and sufficient condition: a language is context-free if and only if it is representable as $h^{-1}(L_0)$, for some h. Another famous example of a representation theorem is the Chomsky–Schützenberger theorem.

The possibility of having hardest sets under homomorphic reductions was also investigated for a few other families of formal languages. Already Greibach [9] showed that for the family of deterministic languages—that is, the languages described by LR(1) grammars—there cannot be a hardest language under homomorphic reductions. Greibach [8] has also proved a variant of her theorem for a certain restricted type of Turing machines. Čulík and Maurer [6] similarly found the same kind of a hardest language in the complexity class NSPACE(n). Boasson and Nivat [5] proved that there is no hardest language in the family described by linear grammars, Autebert [2] proved the same negative result for one-counter automata, whereas Čulík and Maurer [6] showed such a result for the regular languages.

The purpose of this paper is to prove the existence of the hardest language for the family of *conjunctive grammars* [12,16], which allow a conjunction of any syntactic conditions to be expressed in a rule. Consider that a rule $A \to BC$ in an ordinary grammar states that if a string w is representable as BC—that is, as $w = uv$, where u has the property B and v has the property C—then w has the property A. In a conjunctive grammar, one can define a rule of the form $A \to BC \,\&\, DE$, which asserts that every string w representable *both* as BC (with $w = uv$) *and at the same time* as DE (with $w = xy$) therefore has the property A. The importance of conjunctive grammars is justified by two facts: on the one hand, they add useful logical operations to standard inductive definitions of syntax, and these operations allow one to express some syntactic constructs beyond the scope of ordinary grammars. On the other hand, conjunctive grammars have generally the same parsing algorithms as ordinary grammars [1,16], and the same subcubic upper bound on the time complexity of parsing [17]. Among numerous theoretical results on conjunctive grammars, the one particularly relevant for this paper is the closure of the language family described by conjunctive grammars under inverse homomorphisms [11]. For more information on this grammar family, the reader is directed to a recent survey paper [16].

The result presented in this paper is that there exists a hardest language for conjunctive grammars. Unlike much of the previous work on conjunctive grammars, which worked out by directly extending the corresponding results from the case of ordinary grammars, a conjunctive version of Greibach's theorem has to be proved in quite a different way. Consider that Greibach's own proof of her theorem requires a grammar in the *Greibach normal form* [7], with all rules of the form $A \to a\alpha$, where a is a symbol of the alphabet. No analogue of this normal form has been established for conjunctive grammars, and for this reason, the proof of the hardest language theorem presented in this paper has to rely upon an entirely different normal form defined by Okhotin and Reitwießner [18]. Using that normal form instead of the Greibach normal form requires a new construction, which is quite different from the construction used by Greibach [7] for ordinary grammars.

2 Conjunctive Grammars

Rules in ordinary formal grammars may concatenate substrings to each other, and may use *disjunction* of syntactic conditions, represented by multiple rules for a nonterminal symbol. *Conjunctive grammars* extend this logic by allowing conjunction within the same kind of definitions.

Definition 1. *A conjunctive grammar is a quadruple* $G = (\Sigma, N, R, S)$, *in which:*

- Σ *is the* alphabet *of the language being defined;*
- N *is a finite set of symbols for the syntactic categories defined in the grammar ("nonterminal symbols");*
- R *is a finite set of* rules, *each of the form*

$$A \to \alpha_1 \& \ldots \& \alpha_m, \qquad (1)$$

   where $A \in N$, $m \geqslant 1$ *and* $\alpha_1, \ldots, \alpha_m \in (\Sigma \cup N)^*$;
- $S \in N$ *is a symbol representing the property of being a syntactically well-formed sentence of the language ("the initial symbol").*

Each concatenation α_i in a rule (1) is called a *conjunct*. If a grammar has a unique conjunct in every rule ($m = 1$), it is an *ordinary grammar* (Chomsky's "context-free").

Each rule (1) means that any string representable as each concatenation α_i therefore has the property A. These dependencies form *parse trees with shared leaves*, where the subtrees corresponding to different conjuncts in a rule (1) define multiple interpretations of the same substring, and accordingly lead to the same set of leaves, as illustrated in Fig. 1. This understanding can be formalized in several equivalent ways: by term rewriting [12], by logical deduction [15] and by language equations [13].

Consider one of those definitions, which extends Chomsky's definition of ordinary grammars by string rewriting, using terms instead of strings.

Definition 2 [12]. *Let* $G = (\Sigma, N, R, S)$ *be a conjunctive grammar, and consider terms over concatenation and conjunction, with symbols from* $\Sigma \cup N$ *and the empty string* ε *as atomic terms. The relation of one-step rewriting on such terms* $(\Longrightarrow)$ *is defined as follows.*

- *Using a rule* $A \to \alpha_1 \& \ldots \& \alpha_m \in R$, *any atomic subterm* $A \in N$ *of any term may be rewritten by the term* $(\alpha_1 \& \ldots \& \alpha_m)$.

$$\ldots A \ldots \Longrightarrow \ldots (\alpha_1 \& \ldots \& \alpha_m) \ldots$$

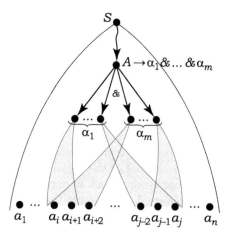

Fig. 1. Parse trees in conjunctive grammars: a subtree with root $A \to \alpha_1 \& \ldots \& \alpha_m$, representing m parses of a substring $a_i \ldots a_j$.

– *A conjunction of several identical strings may be rewritten to one such string.*

$$\ldots (w \& \ldots \& w) \ldots \Longrightarrow \ldots w \ldots \qquad (w \in \Sigma^*)$$

The language generated by a term φ is the set of all strings over Σ obtained from it in a finite number of rewriting steps.

$$L_G(\varphi) = \{\, w \mid w \in \Sigma^*,\ \varphi \Longrightarrow^* w \,\}$$

The language generated by the grammar is the language generated by its initial symbol.

$$L(G) = L_G(S) = \{\, w \mid w \in \Sigma^*,\ S \Longrightarrow^* w \,\}$$

Example 1 [12]. The following conjunctive grammar generates the language $\{\, a^n b^n c^n \mid n \geqslant 0 \,\}$.

$$S \to AB \,\&\, DC$$
$$A \to aA \mid \varepsilon$$
$$B \to bBc \mid \varepsilon$$
$$C \to cC \mid \varepsilon$$
$$D \to aDb \mid \varepsilon$$

The proof of an inverse homomorphic characterization for conjunctive grammars presented in this paper is based upon a normal form that extends the operator normal form for ordinary grammars. This normal form, called the *odd normal form*, derives its name from the fact that all strings generated by all nonterminal symbols (except maybe the initial symbol), are of odd length.

Theorem A (Okhotin and Reitwießner [18]**).** *For every conjunctive grammar there exists and can be effectively constructed a conjunctive grammar*

$G = (\Sigma, N, R, S)$ *generating the same language, which is in the* **odd normal form***, that is, with all rules of the following form.*

$$A \to B_1 a_1 C_1 \& \ldots \& B_m a_m C_m \qquad (m \geqslant 1,\ B_i, C_i \in N,\ a_i \in \Sigma)$$
$$A \to a \qquad\qquad\qquad\qquad\qquad (a \in \Sigma)$$

If the initial symbol S is never used in the right-hand sides of any rules, then the rules of the form $S \to aA$, with $a \in \Sigma$ and $A \in N$, and $S \to \varepsilon$ are also allowed.

For the purposes of this paper, the crucial quality of the odd normal form is that every conjunct contains a symbol of the alphabet, such as a in BaC. The homomorphic image $h(a)$ of that symbol shall, in particular, encode the conjunct BaC. A substring can then be parsed according to that conjunct, beginning from the image $h(a)$, and then proceeding in both directions to parse the appropriate substrings according to B and to C.

3   Hardest Language for Conjunctive Grammars

The goal is to establish the following analogue of Greibach's hardest language theorem for conjunctive grammars.

Theorem 1. *There exists such a conjunctive language L_0 over the alphabet $\Sigma_0 = \{a, b, c, d, \#\}$, that for every conjunctive language L over any alphabet Σ, there is such a homomorphism $h \colon \Sigma \to \Sigma_0^*$, that $L = h^{-1}(L_0)$ if $\varepsilon \notin L$ and $L = h^{-1}(L_0 \cup \{\varepsilon\})$ if $\varepsilon \in L$.*

Let $G = (\Sigma, N, R, X)$ be any conjunctive grammar, in which every conjunct is of the form YsZ, Ys, sZ or s, where $Y, Z \in N$ and $s \in \Sigma$; in particular, every grammar in the odd normal form satisfies this condition. Let $\mathcal{C} = \{\alpha_1, \alpha_2, \ldots, \alpha_{|\mathcal{C}|}\}$, with $\alpha_i \in N\Sigma N \cup \Sigma N \cup N\Sigma \cup \Sigma$, be an enumeration of all distinct conjuncts used in the grammar, so that each rule is of the form

$$A \to \alpha_{i_1} \& \ldots \& \alpha_{i_m}, \quad \text{with } m \geqslant 1 \text{ and } i_1, \ldots, i_m \in \{1, \ldots, |\mathcal{C}|\}. \tag{2}$$

The proposed encoding of G consists of *definitions of conjuncts*. For each conjunct $\alpha_i = YsZ$, its definition is included in the image of the symbol s, so that a substring $h(usv)$ could be parsed according to YsZ beginning from $h(s)$ in the middle. The definition consists of all possible *expansions* of a conjunct, with any rules for Y and for Z substituted instead of Y and Z. Consider any two such rules.

$$Y \to \alpha_{i_1} \& \ldots \& \alpha_{i_m}$$
$$Z \to \alpha_{j_1} \& \ldots \& \alpha_{j_n}$$

The expansion of conjunct number i with this pair of rules represents them as two lists of conjunct numbers: the list of conjuncts $\{i_1, \ldots, i_m\}$ to be used on the left, representing the rule for Y, and the list of conjuncts $\{j_1, \ldots, j_n\}$ on the right. This expansion is then encoded in the image of the symbol s as a triple $(\{i_1, \ldots, i_m\}, i, \{j_1, \ldots, j_n\})$. The image of s contains such triples for every conjunct YsZ and for every choice of rules for Y and for Z.

The encoding uses a five-symbol alphabet $\Sigma_0 = \{a, b, c, d, \#\}$, in which the symbols have the following meaning.

- Symbols a are used to represent any reference to each conjunct α_i as a^i.
- Symbols b are used to mark each expansion of a conjunct $\alpha_i = YsZ$ by b^i.
- The symbol c represents conjunction in the right-hand side of any rule. Each rule (2) has the following symmetric *left* and *right* representations that list all conjuncts used in the rule.

$$\lambda(A \to \alpha_{i_1} \& \ldots \& \alpha_{i_m}) = ca^{i_m} \ldots ca^{i_1}$$
$$\rho(A \to \alpha_{i_1} \& \ldots \& \alpha_{i_m}) = a^{i_1}c \ldots a^{i_m}c$$

Any expansion of a conjunct $\alpha_k = YsZ$ consists of a marker b^k preceded by a left representation of a rule r for Y and followed by a right representation of a rule r' for Z. This is a string $\lambda(r)b^k\rho(r')$. For a conjunct $\alpha_k = Ys$ with Z omitted, its expansion accordingly omits $\rho(r')$ and is of the form $\lambda(r)b^k$. Similarly, a conjunct $\alpha_k = sZ$ is expanded as $b^k\rho(r')$, where r' is a rule for Z. A conjunct $\alpha_k = s$ has a unique expansion b^k.

- The symbol d is used to separate any expansions of conjuncts with the same symbol s in the middle. The definition of a conjunct α_k, denoted by $\sigma(\alpha_k)$, is the following concatenation of all its expansions (using the product notation for concatenations over all rules).

$$\sigma(\alpha_k) = \begin{cases} \prod_{r \text{ is a rule for } Y} \prod_{r' \text{ is a rule for } Z} \lambda(r)b^k\rho(r')d, & \text{if } \alpha_k = YsZ \\ \prod_{r \text{ is a rule for } Y} \lambda(r)b^kd, & \text{if } \alpha_k = Ys \\ \prod_{r' \text{ is a rule for } Z} b^k\rho(r')d, & \text{if } \alpha_k = sZ \\ b^kd, & \text{if } \alpha_k = s \end{cases}$$

- The separator symbol $\# \in \Sigma_0$ concludes the image $h(s)$ of any symbol $s \in \Sigma$.

The image of every symbol $s \in \Sigma$ under h consists of two parts. It begins with the set of all rules for the initial symbol, which will actually be used only in the image of the first symbol of a string. Next, after a double separator dd, there is the list of all expansions of all conjuncts containing the symbol s. The image is concluded with the separator symbol $\#$.

$$h_G(s) = \left(\prod_{r \text{ is a rule for } S} \rho(r)d \right) \cdot d \cdot \left(\prod_{\substack{\alpha_k \in \mathcal{C} \\ \alpha_k \in NsN \cup sN \cup Ns \cup \{s\}}} \sigma(\alpha_k) \right) \cdot \#$$

These definitions are illustrated on the following grammar.

Example 2. Let $\Sigma = \{s, t\}$ and consider a grammar $G = (\Sigma, \{X, Y, Z\}, R, X)$, with the following rules.

$$X \rightarrow tY$$
$$Y \rightarrow YsZ \,\&\, ZsZ \mid t$$
$$Z \rightarrow t$$

The parse tree of the string $ttst$ is presented in Fig. 2.

Let the conjuncts in G be numbered as $\alpha_1 = tY$, $\alpha_2 = YsZ$, $\alpha_3 = ZsZ$ and $\alpha_4 = t$. Then its rules have the following left and right representations.

$$\lambda(X \rightarrow tY) = ac \qquad\qquad \rho(X \rightarrow tY) = ca$$
$$\lambda(Y \rightarrow YsZ \,\&\, ZsZ) = ca^3ca^2 \qquad \rho(Y \rightarrow YsZ \,\&\, ZsZ) = a^2ca^3c$$
$$\lambda(Y \rightarrow t) = a^4c \qquad\qquad \rho(X \rightarrow tY) = ca^4$$
$$\lambda(Z \rightarrow t) = a^4c \qquad\qquad \rho(Z \rightarrow tY) = ca^4$$

The image of the symbol s begins with the rule for the initial symbol X and continues with the expansions of the conjuncts YsZ and ZsZ (that is, of all conjuncts containing the symbol s). The conjunct YsZ has two expansions, one with the first rule for Y and the other with the second rule for Y. The conjunct ZsZ has only one expansion, because Z has a unique rule $Z \rightarrow t$.

$$h_G(s) = \underbrace{ac}_{\rho(X \rightarrow tY)} dd \overbrace{ca^2ca^3}^{\lambda(Y \rightarrow YsZ\,\&\,ZsZ)} b^2 \underbrace{a^4c}_{\rho(Z \rightarrow t)} \overbrace{d}^{\lambda(Y \rightarrow t)} ca^4 b^2 \underbrace{a^4c}_{\rho(Z \rightarrow t)} \overbrace{d}^{\lambda(Z \rightarrow t)} ca^4 b^3 \underbrace{a^4c}_{\rho(Z \rightarrow t)} d\#$$

The image of t begins with the same rule for the initial symbol X. Then, there are two conjuncts to expand: tY and t. The former conjunct has two expansions corresponding to the two rules for Y, whereas the conjunct $\alpha_4 = t$ has a unique expansion not referring to any rules (b^4).

$$h_G(t) = \underbrace{ac}_{\rho(X \rightarrow tY)} dd b \underbrace{a^4c}_{\rho(Y \rightarrow t)} db \overbrace{a^2ca^3c}^{\rho(Y \rightarrow YsZ\,\&\,ZsZ)} db^4d\#,$$

Accordingly, the string $ttst \in L$ has the following image.

$$h_G(ttst) = acddba^4cdba^2ca^3cdb^4d\# \, acddba^4cdba^2ca^3cdb^4d\#$$
$$acddca^2ca^3b^2a^4cdca^4b^2a^4cdca^4b^3a^4cd\# \, acddba^4cdba^2ca^3cdb^4d\#$$

Figure 2 illustrates, how the parse tree of the string $ttst$ for the grammar in Example 2 can be reconstructed by following the pointers inside this image.

In the general case, the goal is to construct a single conjunctive grammar that describes this kind of structure. The desired grammar $G_0 = (\Sigma_0, N_0, R_0, S_0)$ uses the following set of nonterminal symbols; the purpose of each of them is explained below, along with the rules of the grammar.

$$N_0 = \{S_0, A, B, C, D, \overrightarrow{E}, \overrightarrow{E_+}, \overrightarrow{F}, \overleftarrow{E}, \overleftarrow{E_+}, \overleftarrow{F}, \overrightarrow{E_0}, \overrightarrow{F_0}\}$$

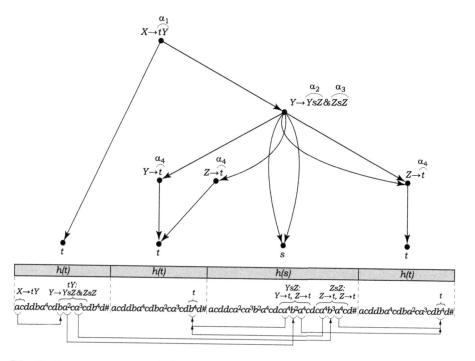

Fig. 2. How a parse tree of the string $w = ttst$ is reconstructed by analyzing its image $h(w)$.

The main task carried out in the grammar G_0 is parsing a substring $w \in \Sigma^*$ according to a rule r of G, given its image $h(w)$. Unless w is the whole string, the matching begins in the image of either of the symbols adjacent to w. Assume that this is the symbol immediately preceding w (the other case is handled symmetrically), and let $t \in \Sigma^*$ be that symbol.

The grammar G_0 implements this by defining a string $\rho(r)dx\#h_G(w)$, where $\rho(r)dx\#$ is a suffix of $h(t)$ that begins with an encoding of the desired rule, and the remaining symbols in $h(t)$ are denoted by $x \in \{a, b, c, d\}^*$. Such a string is defined by the nonterminal symbol $\overrightarrow{E}$, which should verify that each of the conjuncts listed in $\rho(r)$ generates the substring w.

Let the rule r be $A \to \alpha_{i_1} \& \ldots \& \alpha_{i_m}$, so that its right representation is $\rho(r) = a^{i_1}c \ldots a^{i_m}c$. Then, $\overrightarrow{E}$ describes the form of the string $a^{i_1}ca^{i_2}c \ldots a^{i_m}cdx\#h_G(w)$ by checking that w can be parsed as the conjunct α_{i_1}, and at the same time by applying the same test to the string $a^{i_2}c \ldots a^{i_m}cdx\#h_G(w)$, in order to check the rest of the conjuncts of this rule (the self-reference actually uses a variant of $\overrightarrow{E}$, denoted by $\overrightarrow{E_+}$, which is explained later).

Consider how the first conjunct α_{i_1} is checked by $\overrightarrow{E}$. Assume that it is of the form $\alpha_{i_1} = YsZ$, with $Y, Z \in N$ and $s \in \Sigma$. For w to be generated by this conjunct, there should exist a partition $w = usv$, with u generated by Y using some rule r', and with v generated by Z using a rule r''. Then the image of s contains a substring $\lambda(r')b^{i_1}\rho(r'')$ corresponding to this case, with $h(s) = x'\lambda(r')b^{i_1}\rho(r'')x''$. The task for $\overrightarrow{E}$ is to locate this substring and then apply another instance of $\overrightarrow{E}$ to the suffix $\rho(r'')x''h(v)$, to test the generation of v by r''; the prefix $h(u)x'\lambda(r'')$ is similarly checked by a symbol $\overleftarrow{E}$ with symmetrically defined rules, ensuring that u is generated by r'.

$$a^{i_1}ca^{i_2}c\ldots a^{i_m}cdx\# \cdot \overbrace{h(u) \cdot x'\, \lambda(r')\, b^{i_1}\, \overbrace{\rho(r'')\, x'' \cdot h(v)}^{\overrightarrow{E}}}^{\overbrace{}^{\overleftarrow{E}}} \in L_{G_0}(\overrightarrow{E})$$

where the large brace over the whole right part is labeled $\overrightarrow{F}$.

Two rules are defined for $\overrightarrow{E}$. The first rule begins the test for an encoded rule r, as explained above, with the first conjunct using an intermediate symbol $\overrightarrow{F}$ to check the conjunct's number and to apply $\overleftarrow{E}$ to the correct substring, and with the second conjunct skipping $a^{i_1}c$ and invoking $\overrightarrow{E_+}$ to test the rest of the rule r, beginning with the conjunct i_2.

$$\overrightarrow{E} \to \overrightarrow{F}\,\overrightarrow{E} \;\&\; Ac\overrightarrow{E_+} \tag{3a}$$

$$A \to aA \mid a \tag{3b}$$

The other rule for $\overrightarrow{E}$ handles the case when the rule r is entirely missing, and the image of t contains an encoding of a conjunct Xt or t. In this case, $\overrightarrow{E}$ has to ensure that w is empty, which is handled by the following rule.

$$\overrightarrow{E} \to dC\# \tag{3c}$$

$$C \to aC \mid bC \mid cC \mid dC \mid \varepsilon \tag{3d}$$

The rules for $\overrightarrow{F}$ are responsible for matching the conjunct's code a^{i_1} to the corresponding code b^{i_1} for this conjunct's expansion, and then skip the rest of the image $h(t)$ to apply $\overleftarrow{E}$ to $h(u)x'\lambda(r'')$.

$$\overrightarrow{F} \to a\overrightarrow{F}b \mid acC\#\overleftarrow{E}b \tag{3e}$$

The rest of the conjuncts of this rule $(\alpha_{i_2}, \ldots, \alpha_{i_m})$ are checked in exactly the same way, using the symbol $\overrightarrow{E_+}$ with the same main rule. Once all the conjuncts are handled, the second rule for $\overrightarrow{E_+}$ skips the remaining string $dx\#h(w)$.

$$\overrightarrow{E_+} \to \overrightarrow{F}\,\overrightarrow{E} \;\&\; Ac\overrightarrow{E_+} \mid dC\#D \tag{3f}$$

$$D \to C\#D \mid \varepsilon \tag{3g}$$

Symmetrically to $\overrightarrow{E}$, the symbol $\overleftarrow{E}$ describes a string $h_G(w)x\lambda(r)$, with $\lambda(r) = ca^{i_m}\ldots ca^{i_2}ca^{i_1}$ and $x \in \{a,b,c,d\}^*$, as long as w is generated by each of the conjuncts $\alpha_{i_1}, \ldots, \alpha_{i_m}$.

$$\overleftarrow{E} \to \overleftarrow{E}\overleftarrow{F} \mathbin{\&} \overleftarrow{E_+}cA \mid Cd \tag{3h}$$

$$\overleftarrow{E_+} \to \overleftarrow{E}\overleftarrow{F} \mathbin{\&} \overleftarrow{E_+}cA \mid DCd \tag{3i}$$

$$\overleftarrow{F} \to b\overleftarrow{F}a \mid b\overrightarrow{E}Cca \tag{3j}$$

It remains to define the rules for the initial symbol S_0, which should describe an encoding $h_G(w)$ if and only if S generates w in the grammar G. Let $w = a_1 \dots a_\ell$. The image of its first symbol a_1 (just like the image of any subsequent symbol) begins with the list of all the rules for S in G; the rule for S_0 chooses one of these encoded rules and then uses another variant of $\overrightarrow{E}$, called $\overrightarrow{E_0}$, to match the whole string w according to this rule.

$$S_0 \to BdS_0 \mid \overrightarrow{F_0}\overrightarrow{E} \mathbin{\&} Ac\overrightarrow{E_0} \tag{3k}$$

$$B \to aB \mid cB \mid a \mid c \tag{3l}$$

The reason for using a new symbol $\overrightarrow{E_0}$ rather than $\overrightarrow{E_+}$ is that some conjuncts of the encoded rule for S may be of the form sY or just s, and in this case one has to match a rule written in the first part of $h(a_1)$ to its definition included in the second part of the image of the same symbol.

$$\overrightarrow{E_0} \to \overrightarrow{F_0}\overrightarrow{E} \mathbin{\&} Ac\overrightarrow{E_0} \mid dC\#D \tag{3m}$$

$$\overrightarrow{F_0} \to a\overrightarrow{F_0}b \mid acCdb \mid ac\overleftarrow{E}b \tag{3n}$$

This completes the construction of the "hardest" conjunctive grammar G_0.

Lemma 1. *Let $G = (\Sigma, N, R, X)$ be a conjunctive grammar with all conjuncts of the form YsZ, Ys, sZ or s, where $Y, Z \in N$ and $s \in \Sigma$. Then, a string $h(w)$, with $w \in \Sigma^+$, is in $L(G_0)$ if and only if w is in $L(G)$.*

A proof of the lemma can be given by induction on the length of strings, along the lines of the above explanation. The lemma then implies Theorem 1.

Suggested future work on this characterization of conjunctive grammars is to construct a different hardest language that would have a simpler grammar, or perhaps some theoretically interesting explanation, similar to Greibach's "nondeterministic version of the Dyck set" [8].

4   Further Characterizations

A similar hardest language theorem can be established for the family of *Boolean grammars* [14,16], which are a further extension of conjunctive grammars featuring a negation operator. A Boolean grammar allows such rules as $A \to BC \mathbin{\&} \neg DE$, which expresses all strings representable as BC, but not representable as DE; this definition is formalized using language equations [10,14]. The hardest language theorem is proved by a slight extension of the construction in Sect. 3, and will be included in the full version of this paper.

Another noteworthy extension of ordinary formal grammars are the *multi-component grammars* of Vijay-Shanker et al. [20] and of Seki et al. [19]. In these grammars, a nonterminal symbol may deal with several substrings at once, and those substrings may be arbitrarily concatenated in the rules. The maximum number of substrings handled at once is the *dimension* of the grammar, and there is an infinite hierarchy of representable languages with respect to this dimension. Since the language family described by grammars of every fixed dimension d is closed under inverse homomorphisms, there cannot be a single hardest language for all dimensions. However, it is possible that there is a hardest language for each dimension, and checking that is left for future studies.

A similar question can be asked about *grammars with context operators*, recently defined by Barash and Okhotin [3,4], which further extend conjunctive grammars and pertain to implementing the notion of a rule applicable in a context. It is not yet known whether they are closed under inverse homomorphisms, and establishing an analogue of the odd normal form might be a challenging task. Nevertheless, the existence of a hardest language is conjectured.

References

1. Aizikowitz, T., Kaminski, M.: $LR(0)$ conjunctive grammars and deterministic synchronized alternating pushdown automata. In: Kulikov, A., Vereshchagin, N. (eds.) CSR 2011. LNCS, vol. 6651, pp. 345–358. Springer, Heidelberg (2011)
2. Autebert, J.-M.: Non-principalité du cylindre des langages à compteur. Math. Syst. Theory 11(1), 157–167 (1977)
3. Barash, M., Okhotin, A.: An extension of context-free grammars with one-sided context specifications. Inf. Comput. 237, 268–293 (2014)
4. Barash, M., Okhotin, A.: Two-sided context specifications in formal grammars. Theoret. Comput. Sci. 591, 134–153 (2015)
5. Boasson, L., Nivat, M.: Le cylindre des langages linéaires. Math. Syst. Theory 11, 147–155 (1977)
6. Čulík II, K., Maurer, H.A.: On simple representations of language families. RAIRO Informatique Théorique et Appl. 13(3), 241–250 (1979)
7. Greibach, S.A.: A new normal-form theorem for context-free phrase structure grammars. J. ACM 12, 42–52 (1965)
8. Greibach, S.A.: The hardest context-free language. SIAM J. Comput. 2(4), 304–310 (1973)
9. Greibach, S.A.: Jump PDA's and hierarchies of deterministic context-free languages. SIAM J. Comput. 3(2), 111–127 (1974)
10. Kountouriotis, V., Nomikos, C., Rondogiannis, P.: Well-founded semantics for Boolean grammars. Inf. Comput. 207(9), 945–967 (2009)
11. Lehtinen, T., Okhotin, A.: Boolean grammars and GSM mappings. Int. J. Found. Comput. Sci. 21(5), 799–815 (2010)
12. Okhotin, A.: Conjunctive grammars. J. Automata Lang. Comb. 6(4), 519–535 (2001)
13. Okhotin, A.: Conjunctive grammars and systems of language equations. Program. Comput. Sci. 28(5), 243–249 (2002)
14. Okhotin, A.: Boolean grammars. Inf. Comput. 194(1), 19–48 (2004)

15. Okhotin, A.: The dual of concatenation. Theoret. Comput. Sci. **345**(2–3), 425–447 (2005)
16. Okhotin, A.: Conjunctive and Boolean grammars: the true general case of the context-free grammars. Comput. Sci. Rev. **9**, 27–59 (2013)
17. Okhotin, A.: Parsing by matrix multiplication generalized to Boolean grammars. Theoret. Comput. Sci. **516**, 101–120 (2014)
18. Okhotin, A., Reitwießner, C.: Conjunctive grammars with restricted disjunction. Theoret. Comput. Sci. **411**(26–28), 2559–2571 (2010)
19. Seki, H., Matsumura, T., Fujii, M., Kasami, T.: On multiple context-free grammars. Theoret. Comput. Sci. **88**(2), 191–229 (1991)
20. Vijay-Shanker, K., Weir, D.J., Joshi, A.K.: Characterizing structural descriptions produced by various grammatical formalisms. In: 25th Annual Meeting of the Association for Computational Linguistics (ACL 1987), pp. 104–111 (1987)

Low-Rank Approximation of a Matrix: Novel Insights, New Progress, and Extensions

Victor Y. Pan[1,2(✉)] and Liang Zhao[2]

[1] Departments of Mathematics and Computer Science, Lehman College of the City University of New York, Bronx, NY 10468, USA
victor.pan@lehman.cuny.edu

[2] Ph.D. Programs in Mathematics and Computer Science, The Graduate Center of the City University of New York, New York, NY 10036, USA
lzhao1@gc.cuny.edu
http://comet.lehman.cuny.edu/vpan/

Abstract. Empirical performance of the celebrated algorithms for low-rank approximation of a matrix by means of random sampling has been consistently efficient in various studies with various sparse and structured multipliers, but so far formal support for this empirical observation has been missing. Our new insight into this subject enables us to provide such an elusive formal support. Furthermore, our approach promises significant acceleration of the known algorithms by means of sampling with more efficient sparse and structured multipliers. It should also lead to enhanced performance of other fundamental matrix algorithms. Our formal results and our initial numerical tests are in good accordance with each other, and we have already extended our progress to the acceleration of the Fast Multipole Method and the Conjugate Gradient algorithms.

Keywords: Low-rank approximation of a matrix · Random sampling · Derandomization · Fast multipole method · Conjugate gradient algorithms

1 Introduction

Low-rank approximation of a matrix by means of random sampling is an increasingly popular subject area with applications to the most fundamental matrix computations [19] as well as numerous problems of data mining and analysis, "ranging from term document data to DNA SNP data" [20]. See [19,20], and [14, Sect. 10.4.5], for surveys and ample bibliography; see [7,12,16,17,26], for sample early works.

All these studies rely on the proved efficiency of random sampling with Gaussian multipliers and on the empirical evidence that the algorithms work as efficiently with various random sparse and structured multipliers. So far formal support for this empirical evidence has been missing, however.

Our novel insight into this subject provides such an elusive formal support and promises significant acceleration of the computations. Next we outline our progress and then specify it in some detail (see also [23]).

© Springer International Publishing Switzerland 2016
A.S. Kulikov and G.J. Woeginger (Eds.): CSR 2016, LNCS 9691, pp. 352–366, 2016.
DOI: 10.1007/978-3-319-34171-2_25

We recall some basic definitions in the next section and in the Appendix. As this is customary in the study of matrix computations, we use freely the concepts "large", "small", "near", "close", "approximate", "ill-conditioned" and "well-conditioned" quantified in the context, although we specify them quantitatively as needed.

The acronym "i.i.d." stands for "independent identically distributed", and we refer to standard Gaussian random variables just as *Gaussian*.

We call an $m \times n$ matrix *Gaussian* if all its entries are i.i.d. Gaussian variables.

Hereafter *"likely"* means "with a probability close to 1", and *"flop"* stands for "floating point arithmetic operation".

Basic Algorithm; its Efficiency with Gaussian Random Multipliers. nrank(M) denotes *numerical rank* of an $m \times n$ matrix M: a matrix M can be closely approximated by a matrix of rank at most r if and only if $r \geq$ nrank(M) (cf. part 9 of the next section).

The following randomized algorithm (cf. [19, Algorithm 4.1 and Sect. 10]) computes such an approximation by a product FH where F and H are $m \times l$ and $l \times n$ matrices, respectively, $l \geq r$, and in numerous applications of the algorithm, l is small compared to m and n.

Algorithm 1. Low-rank approximation of a matrix via random sampling.

INPUT: *An $m \times n$ matrix M having numerical rank r where $m \geq n > r > 0$.*
INITIALIZATION: *Fix an integer p such that $0 \leq p < n - r$. Compute $l = r + p$.*
COMPUTATIONS: *1. Generate an $n \times l$ matrix B. Compute the matrix MB.*
2. Orthogonalize its columns, producing matrix $Q = Q(MB)$.
3. Output the rank-l matrix $\tilde{M} = QQ^T M \approx M$ and the relative residual norm $\Delta = \frac{\|\tilde{M} - M\|}{\|M\|}$.

The following theorem supports Algorithm 1 with a Gaussian multiplier B.

Theorem 1. The Power of Gaussian random sampling. *Approximation of a matrix M by a rank-l matrix produced by Algorithm 1 is likely to be optimal up to a factor of f having expected value $1 + (1 + \sqrt{m} + \sqrt{l})\frac{e}{p}\sqrt{r}$, provided that $e = 2.71828\ldots$, nrank(M) $\leq r$, and B is an $n \times l$ Gaussian matrix.*

This is [19, Theorem 10.6] for a slightly distinct factor f. For the sake of consistency and completeness of our presentation, we prove the theorem in Sect. 4 and then extend it into a new direction.

By combining Algorithm 1 with the Power Scheme of Remark 1, one can decrease the bound of the theorem dramatically at a low computational cost.

Modifications with SRFT and SRHT Structured Random Sampling. A Gaussian matrix B involves nl random parameters, and we multiply it by M by using $ml(2n - 1)$ flops. They dominate $O(ml^2)$ flops for orthogonalization at Stage 2 if $l \ll n$.

An $n \times n$ matrix B of *subsample random Fourier or Hadamard transform*[1] is defined by n parameters, and we need just $O(mn \log(l))$ flops, $l = O(r \log(r))$, in order to multiply an input matrix M by an $n \times l$ block of such a matrix B (cf. [19, Sects. 4.6 and 11]).

Both SRFT and SRHT multipliers are *universal*, like Gaussian ones: it is proven that Algorithm 1 is likely to produce correct output with them for any matrix M, although the known upper bounds on the failure probability increase. E.g., with SRFT multipliers such a bound grows to $O(1/r)$ from order $3 \exp(-p)$, for $p \geq 4$, with Gaussian ones (cf. [19, Theorems 10.9 and 11.1]).

Empirically the algorithm fails very rarely with SRFT multipliers, $l = r + p$, $p = 20$, and even $p = 4$ (and similarly with SRHT multipliers), but for special inputs M, it is likely to fail if $l = O(r \log(r))$ (cf. [19, Remark 11.2]).

Related Work. Similar empirical behavior has been consistently observed by ourselves and by many other researchers when Algorithm 1 was applied with a variety of sparse and structured multipliers [19,20,22], but so far formal support for such empirical observations has been missing from the huge bibliography on this highly popular subject.

Our Goals and our Progress. In this paper we are going to

(i) fill the void in the bibliography by supplying the missing *formal support*,
(ii) define *new more efficient multipliers* for low-rank approximation,
(iii) compare our formal results with those of *our numerical tests*, and
(iv) *extend our findings* to another important computational area.

Our basic step is the proof of a dual version of Theorem 1, which relies on the following concept.

Definition 1. Factor Gaussian matrices with small expected rank. *For three integers m, n, and r, $m \geq n > r > 0$, define the class $\mathcal{G}(m,n,r)$ of $m \times n$ factor Gaussian matrices $M = UV$ with expected rank r such that U is an $m \times r$ Gaussian matrix and V is a $r \times n$ Gaussian matrix.*

Recall that rectangular Gaussian matrices have full rank with probability 1 and are likely to be well-conditioned (see Theorems 9 and 10), and so the matrices $M \in \mathcal{G}(m,n,r)$ are likely to have rank r indeed, as we state in the definition.

Theorem 2. The Duality Theorem.[2]
The claims of Theorem 1 still hold if we assume that the $m \times n$ input matrix M is a small-norm perturbation of a factor Gaussian matrix with expected numerical rank r and if we allow a multiplier B to be any $n \times l$ well-conditioned matrix of full rank l.

---

[1] Hereafter we use the acronyms *SRFT* and *SRHT*.
[2] It is sufficient to prove the theorem for the matrices in $\mathcal{G}(m,n,r)$. The extension to their small-norm perturbation readily follows from Theorem 3 of the next section.

Theorem 2 implies that Algorithm 1 produces a low-rank approximation to average input matrix M that has numerical rank r under the mildest possible restriction on the choice of a multiplier provided that average matrix is defined under the Gaussian probability distribution. This provision is customary, and it is quite natural in view of the Central Limit Theorem.

Our novel point of view implies formal support for the cited empirical observations and for almost unrestricted choice of efficient multipliers, which can be viewed as *derandomization* of Algorithm 1 and which promises significant improvement of its performance. This promise, based on our formal analysis, turned out to be in good accordance with our initial numerical tests in Sect. 5 and [23].

In Sect. 3 we describe some promising candidate sparse and structured multipliers, in particular those simplifying the matrices of SRHT, and we test our recipes numerically, thus fulfilling our goals (ii) and (iii). As we prove in Sect. 4, the accuracy of low-rank approximations output by Algorithm 1 is likely to be reasonable even where the oversampling integer p vanishes and to increase fast as this integer grows. Moreover we can increase the accuracy dramatically by applying the Power Scheme of Remark 1 at a low computational cost.

Our progress can be extended to various other computational tasks, e.g. (see Sect. 6), to the acceleration of the Fast Multipole Method[3] of [5,15], listed as one of the 10 most important algorithms of the 20th century, widely used, and increasingly popular in Modern Matrix Computations.

The extension to new tasks is valid as long as the definition of their average inputs under the Gaussian probability distribution is appropriate and relevant. For a specific input to a chosen task we can test the validity of extension *by action*, that is, by applying our algorithms and checking the output accuracy.

Organization of the Paper. We organize our presentation as follows. In the next section and in the Appendix we recall some basic definitions. In Sect. 3 we specify our results for low-rank approximation, by elaborating upon Theorems 1 and 2. In Sect. 4 we prove these theorems. Section 5 covers our numerical experiments, which are the contribution of the second author. In Sect. 6 we extend our results to the acceleration of the FMM.

In our report [23], we extend them further to the acceleration of the Conjugate Gradient celebrated algorithms and include a proof of our Theorem 7 and test results omitted due to the limitation on the paper size.

2 Some Definitions and Basic Results

We recall some relevant definitions and basic results for random matrices in the Appendix. Next we list some definitions for matrix computations (cf. [14]).

For simplicity we assume dealing with real matrices, but our study can be readily extended to the complex case.

1. I_g is a $g \times g$ identity matrix. $O_{k,l}$ is the $k \times l$ matrix filled with zeros.

---

[3] Hereafter we use the acronym FMM.

2. $(B_1 \mid B_2 \mid \cdots \mid B_k)$ is a block vector of length k, and $\mathrm{diag}(B_1, B_2, \ldots, B_k)$ is a $k \times k$ block diagonal matrix, in both cases with blocks $B_1, B_2, \ldots, B_k$.

3. $W_{k,l}$ denotes the $k \times l$ leading (that is, northwestern) block of an $m \times n$ matrix W for $k \le m$ and $l \le n$. W^T denotes its transpose.

4. An $m \times n$ matrix W is called *orthogonal* if $W^T W = I_n$ or if $WW^T = I_m$.

5. $W = S_{W,\rho} \Sigma_{W,\rho} T^T_{W,\rho}$ is *compact SVD* of a matrix W of rank ρ with $S_{W,\rho}$ and $T_{W,\rho}$ orthogonal matrices of its singular vectors and $\Sigma_{W,\rho} = \mathrm{diag}(\sigma_j(W))^\rho_{j=1}$ the diagonal matrix of its singular values in non-increasing order; $\sigma_\rho(W) > 0$.

6. $W^+ = T_{W,\rho} \Sigma^{-1}_{W,\rho} S^T_{W,\rho}$ is the *Moore–Penrose pseudo inverse* of the matrix W. ($W^+ = W^{-1}$ for a nonsingular matrix W.)

7. $\|W\| = \sigma_1(W)$ and $\|W\|_F = (\sum^\rho_{j=1} \sigma^2_j(W))^{1/2} \le \sqrt{n}\,\|W\|$ denote its *spectral and Frobenius norms*, respectively. ($\|W^+\| = \frac{1}{\sigma_\rho(W)}$; $\|U\| = \|U^+\| = 1$, $\|UW\| = \|W\|$ and $\|WU\| = \|W\|$ if the matrix U is orthogonal.)

8. $\kappa(W) = \|W\|\,\|W^+\| = \sigma_1(W)/\sigma_\rho(W) \ge 1$ denotes the *condition number* of a matrix W. A matrix is called *ill-conditioned* if its condition number is large in context and is called *well-conditioned* if this number $\kappa(W)$ is reasonably bounded. (An $m \times n$ matrix is ill-conditioned if and only if it has a matrix of a smaller rank nearby, and it is well-conditioned if and only if it has full numerical rank $\min\{m, n\}$.)

9. **Theorem 3.** *Suppose C and $C + E$ are two nonsingular matrices of the same size and $\|C^{-1}E\| = \theta < 1$. Then $\||(C + E)^{-1} - C^{-1}\| \le \frac{\theta}{1-\theta}\|C^{-1}\|$. In particular, $\||(C + E)^{-1} - C^{-1}\| \le 0.5\|C^{-1}\|$ if $\theta \le 1/3$.*

Proof. See [24, Corollary 1.4.19] for $P = -C^{-1}E$.

3   Randomized Low-Rank Approximation of a Matrix

Primal and Dual Versions of Random Sampling. In the next section we prove Theorems 1 and 2 specifying estimates for the output errors of Algorithm 1. They imply that the algorithm is nearly optimal under each of the two randomization policies:

(p) *primal*: if $\mathrm{nrank}(M) = r$ and if the multiplier B is Gaussian and

(d) *dual*: if B is a well-conditioned $n \times l$ matrix of full rank l and if the input matrix M is average in the class $\mathcal{G}(m, n, r)$ up to a small-norm perturbation.

We specify later some multipliers B, which we generate and multiply with a matrix M at a low computational cost, but here is a caveat: we prove that Algorithm 1 produces accurate output when it is applied to average $m \times n$ matrix M with $\mathrm{nrank}(M) = r \ll n \le m$, but (unlike the cases of its applications with Gaussian and SRFT multipliers) does not do this for all such matrices M.

E.g., Algorithm 1 is likely to fail in the case of an $n \times n$ matrix $M = P\,\mathrm{diag}(I_l, O_{n-l})P'$, two random permutation matrices P and P', and sparse and structured orthogonal multiplier $B = (O_{l,n-l} \mid I_l)^T$ of full rank l.

Our study of dual randomization implies, however (cf. Corollary 2), that such "bad" pairs (B, M) are rare, that is, application of Algorithm 1 with sparse and structured multipliers B succeeds for a typical input M, that is, for almost any input with only a narrow class of exceptions.

Managing Rare Failures of Algorithm 1. If the relative residual norm Δ of the output of Algorithm 1 is large, we can re-apply the algorithm, successively or concurrently, for a fixed set of various sparse or structured multipliers B or of their linear combinations. We can also define multipliers as the leftmost blocks of some linear combinations of the products and powers (including inverses) of some sparse or structured square matrices. Alternatively we can re-apply the algorithm with a multiplier chosen at random from a fixed reasonably narrow class of such sparse and structured multipliers (see some samples below).

If the application still fails, we can re-apply Algorithm 1 with Gaussian or SRFT universal multiplier, and this is likely to succeed. With such a policy we would compute a low-rank approximation at significantly smaller average cost than in the case where we apply a Gaussian or even SRFT multiplier.

Sparse Multipliers. ASPH and AH Matrices. For a large class of well-conditioned matrices B of full rank, one can compute the product MB at a low cost. For example, fix an integer h, $1 \leq h \leq n/l$, define an $l \times n$ matrix $H = (I_l \mid I_l \mid \cdots \mid I_l \mid O_{l,n-hl})$ with h blocks I_l, choose a pair of random or fixed permutation matrices P and P', write $B = PH^T P'$, and note that the product MB can be computed by using just $(h-1)ln$ additions. In particular the computation uses no flops if $h = 1$. The same estimates hold if we replace the identity blocks I_l with $l \times l$ diagonal matrices filled with the values ± 1. If instead we replace the blocks I_l with arbitrary diagonal matrices, then we would need up to hln additional multiplications.

In the next example, we define such multipliers by simplifying the SRHT matrices in order to decrease the cost of their generation and multiplication by a matrix M. Like SRHT matrices, our multipliers have nonzero entries spread quite uniformly throughout the matrix B and concentrated in neither of its relatively small blocks.

At first recall a customary version of SRHT matrices, $H = DCP$, where P is a (random or fixed) $n \times n$ permutation matrix, $n = 2^k$, k is integer, D is a (random or fixed) $n \times n$ diagonal matrix, and $C = H_k$ is an $n \times n$ core matrix defined recursively, for $d = k$, as follows (cf. [M11]):

$$H_j = \begin{pmatrix} H_{j-1} & H_{j-1} \\ H_{j-1} & -H_{j-1} \end{pmatrix}, \; j = k, k-1, \ldots, k-d+1; \quad H_{k-d} = \begin{pmatrix} I_{2^{k-d}} & I_{2^{k-d}} \\ I_{2^{k-d}} & -I_{2^{k-d}} \end{pmatrix}.$$

By choosing small integers d (instead of $d = k$) and writing $B = C = H_d$, we arrive at $n \times n$ *Abridged Scaled Permuted Hadamard matrices*. For $D = P = I_n$, they turn into *Abridged Hadamard matrices*.[4] Every column and every row of such a matrix is filled with 0s, except for its 2^d entries. Its generation and multiplication by a vector are greatly simplified versus SRHT matrices.

---

[4] Hereafter we use the acronyms *ASPH* and *AH*.

We have defined ASPH and AH matrices of size $2^h \times 2^h$ for integers h, but we use their $n \times l$ blocks (e.g., the leading blocks) with $2^k \leq n < 2^{k+1}$ and any positive integer l, for *ASPH and AH multipliers B in Algorithm* 1.

The l columns of such a multiplier have at most $l2^d$ nonzero entries. The computation of the product MB, for an $m \times n$ matrix M, involves at most $ml2^d$ multiplications and slightly fewer additions and subtractions. In the case of AH multipliers (filled with ± 1 and 0) no multiplications are needed.

4   Proof of Two Basic Theorems

Two Lemmas and a Basic Estimate for the Residual Norm. We use the auxiliary results of the Appendix and the following simple ones.

Lemma 1. *Suppose that H is an $n \times r$ matrix, $\Sigma = \mathrm{diag}(\sigma_i)_{i=1}^n$, $\sigma_1 \geq \sigma_2 \geq \cdots \geq \sigma_n > 0$, $\Sigma' = \mathrm{diag}(\sigma_i')_{i=1}^r$, $\sigma_1' \geq \sigma_2' \geq \cdots \geq \sigma_r' > 0$. Then*

$$\sigma_j(\Sigma H \Sigma') \geq \sigma_j(H)\sigma_n \sigma_r' \text{ for all } j.$$

Lemma 2. *(Cf. [14, Theorem 2.4.8].) For an integer r and an $m \times n$ matrix M where $m \geq n > r > 0$, set to 0 the singular values $\sigma_j(M)$, for $j > r$, and let M_r denote the resulting matrix. Then*

$$||M - M_r|| = \sigma_{r+1}(M) \text{ and } ||M - M_r||_F^2 = \sum_{j=r+1}^{n} \sigma_j^2.$$

Next we estimate the relative residual norm Δ of the output of Algorithm 1 in terms of the norm $||(M_r B)^+||$; then we estimate the latter norm.

Suppose that B is a Gaussian $n \times l$ matrix. Apply part (ii) of Theorem 8, for $A = M_r$ and $H = B$, and deduce that $\mathrm{rank}(M_r B) = r$ with probability 1.

Theorem 4. Estimating the relative residual norm of the output of Algorithm 1 in terms of the norm $||(M_r B)^+||$.

Suppose that B is an $n \times l$ matrix, $\Delta = \frac{||\tilde{M} - M||}{||M||} = \frac{||M - Q(MB)Q^T(MB)M||}{||M||}$ denotes the relative residual norm of the output of Algorithm 1, M_r is the matrix of Lemma 2, $E' = (M - M_r)B$, and so $||E'||_F \leq ||B||_F ||M - M_r||_F$. Then

$$||M - M_r||_F \leq \sigma_{r+1}(M) \sqrt{n-r}$$

and

$$\Delta \leq \frac{\sigma_{r+1}(M)}{\sigma_1(M)} + \sqrt{8} \, ||(M_r B)^+|| \, ||E'||_F + O(||E'||_F^2).$$

Proof. Recall that $||M - M_r||_F^2 = \sum_{j=n-r+1}^{n} \sigma_j(M)^2 \leq \sigma_{r+1}^2(M) \, (n-r)$, and this implies the first claim of the theorem.

Now let $M_r = S_r \Sigma_r T_r^T$ be compact SVD. Then $Q(M_r B)Q(M_r B)^T M_r = M_r$. Therefore (cf. Lemma 2)

$$||M - Q(M_r B)Q(M_r B)^T M|| = ||M - M_r|| = \sigma_{r+1}(M). \tag{1}$$

Apply [22, Corollary C.1], for $A = M_r B$ and $E = E' = (M - M_r)B$, and obtain

$$||Q(MB)Q(MB)^T - Q(M_r B)Q(M_r B)^T|| \leq \sqrt{8}||(M_r B)^+|| \; ||E'||_F + O(||E'||_F^2).$$

Combine this bound with Eq. (1) and obtain

$$||M - Q(MB)Q^T(MB)M|| \leq \sigma_{r+1}(M) + \sqrt{8}||M|| \; ||(M_r B)^+|| \; ||E'||_F + O(||E'||_F^2).$$

Divide both sides of this inequality by $||M||$ and substitute $||M|| = \sigma_1(M)$.

Assessing Some Immediate Implications. By ignoring the smaller order term $O(||E'||^2)$, deduce from Theorem 4 that

$$\Delta \leq \frac{\sigma_{r+1}(M)}{\sigma_r(M)}(1 + \sqrt{8(n-r)} \; \sigma_r(M) \; ||B||_F ||(M_r B)^+||). \tag{2}$$

The norm $||B||_F$ is likely to be reasonably bounded, for Gaussian, SRFT, SRHT, and various other classes of sparse and structured multipliers B, and the ratio $\frac{\sigma_{r+1}(M)}{\sigma_r(M)}$ is presumed to be small. Furthermore $\sigma_r(M) \leq ||M||$. Hence the random variable Δ, representing the relative residual norm of the output of Algorithm 1, is likely to be small unless the value $||(M_r B)^+||$ is large.

For some *bad pairs* of matrices M and B, however, the matrix $M_r B$ is ill-conditioned, that is, the norm $||(M_r B)^+||$ is large, and then Theorem 4 only implies a large upper bound on the relative residual norm $||\tilde{M} - M||/||M||$.

If $l < n$, then, clearly, every matrix M belongs to such a bad pair, and so does every matrix B as well. Our quantitative specification of Theorems 1 and 2 (which we obtain by estimating the norm $||(M_r B)^+||$) imply, however, that the class of such bad pairs of matrices is narrow, particularly if the oversampling integer $p = l - r$ is not close to 0.

Next we estimate the norm $||(M_r B)^+$ in two ways, by randomizing either the matrix M or the multiplier B. We call these two randomization policies *primal* and *dual*, respectively.

(i) **Primal Randomization.** Let us specify bound (2), for $B \in \mathcal{G}^{m \times l}$. Recall Theorem 9 and obtain $||B||_F = \nu_{F,m,l} \leq \nu_{m,l} \sqrt{l}$, and so $\mathbb{E}(||B||_F) < (1 + \sqrt{m} + \sqrt{l})\sqrt{l}$.

Next estimate the norm $||(M_r B)^+||$.

Theorem 5. *Suppose that B is an $n \times l$ Gaussian matrix. Then*

$$||(M_r B)^+|| \leq \nu_{r,l}^+ / \sigma_r(M), \tag{3}$$

for the random variable $\nu_{r,l}^+$ of Theorem 10.

Proof. Let $M_r = S_r \Sigma_r T_r^T$ be compact SVD.

By applying Lemma 3, deduce that $G_{r,l} = T_r^T B$ is a $r \times l$ Gaussian matrix.

Hence $M_r B = S_r \Sigma_r T_r^T B = S_r \Sigma_r G_{r,l}$.

Write $H = \Sigma_r G_{r,l}$ and let $H = S_H \Sigma_H T_H^T$ be compact SVD where S_H is a $r \times r$ orthogonal matrix.

It follows that $S = S_r S_H$ is an $m \times r$ orthogonal matrix.

Hence $M_r B = S \Sigma_H T_H^T$ and $(M_r B)^+ = T_H (\Sigma_H)^+ S^T$ are compact SVDs of the matrices $M_r B$ and $(M_r B)^+$, respectively.

Therefore $||(M_r B)^+|| = ||(\Sigma_H)^+|| = ||(\Sigma_r G_{r,l})^+|| \le ||G_{r,l}^+||\ ||\Sigma_r^+||$.

Substitute $||G_{r,l}^+|| = \nu_{r,l}^+$ and $||\Sigma_r^+|| = 1/\sigma_r(M)$ and obtain the theorem.

Substitute our estimates for the norms $||B||_F$ and $||(M_r B)^+)||$ into bound (2) and obtain the following result.

Corollary 1. Relative Residual Norm of Primal Gaussian Low-Rank Approximation. *Suppose that Algorithm 1 has been applied to a matrix M having numerical rank r and to a Gaussian multiplier B.*

Then the relative residual norm, $\Delta = \frac{||\tilde{M} - M||}{||M||}$, of the output approximation $\tilde{M}$ to M is likely to be bounded from above by $f \frac{\sigma_r(M)}{\sigma_{r+2}(M)}$, for a factor of f having expected value $1 + (1 + \sqrt{m} + \sqrt{l}) \frac{e}{p} \sqrt{r}$ and for $e = 2.71828 \dots$.

Hence Δ is likely to have optimal order $\frac{\sigma_{r+1}(M)}{\sigma_r(M)}$ up to this factor f.

(ii) **Dual Randomization.** Next we extend Theorem 5 and Corollary 1 to the case where $M = UV + E$ is a small-norm perturbation of an $m \times n$ factor Gaussian matrix of rank r (cf. Definition 1) and B is any $n \times l$ matrix having full numerical rank l. At first we readily extend Theorem 4.

Theorem 6. *Suppose that we are given an $n \times l$ matrix B and an $m \times r$ matrix U such that $m \ge n > l \ge r > 0$, nrank$(B) = l$, and nrank$(U) = r$. Let $M = UV + E$ where $V \in \mathcal{G}^{r \times n}$, $||(UV)^+|| = \frac{1}{\sigma_r(UV)} = \frac{1}{\sigma_r(M)}$, and $||E||_F = \sigma_{r+1}(M) \sqrt{n - r}$. Write $E' = EB$ and $\tilde{M} = Q(MB)Q^T(MB)M$. Then*

$$\frac{||\tilde{M} - M||}{||M||} \le \frac{\sigma_{r+1}(M)}{\sigma_1(M)} + \sqrt{8}\ ||(UVB)^+||\ ||E'||_F + O(||E'||_F^2). \quad (4)$$

By extending estimate (3), the following theorem, proved in [23], bounds the norm $||(UVB)^+||$ provided that U is an $m \times r$ matrix that has full rank r and is well-conditioned.

Theorem 7. *Suppose that*

(i) *an $m \times r$ matrix U has full numerical rank r,*

(ii) *$V = G_{r,n}$ is a $r \times n$ Gaussian matrix, and*

(iii) *B is a well-conditioned $n \times l$ matrix of full rank l such that $m \ge n > l \ge r$ and $||B||_F = 1$.*

Then $||(UVB)^+|| \leq ||U^+|| \ \nu_{r,l}^+ \ ||B^+|| = ||(UV)^+|| \ ||B^+||.$

In particular the theorem is likely to hold where U is an $m \times r$ Gaussian matrix because such a matrix is likely to have full rank r and to be well-conditioned, by virtue of Theorems 8 and 10, respectively.

Now we combine and slightly expand the assumptions of Theorems 6 and 7 and then extend Corollary 1 to a small-norm perturbation $M + E$ of a factor Gaussian matrix M with expected rank r as follows.

Corollary 2. Relative Residual Norm of Dual Gaussian Low-Rank Approximation. *Suppose that* $m \geq n > l \geq r > 0$, *all the assumptions of Theorem 7 hold,* $M = UV + E$, $\tilde{M} = Q(MB)Q^T(MB)M$, $||(UV)^+|| = \frac{1}{\sigma_r(UV)} = \frac{1}{\sigma_r(M)}$, $||E||_F = \sigma_{r+1}(M) \sqrt{n-r}$, *and* $E' = EB$. *Then*

$$\frac{||\tilde{M} - M||}{||M||} \leq \frac{\sigma_{r+1}(M)}{\sigma_r(M)}(1 + \kappa(B) \sqrt{8(n-r)l} \ ) + O(\sigma_{r+1}(M)^2).$$

Proof. Note that $||E'|| \leq ||B||_F ||B||_F \leq ||B|| \ ||E|| \ \sqrt{(n-r)l}$, and so $||E'|| \leq ||B|| \ \sigma_{r+1}(M) \ \sqrt{(n-r)l}$.
By combining Theorem 7 and equation $||(UV)^+|| = \frac{1}{\sigma_r(M)}$, obtain $||(UVB)^+|| \leq ||B^+||/\sigma_r(M)$.
Substitute these bounds on the norms $||E'||$ and $||(UVB)^+||$ into estimate (4).

Remark 1. The Power Scheme of increasing the output accuracy of Algorithm 1. Define the Power Iterations $M_i = (M^T M)^i M$, $i = 1, 2, \ldots$. Then $\sigma_j(M_i) = (\sigma_j(M))^{2i+1}$ for all i and j [19, equation (4.5)]. Therefore, at a reasonable computational cost, one can dramatically decrease the ratio $\frac{\sigma_{r+1}(M)}{\sigma_r(M)}$ and thus decrease accordingly the bounds of Corollaries 1 and 2.

5   Numerical Tests

We have tested Algorithm 1, with both AH and ASPH multipliers, applied on one side and both sides of the matrix M, as well as with one-sided dense multipliers $B = B(\pm 1, 0)$ that have i.i.d. entries ± 1 and 0, each value chosen with probability $1/3$. We generated the input matrices M for these tests by extending the customary recipes of [18, Sect. 28.3]: at first, we generated two matrices S_M and T_M by orthogonalizing a pair of $n \times n$ Gaussian matrices, then wrote $\Sigma_M = \text{diag}(\sigma_j)_{j=1}^n$, for $\sigma_j = 1/j$, $j = 1, \ldots, r$, $\sigma_j = 10^{-10}$, $j = r + 1, \ldots, n$, and finally computed the $n \times n$ matrices M defined by their compact SVDs, $M = S_M \Sigma_M T_M^T$. (In this case $||M|| = 1$ and $\kappa(M) = ||M|| \ ||M^{-1}|| = 10^{10}$).

Table 1 represents the average relative residuals norms Δ of low-rank approximation of the matrices M over 1000 tests for each pair of n and r, $n = 256, 512, 1024$, $r = 8, 32$, and various multipliers B of the five classes B above. For all classes and all pairs of n and r, average relative residual norms ranged from 10^{-7} to about 10^{-9} in these tests.

In [23] we present similar results of our tests with matrices M involved in the discrete representation of PDEs and data analysis.

6   An Extension

Our acceleration of low-rank approximation implies acceleration of various related popular matrix computations for average input matrices, and thus statistically for most of the inputs, although possibly not for all inputs of practical interest. Next, for a simple example, we accelerate the Fast Multipole Method for average input matrix. In [23] we further extend the resulting algorithm to the acceleration of the Conjugate Gradient algorithms.

In order to specify the concept of "average" to the case of FMM applications, we recall the definitions and basic results for the computations with HSS matrices[5], which naturally extend the class of banded matrices and their inverses, are closely linked to FMM, and have been intensively studied for decades (cf. [1,3,5,10,13,15,26–33], and the bibliography therein).

Definition 2. *(Cf. [21].) A neutered block of a block diagonal matrix is the union of a pair of its off-block-diagonal blocks sharing their column sets.*

Definition 3. *(Cf. [1,5,15,30–32].)*
   An $m \times n$ matrix M is called a r-HSS matrix, for a positive integer r, if

(i) this is a block diagonal matrix whose diagonal blocks consist of $O((m+n)r)$ entries and
(ii) r is the maximum rank of its neutered blocks.

Remark 2. Many authors work with (l, u)-HSS rather than r-HSS matrices M where l and u are the maximum ranks of the sub- and super-block-diagonal blocks, respectively. The (l, u)-HSS and r-HSS matrices are closely related. Indeed, if a neutered block N is the union of a sub-block-diagonal block B_- and a super-block-diagonal block B_+, then $\text{rank}(N) \leq \text{rank}(B_-) + \text{rank}(B_+)$, and so an (l, u)-HSS matrix is a p-HSS matrix, for $p \leq l + u$, while clearly a r-HSS matrix is a (q, s)-HSS matrix, for $q \leq r$ and $s \leq r$.

Table 1. Low-rank approximation: residual norms with AH, ASPH, and $B(\pm 1, 0)$ multipliers.

n	r	Pre- and Post-multiplication		Pre-multiplication only		
		AH	ASPH	AH	ASPH	$B(\pm 1, 0)$
256	8	8.43e-09	4.89e-08	2.25e-08	2.70e-08	2.52e-08
256	32	3.53e-09	5.47e-08	5.95e-08	1.47e-07	3.19e-08
512	8	7.96e-09	3.16e-09	4.80e-08	2.22e-07	4.76e-08
512	32	1.75e-08	7.39e-09	6.22e-08	8.91e-08	6.39e-08
1024	8	6.60e-09	3.92e-09	5.65e-08	2.86e-08	1.25e-08
1024	32	7.50e-09	5.54e-09	1.94e-07	5.33e-08	4.72e-08

---

[5] We use the acronym for "hierarchically semiseparable".

The FMM enables us to exploit the r-HSS structure of a matrix as follows (cf. [1, 10, 28]).

(i) At first we should cover all its off-block-diagonal entries with a set of neutered blocks that pairwise have no overlaps and then

(ii) express every $h \times k$ block N of this set as the product $N = FG^T$ of two *generator matrices*, F of size $h \times r$ and G of size $r \times k$. Call such a pair of F and G a *length r generator* of the neutered block N.

(iii) Suppose that, for an r-HSS matrix M of size $m \times n$ having s diagonal blocks, such an HSS representation via generators of length at most r has been computed. Then we can readily multiply the matrix M by a vector by using $O((m + n)r \log(s))$ flops and

(iv) in a more advanced application of FMM we can solve a nonsingular r-HSS linear system of n equations by using $O(nr \log^3(n))$ flops under some mild additional assumptions on the input.

This approach is readily extended to (r, ϵ)-HSS matrices, that is, matrices approximated by r-HSS matrices within perturbation norm ϵ where a positive tolerance ϵ is small in context (e.g., is the unit round-off). Likewise, one defines an (r, ϵ)-*HSS representation* and (r, ϵ)-*generators*. (r, ϵ)-HSS matrices (for r small in context) appear routinely in modern computations, and computations with such matrices are performed efficiently by using the above techniques.

The computation of (r, ϵ)-generators for a (r, ϵ)-HSS representation of a (r, ϵ)-HSS matrix M (that is, for low-rank approximation of the blocks in that representation) turned out to be the bottleneck stage of such applications of FMM.

Indeed, suppose one applies random sampling Algorithm 1 at this stage. Multiplication of a $k \times k$ block by $k \times r$ Gaussian matrix requires $(2k - 1)kr$ flops, while standard HSS-representation of an $n \times n$ HSS matrix includes $k \times k$ blocks for $k \approx n/2$. Therefore the cost of computing such a representation of the matrix M is at least quadratic in n and thus dramatically exceeds the above estimate of $O(rn \log(s))$ flops at the other stages of the computations if $r \ll n$.

Alternative customary techniques for low-rank approximation rely on computing SVD or rank-revealing factorization of an input matrix and are at least as costly as the computations by means of random sampling.

Can we fix such a mishap? Yes, by virtue of Corollary 2, we can perform this stage at the dominated randomized arithmetic cost $O((k + l)r)$ in the case of average (r, ϵ)-HSS input matrix of size $k \times l$, if we just apply Algorithm 1 with AH, ASPH, or other sparse multipliers.

By saying "average", we mean that Corollary 2 can be applied to low-rank approximation of all the off-block diagonal blocks in a (r, ϵ)-HSS representation of a (r, ϵ)-HSS matrix.

Acknowledgements. Our research has been supported by NSF Grant CCF-1116736 and PSC CUNY Awards 67699-00 45 and 68862-00 46. We are also grateful to the reviewers for valuable comments.

Appendix

A   Gaussian Matrices

Theorem 8. *Assume a nonsingular $n \times n$ matrix A and an $n \times n$ matrix H whose entries are linear combinations of finitely many i.i.d. Gaussian variables.*

Let $\det((AH)_{l,l})$ vanish identically in them for neither of the integers l, $l = 1, \ldots, n$. Then the matrices $(AH)_{l,l}$, for $l = 1, \ldots, n$, are nonsingular with probability 1.

Proof. The theorem follows because the equation $\det((AH)_{l,l})$ for any integer l in the range from 1 to n defines an algebraic variety of a lower dimension in the linear space of the input variables (cf. [2, Proposition 1]).

Lemma 3. *(Rotational invariance of a Gaussian matrix.) Suppose that k, m, and n are three positive integers, G is an $m \times n$ Gaussian matrix, and S and T are $k \times m$ and $n \times k$ orthogonal matrices, respectively.*

Then SG and GT are Gaussian matrices.

We keep stating all results and estimates for real matrices, but estimates similar to the ones of the next theorems in the case of complex matrices can be found in [4,6,9], and [11].

Write $\nu_{m,n} = ||G||$, $\nu_{m,n}^+ = ||G^+||$, and $\nu_{m,n,F}^+ = ||G^+||_F$, for a Gaussian $m \times n$ matrix G, and write $\mathbb{E}(v)$ for the expected value of a random variable v.

Theorem 9. *(Cf. [8, Theorem II.7].) Suppose that m and n are positive integers, $h = \max\{m, n\}$, $t \geq 0$. Then*

(i) Probability$\{\nu_{m,n} > t + \sqrt{m} + \sqrt{n}\} \leq \exp(-t^2/2)$ *and*
(ii) $\mathbb{E}(\nu_{m,n}^+) < 1 + \sqrt{m} + \sqrt{n}$.

Theorem 10. *Let $\Gamma(x) = \int_0^\infty \exp(-t) t^{x-1} dt$ denote the Gamma function and let $x > 0$. Then*

(i) Probability $\{\nu_{m,n}^+ \geq m/x^2\} < \frac{x^{m-n+1}}{\Gamma(m-n+2)}$ *for $m \geq n \geq 2$,*
(ii) Probability $\{\nu_{n,n}^+ \geq x\} \leq 2.35\sqrt{n}/x$ *for $n \geq 2$,*
(iii) $\mathbb{E}((\nu_{F,m,n}^+)^2) = m/|m - n - 1|$, *provided that $|m - n| > 1$, and*
(iv) $\mathbb{E}(\nu_{m,n}^+) \leq e\sqrt{m}/|m - n|$, *provided that $m \neq n$ and $e = 2.71828\ldots$.*

Proof. See [4, Proof of Lemma 4.1] for part (i), [25, Theorem 3.3] for part (ii), and [19, Proposition 10.2] for parts (iii) and (iv).

Theorem 10 provides probabilistic upper bounds on $\nu_{m,n}^+$. They are reasonable already for square matrices, for which $m = n$, but become much stronger as the difference $|m - n|$ grows large.

Theorems 9 and 10 combined imply that an $m \times n$ Gaussian matrix is well-conditioned unless the integer $m + n$ is large or the integer $m - n$ is close to 0 and that such a matrix can still be considered well-conditioned (possibly with some grain of salt) if the integer m is not large and if the integer $|m - n|$ is small or even vanishes. These properties are immediately extended to all submatrices because they are also Gaussian.

References

1. Börm, S.: Efficient Numerical Methods for Non-local Operators: H2-Matrix Compression. Algorithms and Analysis. European Mathematical Society, Zürich (2010)
2. Bruns, W., Vetter, U.: Determinantal Rings. LNM, vol. 1327. Springer, Heidelberg (1988)
3. Barba, L.A., Yokota, R.: How will the fast multipole method fare in exascale era? SIAM News **46**(6), 1–3 (2013)
4. Chen, Z., Dongarra, J.J.: Condition numbers of Gaussian random matrices. SIAM J. Matrix Anal. Appl. **27**, 603–620 (2005)
5. Carrier, J., Greengard, L., Rokhlin, V.: A fast adaptive algorithm for particle simulation. SIAM J. Sci. Comput. **9**, 669–686 (1988)
6. Demmel, J.: The probability that a numerical analysis problem is difficult. Math. Comput. **50**, 449–480 (1988)
7. Drineas, P., Kannan, R., Mahoney, M.W.: Fast Monte Carlo algorithms for matrices I-III. SIAM J. Comput. **36**(1), 132–206 (2006)
8. Davidson, K.R., Szarek, S.J.: Local operator theory, random matrices, and Banach spaces. In: Johnson, W.B., Lindenstrauss, J. (eds.) Handbook on the Geometry of Banach Spaces, pp. 317–368. North Holland, Amsterdam (2001)
9. Edelman, A.: Eigenvalues and condition numbers of random matrices. SIAM J. Matrix Anal. Appl. **9**(4), 543–560 (1988)
10. Eidelman, Y., Gohberg, I., Haimovici, I.: Separable Type Representations of Matrices and Fast Algorithms Volume 1: Basics Completion Problems. Multiplication and Inversion Algorithms, Volume 2: Eigenvalue Methods. Birkhauser, Basel (2013)
11. Edelman, A., Sutton, B.D.: Tails of condition number distributions. SIAM J. Matrix Anal. Appl. **27**(2), 547–560 (2005)
12. Frieze, A., Kannan, R., Vempala, S.: Fast Monte-Carlo algorithms for finding low-rank approximations. J. ACM **51**, 1025–1041 (2004). (Proceedings version in 39th FOCS, pp. 370–378. IEEE Computer Society Press (1998))
13. Grasedyck, L., Hackbusch, W.: Construction and arithmetics of h-matrices. Computing **70**(4), 295–334 (2003)
14. Golub, G.H., Van Loan, C.F.: Matrix Computations. The Johns Hopkins University Press, Baltimore, Maryland (2013)
15. Greengard, L., Rokhlin, V.: A fast algorithm for particle simulation. J. Comput. Phys. **73**, 325–348 (1987)
16. Goreinov, S.A., Tyrtyshnikov, E.E., Zamarashkin, N.L.: A theory of pseudo-skeleton approximations. Linear Algebra Appl. **261**, 1–21 (1997)
17. Goreinov, S.A., Zamarashkin, N.L., Tyrtyshnikov, E.E.: Pseudo-skeleton approximations by matrices of maximal volume. Math. Notes **62**(4), 515–519 (1997)
18. Higham, N.J.: Accuracy and Stability in Numerical Analysis, 2nd edn. SIAM, Philadelphia (2002)
19. Halko, N., Martinsson, P.G., Tropp, J.A.: Finding Structure with randomness: probabilistic algorithms for constructing approximate matrix decompositions. SIAM Rev. **53**(2), 217–288 (2011)
20. Mahoney, M.W.: Randomized algorithms for matrices and data. Found. Trends Mach. Learn. **3**(2) (2011). (In: Way, M.J., et al. (eds.) Advances in Machine Learning and Data Mining for Astronomy (abridged version), pp. 647-672. NOW Publishers (2012))
21. Martinsson, P.G., Rokhlin, V., Tygert, M.: A fast algorithm for the inversion of general Toeplitz matrices. Comput. Math. Appl. **50**, 741–752 (2005)

22. Pan, V.Y., Qian, G., Yan, X.: Random multipliers numerically stabilize Gaussian and block Gaussian elimination: proofs and an extension to low-rank approximation. Linear Algebra Appl. **481**, 202–234 (2015)
23. Pan, V.Y., Zhao, L.: Low-rank approximation of a matrix: novel insights, new progress, and extensions. arXiv:1510.06142 [math.NA]. Submitted on 21 October 2015, Revised April 2016
24. Stewart, G.W.: Matrix Algorithms: Basic Decompositions, vol. I. SIAM, Philadelphia (1998)
25. Sankar, A., Spielman, D., Teng, S.-H.: Smoothed analysis of the condition numbers and growth factors of matrices. SIAM J. Matrix Anal. Appl. **28**(2), 46–476 (2006)
26. Tyrtyshnikov, E.E.: Incomplete cross-approximation in the mosaic-skeleton method. Computing **64**, 367–380 (2000)
27. Vandebril, R., van Barel, M., Golub, G., Mastronardi, N.: A bibliography on semiseparable matrices. Calcolo **42**(3–4), 249–270 (2005)
28. Vandebril, R., van Barel, M., Mastronardi, N.: Matrix Computations and Semiseparable Matrices: Linear Systems, vol. 1. The Johns Hopkins University Press, Baltimore, Maryland (2007)
29. Vandebril, R., van Barel, M., Mastronardi, N.: Matrix Computations and Semiseparable Matrices: Eigenvalue and Singular Value Methods, vol. 2. The Johns Hopkins University Press, Baltimore, Maryland (2008)
30. Xia, J.: On the complexity of some hierarchical structured matrix algorithms. SIAM J. Matrix Anal. Appl. **33**, 388–410 (2012)
31. Xia, J.: Randomized sparse direct solvers. SIAM J. Matrix Anal. Appl. **34**, 197–227 (2013)
32. Xia, J., Xi, Y., Cauley, S., Balakrishnan, V.: Superfast and stable structured solvers for Toeplitz least squares via randomized sampling. SIAM J. Matrix Anal. Appl. **35**, 44–72 (2014)
33. Xia, J., Xi, Y., Gu, M.: A superfast structured solver for Toeplitz linear systems via randomized sampling. SIAM J. Matrix Anal. Appl. **33**, 837–858 (2012)

Representations of Analytic Functions and Weihrauch Degrees

Arno Pauly[1,2(✉)] and Florian Steinberg[3]

[1] Computer Laboratory, University of Cambridge, Cambridge, UK
Arno.Pauly@cl.cam.ac.uk
[2] Département d'Informatique, Université Libre de Bruxelles, Brussels, Belgium
[3] Fachbereich Mathematik, TU-Darmstadt, Darmstadt, Germany
steinberg@mathematik.tu-darmstadt.de

Abstract. This paper considers several representations of the analytic functions on the unit disk and their mutual translations. All translations that are not already computable are shown to be Weihrauch equivalent to closed choice on the natural numbers. Subsequently some similar considerations are carried out for representations of polynomials. In this case in addition to closed choice the Weihrauch degree LPO* shows up as the difficulty of finding the degree or the zeros.

Keywords: Computable analysis · Analytic function · Weihrauch reduction · Polynomials · Closed choice · LPO*

1 Introduction

In order to make sense of computability questions in analysis, the spaces of objects involved have to be equipped with representations: A representation determines the kind of information that is provided (or has to be provided) when computing on these objects. When restricting from a more general to more restrictive setting, there are two options: Either to merely restrict the scope to the special objects and retain the representation, or to actually introduce a new representation containing more information.

As a first example of this, consider the closed subsets of $[0,1]^2$ and the closed convex subsets of $[0,1]^2$ (following [8]). The former are represented by an enumeration of open balls exhausting their complement. The latter are represented as the intersection of a decreasing sequence of rational polygons. Thus, prima facie the notion of *closed set which happens to be convex* and *convex closed set* are different. In this case it turns out they are computably equivalent after all (the proof, however, uses the compactness of $[0,1]^2$).

This paper focuses on a different example of the same phenomenon: The difference between an *analytic function* and a *continuous function that happens to be analytic*. It is known that these actually are different notions. Sections 3.1 and 3.2 quantify how different they are using the framework of Weihrauch reducibility. As a further example Sects. 3.3 and 3.4 consider *continuous functions that*

© Springer International Publishing Switzerland 2016
A.S. Kulikov and G.J. Woeginger (Eds.): CSR 2016, LNCS 9691, pp. 367–381, 2016.
DOI: 10.1007/978-3-319-34171-2_26

happen to be polynomials versus *analytic functions that happen to be polynomials* versus *polynomials*. All translations turn out to be either computable, or Weihrauch equivalent to one of the two well-studied principles C_N and LPO*. The results are summarized in Fig. 3 on Page 377 and Fig. 5 on Page 379.

The additional information one needs about an analytic function over a continuous function can be expressed by a single natural number – the same holds for the other examples studied. Thus, this can be considered as an instance of computation with discrete advice as introduced in [20]. That finding this number is Weihrauch equivalent to C_N essentially means that while the number can be chosen to be verifiable (i.e. wrong values can be detected eventually), this is the only computationally relevant restriction on how complicated the relationship between object and associated number can be.

Before ending this introduction, we shall briefly mention two alternative perspectives on the phenomenon: Firstly, recall that in intuitionistic logic a negative translated statement behaves like a classical one, and that double negations generally do not cancel. In this setting the difference boils down to considering either analytic functions or continuous functions that are not analytic. Secondly, from a topological perspective, Weihrauch equivalence of a translation to C_N implies that the topologies induced by the representations differ. Indeed, the suitable topology on the space of analytic functions is not just the subspace topology inherited from the space of continuous functions but in fact obtained as a direct limit.

A version of this paper containing all the proofs can be found on the arXiv [17].

2   Background

This section provides a very brief introduction to the required concepts from computable analysis, Weihrauch reducibility, and then in more detailed introduction of the representations of analytic functions that are considered. For a more in depth introduction into computable analysis and further information, the reader is pointed to the standard textbook in computable analysis [19], and to [14]. Also, [18] should be mentioned as an excellent source, even though the approach differs considerably from the one taken here. The research programme of Weihrauch reducibility was formulated in [2], a more up-to-date introduction to Weihrauch reducibility can be found in the introduction of [3].

2.1   Represented Spaces

Recall that a **represented space** $\mathbf{X} = (X, \delta_{\mathbf{X}})$ is given by a set X and a partial surjection $\delta_{\mathbf{X}} :\subseteq \mathbb{N}^{\mathbb{N}} \to X$ from Baire space onto it. The elements of $\delta_{\mathbf{X}}^{-1}(x)$ should be understood as encodings of x and are called the **X-names** of x. Since Baire space inherits a topology, each represented space can be equipped with a topology: The final topology of the chosen representation. We usually refrain from mentioning the representation of a represented space in the same way as the topology of a topological space is usually not mentioned. For instance the set of natural numbers is regarded as a represented space with the representation $\delta_{\mathbb{N}}(p) := p(0)$. Therefore, from now on denote by $\mathbb{N}$ not only the set or the topological space, but the **represented space of natural numbers**. If the set that is to be represented already inherits a topology, we always choose the representation such that it fits the topology. This can be checked easily for the case $\mathbb{N}$ above, where the final topology of the representation is the discrete topology.

If $\mathbf{X}$ is a represented space and Y is a subset of $\mathbf{X}$, then Y can be turned into a represented space by considering the range restriction of the representation of $\mathbf{X}$ on it. We denote the represented space arising in this way by $\mathbf{X}|_Y$. Note that here only set inclusion is considered. The set Y may be a subset of many different represented spaces and the restrictions need not coincide. They often turn out to be inappropriate. We use the same notation $\mathbf{X}|_\mathbf{Y}$ if $\mathbf{Y}$ is a represented space already. In this case, however, no information about the representation of $\mathbf{Y}$ is carried over to $\mathbf{X}|_\mathbf{Y}$.

The remainder of this section introduces the represented spaces that are needed for the content of the paper.

Sets of Natural Numbers. Let $\mathcal{O}(\mathbb{N})$ resp. $\mathcal{A}(\mathbb{N})$ denote the **represented spaces of open** resp. **closed subsets of** $\mathbb{N}$. The underlying set of both $\mathcal{O}(\mathbb{N})$ and $\mathcal{A}(\mathbb{N})$ is the power set of $\mathbb{N}$. The representation of $\mathcal{O}(\mathbb{N})$ is defined by

$$\delta_{\mathcal{O}(\mathbb{N})}(p) = O \quad \Leftrightarrow \quad O = \{p(n) - 1 \mid p(n) > 0\}.$$

That is: A name of an open set is an enumeration of that set, however, to include the empty set, the enumeration is allowed to not return an element of the set in each step. The closed sets $\mathcal{A}(\mathbb{N})$ are represented as complements of open sets:

$$\delta_{\mathcal{A}(\mathbb{N})}(p) = A \quad \Leftrightarrow \quad \delta_{\mathcal{O}(\mathbb{N})}(p) = A^c.$$

Normed Spaces, $\mathbb{R}$, $\mathbb{C}$, $\mathcal{C}(D)$. Given a triple $\mathcal{M} = (M, d, (x_n)_{n \in \mathbb{N}})$ such that (M, d) is a separable metric space and x_n is a dense sequence, $\mathcal{M}$ can be turned into a represented space by equipping it with the representation

$$\delta_{\mathcal{M}}(p) = x \quad \Leftrightarrow \quad \forall n \in \mathbb{N} : d(x, x_{p(n)}) < 2^{-n}.$$

In this way $\mathbb{R}$, $\mathbb{R}^d$, $\mathbb{C}$ (where the dense sequences are standard enumerations of the rational elements) and $C([0,1])$, $\mathcal{C}(D)$ (where D is a compact subset of $\mathbb{R}^d$ and the dense sequences are standard enumerations of the polynomials with rational coefficients) can be turned into represented spaces.

Sequences in a Represented Space. For a represented space $\mathbf{X}$ there is a canonical way to turn the set of sequences in $\mathbf{X}$ into a **represented space** $\mathbf{X}^{\mathbb{N}}$: Let $\langle \cdot, \cdot \rangle : \mathbb{N} \times \mathbb{N} \to \mathbb{N}$ be a standard paring function (i.e. bijective, recursive with recursive projections). Define a function $\langle \cdot \rangle : \left(\mathbb{N}^{\mathbb{N}} \right)^{\mathbb{N}} \to \mathbb{N}^{\mathbb{N}}$ by

$$\langle (p_k)_{k \in \mathbb{N}} \rangle(\langle m, n \rangle) := p_m(n).$$

For a represented space $\mathbf{X}$ define a representation of the set $X^{\mathbb{N}}$ of the sequences in the set X underlying $\mathbf{X}$ by

$$\delta_{\mathbf{X}^{\mathbb{N}}}(\langle (p_k)_{k \in \mathbb{N}} \rangle) = (x_k)_{k \in \mathbb{N}} \quad \Leftrightarrow \quad \forall m \in \mathbb{N} : \delta_{\mathbf{X}}(p_m) = x_m.$$

In particular the spaces $\mathbb{R}^{\mathbb{N}}$ and $\mathbb{C}^{\mathbb{N}}$ of real and complex sequences are considered represented spaces in this way. Also $\mathcal{C}(D)^{\mathbb{N}}$ briefly shows up in Sect. 3.2.

2.2 Weihrauch Reducibility

Recall that a multivalued function f from X to Y (or $\mathbf{X}$ to $\mathbf{Y}$) is an assignment that assigns to each element x of its domain a set $f(x)$ of acceptable return values. Multivaluedness of a function is indicated by $f : \mathbf{X} \rightrightarrows \mathbf{Y}$. The domain of a multivalued function is the set of elements such that the image is not empty. Furthermore, recall that we write $f :\subseteq \mathbf{X} \to \mathbf{Y}$ if the function f is allowed to be partial, that is if its domain can be a proper subset of $\mathbf{X}$.

Definition 1. *A partial function* $F :\subseteq \mathbb{N}^{\mathbb{N}} \to \mathbb{N}^{\mathbb{N}}$ *is a **realizer** of a multivalued function* $f :\subseteq \mathbf{X} \rightrightarrows \mathbf{Y}$ *if* $\delta_{\mathbf{Y}}(F(p)) \in f(\delta_{\mathbf{X}}(p))$ *for all* $p \in \delta_{\mathbf{X}}^{-1}(\mathrm{dom}(f))$ *(compare Fig. 1).*

A function between represented spaces is called computable if it has a computable realizer, where computability on Baire space is defined via oracle Turing machines (as in e.g. [6]) or via Type-2 Turing machines (as in e.g. [19]). The computable Weierstraß approximation theorem can be interpreted to state that an element of $\mathcal{C}([0,1])$ is computable if and only if it has a computable realizer as function on the represented space $\mathbb{R}$.

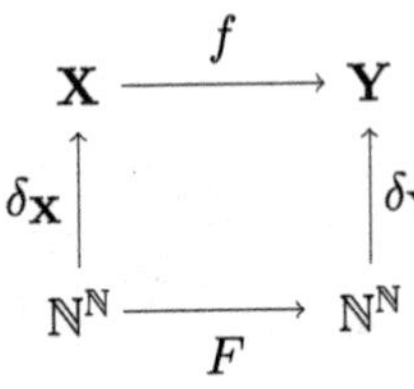

Fig. 1. Realizer.

Every multivalued function $f :\subseteq \mathbf{X} \rightrightarrows \mathbf{Y}$ corresponds to a computational task. Namely: 'given information about x and the additional assumption $x \in \mathrm{dom}(f)$ find suitable information about some $y \in f(x)$'. What information about x resp. $f(x)$ is provided resp. asked for is reflected in the choice of the representations for $\mathbf{X}$ and $\mathbf{Y}$. The following example of this is very relevant for the content of this paper:

Definition 2. *Let **closed choice on the integers** be the multivalued function* $C_{\mathbb{N}} :\subseteq \mathcal{A}(\mathbb{N}) \rightrightarrows \mathbb{N}$ *defined on nonempty sets by*

$$y \in C_{\mathbb{N}}(A) \Leftrightarrow y \in A.$$

The corresponding task is 'given an enumeration of the complement of a set of natural numbers and provided that it is not empty, return an element of the set'. $C_{\mathbb{N}}$ does not permit a computable realizer: Whenever a machine decides that the name of the element of the set should begin with n, it has only read a finite beginning segment of the enumeration. The next value might as well be n.

From the point of view of multi-valued functions as computational tasks, it makes sense to compare their difficulty by comparing the corresponding multivalued functions. This paper uses Weihrauch reductions as a formalization of such a comparison. Weihrauch reductions define a rather fine pre-order on multivalued functions between represented spaces.

Definition 3. *Let f and g be partial, multivalued functions between represented spaces. Say that f is **Weihrauch reducible** to g, in symbols $f \leq_W g$, if there are computable functions $K :\subseteq \mathbb{N}^{\mathbb{N}} \times \mathbb{N}^{\mathbb{N}} \to \mathbb{N}^{\mathbb{N}}$ and $H :\subseteq \mathbb{N}^{\mathbb{N}} \to \mathbb{N}^{\mathbb{N}}$ such that whenever G is a realizer of g, the function $F := (p \mapsto K(p, G(H(p))))$ is a realizer for f.*

H is called the **pre-processor** and K the **post-processor** of the Weihrauch reduction. This definition and the nomenclature is illustrated in Fig. 2. The relation $\leq_W$ is reflexive and transitive. We use $\equiv_W$ to denote that reductions in both directions exist and $<_W$ the other reduction does not exist. The equivalence class of a multivalued function with respect to the equivalence relation $\equiv_W$ is called the **Weihrauch degree** of the function. A Weihrauch degree is called non-computable if it contains no computable function.

The Weihrauch degree corresponding to $C_{\mathbb{N}}$ has received significant attention (see for instance [1–3, 10–13]). In particular, as shown in [15], a function between computable Polish spaces is Weihrauch reducible to $C_{\mathbb{N}}$ if and only if it is piecewise computable or equivalently is effectively Δ^0_2-measurable.

For the purposes of this paper, the following representatives of this degree are also relevant:

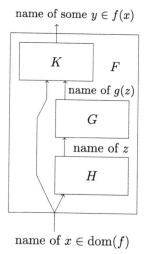

name of some $y \in f(x)$

K F

name of $g(z)$

G

name of z

H

name of $x \in \mathrm{dom}(f)$

Fig. 2. Weihrauch reduction.

Lemma 1 [16]. *The following are Weihrauch equivalent:*

- $C_{\mathbb{N}}$, *that is closed choice on the natural numbers.*
- $\max :\subseteq \mathcal{O}(\mathbb{N}) \to \mathbb{N}$ *defined on the bounded sets in the obvious way.*
- Bound $:\subseteq \mathcal{O}(\mathbb{N}) \rightrightarrows \mathbb{N}$, *where $n \in \mathrm{Bound}(U)$ iff $\forall m \in U : n \geq m$.*

In the later chapters of this paper another non-computable Weihrauch degree is encountered: LPO*. Here, LPO is short for 'limited principle of omniscience'. We refrain from stating LPO* explicitly as it would need more machinery than we introduced. Instead we characterize it by specifying the representative that is used in the proofs: Consider the function

$$\min_B : \mathbb{N}^{\mathbb{N}} \to \mathbb{N}, \quad p \mapsto \min\{p(n) \mid n \in \mathbb{N}\}.$$

Here, the index B is for Baire space and to distinguish the function from the integer minimum function used on the right hand side of the definition.

Proposition 1. $\min_B$ *is a representative of the Weihrauch degree* LPO^*.

LPO^* is also called the Weihrauch degree of finitely many mind changes: To obtain the minimum of an element of Baire space you may guess that it is the smallest value assumed on arguments up to n, and you will only be wrong a finite number of times.

To give a little more intuition as to why this Weihrauch degree shows up in this paper, note the following: LPO^* is derived from the maybe simplest non-computable Weihrauch degree $LPO : \mathbb{N}^{\mathbb{N}} \to \{0,1\}$ defined via

$$LPO(p) := \begin{cases} 1 & \text{if } p \text{ is the zero function, i.e. } \forall n : p(n) = 0. \\ 0 & \text{otherwise.} \end{cases}$$

In computable analysis LPO shows up as the Weihrauch degree of the equality test for real (or complex) numbers $\neq : \mathbb{R} \times \mathbb{R}$. Now, LPO^* corresponds to carrying out a fixed finite but arbitrary high number of equality tests on the real or complex numbers. It is known that $LPO <_W LPO^* <_W C_{\mathbb{N}}$.

2.3   Representations of Analytic Functions

Recall that a function is analytic if it is locally given by a power series:

Definition 4. *Let* $D \subseteq \mathbb{C}$ *be a set. A function* $f : D \to \mathbb{C}$ *is called* **analytic**, *if for every* $x_0 \in D$ *there is a neighborhood* U *of* x_0 *and a sequence* $(a_k)_{k\in\mathbb{N}} \in \mathbb{C}^{\mathbb{N}}$ *such that for each* $x \in U \cap D$

$$f(x) = \sum_{k\in\mathbb{N}} a_k x^k.$$

The set of analytic functions is denoted by $\mathcal{C}^{\omega}(D)$. Each analytic function is continuous, that is $\mathcal{C}^{\omega}(D) \subseteq \mathcal{C}(D)$. If D is open, the analytic functions on D are smooth, i.e. infinitely often differentiable. An analytic function can be analytically extended to an open superset of its domain.

Definition 5. *A pair* $(x, (a_k)_{k\in\mathbb{N}})$ *is called* **germ** *of* $f \in \mathcal{C}^{\omega}(D)$ *if* x *is an element of* D *and* $(a_k)_{k\in\mathbb{N}} \in \mathbb{C}^{\mathbb{N}}$ *is a series expansion of* f *around* x.

As long as the domain is connected, an analytic function is uniquely determined by each of its germs. The one to one correspondence of germs and analytic functions only partially carries over to the computability and complexity realm: It is well known that an analytic function on the unit disk is computable if and only if the germ around any computable point of the domain is computable [5]. However, the proofs of these statements are inherently non-uniform. The operations of obtaining a germ from a function and a function from a germ are discontinuous and therefore not computable [9]. This paper classifies them to be Weihrauch equivalent to closed choice on the naturals in Theorems 3 and 4.

There is a more suitable representation for the analytic functions than the restriction of the representation of continuous functions. This representation has been investigated by different authors for instance in [4,7,9]. For simplicity we restrict to the case of analytic functions on the unit disk. Thus, let D denote the closed unit disk from now on. And let U_m denote the open ball $B_{r_m}(0)$ of radius $r_m := 2^{\frac{1}{m+1}}$ around zero. Recall from the introduction that the space $\mathcal{C}(D)$ of continuous functions is represented as a metric space (where $\mathbb{C}$ is identified with $\mathbb{R}^2$).

Definition 6. *Let $\mathcal{C}^\omega(D)$ denote the **represented space of analytic functions on** D, where the representation is defined as follows: A $q \in \mathbb{N}^\mathbb{N}$ is a name of an analytic function f on D, if and only if f extends analytically to the closure of $U_{q(0)}$, the extension is bounded by $q(0)$ and $n \mapsto q(n+1)$ is a name of $f \in \mathcal{C}(D)$.*

Note that the representation of $\mathcal{C}^\omega(D)$ arises from the restriction of the representation of continuous functions by adding discrete additional information. This information is quantified by the advice function $\mathrm{Adv}_{\mathcal{C}^\omega} :\subseteq \mathcal{C}(D) \to \mathbb{N}$ whose domain are the analytic functions and that on those is defined by

$$\mathrm{Adv}_{\mathcal{C}^\omega}(f) := \{q(0) \mid q \text{ is a } \mathcal{C}^\omega(D)\text{-name of } f)\}$$
$$= \{m \in \mathbb{N} \mid f \text{ has an analytic cont. to } U_m \text{ bounded by } m\}. \tag{1}$$

This function turns up in the results of this paper. In the terminology of [4], one would say that $\mathcal{C}^\omega(D)$ arises from the restriction $\mathcal{C}(D)|_{\mathcal{C}^\omega(D)}$ by enriching with the discrete advice $\mathrm{Adv}_{\mathcal{C}^\omega}$.

The topology induced by the representation of $\mathcal{C}^\omega(D)$ is well known and used in analysis: It can be constructed as a direct limit topology and makes $\mathcal{C}^\omega(D)$ a so called Silva-Space. For more information on this topology and its relation to computability and complexity theory also compare [7].

Consider the set of germs around zero, i.e. of power series with radius of convergence strictly larger than 1. Since the base point 0 is fixed, it is often omitted and the germ identified with a sequence. This set may be represented as follows:

Definition 7. *Let $\mathcal{O}$ denote the represented space of germs around zero, where the representation is defined as follows: A $q \in \mathbb{N}^\mathbb{N}$ is a name of a germ $(0, (a_k)_{k \in \mathbb{N}})$, if and only if*

$$\forall k \in \mathbb{N} : |a_k| \le 2^{-\frac{k}{q(0)+1}} q(0)$$

and $n \mapsto q(n+1)$ is a name of the sequence $(a_k)_{k \in \mathbb{N}}$ as element of $\mathbb{C}^\mathbb{N}$.

As above, this representation is related to the restriction of the representation of $\mathbb{C}^\mathbb{N}$ by means of the advice function $\mathrm{Adv}_{\mathcal{O}} :\subseteq \mathbb{C}^\mathbb{N} \rightrightarrows \mathbb{N}$ whose domain are the sequences with radius of convergence strictly larger than one and that is defined on those by

$$\mathrm{Adv}_{\mathcal{O}}((a_k)_{k \in \mathbb{N}}) := \{q(0) \mid q \text{ is a } \mathcal{O}\text{-name of } (a_k)_{k \in \mathbb{N}}\}$$
$$= \{n \in \mathbb{N} \mid \forall k \in \mathbb{N} : |a_k| \le 2^{-\frac{k}{n+1}} \cdot n\} \tag{2}$$

Again, the topology induced by this representation is well known and used in analysis: It is the standard choice of a topology on the set of germs and can be introduced as a direct limit topology.

Proofs that the following holds can be found in [4] or [9]:

Theorem 1 (Computability of Summation). *The assignment*

$$\mathcal{O} \to \mathcal{C}^{\omega}(D), \quad (a_k)_{k \in \mathbb{N}} \mapsto \left(x \mapsto \sum_k a_k x^k \right)$$

is computable.

A proof of the following can be found in [4]:

Theorem 2. *Differentiation is computable as mapping from $\mathcal{C}^{\omega}(D)$ to $\mathcal{C}^{\omega}(D)$.*

3   The Results

We open this chapter with an addition to Lemma 1. Given $p \in \mathbb{N}^{\mathbb{N}}$ denote the support of this function by $\mathrm{supp}(p) := \{n \in \mathbb{N} \mid p(n) > 0\}$. Furthermore, for a set A denote the number of elements of that set by $\#A$.

Lemma 2. *The function* $\mathrm{Count} :\subseteq \mathbb{N}^{\mathbb{N}} \to \mathbb{N}$, *defined via*

$$\mathrm{dom}(\mathrm{Count}) = \{p \in \mathbb{N}^{\mathbb{N}} \mid \mathrm{supp}(p) \text{ is finite}\} \quad \mathrm{Count}(p) = \#\mathrm{supp}(p)$$

is Weihrauch equivalent to $C_{\mathbb{N}}$, that is: Closed choice on the naturals.

3.1   Summing Power Series

In Sect. 2.3 it was mentioned that the operation of summing a power series is not computable on $\mathbb{C}^{\mathbb{N}}$. Recall that $\mathrm{Adv}_{\mathcal{O}}$ was the advice function of the representation of the represented space $\mathcal{O}$ of germs around zero of analytic functions on the unit disk. The computational task corresponding to this multivalued function is to find from a sequence that is guaranteed to have radius of convergence bigger than one a constant witnessing the exponential decay of the absolute value of the coefficients (compare Eq. 2 on page 373). Theorem 1 states that summation is computable on $\mathcal{O}$. Therefore, the advice function $\mathrm{Adv}_{\mathcal{O}}$ cannot be computable. The following theorem classifies the difficulty of summing power series and $\mathrm{Adv}_{\mathcal{O}}$ in the sense of Weihrauch reductions.

Theorem 3. *The following are Weihrauch-equivalent:*

- *$C_{\mathbb{N}}$, that is: Closed choice on the naturals.*
- *Sum, that is: The partial mapping from $\mathbb{C}^{\mathbb{N}}$ to $\mathcal{C}(D)$ defined on the sequences with radius of convergence strictly larger than one by*

$$\mathrm{Sum}((a_k)_{k \in \mathbb{N}})(x) := \sum_{k \in \mathbb{N}} a_k x^k.$$

I.e. summing a power series.

– Adv$_\mathcal{O}$, *that is: The function from Eq.(2) on page 373. I.e. obtaining the constant from the series.*

Proof (Ideas). Build a Weihrauch reduction circle:

$C_\mathbb{N} \leq_W$ Sum: Lemma 2 permits to replace $C_\mathbb{N}$ by Count. Let the pre-processor assign to $p \in \mathbb{N}^\mathbb{N}$ the sequence

$$a_k := \begin{cases} 1 & \text{if } p(k) > 0 \\ 0 & \text{if } p(k) = 0 \end{cases}.$$

For the post-processor use a realizer of the evaluation in 1.
Sum $\leq_W$ Adv$_\mathcal{O}$: Follows from Theorem 1.
Adv$_\mathcal{O} \leq_W C_\mathbb{N}$: Let the pre-processor be the function that maps a given name p of $(a_k)_{k\in\mathbb{N}} \in \mathbb{C}^\omega$ to an $\mathcal{A}(\mathbb{N})$-name of the set Adv$_\mathcal{O}((a_k)_{k\in\mathbb{N}})$. Note that an enumeration of the complement of this set can be extracted from p as follows: For all k and $m \in \mathbb{N}$ dovetail the test $|a_k| > 2^{-\frac{k}{m+1}} m$. If it holds for some k, return m as an element of the complement. Applying closed choice to this set will give result in a valid return value.

3.2 Differentiating Analytic Functions

In Sect. 2.3 it was remarked that it is not possible to compute the germ of an analytic function just from a name as continuous function. The proof that this is in general impossible from [9], however, argues about analytic functions on an interval. The first lemma of this chapter proves that for analytic functions on the unit disk it is possible to compute a germ if its base point is well inside of the domain. We only consider the case where the base point is zero, but the proof works whenever a lower bound on the distance of the base point to the boundary of the disk is known.

Lemma 3. Germ, *that is: The partial mapping from $C(D)$ to $\mathbb{C}^\mathbb{N}$ defined on analytic functions by mapping them to their series expansion around zero, is computable.*

Proof (Sketch). Use the Cauchy integral Formula.

The next theorem is very similar to Theorem 3. Both the advice function Adv$_{C^\omega}$ and computing a germ around a boundary point are shown to be Weihrauch equivalent to $C_\mathbb{N}$. Note that the coefficients of the series expansion $(a_k)_{k\in\mathbb{N}}$ of an analytic function f around a point x_0 are related to the derivatives $f^{(k)}$ of the function via $k!a_k = f^{(k)}(x_0)$. Therefore, computing a series expansion around a point is equivalent to computing all the derivatives in that point.

Theorem 4. *The following are Weihrauch equivalent:*

– $C_\mathbb{N}$, *that is closed choice on the naturals.*

- Diff, *that is the partial mapping from* $\mathcal{C}(D)$ *to* $\mathbb{C}$ *defined on analytic functions by*

$$\text{Diff}(f) := f'(1).$$

I.e. evaluating the derivative of an analytic function in 1.
- $\text{Adv}_{\mathcal{C}^\omega}$, *that is the function from Eq.* (1). *I.e. obtaining the constant from the function.*

Proof (Outline). By building a circle of Weihrauch reductions:

$\mathbf{C_N} \leq_{\mathbf{W}}$ Diff: Use Lemma 2 and show Count $\leq_{\mathbf{W}}$ Diff instead. For the preprocessor fix a computable sequence of analytic functions $f_n : D \to \mathbb{C}$ such that $f_n'(1) = 1$ and $|f_n(x)| < 2^{-n}$ for all $x \in D$ (compare Fig. 4). For $p \in \mathbb{N}^{\mathbb{N}}$ consider the function

$$f(x) := \sum_{n \in \text{supp}(p)} f_n(x).$$

Note that applying Diff to the function f results in

$$\text{Diff}(f) = f'(1) = \sum_{n \in \text{supp}(p)} f_n'(1) = \#\text{supp}(p).$$

Therefore, the post-processor $K(p, q) := q$ results in a Weihrauch reduction.
Diff $\leq_{\mathbf{W}} \text{Adv}_{\mathcal{C}^\omega}$: Use Theorem 2.
$\text{Adv}_{\mathcal{C}^\omega} \leq_{\mathbf{W}} \mathbf{C_N}$: Theorem 3 proved that $\text{Adv}_{\mathcal{O}} \equiv_{\mathbf{W}} \mathbf{C_N}$.
Let the pre-processor be a realizer of the function Germ from Lemma 3. Applying $\text{Adv}_{\mathcal{O}}$ will return a constant n for the sequence. Set $m := 4(n+1)^2$, then for $|x| \leq 2^{\frac{1}{m+1}} \leq 2^{\frac{1}{2(n+1)}}$

$$\left| \sum_{k \in \mathbb{N}} a_k x^k \right| \leq \sum_{k \in \mathbb{N}} 2^{-\frac{k}{2(n+1)}} n = \frac{1}{1 - 2^{-\frac{1}{2(n+1)}}} n \leq 4(n+1)^2 = m.$$

Therefore, the sum can be evaluated to an analytic function bounded by m on $B_{\frac{1}{2^{m+1}}}(0)$ and m is a valid value for the post-processor.

Recall from the introduction that $\mathcal{C}(D)|_{\mathcal{C}^\omega(D)}$ resp. $\mathbb{C}^{\mathbb{N}}|_{\mathcal{O}}$ denote the represented spaces obtained by restricting the representation of $\mathcal{C}(D)$ resp. $\mathbb{C}^{\mathbb{N}}$ to $\mathcal{C}^\omega(D)$, resp. $\mathcal{O}$. Theorems 1, 3 and 4 and Lemma 3 are illustrated in Fig. 3.

3.3 Polynomials as Finite Sequences

Consider the set $\mathbb{C}[X]$ of polynomials with complex coefficients in one variable X. There are several straightforward ways to represent polynomials. The first one that comes to mind is to represent a polynomial by a finite list of complex numbers. One can either demand the length of the list to equal the degree of the polynomial or just to be big enough to contain all of the non-zero coefficients. The first option fails to make operations like addition of polynomials computable.

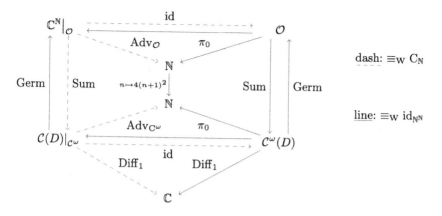

Fig. 3. The results of Theorems 1, 3 and 4 and Lemma 3.

Definition 8. *Let* $\mathbb{C}[X]$ *denote the* **represented space of polynomials**, *where* $p \in \mathbb{N}^{\mathbb{N}}$ *is a* $\mathbb{C}[X]$*-name of* P *if* $p(0) \geq \deg(P)$ *and* $n \mapsto p(n+1)$ *is a* $\mathbb{C}^{p(0)}$*-name of the first* $p(0)$ *coefficients of* P.

Let $\mathbb{C}_m[X]$ denote the set of monic polynomials over $\mathbb{C}$, i.e. the polynomials with leading
coefficient equal to one. Make $\mathbb{C}_m[X]$ a represented space by restricting the representation of
$\mathbb{C}[X]$. Monic polynomials are important because
it is possible to compute their roots – albeit in an
unordered way. To formalize this define a repre-
sentation of the disjoint union $\mathbb{C}^\times := \coprod_{n \in \mathbb{N}} \mathbb{C}^n$
as follows: A $p \in \mathbb{N}^{\mathbb{N}}$ is a name of $x \in \mathbb{C}^\times$ if
and only if $x \in \mathbb{C}^{p(0)}$ and $n \mapsto p(n+1)$ is a
$\mathbb{C}^{p(0)}$ name of x. Note that the construction of

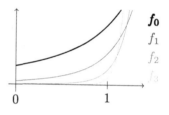

Fig. 4. $f_n(x) := (x - x_n)^{-2^{n+1}}$ for appropriate x_n (Color figure online).

the representation of $\mathbb{C}[X]$ is very similar. The only difference being that vectors
with leading zeros are not identified with shorter vectors.

Now, the task of finding the zeros in an unordered way can be formalized by
computing the multivalued function that maps a polynomial to the set of lists
of its zeros, each appearing according to its multiplicities:

$$\text{Zeros} : \mathbb{C}[X] \rightrightarrows \mathbb{C}^\times, P \mapsto \left\{ (a_1, \dots, a_{\deg(P)}) \mid \exists \lambda : P = \lambda \prod_{k=1}^{\deg(P)} (X - a_k) \right\} \quad (3)$$

The importance of $\mathbb{C}_m[X]$ is reflected in the following well known lemma:

Lemma 4. *Restricted to* $\mathbb{C}_m[X]$ *the mapping* Zeros *is computable.*

The main difficulty in computing the zeros of an arbitrary polynomial is to
find its degree. A polynomial of known degree can be converted to a monic polynomial with the same zeros by scaling. On $\mathbb{C}[X]$ consider the following functions:

- deg: The function assigning to a polynomial its degree.
- Dbnd: The multivalued function where an integer is a valid return value if and only if it is an upper bound of the degree of the polynomial.

Dbnd is computable by definition of the representation of $\mathbb{C}[X]$. The mapping deg, in contrast, is not computable on the polynomials, however, the proof of Lemma 4 includes a proof of the following:

Lemma 5. *On $\mathbb{C}_m[X]$ the degree mapping is computable.*

The next result classifies finding the degree, turning a polynomial into a monic polynomial and finding the zeros to be Weihrauch equivalent to LPO^*.

Proposition 2. *The following are Weihrauch-equivalent to LPO^*:*

- deg, *that is the mapping from $\mathbb{C}[X]$ to $\mathbb{N}$ defined in the obvious way.*
- Monic, *that is the mapping from $\mathbb{C}[X]$ to $\mathbb{C}_m[X]$ defined on the non-zero polynomials by*

$$P = \sum_{k=0}^{\deg(P)} a_k X^k \mapsto \sum_{k=0}^{\deg(P)} \frac{a_k}{a_{\deg(P)}} X^k.$$

- Zeros $: \subseteq \mathbb{C}[X] \rightrightarrows \mathbb{C}^\times$, *mapping a non-zero polynomial to the set of its zeros, each appearing according to its multiplicity (compare Eq. (3)).*

3.4   Polynomials as Functions

As polynomials induce analytic functions on the unit disk, the representations of $C^\omega(D)$ and $C(D)$ can be restricted to the polynomials. The represented spaces that result from this are $C^\omega(D)|_{\mathbb{C}[X]}$, resp. $C(D)|_{\mathbb{C}[X]}$. Here, the choice of the unit disk D as domain seems arbitrary: A polynomial defines a continuous resp. analytic function on the whole space. The following proposition can easily be checked to hold whenever the domain contains an open neighborhood of zero and, since translations are computable with respect to all the representations we consider, if it contains any open set.

Denote the versions of the degree resp. degree bound functions that take continuous resp. analytic functions by $\deg_{C(D)}, \mathrm{Dbnd}_{C(D)}$ resp. $\deg_{C^\omega(D)}$, $\mathrm{Dbnd}_{C^\omega(D)}$. When polynomials are regarded as functions, resp. analytic functions, these maps become harder to compute.

Theorem 5. *The following are Weihrauch-equivalent:*

- $C_\mathbb{N}$, *that is: Closed choice on the naturals.*
- $\mathrm{Dbnd}_{C^\omega(D)}$, *that is: Given an analytic function which is a polynomial, find an upper bound of its degree.*
- $\deg_{C^\omega(D)}$: *Given an analytic function which is a polynomial, find its degree.*

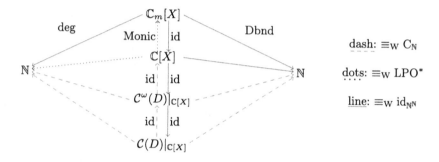

Fig. 5. The result of Lemma 4, Proposition 1 and Theorems 5 and 6.

Proof. $C_{\mathbb{N}} \leq_{\mathbf{W}} \text{Dbnd}_{\mathcal{C}^\omega(D)}$: Use Lemma 1 and reduce to Bound instead. For an enumeration p of a bounded set consider $P(X) := \sum 2^{-\max\{n, p(n)\}} X^{p(n)}$. A $\mathcal{C}^\omega(D)$-name of the function f corresponding to P can be computed from p. Let the pre-processor H be a realizer of this assignment. Set $K(p, q) := q$.
$\text{Dbnd}_{\mathcal{C}^\omega(D)} \leq_{\mathbf{W}} \deg_{\mathcal{C}^\omega(D)}$: Is trivial.
$\deg_{\mathcal{C}^\omega(D)} \leq_{\mathbf{W}} C_{\mathbb{N}}$: By Lemma 1 replace $C_{\mathbb{N}}$ with max. Let p be a $\mathcal{C}^\omega(D)$-name of the function corresponding to some polynomial P. Use Lemma 3 to extract a $\mathbb{C}^{\mathbb{N}}$-name q of the series of coefficients. Define the pre-processor by $H(p)(\langle m, n \rangle) := n + 1$ if the dyadic number encoded by $q(\langle m, n \rangle)$ is bigger than 2^{-m} and 0 otherwise. Set $K(p, q) := q$.

From the proof of the previous theorem it can be seen, that stepping down from analytic to continuous functions is not an issue. For sake of completeness we add a slight tightening of the third item of Theorem 4 and state this as theorem:

Theorem 6. *The following are Weihrauch-equivalent to $C_{\mathbb{N}}$:*

- $\deg_{\mathcal{C}(D)}$: *Given a continuous function which is a polynomial, find its degree.*
- $\text{Dbnd}_{\mathcal{C}(D)}$: *Given an analytic function which is a polynomial, find an upper bound of its degree.*
- $\text{Adv}_{\mathcal{C}^\omega}|_{\mathbb{C}[X]}$: *Given a continuous function which happens to be a polynomial, find the constant needed to represent it as analytic function.*

$\text{Dbnd}_{\mathcal{C}^\omega(D)}$ may be regarded as the advice function of $\mathbb{C}[X]$ over $\mathcal{C}^\omega(D)$: The representation where p is a name of a polynomial P if and only if $p(0) = \text{Dbnd}_{\mathcal{C}^\omega(D)}$ and $n \mapsto p(n+1)$ is a $\mathcal{C}^\omega(D)$-name of P is computationally equivalent to the representation of $\mathbb{C}[X]$. The same way, $\text{Dbnd}_{\mathcal{C}(D)}$ can be considered an advice function of $\mathbb{C}[X]$ over $\mathcal{C}(D)$.

Figure 5 illustrates Lemma 4, Proposition 1 and Theorems 5 and 6.

4 Conclusion

Many of the results proved in Sect. 3 work for more general domains: Lemma 3 generalizes to any computable point of the interior of an arbitrary domain. It can be made a uniform statement by including the base point of a germ. In this case for the proof to go through computability of the distance function of the complement of the domain of the analytic function is needed.

Another example is the part of Theorem 4 that says finding a germ on the boundary is difficult. In this case a disc of finite radius touching the boundary in a computable point is needed. Alternatively, a simply connected bounded Lipshitz domain with a computable point in the boundary can be used. Also in this case it seems reasonable to assume that a uniform statement can be proven.

Furthermore, after considering polynomials and analytic functions [7] also investigates representations for the set of distributions with compact support. In the same vain as in this paper one could compare these representation and the representation of distributions as functions on the spaces of test functions.

Acknowledgements. The work has benefited from the Marie Curie International Research Staff Exchange Scheme *Computable Analysis*, PIRSES-GA-2011- 294962. The first author was supported partially by the ERC inVEST (279499) project, the second by the International Research Training Group 1529 'Mathematical Fluid Dynamics' funded by the DFG and JSPS.

References

1. Brattka, V., de Brecht, M., Pauly, A.: Closed choice and a uniform low basis theorem. Ann. Pure Appl. Logic **163**(8), 986–1008 (2012). http://dx.doi.org/10.1016/j.apal.2011.12.020
2. Brattka, V., Gherardi, G.: Effective choice and boundedness principles in computable analysis. Bull. Symbolic Logic **17**(1), 73–117 (2011). http://dx.doi.org/10.2178/bsl/1294186663
3. Brattka, V., Gherardi, G., Hölzl, R.: Probabilistic computability and choice. Inform. Comput. **242**, 249–286 (2015). http://dx.doi.org/10.1016/j.ic.2015.03.005
4. Kawamura, A., Müller, N., Rösnick, C., Ziegler, M.: Computational benefit of smoothness: parameterized bit-complexity of numerical operators on analytic functions and Gevrey's hierarchy. J. Complex. **31**(5), 689–714 (2015). http://dx.doi.org/10.1016/j.jco.2015.05.001
5. Ko, K.: Complexity Theory of Real Functions. Progress in Theoretical Computer Science. Birkhäuser Boston Inc., Boston (1991). http://dx.doi.org/10.1007/978-1-4684-6802-1
6. Ko, K.: Polynomial-time computability in analysis. In: Handbook of Recursive Mathematics. Stud. Logic Found. Math., vols. 2 and 139, pp. 1271–1317. North-Holland, Amsterdam (1998). http://dx.doi.org/10.1016/S0049-237X(98)80052-9
7. Kunkle, D., Schröder, M.: Some examples of non-metrizable spaces allowing a simple type-2 complexity theory. In: Proceedings of the 6th Workshop on Computability and Complexity in Analysis (CCA 2004). Electron. Notes Theor. Comput. Sci., vol. 120, pp. 111–123. Elsevier, Amsterdam (2005). http://dx.doi.org/10.1016/j.entcs.2004.06.038
8. Le Roux, S., Pauly, A.: Finite choice, convex choice and finding roots. Logical Methods Comput. Sci. (2015, to appear). http://arxiv.org/abs/1302.0380

9. Müller, N.T.: Constructive aspects of analytic functions. In: Ko, K.I., Weihrauch, K. (eds.) Computability and Complexity in Analysis, CCA Workshop. Informatik Berichte, vol. 190, pp. 105–114. FernUniversität Hagen, Hagen (1995)
10. Mylatz, U.: Vergleich unstetiger funktionen in der analysis. Diplomarbeit, Fachbereich Informatik, FernUniversität Hagen (1992)
11. Mylatz, U.: Vergleich unstetiger funktionen: "Principle of Omniscience" und Vollständigkeit in der C-Hierarchie. Ph.D. thesis, Fernuniversität, Gesamthochschule in Hagen (Mai 2006)
12. Pauly, A.: Methoden zum Vergleich der Unstetigkeit von Funktionen. Masters thesis, FernUniversität Hagen (2007)
13. Pauly, A.: Infinite oracle queries in type-2 machines (extended abstract). arXiv:0907.3230v1, July 2009
14. Pauly, A.: On the topological aspects of the theory of represented spaces. Computability (201X, accepted for publication). http://arxiv.org/abs/1204.3763
15. Pauly, A., de Brecht, M.: Non-deterministic computation and the Jayne Rogers theorem. In: DCM 2012. Electronic Proceedings in Theoretical Computer Science, vol. 143 (2014)
16. Pauly, A., Davie, G., Fouché, W.: Weihrauch-completeness for layerwise computability. arXiv:1505.02091 (2015)
17. Pauly, A., Steinberg, F.: Representations of analytic functions and Weihrauch degrees. arXiv:1512.03024
18. Pour-El, M.B., Richards, J.I.: Computability in Analysis and Physics. Perspectives in Mathematical Logic. Springer, Berlin (1989)
19. Weihrauch, K.: Computable Analysis. An Introduction. Texts in Theoretical Computer Science, An EATCS Series. Springer, Berlin (2000). http://dx.doi.org/10.1007/978-3-642-56999-9
20. Ziegler, M.: Real computation with least discrete advice: a complexity theory of nonuniform computability with applications to effective linear algebra. Ann. Pure Appl. Logic 163(8), 1108–1139 (2012). http://dx.doi.org/10.1016/j.apal.2011.12.030

On Expressive Power of Regular Expressions over Infinite Orders

Alexander Rabinovich[✉]

The Blavatnik School of Computer Science, Tel Aviv University, Tel Aviv, Israel
rabinoa@post.tau.ac.il

Abstract. Two fundamental results of classical automata theory are the Kleene theorem and the Büchi-Elgot-Trakhtenbrot theorem. Kleene's theorem states that a language of finite words is definable by a regular expression iff it is accepted by a finite state automaton. Büchi-Elgot-Trakhtenbrot's theorem states that a language of finite words is accepted by a finite-state automaton iff it is definable in the weak monadic second-order logic. Hence, the weak monadic logic and regular expressions are expressively equivalent over finite words. We generalize this to words over arbitrary linear orders.

1 Definitions and Result

A linear ordering $(L, <)$ is a non-empty set L equipped with a total order. A subset I of a linear order $(L, <)$ is convex, if for all $x < y < z$ with $x, z \in I$ also $y \in I$. We use "*interval*" as a synonym for "convex subset."

A linear order $(A, <)$ is *Dedekind complete* if every non-empty subset (of the domain) which has an upper bound has a least upper bound. For example, finite orders, the naturals and reals are Dedekind complete, while the order of the rationals is not.

In this paper a *cut* of a linearly ordered set $(A, <)$ is a downward closed set $C \subseteq A$. A cut C is non-trivial if it is not empty and is a proper subset of A. If $(A, <)$ is Dedekind complete and C is its nontrivial cut, then there is $a \in A$ such that $C := \{c \in A \mid c \leq a\}$ or $C := \{c \in A \mid c < a\}$.

1.1 Extended Regular Expression

We use a generalized notion of a word, which coincides with the notion of a labeled linear ordering. Given a finite alphabet Σ, a word over Σ or Σ-labeled chain is a linear order $(L, <)$ equipped with a function lab from L into Σ. A language over Σ is a class of words over Σ. Whenever Σ is clear from the context or unimportant we will use "word" for "word over Σ" and "language" for "language over Σ."

The concatenation (the lexicographical sum) of two words $w_1 = (L_1, <_1, lab_1)$ and $w_2 = (L_2, <_2, lab_2)$ over the same alphabet (up to renaming, assume that L_1 and L_2 are disjoint) is a word $(L_1 \cup L_2, <, lab)$, where (1) lab coincides with lab_1 on L_1, with lab_2 on L_2, and (2) $<$ coincides with $<_1$ on L_1, with $<_2$ on L_2,

© Springer International Publishing Switzerland 2016
A.S. Kulikov and G.J. Woeginger (Eds.): CSR 2016, LNCS 9691, pp. 382–393, 2016.
DOI: 10.1007/978-3-319-34171-2_27

and if $a \in L_1$ and $b \in L_2$ then $a < b$. The concatenation of words w_1 and w_2 is denoted[1] by $w_1 + w_2$.

For languages C_1 and C_2, their concatenation is defined as $\{w_1 + w_2 \mid w_1 \in C_1$ and $w_2 \in C_2\}$ and is denoted by $C_1; C_2$.

The Kleene iteration or the positive concatenation closure of a language C is denoted by C^+ and is defined as $\cup_{k=1}^{\infty}\{w_1 + w_2 + \cdots + w_k \mid w_i \in C\}$.

Extended regular expressions over an alphabet Σ are defined by the following grammar: $E := \emptyset \mid \sigma \mid E \cup E \mid E; E \mid E^+ \mid \neg E$, where $\sigma \in \Sigma$. The semantics assigns to such an expression a language over Σ, as follows: (1) The empty language is assigned to $\emptyset$. (2) A language consisting of one element order labeled by σ is assigned to σ. (3) $\cup$ is interpreted as the union and $\neg$ as the complementation with respect to the class of all words over Σ. (4) $E_1; E_2$ is the concatenation of the languages assigned to E_1 and E_2, and (5) E^+ is the positive concatenation closure of the language assigned to E.

A *regular expression* is an extended regular expression without negation. Note that the semantics assigns to a regular expression only a set of finite words. Usually, in classical automata theory the complementation is taken only with respect to the set of finite words. Clearly, under such finite-words interpretation of complementation only languages of finite words are defined by extended regular expressions.

We conclude this section with examples which illustrate the expressive power of extended regular expressions.

All expressions below are over unary alphabet $\{1\}$; a word over a unary alphabet can be identified with the underlying linear order.

- $All := \neg\emptyset$ - defines the class of all linear orders.
- $Max := 1 \cup All; 1$ - defines the linear orders with a maximal element.
- $Min := 1 \cup 1; All$ - defines the linear orders with a minimal element.
- Dense:= $\neg(Max; Min)$ - defines the dense linear orders.
- Dedekind:=$1 \cup \neg((\neg Max); (\neg Min))$ - defines the Dedekind complete linear orders.
- Dense$^+$ - defines the orders which can be partitioned in a finite set of dense intervals; equivalently the linear order with a finite set of a successor elements where a is a successor if there is $b < a$ such that no element exists between b and a.

1.2   Fragments of MSO

The Monadic second-order logic (MSO) is an extension of first-order logic that allows to quantify over elements as well as over subsets of the domain of the structure.

The structures considered in this paper are expansions of nonempty linear orderings $(A, <^A)$ by subsets $P_1^A, \ldots, P_l^A$. When no confusion arises we

---

[1] In algebraic framework to formal languages the concatenation of w_1 and w_2 is called "the product" and is denoted by $w_1 \cdot w_2$.

cancel the superscript A, use the abbreviating notation $\overline{P}$ for the set tuple $(P_1^A, \ldots, P_l^A)$, and write $(A, \overline{P})$.

Such a structure is called l-chain. It can be regarded as a labeled ordering (or generalized word) with labels in $\{0,1\}^l$: the element $a \in A$ has the label $(b_1, \ldots, b_l)$ defined by $b_i := 1$ iff $a \in P_i$. When $P_1, \ldots, P_l$ partitions the domain of a linear ordering $(A, <^A)$, such a structure can be regarded as a word with labels in $\{1, \ldots, l\}$: the element $a \in A$ is labeled by i iff $a \in P_i$.

The standard language of MSO for structures of this signature is built up as follows, using the relation symbols $<$ and $P_1, \ldots P_m$. We have first-order variables $x, y, \ldots$ for elements of structures, monadic second order variables $X, Y \ldots$ for sets of elements of structures, and the atomic formulas are of the form $x = y$, $x < y$, $P_i(x)$, and $Y(x)$, with the canonical interpretation. Formulas are constructed from atomic formulas by the Boolean connectives, and by applying the first-order quantifier $\exists x$ "there is an element x" to first-order variables, and the monadic second-order quantifier $\exists X$ - "there is a set X" to monadic variables.

The Weak Monadic Second-Order logic is an extension of first-order logic that allows to quantify over elements as well as over finite subsets of the domain of the structure. So, it has the first-order quantifiers, and the quantifier $\exists^{fin} X$ - "there is a finite set X". We denote this logic by $MSO[\exists^{fin}]$.

The logic we are going to consider is denoted by $MSO[\exists^{fin}, \exists^{cut}]$ and it extends the weak monadic logic by the quantifier over cuts: $\exists^{cut} X$ - "there is a cut X."

A language (or a class of chains) $definable$ by a formula φ is the class of all chains that satisfy φ.

Note that over Dedekind complete chains $MSO[\exists^{fin}]$ is expressively equivalent to $MSO[\exists^{fin}, \exists^{cut}]$. Both $MSO[\exists^{fin}]$ and $MSO[\exists^{fin}, \exists^{cut}]$ are equivalent to MSO over the class of finite words. McNaughton's theorem [10] implies that an ω-language is definable in MSO iff it is accepted by a deterministic Muller automaton. For a deterministic automaton "the run on an ω-word is accepting" can be formalized in $MSO[\exists^{fin}]$. Hence, $MSO[\exists^{fin}]$, $MSO[\exists^{fin}, \exists^{cut}]$ and MSO are expressively equivalent on the class of ω-words.

1.3   Result

Kleene [7] introduced regular expressions and proved that a language is definable by a regular expression iff it is accepted by a finite state automaton, and that the transformations from expressions to automata and vice versa are computable. The Büchi-Elgot-Trakhtenbrot theorem states that finite-state automata and the Weak Monadic Second-Order Logic (interpreted over finite words) have the same expressive power, and that the transformations from formulas to automata and vice versa are computable [1,4,17]. Hence, the classical theorem is:

Theorem 1.1 (Kleene, Büchi, Elgot, Trakhtenbrot). *The following are equivalent for languages of finite words:*

1. A language is definable by a regular expression.
2. A language is accepted by a finite state automaton.
3. A language is definable in $MSO[\exists^{fin}]$.

We generalize the equivalence between (1) and (3) of this classical result to arbitrary words, as follows:

Theorem 1.2 (Main). *A language of labelled orderings is definable by an extended regular expression iff it is definable in $MSO[\exists^{fin}, \exists^{cut}]$.*

Hence, extended regular expressions and $MSO[\exists^{fin}, \exists^{cut}]$ have the same expressive power over the class of all words. The transformations from formulas to extended regular expressions and vice versa are computable and can be easily extracted from the proof.

The paper is organized as follows. The next section provides a logical background and summarizes elements of the composition method. In Sect. 3 we prove that every $MSO[\exists^{fin}, \exists^{cut}]$ formula is equivalent to an extended regular expression. In Sect. 4 we prove that every extended regular expression is equivalent to a $MSO[\exists^{fin}, \exists^{cut}]$ formula. Section 5 presents a conclusion and further results.

2 Logical Background

2.1 A Variant of $MSO[\exists^{fin}, \exists^{cut}]$

It will be convenient to work with a slightly modified (but expressively equivalent) set-up, in which the first-order variables are canceled. We allow only monadic second-order variables and take as atomic formulas of $MSO[\exists^{fin}, \exists^{cut}]$ the following: $Empty(X)$, $X \subseteq Y$, $Sing(X)$, $X < Y$, $All(X)$, $Finite(X)$ and $Cut(X)$. These are interpreted, respectively, as "X is empty," "X is a subset of Y," "X contains one element," "X contains one element and Y contains one element and the element of X is smaller than the element of Y," "X is the universe," "X is finite," and "X is a cut."

Formulas are constructed from atomic formulas by the Boolean connectives, and by the quantifiers $\exists^{fin}$ and $\exists^{cut}$.

The use of the unary relation symbols P_i will be avoided by taking free set variables X_i instead. Thus, we shall use labeled chains $(A, <, \overline{P})$ as interpretations of monadic formulas $\varphi(\overline{X})$.

The quantifier rank of a formula φ, denoted $\mathrm{qr}(\varphi)$, is the maximum depth of nesting of quantifiers in φ. For $r, l \in \mathbb{N}$ we denote by $\mathfrak{Form}_l^r$ the set of formulas of quantifier rank $\leq r$ and with free variables among $X_1, \ldots, X_l$.

2.2 Elements of the Composition Method

Our proofs use a technique known as the composition method [9,14]. To fix notations and to aid the reader unfamiliar with this technique, we briefly review those definitions and results that we require. A more detailed presentation can be found in [16] or in [5].

2.2.1   Hintikka Formulas and r-types
Definition 2.1 *Let* $r, l \in \mathbb{N}$ *and* $\mathfrak{A}, \mathfrak{B}$ l-*chains. The* r-*theory of* $\mathfrak{A}$ *is*

$$\mathrm{Th}^r(\mathfrak{A}) := \{\varphi \in \mathfrak{Form}_l^r \mid \mathcal{M} \models \varphi\}.$$

If $\mathrm{Th}^r(\mathfrak{A}) = \mathrm{Th}^r(\mathfrak{B})$, *we say that* $\mathfrak{A}$ *and* $\mathfrak{B}$ *are* r-*equivalent and write* $\mathfrak{A} \equiv^r \mathfrak{B}$.

Clearly, $\equiv^r$ is an equivalence relation. For any $r, l \in \mathbb{N}$, the set $\mathfrak{Form}_l^r$ is infinite. However, it contains only finitely many semantically distinct formulas. So, there are finitely many $\equiv^r$-classes of l-chains. In fact, we can compute "representatives" for these classes:

Lemma 2.2 (Hintikka Lemma). *For* $r, l \in \mathbb{N}$, *we can compute a finite set* $H_l^r \subseteq \mathfrak{Form}_l^r$ *such that:*

(a) *For distinct* $\tau, \tau' \in H_l^r$, $\tau \wedge \tau'$ *is not satisfiable.*
(b) *If* $\tau \in H_l^r$ *and* $\varphi \in \mathfrak{Form}_l^r$, *then either* $\tau \models \varphi$ *or* $\tau \models \neg\varphi$. *Furthermore, there is an algorithm that, given such* τ *and* φ, *decides which of these two possibilities holds.*
(c) *For every* l-*structure* $\mathfrak{A}$, *there is a* unique $\tau \in H_l^r$ *such that* $\mathfrak{A} \models \tau$.

Any member of H_l^r *we call an* (r, l)-*Hintikka formula*[2] *or a formal* (r, l)-*type.*

Definition 2.3 (r-type). *For* $r, l \in \mathbb{N}$ *and* $\mathfrak{A}$ *an* l-*chain, we denote by* $\mathrm{Tp}^r(\mathfrak{A})$ *the unique member of* H_l^r *satisfied by* $\mathfrak{A}$ *and call it the* r-*type of* $\mathfrak{A}$.

Thus, $\mathrm{Tp}^r(\mathfrak{A})$ determines $\mathrm{Th}^r(\mathfrak{A})$ and, indeed, $\mathrm{Th}^r(\mathfrak{A})$ is computable from $\mathrm{Tp}^r(\mathfrak{A})$.

Lemma 2.4 (Projection). *For* $r, l \in \mathbb{N}$, *there is an operation* Pr_l^r *from* H_l^r *into* H_{l-1}^r *such that if* $\mathrm{Tp}_l^r(A, <, P_1^{\mathfrak{A}}, \cdots, P_{l-1}^{\mathfrak{A}}, P_l^{\mathfrak{A}}) = \tau$, *then* $\mathrm{Tp}_{l-1}^r(A, <, P_1^{\mathfrak{A}}, \cdots, P_{l-1}^{\mathfrak{A}}) = \mathrm{Pr}_l^r(\tau)$.

2.2.2   The Lexicographical Sum of Chains and of r-types
Let $\mathfrak{A} := (A, <^{\mathfrak{A}}, P_1^{\mathfrak{A}}, \ldots, P_l^{\mathfrak{A}})$ and $\mathfrak{B} := (B, <^{\mathfrak{B}}, , P_1^{\mathfrak{B}}, \ldots, P_l^{\mathfrak{B}})$ be l-chains with disjoint domains. The lexicographical sum (or concatenation) of $\mathfrak{A}$ and $\mathfrak{B}$ is denoted $\mathfrak{A} + \mathfrak{B}$ and is defined as the l-chain $(A \cup B, <, P_1^{\mathfrak{A}} \cup P_1^{\mathfrak{B}}, \ldots, P_l^{\mathfrak{A}} \cup P_l^{\mathfrak{B}})$ where $a < b$ if $a \in A$ and $b \in B$ or $a, b \in A$ and $a <_{\mathfrak{A}} b$ or $a, b \in B$ and $a <_{\mathfrak{B}} b$.

As usual, we do not distinguish between isomorphic structures. So, if the domains of $\mathfrak{A}$ and $\mathfrak{B}$ are not disjoint, replace them with isomorphic l-chains that have disjoint domains, and proceed as before.

It is clear that the sum of chains is associative. We will use the notation $\mathfrak{A}_1 + \mathfrak{A}_2 + \cdots + \mathfrak{A}_k$ for the sum of k chains.

The next Lemma says that $\equiv^r$ is a congruence with respect to the sum.

Lemma 2.5 *The* r-*types of* l-*chains* $\mathfrak{A}$, $\mathfrak{B}$ *determine the* r-*type of* $\mathfrak{A} + \mathfrak{B}$.

---

[2] Hintikka formulas made their first appearance in [6], in the framework of first-order logic.

The Lemma justifies the notation $\tau_1 + \tau_2$ for the r-type of an l-chain which is the sum of two l-chains of r-types τ_1 and τ_2, respectively. The composition theorem states that $+$ can be extended to a (uniformly) computable operation on the formal types.

Theorem 2.6 (Composition Theorem). *For $r, l \in \mathbb{N}$, there is an associative operation $+ : H_l^r \times H_l^r \to H_l^r$ such that for every l-chains $\mathfrak{A}, \mathfrak{B}$ if $\mathrm{Tp}^r(\mathfrak{A}) = \tau_1$ and $\mathrm{Tp}^r(\mathfrak{B}) = \tau_2$ then $\mathrm{Tp}^r(\mathfrak{A} + \mathfrak{B}) = \tau_1 + \tau_2$. Furthermore, the sum of (r, l)-formal types is (uniformly) computable.*

The reader may wonder why we do not say: "$\tau_1 + \tau_2$ is the *unique* element of H_l^r such that ...". The reason is that by Hintikka's construction [6] there are in H_l^r formulas that are not satisfied in any structure.

3   From Logic to Expressions

In this section we prove that for every formula φ in $MSO[\exists^{fin}, \exists^{cut}]$ there is an equivalent extended regular expression E_φ.

We proceed by induction on the quantifier rank of formulas.

For a quantifier free formula the corresponding equivalent expression is easily constructed.

If φ_1 is equivalent to E_{φ_i} for $i = 1, 2$, then $\varphi_1 \vee \varphi_2$ is equivalent to $E_{\varphi_1} \cup E_{\varphi_1}$, and $\neg \varphi_1$ is equivalent to $\neg E_{\varphi_1}$

The only interesting case is for quantifiers:

3.1   Translation for $\exists^{cut}$ Quantifier

Assume that the inductive assumption holds for r. In particular, for every Hintikka formula τ of quantifier rank r there is an equivalent expression E_τ.

Let $\varphi(X_1, \cdots, X_l)$ be a formula and assume that $\mathrm{qr}(\varphi) = r$.
$\exists^{cut} X_l \varphi$ is equivalent to a disjunction of

1. $\varphi_0 := \exists^{cut} X_l Empty(X_l) \wedge \varphi$
2. $\varphi_1 := \exists^{cut} X_l All(X_l) \wedge \varphi$
3. $\varphi_2 := \exists^{cut} X_l \neg Empty(X_l) \wedge \neg All(X_l) \wedge \varphi$

Let $S_0 \subseteq H_{l-1}^r$ be defined as $\{\mathrm{Pr}_l^r(\tau_1) \mid \tau_1 \in H_l^r$ and $\tau_1 \models \varphi \wedge Empty(X_l)\}$, where Pr_l^r was defined in Lemma 2.4. Then $\mathfrak{A} \models \varphi_0$ iff $\mathrm{Tp}_{l-1}^r(\mathfrak{A}) \in S_0$. Therefore, φ_0 is equivalent to $\cup_{\tau \in S_0} E_\tau$ (where E_τ are defined by the inductive assumption). For φ_1 an equivalent expression E_{φ_1} is defined in a similar way as $E_{\varphi_1} := \cup_{\tau \in S_1} E_\tau$, where $S_1 := \{\mathrm{Pr}_l^r(\tau_1) \mid \tau_1 \in H_l^r$ and $\tau_1 \models \varphi \wedge All(X_l)\}$.

In order to translate φ_2 into an equivalent expression we will use the composition theorem and an observation that every non-empty proper downward closed subset P of the domain of $\mathfrak{A}$ induces a representation of $\mathfrak{A}$ as the sum $\mathfrak{A}_1 + \mathfrak{A}_2$ where $\mathfrak{A}_1$ (respectively, $\mathfrak{A}_2$) is the substructure of $\mathfrak{A}$ over P (respectively, the complement of P).

Set $\psi_2 := \neg Empty(X_l) \wedge \neg All(X_l) \wedge Cut(X_l) \wedge \varphi$. Hence, $\varphi_2 := \exists^{cut} X_l \psi_2$.

Claim 1. *Let* $\mathfrak{B}$ *be an l-chain.* $\mathfrak{B} \models \psi_2$ *iff there are* $\tau_1, \tau_2 \in H_l^r$ *and* $\mathfrak{B}_1$ *and* $\mathfrak{B}_2$ *such that*

1. $\mathfrak{B} = \mathfrak{B}_1 + \mathfrak{B}_2$ *and* $\tau_i = \mathrm{Tp}^r(\mathfrak{B}_1)$ *for* $i = 1, 2$.
2. $\tau_1 + \tau_2 \models \psi_2$.
3. $\tau_1 \models All(X_l)$ *and* $\tau_2 \models Empty(X_l)$.

Proof. $\Leftarrow$ is immediate.
    $\Rightarrow$ Take as $\mathfrak{B}_1$ (respectively, $\mathfrak{B}_2$) the substructure of $\mathfrak{B}$ over P_l (respectively, over the complement of P_l), and as τ_i the r-type of $\mathfrak{B}_i$. $\square$

Let S be the set of pairs $\langle \tau_1, \tau_2 \rangle$ of H_l^r formulas, which satisfy conditions (2) and (3) of Claim 1.
    Define $\widehat{S} \subseteq H_{l-1}^r \times H_{l-1}^r$ as $\widehat{S} := \{ \langle \mathrm{Pr}_l^r(\tau_1), \mathrm{Pr}_l^r(\tau_2) \rangle \mid \langle \tau_1, \tau_2 \rangle \in S \}$, where Pr_l^r was defined in Lemma 2.4. Thus we obtain:

Claim 2. $\mathfrak{A} \models \varphi_2$ *if and only if there are* $\mathfrak{A}_1$ *and* $\mathfrak{A}_2$ *such that* $\mathfrak{A} = \mathfrak{A}_1 + \mathfrak{A}_2$ *and* $\langle \mathrm{Tp}^r(\mathfrak{A}_1), \mathrm{Tp}^r(\mathfrak{A}_2) \rangle \in \widehat{S}$.

By the inductive assumption each formula of quantifier rank r is equivalent to an expression. In particular, each Hintikka formula τ of quantifier rank r is equivalent to an expression E_τ. Finally, Claim 2 implies that φ_2 is equivalent to
$$\cup_{\langle \tau_1, \tau_2 \rangle \in \widehat{S}} E_{\tau_1} ; E_{\tau_2}.$$

3.2   Translation for $\exists^{fin}$ Quantifier

In order to translate $\exists^{fin} X_l \varphi$ into an equivalent expression we will use the composition theorem and an observation that every finite subset of the domain of $\mathfrak{A}$ induces a natural representation of $\mathfrak{A}$ as a finite sum of its subchains.

Claim 3. $\mathfrak{B} \models \varphi \wedge Finite(X_l)$ *iff there is a sequence* $\tau_1, \ldots, \tau_k$ *of* H_l^r *formulas and a sequence* $\mathfrak{B}_1 \ldots, \mathfrak{B}_k$ *of l-chains such that*

1. $\mathfrak{B} = \mathfrak{B}_1 + \mathfrak{B}_2 + \cdots + \mathfrak{B}_k$ *and* $\tau_i = \mathrm{Tp}^r(\mathfrak{B}_i)$ *for* $i = 1, \ldots, k$.
2. $\tau_1 + \tau_2 + \cdots + \tau_k \models \varphi$ *and*
3. *if* $\tau_i \models \neg Empty(X_l)$ *then* $\tau_i \models Sing(X_l) \wedge All(X_l)$, *i.e.,* τ_i *holds only on singleton chains.*

Proof. $\Leftarrow$ is immediate.
    $\Rightarrow$ Assume $\mathfrak{B} \models \varphi \wedge Finite(X_l)$. Hence, P_l is finite. Define an equivalence $\sim$ as follows: $a_1 \sim a_2$ iff either $a_1 = a_2 \in P_l$ or there is no element of P_l in the interval $[\min(a_1, a_2), \max(a_1, a_2)]$. It is clear that $\sim$ is an equivalence relation. It has finitely many equivalence classes, and each $\sim$ equivalence class is an interval of the domain of $\mathfrak{B}$. Let $I_1 < \cdots < I_k$ be the $\sim$-equivalence classes. For $j = 1, \ldots, k$, define $\mathfrak{B}_j$ as the substructure of $\mathfrak{B}$ over I_j and $\tau_j := \mathrm{Tp}^r(\mathfrak{B}_j)$. It is clear that $\mathfrak{B}_j$ and τ_j satisfy the requirements of the claim. $\square$

Let S be the set of finite sequences of H_l^r formulas, which satisfy conditions (2) and (3) of Claim 3.

Define a set $\widehat{S}$ of finite sequences of H_{l-1}^r formulas as $\widehat{S} := \{\langle \mathrm{Pr}_l^r(\tau_1), \ldots, \mathrm{Pr}_l^r(\tau_k) \rangle \mid \langle \tau_1, \ldots, \tau_k \rangle \in S\}$. Therefore, Claim 3 implies:

Claim 4. $\mathfrak{A} \models \exists^{fin} X_l \varphi$ *iff there is a sequence* $\langle \tau_1, \ldots, \tau_k \rangle \in \widehat{S}$ *and a sequence* $\mathfrak{A}_1 \ldots, \mathfrak{A}_k$ *of* $(l-1)$*-chains such that* $\mathfrak{A} = \mathfrak{A}_1 + \mathfrak{A}_2 + \cdots + \mathfrak{A}_k$ *and* $\tau_i = \mathrm{Tp}^r(\mathfrak{A}_i)$ *for* $i = 1, \ldots, k$.

Claim 5. *There is a regular expression* E *which defines* $\widehat{S}$.

Proof. We will construct a finite state automaton $\mathcal{A}$ which accepts $\widehat{S}$. The set $Q_{\mathcal{A}}$ of its states is $Q_{\mathcal{A}} := \{q_i\} \cup H_l^r$, where $q_i \notin H_l^r$ is a fresh state.

q_i is the initial state of $\mathcal{A}$. The set Acc of accepting states is defined as $Acc := \{\tau \in H_l^r \mid \tau \models \varphi \wedge Finite(X_l)\}$.

For every $\tau \in H_{l-1}^r$ define two sets $D(\tau), F(\tau) \subseteq H_l^r$ as $D(\tau) := \{\tau' \in H_l^r \mid \tau' \models \tau \wedge Empty(X_l)\}$ and $F(\tau) := \{\tau' \in H_l^r \mid \tau' \models \tau \wedge All(X_l) \wedge Sing(X_l)\}$. The transition relation $\rightarrow_{\mathcal{A}} \subseteq Q_{\mathcal{A}} \times H_{l-1}^r \times Q_{\mathcal{A}}$ is defined as follows:

1. $\langle q_i, \tau, \tau' \rangle \in \rightarrow_{\mathcal{A}}$ iff $\tau' \in D(\tau) \cup F(\tau)$.
2. $\langle \tau_1, \tau, \tau_2 \rangle \in \rightarrow_{\mathcal{A}}$ iff there is $\tau' \in D(\tau) \cup F(\tau)$ such that $\tau_2 = \tau_1 + \tau'$.

It is straightforward to check that $\mathcal{A}$ accepts $\widehat{S}$. Therefore, by Theorem 1.1, $\widehat{S}$ is definable by a regular expression. □

By the inductive assumption each formula of quantifier rank r is equivalent to an expression. In particular, each Hintikka formula τ of quantifier rank r is equivalent to an expression E_τ.

Finally, let E_φ be obtained from a regular (complementation free) expression E of Claim 5, by replacing each letter $\tau \in H_{l-1}^r$ with an equivalent extended regular expression E_τ. Claims 4 and 5 imply that φ is equivalent to E_φ.

4   From Expressions to Logic

We are going to prove that for every expression E over an alphabet Σ there is an equivalent $MSO[\exists^{fin}, \exists^{cut}]$ formula φ.
We proceed by the structural induction on expressions.
It is straightforward to write a formula for $\emptyset$ and for a letter $\sigma \in \Sigma$.
If E_i are equivalent to φ_i for $i = 1, 2$, then $E_1 \cup E_2$ is equivalent to $\varphi_1 \vee \varphi_2$ and $\neg E_1$ is equivalent to $\neg \varphi_1$.
Below we will treat concatenation and iteration.
First, let us introduce notations and state a standard "relativization" lemma which will be used several times.

Notation 4.1. Let $l \in \mathbb{N}$, $\mathfrak{A} := (A, <, P_1, \ldots, P_l)$ an l-chain and D a non-empty subset of A. The *restriction* of $\mathfrak{A}$ to D is the l-chain $\mathfrak{A}_{\restriction D}$ defined as $\mathfrak{A}_{\restriction D} := (D, <, P_1 \cap D, \ldots, P_l \cap D)$.

Lemma 4.2. (Relativization). *Let $\varphi(\overline{Y})$ be a formula, U a variable not appearing in φ. There is a formula $\varphi_{\restriction U}(\overline{Y}, U)$ such that for every chain $(A, <, \overline{P})$ and every non-empty $D \subseteq A$,*

$$(A, <, \overline{P}, D) \models \varphi_{\restriction U}(\overline{Y}, U) \text{ iff } (A, <, \overline{P})_{\restriction D} \models \varphi(\overline{Y}).$$

When this is the case, we say that φ holds in $(A, <, \overline{P})$ relativized to D.

4.1   Concatenation

Assume φ_i is equivalent to E_i for $i = 1, 2$.

Then $E_1; E_2$ is equivalent to $\exists^{cut} X \varphi$ where φ is the conjunction of the following:

1. X is a non-empty proper downward closed subset of the domain of $\mathfrak{A}$.
2. φ_1 holds in $\mathfrak{A}$ relativized to X.
3. φ_2 holds in $\mathfrak{A}$ relativized to the complement of X.

(1)–(3) are easily formalized in $MSO[\exists^{fin}, \exists^{cut}]$. Moreover, if φ_1 and φ_2 are $MSO[\exists^{cut}]$ formulas, then (1)–(3) are easily formalized in $MSO[\exists^{cut}]$.

4.2   Kleene Iteration

Assume that E is equivalent to φ.

Recall that $\mathfrak{A}$ is in E^+ iff there is $k > 0$ and a partition of the domain of $\mathfrak{A}$ into intervals $I_1, \ldots, I_k$ such that $\mathfrak{A}_{\restriction I_j}$ are in E. In the case when all I_j are intervals with endpoints in $\mathfrak{A}$ this can be easily formalized. However, $\mathfrak{A}$ is not necessarily Dedekind complete, and not all intervals have end-points in $\mathfrak{A}$. To overcome this problem we use the following Lemma:

Lemma 4.3. *Let $\varphi(\overline{X})$ be a formula. Then there are formulas $\psi^i_{\leq}(\overline{X})$ and $\psi^i_{\geq}(\overline{X})$ $(i = 0, \ldots, m)$ such that for every $\mathfrak{A}$, element $a \in \mathfrak{A}$, and intervals $I_{\leq a} := \{b \in \mathfrak{A} \mid b \leq a\}$ and $I_{\geq a} := \{b \in \mathfrak{A} \mid b \geq a\}$:*

$$\mathfrak{A} \models \varphi \text{ iff there is } i \text{ such that } \mathfrak{A}_{\restriction I_{\leq a}} \models \psi^i_{\leq} \text{ and } \mathfrak{A}_{\restriction I_{\geq a}} \models \psi^i_{\geq}$$

The Lemma is easily obtained from Lemma 2.5 (one can take as $\psi^i_{\leq}(\overline{X})$, $\psi^i_{\geq}(\overline{X})$ formulas of quantifier rank smaller than $\mathrm{qr}(\varphi) + 3$).

The Lemma implies that "$\mathfrak{A}$ is in E^+" can be rephrased as:

there is a partition of the domain of $\mathfrak{A}$ into intervals $I_1, \ldots, I_k$ and there are $a_j \in I_j$ and a function $F : \{1, \ldots, k\}$ into $\{0, \ldots, m\}$ such that for every $j \in \{1, \ldots, k\}$ and $s := F(j)$
1. the substructure of $\mathfrak{A}$ over the interval $\{b \in I_j \mid b \geq a_j\}$ satisfies $\psi^s_{\geq}$
2. the substructure of $\mathfrak{A}$ over the interval $\{b \in I_j \mid b \leq a_j\}$ satisfies $\psi^s_{\leq}$

The above is equivalent to

there is a non-empty finite subset P of the domain of $\mathfrak{A}$ and a function
$F : P \to \{0, \ldots, m\}$ such that
1. if a is the maximal element of P and $s = F(a)$ then the substructure
   of $\mathfrak{A}$ over the interval $\{b \mid b \geq a\}$ satisfies $\psi_{\geq}^{s}$
2. if a is the minimal element of P and $s = F(a)$ then the substructure
   of $\mathfrak{A}$ over the interval $\{b \mid b \leq a\}$ satisfies $\psi_{\leq}^{s}$, and
3. If $a < c$ are successive elements of P and $s = F(a)$ and $p = F(c)$, then
   there is a downward closed set D such that
   (a) $a \in D$, $c \notin D$
   (b) the substructure of $\mathfrak{A}$ over the interval $\{b \in D \mid b \geq a\}$ satisfies $\psi_{\geq}^{s}$
   (c) the substructure of $\mathfrak{A}$ over the interval $\{b \notin D \mid b \leq c\}$ satisfies $\psi_{\leq}^{p}$

Observe that F cannot be represented by a single monadic predicate. However, since F is a mapping from a finite set P to a set of size $m + 1$ (m is defined in Lemma 4.3 depens on φ, but is independent of P), it can be represented by a tuple of finite sets and the conditions (1)–(3) can be easily formalized in $MSO[\exists^{fin}, \exists^{cut}]$.

5   Conclusion

The classical automata theory establishes equivalence (over finite words) between three fundamental formalisms: the monadic second-order logic, regular expressions and finite state automata. The cornerstones of automata theory on infinite objects are Büchi's and Rabin's theorems. The Büchi theorem states that MSO and finite automata are equivalent over ω-words [2] and the Rabin theorem states that MSO and finite automata are equivalent over labeled binary trees [12].

MSO and its fragment have a natural interpretation over arbitrary (even partial) orders. Regular expressions have a natural interpretation over arbitrary linear orders. We proved expressive equivalence (over arbitrary words) between the extended regular expressions and $MSO[\exists^{fin}, \exists^{cut}]$. It seems that there is no natural notion of automata which has the same expressive power as the above formalisms. Usually, automata correspond to logical formulas of a fixed quantifier alternation depth. However, Thomas Colcombet pointed out that the quantifier alternation hierarchy does not collapse for $MSO[\exists^{fin}, \exists^{cut}]$.

Below we comment about some extensions of our results.

5.1   Words over Linear Orders of a Bounded Cardinality

Let $\aleph$ be an infinite cardinal. A linear order $(L, <)$ is an $\aleph^{<}$-order if the cardinality of L is less than $\aleph$. Given a finite alphabet Σ, an $\aleph^{<}$-word over Σ or Σ-labeled $\aleph^{<}$-chain is an $\aleph^{<}$-linear order $(L, <)$ equipped with a function lab from L into Σ. A $\aleph^{<}$-language over Σ is a set of $\aleph^{<}$-words over Σ. Whenever Σ is clear from the context or unimportant we will use "$\aleph^{<}$-word" for "$\aleph^{<}$-word over Σ" and "$\aleph^{<}$-language" for "$\aleph^{<}$-language over Σ."

For an extended regular expression E over Σ the $\aleph^<$-semantics assigns an $\aleph^<$-language over Σ. The $\aleph^<$-semantics is defined exactly like the semantics of extended regular expressions in Sect. 1.1 with the only exception that complementation is taken with respect to the set of $\aleph^<$-word over Σ. Namely, the $\aleph^<$-semantics is defined as follows: (1) The empty language is assigned to $\emptyset$. (2) A language consisting of one element order labeled by σ is assigned to σ. (3) $\cup$ is interpreted as the union and $\neg$ as the complementation with respect to the set of all $\aleph^<$-words over Σ. (4) $E_1; E_2$ is the concatenation of the languages assigned to E_1 and E_2. (5) E^+ is the positive concatenation closure of the language assigned to E.

Note

(1) For the first infinite cardinal $\aleph_0$, the $\aleph_0^<$-semantics assigns to an extended regular expression the same language (of finite words) as the classical semantics does.

(2) If C is the class of words assigned to E by the semantics defined in Sect. 1.1, then $\aleph^<$-semantics assigns to E the set of all $\aleph^<$-words in C.

We say that a language C is $\aleph^<$-*definable* by an expression E if $\aleph^<$-semantics assigns C to E. We say that an $\aleph^<$-language is *definable* by an MSO formula φ iff it is the set of all $\aleph^<$-words that satisfy φ.

Our main theorem and (2) imply the following Theorem:

Theorem 5.1. *Let $\aleph$ be an infinite cardinal. An $\aleph^<$-language is definable by an extended regular expression iff it is definable by an $MSO[\exists^{fin}, \exists^{cut}]$ formula.*

From our proof it is also easy to extract that a language of labelled Dedekind complete orderings is definable by an extended regular expression iff it is definable by an $MSO[\exists^{fin}]$ formula.

5.2   Star-Free Expressions

McNaughton and Papert introduced star-free regular expressions. These are extended regular expressions without the Kleene iteration. Namely, given an alphabet Σ, the star-free expressions over Σ are built up from $\emptyset$ and the letters in Σ by union, concatenation and complementation. A famous theorem of McNaughton and Papert [11] states that a language of finite words is definable by a star-free expression if and only if it is definable in first-order logic. This theorem was extended to ω-languages in Ladner [8] and Thomas [15], and to languages over the real order by Rabinovich [13]. The following generalization to Dedekind complete orders was proved in [13]:

Theorem 5.2. *A language of labelled Dedekind complete orderings is definable by a star-free regular expression iff it is definable by a first-order formula.*

Our proof of Theorem 1.2 can be easily modified to show that:

Theorem 5.3. *A language of labelled orderings is definable by a star-free regular expression iff it is definable by an $MSO[\exists^{cut}]$ formula.*

References

1. Büchi, J.R.: Weak second-order arithmetic and finite automata. Zeit. Math. Logik und Grundl. Math. **6**, 66–92 (1960)
2. Büchi, J.R.: On a decision method in restricted second order arithmetic. In: Logic, Methodology and Philosophy of Science, pp. 1–11. Stanford University Press (1962)
3. Colcombet, T.: Personal communication, September 2012
4. Elgot, C.: Decision problems of finite-automata design and related arithmetics. Trans. Am. Math. Soc. **98**, 21–51 (1961)
5. Gurevich, Y.: Monadic second order theories. In: Barwise, J., Feferman, S. (eds.) Model Theoretic Logics, pp. 479–506. Springer, Heidelberg (1986)
6. Hintikka, J.: Distributive normal forms in the calculus of predicates. Acta Philos. Fennica **6** (1953)
7. Kleene, S.: Representation of events in nerve nets and finite automata. Automata Studies, pp. 3–41. Princeton University Press (1956)
8. Ladner, R.E.: Application of model theoretical games to linear orders and finite automata theory. Inf. Control **9**, 521–530 (1977)
9. Läuchli, H., Leonard, J.: On the elementary theory of linear order. Fund. Math. **59**, 109–116 (1966)
10. McNaughton, R.: Testing and generating infinite sequences by a finite automaton. Inf. Control **9**, 521–530 (1966)
11. McNaughton, R., Papert, S.: Counter-Free Automata. The MIT Press, Cambridge (1971)
12. Rabin, M.O.: Decidability of second-order theories and automata on infinite trees. Trans. Am. Math. Soc. **141**, 1–35 (1969)
13. Rabinovich, A.: Star free expressions over the reals. Theoret. Comput. Sci. **233**, 233–245 (2000)
14. Shelah, S.: The monadic theory of order. Ann. Math. **102**, 379–419 (1975). Ser. 2
15. Thomas, W.: Star free regular sets of ω-sequences. Inf. Control **42**, 148–156 (1979)
16. Thomas, W.: Ehrenfeucht games, the composition method, and the monadic theory. In: Mycielski, J., Rozenberg, G., Salomaa, A. (eds.) Structures in Logic and Computer Science. LNCS, vol. 1261, pp. 118–143. Springer, Heidelberg (1997)
17. Trakhtenbrot, B.A.: The synthesis of logical nets whose operators are described in terms of one-place predicate calculus. Doklady Akad. Nauk SSSR **118**(4), 646–649 (1958)

Prediction of Infinite Words with Automata

Tim Smith[1,2(✉)]

[1] Northeastern University, Boston, MA, USA
[2] Université Paris-Est Marne-la-Vallée, Champs-sur-Marne, France
tim.smith@u-pem.fr

Abstract. In the classic problem of sequence prediction, a predictor receives a sequence of values from an emitter and tries to guess the next value before it appears. The predictor masters the emitter if there is a point after which all of the predictor's guesses are correct. In this paper we consider the case in which the predictor is an automaton and the emitted values are drawn from a finite set; i.e., the emitted sequence is an infinite word. We examine the predictive capabilities of finite automata, pushdown automata, stack automata (a generalization of pushdown automata), and multihead finite automata. We relate our predicting automata to purely periodic words, ultimately periodic words, and multilinear words, describing novel prediction algorithms for mastering these sequences.

1 Introduction

One motivation for studying prediction of infinite words comes from its position as a kind of underlying "simplest case" of other prediction tasks. For example, take the problem of designing an intelligent agent, a purposeful autonomous entity able to explore and interact with its environment. At each moment, it receives data from its sensors, which it stores in its memory. We would like the agent to analyze the data it is receiving, so that it can make predictions about future data and carry out actions in the world on the basis of those predictions. That is, we would like the agent to discover the laws of nature governing its environment.

Without any constraints on the problem, this is a formidable task. The data being received by the agent might be present in multiple channels, corresponding to sight, hearing, touch, and other senses, and in each channel the data given at each instant could have a complex structure, e.g. a visual field or tactile array. The data source could be nondeterministic or probabilistic, and furthermore could be sensitive to actions taken by the agent, leading to a feedback loop between the agent and its environment. The laws governing the environment could be mathematical in nature or arise from intensive computational processing.

Due to space constraints, some proofs are only sketched. The full version is available at http://arxiv.org/abs/1603.02597.

© Springer International Publishing Switzerland 2016
A.S. Kulikov and G.J. Woeginger (Eds.): CSR 2016, LNCS 9691, pp. 394–408, 2016.
DOI: 10.1007/978-3-319-34171-2_28

A natural approach to tackling such a complex problem is to start with the easiest case. How, then, can we simplify the above scenario? First, say that instead of receiving data through multiple channels, the agent has only a single channel of data. And say that instead of the data having a complex structure like a visual field, it simply consists of a succession of symbols, and that the set of possible symbols is finite. Say that the data source is completely deterministic, and moreover that the data is not sensitive to the actions or predictions of the agent, but is simply output one symbol at a time without depending on any input.

Under these simplifying assumptions, the problem we are left with is that of predicting an infinite word. That is, the agent's environment now consists of some infinite word, which it is the agent's task to predict on the basis of the symbols it has seen so far. We hope that by exploring and making progress in this simple setting, we can develop techniques which may help with the more general prediction problems encountered in the original scenario.

1.1 Our Contributions

In this paper, we consider the case in which the predictor in the above setting is an automaton. In our model, a predicting automaton M takes as input an infinite word α and produces as output an infinite word $M(\alpha)$, with the restriction that for each $i \geq 1$, M must output the ith symbol of $M(\alpha)$ before it can read beyond the $i - 1$th symbol of α. If there is an $n \geq 1$ such that for every $i \geq n$, the ith symbol of $M(\alpha)$ equals the ith symbol of α, then we say that M **masters** α.

We consider three classes of infinite words. The first are the purely periodic words, those of the form $xxx \cdots$ for some string x. Next are the ultimately periodic words, those of the form $xyyy \cdots$ for strings x, y. Finally we consider the multilinear words [21], which consist of an initial string followed by strings that repeat in a way governed by linear polynomials, for example abaabaaab $\cdots$.

All of the automata we consider are deterministic automata with a one-way input tape. We first examine DFAs (deterministic finite automata), showing that no DFA predictor masters every purely periodic word. We then consider DPDAs (deterministic pushdown automata), showing that no DPDA predictor masters every purely periodic word. We next turn to DSAs (deterministic stack automata). Stack automata are a generalization of pushdown automata whose stack head, in addition to pushing and popping when at the top of the stack, can move up and down the stack in read-only mode [10]. We show that there is a DSA predictor which masters every purely periodic word, and we provide an algorithm by which it can do so.

Next, we consider multi-DFAs (multihead deterministic finite automata), finite automata with one or more input heads [13]. We show that there is a multi-DFA predictor which masters every ultimately periodic word, and we provide an algorithm by which it can do so. Finally, we consider sensing multi-DFAs, multihead DFAs extended with the ability to sense, for each pair of heads, whether those two heads are at the same position on the input tape [14]. We show that

Table 1. Prediction of classes of infinite words. A checkmark means that there is a predictor in that row which masters every infinite word in that column. A cross means that this is not the case.

$\exists \xrightarrow{\ masters\ } \forall$	purely periodic	ultimately periodic	multilinear
DFA	✗	✗	✗
DPDA	✗	✗	✗
DSA	✓	?	?
multi-DFA	✓	✓	?
sensing multi-DFA	✓	✓	✓

there is a sensing multi-DFA predictor which masters every multilinear word, and we provide an algorithm by which it can do so. Our results are depicted in Table 1.

1.2   Related Work

A classic survey of inductive inference, including the problem of sequence prediction, can be found in [2]. The concept of "mastering" an infinite word is a form of "learning in the limit", a concept which originates with the seminal paper of Gold [11], where it is applied to language learnability. Turing machines are considered as sequence extrapolators in [4]. An early work on prediction of periodic sequences is [20], where these sequences appear in the setting of two-player emission-prediction games. Inference of ultimately periodic sequences is treated in [15] in an "offline" setting, where the input is a finite string and the output is a description of an ultimately periodic sequence. An algorithm is presented which computes the shortest possible description of an ultimately periodic sequence when given a long enough prefix of that sequence, and can be implemented in time and space linear in the size of the input, using techniques from string matching. The algorithm works by finding the LRS (longest repeated suffix) of the input and predicting the symbol which followed that suffix on its previous occurrence.

In [18], finite-state automata are considered as predicting machines and the question of which sequences appear "random" to these machines is answered. A binary sequence is said to appear random to a predicting machine if no more than half of the predictions made of the sequence's terms by that machine are correct. Further work on this concept appears in [5]. In [9] the finite-state predictability of an infinite sequence is defined as the minimum fraction of prediction errors that can be made by an finite-state predictor, and it is proved that finite-state predictability can be obtained by an efficient prediction procedure using techniques from data compression. In [3] a random prediction method for binary sequences is given which ensures that the proportion of correct predictions approaches the frequency of the more common symbol (0 or 1) in the sequence. In [16], "inverse problems" for D0L systems are discussed (in the title and throughout the paper, the term "finite automata" refers to morphisms). These problems ask, given a word, to find a morphism and initial string which generate that word (bounds are assumed on the size of the morphism and initial string). An approach is given for solving this problem by trying different string lengths for the righthand side of the morphism until a combination is found which is compatible with the input. A genetic algorithm is described to search the space of word lengths. In [6], an evolutionary algorithm is used to search for the finite-state machine with the highest prediction ratio for a given purely periodic word, in the space of all automata with a fixed number of states. In [7], the problem of successfully predicting a single 0 in an infinite binary word being revealed sequentially to the predictor is considered; only one prediction may be made, but at a time of the predictor's choosing. Learning of languages consisting of infinite words has also been studied; see [1] for recent work.

An early and influential approach to predicting infinite sequences is that of program-size complexity [22]. Unfortunately this model is incomputable, and in [17] it is shown furthermore that some sequences can only be predicted by very complex predictors which cannot be discovered mathematically due to problems of Gödel incompleteness. [17] concludes that "perhaps the only reasonable solution would be to add additional restrictions to both the algorithms which generate the sequences to be predicted, and to the predictors." This suggestion is akin to the approach followed in the present paper, where the automata and infinite words considered are of various restricted classes. Following on from [17], in [12] the formalism of sequence prediction is extended to a competition between two agents, which is shown to be a computational resources arms race.

1.3 Outline of Paper

The rest of the paper is organized as follows. Section 2 gives definitions for infinite words and predicting automata. Section 3 studies prediction of purely periodic and ultimately periodic words. Section 4 studies prediction of multilinear words. Section 5 gives our conclusions.

2 Preliminaries

2.1 Words

Where X is a set, we denote the cardinality of X by $|X|$. For a list or tuple v, $v[i]$ denotes the ith element of v; indexing starts at 1. An **alphabet** A is a finite set of symbols. A **word** is a concatenation of symbols from A. We denote the set of finite words by A^* and the set of infinite words by A^ω. We call finite words **strings** and infinite words **streams** or ω-**words**. The length of x is denoted by $|x|$. We denote the empty string by λ. A **language** is a subset of A^*. A (symbolic) **sequence** S is an element of $A^* \cup A^\omega$. A **prefix** of S is a string x such that $S = xS'$ for some sequence S'. The ith symbol of S is denoted by $S[i]$; indexing starts at 1. For a non-empty string x, x^ω denotes the infinite word

$xxx \cdots$. Such a word is called **purely periodic**. An infinite word of the form xy^ω, where x and y are strings and $y \neq \lambda$, is called **ultimately periodic**. An infinite word is **multilinear** if it has the form

$$q \prod_{n \geq 0} r_1^{a_1 n + b_1} r_2^{a_2 n + b_2} \cdots r_m^{a_m n + b_m},$$

where $\prod$ denotes concatenation, q is a string, m is a positive integer, and for each $1 \leq i \leq m$, r_i is a non-empty string and a_i and b_i are nonnegative integers such that $a_i + b_i > 0$. For example, $\prod_{n \geq 0} \mathsf{a}^{n+1}\mathsf{b} = \mathsf{abaabaaab} \cdots$ is a multilinear word. The class of multilinear words appears in [21] and also in [8] (as the reducts of the "prime" stream Π). Clearly the multilinear words properly include the ultimately periodic words. Any multilinear word which is not ultimately periodic we call **properly multilinear**.

2.2 Predictors

We now define predictors based on various types of automata. (See [23] for results on the original automata, which are language recognizers rather than predictors.) Each predictor M takes as input an infinite word α and produces as output an infinite word $M(\alpha)$, with the restriction that for each $i \geq 1$, M must output the ith symbol of $M(\alpha)$ before it can read beyond the $i - 1$th symbol of α. We call $M(\alpha)[i]$ M's **guess** about position i of α. If $M(\alpha)[i] = \alpha[i]$ then we say that the guess is correct; otherwise we say that it is incorrect. If there is an $n \geq 1$ such that for every $i \geq n$, $M(\alpha)[i] = \alpha[i]$, then we say that M **masters** α. (If M outputs only a finite number of symbols when given α, then we say that $M(\alpha)$ is undefined and M does not master α.)

DFA Predictors. A **DFA predictor** is a tuple $M = (Q, A, T, \triangleright, q_s)$, where Q is the set of states, A is the input alphabet, $\triangleright$ is the start-of-input marker, $q_s \in Q$ is the initial state, and T is a transition function of the form $[Q \times (A \cup \{\triangleright\})] \to [Q \times A]$.

To perform a computation, M is given an input consisting of the symbol $\triangleright$ followed by an infinite word α. M starts in state q_s with its input head positioned at $\triangleright$. M then makes transitions based on its current state and input symbol. At each transition, M changes state, moves its head to the right, and makes a guess about what the next symbol will be. The sequence of these guesses constitutes $M(\alpha)$. More formally, let $C = [C_1, C_2, C_3, \ldots]$ where $C_i = \{[q_i, c_i, g_i]$ with $q_i \in Q$, $c_i \in (A \cup \{\triangleright\}), g_i \in A$ such that $q_1 = q_s$ and for each $i \geq 1$, $c_i = (\triangleright\alpha)[i]$ and $T(q_i, c_i) = [q_{i+1}, g_i]$. Notice that there is only one possible C, given M and α. Now for $i \geq 1$, set $M(\alpha)[i] = g_i$.

DPDA Predictors. A **DPDA predictor** is a tuple $M = (Q, A, F, T, \triangleright, \Delta, q_s)$, where Q is the set of states, A is the input alphabet, F is the stack alphabet,

$\triangleright$ is the start-of-input marker, $\triangle$ is the bottom-of-stack marker, $q_s \in Q$ is the initial state, and T is a transition function of the form

$$[Q \times (A \cup \{\triangleright\}) \times (F \cup \{\triangle\})] \rightarrow [Q \times (A \cup \{\mathsf{stay}\}) \times (F \cup \{\mathsf{pop}, \mathsf{keep}\})].$$

To perform a computation, M is given an input consisting of the symbol $\triangleright$ followed by an infinite word α. M starts in state q_s with stack $\triangle$ and with its input head positioned at $\triangleright$. M then makes transitions based on its current state, input symbol, and stack symbol. At each transition, M (1) changes state, (2) either moves its input head to the right and guesses what the next symbol will be, or else keeps it in place (using stay), and (3) either pushes a symbol to the stack, pops the stack, or leaves it alone (using keep). It is illegal for M to pop $\triangle$. The sequence of guesses made by M constitutes $M(\alpha)$.

DSA Predictors. A **DSA predictor** is a tuple $M = (Q, A, F, T, \triangleright, \triangle, q_s)$, where Q is the set of states, A is the input alphabet, F is the stack alphabet, $\triangleright$ is the start-of-input marker, $\triangle$ is the bottom-of-stack marker, $q_s \in Q$ is the initial state, and T is a transition function of the form

$$[Q \times (A \cup \{\triangleright\}) \times (F \cup \{\triangle\}) \times \{\mathsf{top}, \mathsf{inside}\}] \rightarrow$$
$$[Q \times (A \cup \{\mathsf{stay}\}) \times (F \cup \{\mathsf{pop}, \mathsf{keep}, \mathsf{up}, \mathsf{down}\})].$$

To perform a computation, M is given an input consisting of the symbol $\triangleright$ followed by an infinite word α. M starts in state q_s with stack $\triangle$ and with its input head positioned at $\triangleright$. M then makes transitions based on its current state, input symbol, stack symbol, and whether or not the stack head is at the top of the stack (top means the stack head is at the top; inside means it is not). At each transition, M (1) changes state, (2) either moves its input head to the right and guesses what the next symbol will be, or else keeps it in place (using stay), and (3) either pushes a symbol to the stack, pops the stack, leaves it alone (using keep), or moves its stack head up or down. It is illegal for M to push or pop the stack when the stack head is not at the top of the stack, or to move it up when it is already at the top or down when it is already at the bottom. The sequence of guesses made by M constitutes $M(\alpha)$.

Multi-DFA Predictors. A **multi-DFA predictor** is a tuple of the form $M = (Q, A, k, T, \triangleright, q_s)$, where Q is the set of states, A is the input alphabet, $k \geq 1$ is the number of input heads, $\triangleright$ is the start-of-input marker, $q_s \in Q$ is the initial state, and T is a transition function of the form

$$[Q \times (A \cup \{\triangleright\})^k] \rightarrow [Q \times \{\mathsf{stay}, \mathsf{right}\}^k \times A].$$

To perform a computation, M is given an input consisting of the symbol $\triangleright$ followed by an infinite word α. M starts in state q_s with its k input heads all positioned at $\triangleright$. M then makes transitions based on its current state and the input symbols it sees under each of its heads. At each transition, M (1) changes state, (2) for each head either moves it to the right or keeps it in place (using stay), and (3) makes a guess about what the next symbol will be. If in a given transition, M does not reach a new input position (one which had not previously been reached by any head), M's guess at that transition is disregarded (i.e., it is not included in $M(\alpha)$). That is, $M(\alpha)[i]$ is the guess of the first transition which moves any head to $\alpha[i]$.

A **sensing multi-DFA predictor** is a multi-DFA predictor extended so that its transition function takes an additional argument indicating, for each pair of heads, whether those two heads are at the same input position.

3   Prediction of Periodic Words

In this section we study finite automata, pushdown automata, stack automata, and multihead finite automata as predictors of purely periodic and ultimately periodic words.

3.1   Prediction by DFAs

Theorem 1. *Let A be an alphabet such that $|A| \geq 2$. Then no DFA predictor masters every purely periodic word over A.*

Proof. Suppose some DFA predictor M masters every purely periodic word over A. M has some number of states p. Take any $a, b \in A$ such that $a \neq b$. Let α be the purely periodic word $(a^{p+1}b)^\omega$. Then there is an $n \geq 1$ such that for every $i \geq n$, $M(\alpha)[i] = \alpha[i]$. Take the first segment of $p + 1$ consecutive as after the position n. At two of these as, M is in the same state. Then M will repeat the guesses it made between those two as for as long as it keeps reading as. But then M will guess a for the next b, a contradiction. So M does not master α. $\qquad\square$

3.2   Prediction by DPDAs

Theorem 2. *Let A be an alphabet such that $|A| \geq 2$. Then no DPDA predictor masters every purely periodic word over A.*

Proof (Sketch). Suppose some DPDA predictor $M = (Q, A, F, T, \triangleright, \Delta, q_s)$ masters every purely periodic word over A. We set p to be very large with respect to $|Q|$ and $|F|$. Take any $a, b \in A$ such that $a \neq b$. Let α be the purely periodic word $(a^p b)^\omega$. Then there is some position $m \geq 0$ after which all of M's guesses about α are correct. Now, between each two segments of p consecutive a's, there is only one symbol (a single b), so the stack can grow by at most $|Q| \cdot |F|$ between each two segments. It follows that in some segment of p consecutive a's occurring after m, the stack height does not decrease by more than $|Q| \cdot |F|$, since otherwise it would eventually become negative. We show that in such a segment, because p is so large with respect to $|Q|$ and $|F|$, there are two configurations C_i and C_j of M occurring at different input positions with the same state and stack symbol, such that the stack below the top symbol at C_i is not accessed between C_i and

C_j. Then since all of M's guesses between C_i and C_j are a's, M will continue to guess a's for as long as it continues to read a's. But then M will guess a for the b at the end of the segment, contradicting the supposition that all of M's guesses about α after m are correct. Therefore M does not master every purely periodic word over A. $\qquad\square$

3.3   Prediction by DSAs

We give two results about the predictive capabilities of DSAs: first, that some DSA predictor masters every purely periodic word, and second, that no DSA predictor can master any infinite word which is not multilinear.

Algorithm 1. A DSA predictor which masters every purely periodic word. The input head is denoted by h_i and the stack head is denoted by h_s. The input consists of the symbol $\triangleright$ followed by an infinite word α. Wherever a guess is not specified, it may be taken to be arbitrary.

1: **loop**
2:     move h_i
3:     push $\alpha[h_i]$
4:     *recovering* $\leftarrow$ false
5:     **loop**
6:         move h_s down **until** $stack[h_s] = \triangle$
7:         *matched* $\leftarrow$ true
8:         **loop**
9:             move h_s up
10:            move h_i, guessing $stack[h_s]$
11:            *matched* $\leftarrow$ false **if** $\alpha[h_i] \neq stack[h_s]$
12:            **break if** top
13:        *recovering* $\leftarrow$ true **if not** *matched*
14:        **break if** *recovering* **and** *matched*

Theorem 3. *Let A be an alphabet. Then some DSA predictor masters every purely periodic word over A.*

Proof. Let M be a DSA predictor which implements Algorithm 1. (The boolean variables *recovering* and *matched* can be accommodated using M's finite state control.) The idea is that M will gradually build up its stack until the stack consists of the period (or a cyclic shift thereof) of the purely periodic word to be mastered. Following Algorithm 1, M begins by pushing the first symbol of the input after $\triangleright$ onto its stack, and then enters the loop spanning lines 5–14. This loop moves the stack head to the bottom of the stack and then moves it up symbol by symbol, predicting that the input will match the stack. Call each iteration of the loop spanning lines 5–14 a "pass", and call a pass successful if *matched* is true at line 14 and unsuccessful otherwise. Observe that if a pass is successful, then all of the guesses made during it (on line 10) are correct, and that if eventually there are no more unsuccessful passes, then M masters its input.

Now take any purely periodic word $\alpha = x^\omega$. To show that M masters α, we first show that every unsuccessful pass will eventually be followed by a successful pass. Observe that there must be at least one successful pass, since M begins the passes with only one symbol on the stack, and that symbol will eventually reappear in the input. So take any unsuccessful pass after the first successful pass. Now take the most recent successful pass prior to that unsuccessful pass. Let i be the position of the input head in x (counting from zero, so $0 \leq i < |x|$) at the beginning of this most recent successful pass and let h be the height of the stack. Then the position of the input head in x after the successful pass is $(i + h) \bmod |x|$. Then after $|x| - 1$ unsuccessful passes, the position of the input head in x will be $(i + h|x|) \bmod |x| = i$. So the next pass after that will be successful. Hence every unsuccessful pass will eventually be followed by a successful pass.

Since each unsuccessful pass sets *recovering* to true, the next successful pass after it will break at line 14, causing M to push another symbol onto the stack. If the height of the stack never reaches $|x|$, then after some point, every pass is successful and M masters α. So say the height of the stack eventually reaches $|x|$. Then since the last pass before the stack reached that height was successful, and the input symbol following that pass is now at the top of the stack, the previous $|x|$ symbols of the input match the stack. Then every subsequent pass will be successful, and M masters α. $\qquad\square$

Theorem 4. *Every infinite word mastered by a DSA predictor is multilinear.*

Proof. Let M be a DSA predictor and let α be any infinite word mastered by M. We will show that there is a DSA recognizer for Prefix(α), the set of all prefixes of α. Since M masters α, there is an $n \geq 1$ such that for every $i \geq n$, $M(\alpha)[i] = \alpha[i]$. Take any such n. Let $C = (q, s, i)$ be the configuration of M upon reaching position n of α, where q is the state of M, s is the stack, and i is the position of the stack head within s. Let M_α be a DSA recognizer which operates as follows. First M_α uses its finite control to check that the first n symbols of its input match the first n symbols of α. Then M_α uses its finite control to push s onto its stack and move its stack head to position i within s. Next M_α simulates M, starting from C. Whenever M would make a guess, M_α instead checks that the next symbol of the input matches M's guess. If any check fails, then M_α rejects its input; otherwise, when M_α reaches end-of-input, it accepts. Since all of M's guesses after n are correct, M_α now recognizes Prefix(α), and hence M_α determines α in the sense of [21]. Then by Theorem 8 of [21], α is multilinear. $\square$

3.4   Prediction by Multi-DFAs

We next consider multi-DFA predictors. We leave their more powerful cousins, sensing multi-DFA predictors, to Sect. 4.

Algorithm 2. A 2-head DFA predictor which masters every ultimately periodic word. The heads are denoted by t and h. The input consists of the symbol $\triangleright$ followed by an infinite word α. Wherever a guess is not specified, it may be taken to be arbitrary.

> move h
> **loop**
>   move t
>   move h, guessing $\alpha[t]$
>   move h **if** $\alpha[h] \neq \alpha[t]$

Theorem 5. *Let A be an alphabet. Then some multi-DFA predictor masters every ultimately periodic word over A.*

Proof. We employ a variation of the "tortoise and hare" cycle detection algorithm [19], adapted to our setting. Let M be a 2-head DFA predictor which implements Algorithm 2. Take any ultimately periodic word $\alpha = xy^\omega$. Following the algorithm, the two heads t (for "tortoise") and h (for "hare") begin at the start of the input. M moves h one square to the right (making an arbitrary guess) and then enters the loop. In the loop, M guesses that h will match t. After each missed guess, h moves ahead an extra square (making an arbitrary guess), so the distance between the two heads increases by 1. If this distance stops growing, then there are no more missed guesses, so M masters α. Otherwise, both heads will reach the periodic part y^ω of α and the distance between them will reach a multiple of $|y|$. Then each head will point to the same position in y as the other, so all guesses will be correct from that point on. So again M masters α. $\square$

4   Prediction of Multilinear Words

We turn now to prediction of the class of multilinear words. We give an algorithm by which a sensing multi-DFA can master every multilinear word.

Theorem 6. *Let A be an alphabet. Then some sensing multi-DFA predictor masters every multilinear word over A.*

Proof (Sketch). Let M be a sensing 10-head DFA predictor which implements Algorithm 3. The idea of the algorithm is as follows. Any properly multilinear word α can be written as $q \prod_{n \geq 1} \prod_{i \geq 1}^{m} p_i s_i^n$ for some $m \geq 1$ and strings q, p_i, s_i subject to certain conditions. That is, α can be broken into "blocks", each block consisting of m "segments" of the form $p_i s_i^n$. To master α, M will alternate between two procedures, CORRECTION and MATCHING. CORRECTION attempts to position h_1, h_2, h_3, and h_4 so that each head is at the beginning of a segment, h_2 is ahead of h_1 by a given number of segments, h_3 is ahead of h_2 by the same number of segments, and h_4 is ahead of h_3 by the same number of segments. Each time CORRECTION is entered, the given number of segments used to separate the heads is increased by one. MATCHING attempts to master α on the assumption that CORRECTION has successfully positioned h_1, h_2, h_3, and h_4 at the beginning of segments and that the number of segments separating the heads is a multiple of m (meaning that the segments share the same p_i and s_i). If any problem is detected, MATCHING is exited and CORRECTION is entered again.

The number of segments used to separate the heads is given by $r - l$. Before each call to CORRECTION, r is moved forward, increasing this number by one. CORRECTION works by first moving h_1 forward to h_4 and then calling ADVANCEONE(1), which tries to move h_1 to the beginning of the next segment. Then CORRECTION moves h_2 to h_1 and calls ADVANCEMANY(2), which tries to move h_2 forward by $r - l$ segments. CORRECTION then moves h_3 to h_2 and calls ADVANCEMANY(3), which tries to move h_3 forward by $r - l$ segments. Finally, CORRECTION moves h_4 to h_3 and calls ADVANCEMANY(4), which tries to move h_4 forward by $r - l$ segments. If everything worked as intended, the four heads are now at the beginning of segments and each pair of heads h_i and h_{i+1} are separated by the same number of segments, $r - l$.

MATCHING works by using h_1, h_2, and h_3 to predict h_4. If the four heads are separated by the same number of segments, and if this number is a multiple of m, then the heads share the same p_i and s_i. In this case, the later heads have extra copies of s_i: for some $d \geq 1$, in each segment i, h_4 will see d more copies of s_i than h_3, which will see d more than h_2, which will see d more than h_1. MATCHING moves the heads together, using the earlier heads to predict h_4 and detecting when each head passes its last copy of s_i by comparing the heads with each other. By use of a normal form for properly multilinear words, we guarantee that the first symbol of p_{i+1} differs from the first symbol of s_i, ensuring that the next segment can be detected. The supplemental head h_{3a} is used to predict h_4's last d copies of s_i by using h_3's last d copies a second time. Once all heads are at the beginning of the next segment, MATCHING repeats from the start. If any guess is incorrect, then the heads were not separated by a multiple of m segments when MATCHING was entered. Upon making an incorrect guess, MATCHING exits, $r - l$ is increased, and CORRECTION is entered again.

The fact that M is sensing allows it to perform operations a designated number of times, a technique used in the procedures ADVANCEMANY and ADVANCEONE called by CORRECTION. This technique works in the following way. Let n be the distance between the heads l and r at a given point in the computation. To perform an operation n times, we first move another head, say *inner*, to r. Then we move l and r together until l reaches *inner*, performing the operation after each step. Now the operation has been performed n times, and we can repeat this process to perform it another n times. Further, by increasing the distance between l and r, we can increase n. It is also possible to nest this process, by moving another head, say *outer*, to r, keeping *outer*'s position constant relative to l and r during the inner process, and moving l and r, but not *outer*, each time the inner process is completed. When l reaches *outer*, the inner process has been executed n times, each time performing its operation n times. In ADVANCEMANY and ADVANCEONE, this technique is used to advance

Algorithm 3. A sensing 10-head DFA predictor which masters every multilinear word. The heads are denoted by h_1, h_2, h_{3a}, h_3, h_4, t, l, r, $inner$, and $outer$. The input consists of the symbol $\triangleright$ followed by an infinite word α. Wherever a guess is not specified, it may be taken to be arbitrary.

loop
  move r
  CORRECTION
  MATCHING

procedure MATCHING
  loop
    move h_{3a} **until** $h_{3a} = h_3$

    while $\alpha[h_1] = \alpha[h_2] = \alpha[h_3] = \alpha[h_4]$ **do**
      move h_1, h_2, h_{3a}, h_3
      move h_4, guessing $\alpha[h_2]$
    break unless $\alpha[h_2] = \alpha[h_4]$

    while $\alpha[h_2] = \alpha[h_3] = \alpha[h_4]$ **do**
      move h_2, h_3
      move h_4, guessing $\alpha[h_3]$
    break unless $\alpha[h_3] = \alpha[h_4]$

    while $\alpha[h_{3a}] = \alpha[h_3] = \alpha[h_4]$ **do**
      move h_{3a}, h_3
      move h_4, guessing $\alpha[h_{3a}]$
    break unless $\alpha[h_{3a}] = \alpha[h_4]$

    while $h_{3a} \neq h_3$ **and** $\alpha[h_{3a}] = \alpha[h_4]$ **do**
      move h_{3a}
      move h_4, guessing $\alpha[h_{3a}]$
    break unless $\alpha[h_{3a}] = \alpha[h_4]$

procedure CORRECTION
  move h_1 **until** $h_1 = h_4$
  ADVANCEONE(1)

  move h_2 **until** $h_2 = h_1$
  ADVANCEMANY(2)

  move h_3 **until** $h_3 = h_2$
  ADVANCEMANY(3)

  move h_4 **until** $h_4 = h_3$
  ADVANCEMANY(4)

procedure ADVANCEMANY(i)
  move $outer$ **until** $outer = r$
  while $l \neq outer$ **do**
    ADVANCEONE(i)
    move l, r

procedure ADVANCEONE(i)
  move t **until** $t = h_i$
  move h_i
  move $inner$ **until** $inner = r$
  while $l \neq inner$ **do**
    if $\alpha[t] = \alpha[h_i]$ **then**
      move $l, r, outer$
    else
      move $inner$ **until** $inner = r$
      move h_i
  move t
  move h_i
  while $\alpha[t] = \alpha[h_i]$ **do**
    move t
    move h_i, guessing $\alpha[t]$

a given h_i by n segments, using within each segment a threshold based on n to detect the beginning of the next segment.

To show that M masters every multilinear word α, we first show that if either MATCHING or CORRECTION gets "stuck", i.e. is entered and does not end, then in its stuck state it will continue to make guesses, all of which are correct, and so M masters α. In particular, we show that the first **while** loop of ADVANCEONE will always end. This loop implements the "tortoise and hare" routine of Algorithm 2 on α, waiting for a streak of $r-l$ consecutive matches. Such a streak will eventually be obtained, because if α is ultimately periodic, then by the proof of Theorem 5, the "tortoise and hare" algorithm masters α, and if α is properly multilinear, then we show that the "tortoise and hare" algorithm will eventually achieve k consecutive matches on α for any $k \geq 1$, and so the loop will end.

So we are left with the case in which MATCHING and CORRECTION always end. Since r is moved at the beginning of each iteration of the main loop, and since CORRECTION and MATCHING leave $r-l$ unchanged, $r-l$ will grow. If α is ultimately periodic, then eventually $r-l$ will be large enough for ADVANCEONE to "line up" the heads h_i and t with respect to the periodic part of α, so that M masters α. If α is properly multilinear, then eventually $r-l$ will be large enough for ADVANCEONE to always advance h_i by at least one segment. We show further that $r-l$ will grow slowly enough with respect to the segment length that eventually whenever h_i is at the beginning of a segment, ADVANCEONE will move it to the beginning of the next segment and not farther. As a result, eventually CORRECTION will always end with the four heads h_1, h_2, h_3, and h_4 at the beginning of segments, with the heads separated by $r-l$ segments as desired. When $r-l$ next reaches a multiple of m, the segments of the four heads will share the same p_i and s_i. We show that then MATCHING can make use of h_1, h_2, and h_3 to correctly predict h_4 as intended. Thus M masters α.   □

5   Conclusion

In this paper, we studied the classic problem of sequence prediction from the angle of automata and infinite words. We examined several types of automata and sought to find out which classes of infinite words they could master. In doing so we described novel prediction algorithms for the classes of purely periodic, ultimately periodic, and multilinear words. Open questions in our investigation include whether there is a DSA predictor which masters every ultimately periodic word, and whether there is a multi-DFA predictor without sensing which masters every multilinear word. Other directions for further research would be to consider other types of automata as predictors, e.g. automata with two-way input tapes, and to attempt prediction of other classes of infinite words, e.g. morphic words. It would also be interesting to consider questions of computational tractability, e.g. how many guesses and how much time is required to achieve mastery.

Acknowledgments. I would like to thank my Ph.D. advisor at Northeastern, Rajmohan Rajaraman, for his helpful comments and suggestions. The continuation of this work at Marne-la-Vallée was supported by the Agence Nationale de la Recherche (ANR) under the project EQINOCS (ANR-11-BS02-004).

References

1. Angluin, D., Fisman, D.: Learning regular omega languages. In: Auer, P., Clark, A., Zeugmann, T., Zilles, S. (eds.) ALT 2014. LNCS, vol. 8776, pp. 125–139. Springer, Heidelberg (2014)
2. Angluin, D., Smith, C.H.: Inductive inference: theory and methods. ACM Comput. Surv. **15**(3), 237–269 (1983). http://doi.acm.org/10.1145/356914.356918
3. Blackwell, D.: Minimax vs. Bayes prediction. Probab. Eng. Inf. Sci. **9**(1), 53–58 (1995). http://journals.cambridge.org/article_S0269964800003685
4. Blum, L., Blum, M.: Toward a mathematical theory of inductive inference. Inf. Control **28**(2), 125–155 (1975). http://www.sciencedirect.com/science/article/pii/S0019995875902612
5. Broglio, A., Liardet, P.: Predictions with automata. In: Symbolic Dynamics and Its Applications. Contemporary Mathematics, vol. 135, pp. 111–124. American Mathematical Society (1992)
6. Cerruti, U., Giacobini et al., M., Liardet, P.: Prediction of binary sequences by evolving finite state machines. In: Collet, P., Fonlupt, C., Hao, J.-K., Lutton, E., Schoenauer, M. (eds.) EA 2001. LNCS, vol. 2310, pp. 42–53. Springer, Heidelberg (2002)
7. Drucker, A.: High-confidence predictions under adversarial uncertainty. TOCT **5**(3), 12 (2013). http://doi.acm.org/10.1145/2493252.2493257
8. Endrullis, J., Hendriks, D., Klop, J.W.: Degrees of streams. In: Integers, Electronic Journal of Combinatorial Number Theory 11B(A6), 1–40. 2010 Proceedings of the Leiden Numeration Conference (2011)
9. Feder, M., Merhav, N., Gutman, M.: Universal prediction of individual sequences. IEEE Trans. Inf. Theory **38**, 1258–1270 (1992)
10. Ginsburg, S., Greibach, S.A., Harrison, M.A.: One-way stack automata. J. ACM **14**(2), 389–418 (1967)
11. Gold, E.M.: Language identification in the limit. Inf. Control **10**(5), 447–474 (1967). http://groups.lis.illinois.edu/amag/langev/paper/gold67limit.html
12. Hibbard, B.: Adversarial sequence prediction. In: Proceedings of the 2008 Conference on Artificial General Intelligence 2008: Proceedings of the First AGI Conference. pp. 399–403. IOS Press, Amsterdam, The Netherlands (2008). http://dl.acm.org/citation.cfm?id=1566174.1566212
13. Holzer, M., Kutrib, M., Malcher, A.: Complexity of multi-head finite automata: origins and directions. Theor. Comput. Sci. **412**(1–2), 83–96 (2011)
14. Hromkovič, J.: One-way multihead deterministic finite automata. Acta Informatica **19**(4), 377–384 (1983). http://dx.doi.org/10.1007/BF00290734
15. Johansen, P.: Inductive inference of ultimately periodic sequences. BIT Numer. Math. **28**(3), 573–580 (1988). http://dx.doi.org/10.1007/BF01941135
16. Leblanc, B., Lutton, E., Allouche, J.-P.: Inverse problems for finite automata: a solution based on genetic algorithms. In: Hao, J.-K., Lutton, E., Ronald, E., Schoenauer, M., Snyers, D. (eds.) AE 1997. LNCS, vol. 1363, pp. 157–166. Springer, Heidelberg (1998)
17. Legg, S.: Is there an elegant universal theory of prediction? In: Balcázar, J.L., Long, P.M., Stephan, F. (eds.) ALT 2006. LNCS (LNAI), vol. 4264, pp. 274–287. Springer, Heidelberg (2006)
18. O'Connor, M.G.: An unpredictability approach to finite-state randomness. J. Comput. Syst. Sci. **37**(3), 324–336 (1988). http://dx.doi.org/10.1016/0022-0000(88)90011-6

19. Sedgewick, R., Szymanski, T.G., Yao, A.C.: The complexity of finding cycles in periodic functions. SIAM J. Comput. **11**(2), 376–390 (1982)
20. Shubert, B.: Games of prediction of periodic sequences. Technical report, United States Naval Postgraduate School (1971)
21. Smith, T.: On infinite words determined by stack automata. In: FSTTCS 2013. Leibniz International Proceedings in Informatics (LIPIcs), vol. 24, pp. 413–424. Schloss Dagstuhl-Leibniz-Zentrum fuer Informatik, Dagstuhl, Germany (2013)
22. Solomonoff, R.: A formal theory of inductive inference. part i. Inf. Control **7**(1), 1–22 (1964). http://www.sciencedirect.com/science/article/pii/S0019995864902232
23. Wagner, K., Wechsung, G.: Computational complexity. Mathematics and its Applications. Springer (1986)

Fourier Sparsity of GF(2) Polynomials

Hing Yin Tsang[1(✉)], Ning Xie[2], and Shengyu Zhang[3]

[1] University of Chicago, Chicago, USA
hytsang@uchicago.edu
[2] Florida International University, Miami, USA
nxie@cis.fiu.edu
[3] The Chinese University of Hong Kong, Hong Kong, China
syzhang@cse.cuhk.edu.hk

Abstract. We study a conjecture called "linear rank conjecture" recently raised in (Tsang *et al.* [16]), which asserts that if many linear constraints are required to lower the degree of a GF(2) polynomial, then the Fourier sparsity (i.e. number of non-zero Fourier coefficients) of the polynomial must be large. We notice that the conjecture implies a surprising phenomenon that, if the highest degree monomials of a GF(2) polynomial satisfy a certain condition (Specifically, the highest degree monomials do not vanish under a small number of linear restrictions.), then the Fourier sparsity of the polynomial is large regardless of the monomials of lower degrees—whose number is generally much larger than that of the highest degree monomials. We develop a new technique for proving lower bound on the Fourier sparsity of GF(2) polynomials, and apply it to certain special classes of polynomials to showcase the above phenomenon (A full version of this paper is available at http://arxiv.org/abs/1508.02158).

1 Introduction

The study of *communication complexity*, introduced by Yao [17] in 1979, aims at investigating the minimum amount of information exchange required for computing functions whose inputs are distributed among multiple parties [7]. In the standard two-party setting, Alice holds an input x, Bob holds an input y, and they wish to compute a function F on (x, y) by as little communication as possible. Perhaps the most important open problem in communication complexity is the so-called *Log-rank Conjecture* proposed by Lovász and Saks [11], which states that the *deterministic communication complexity* of any $F : \{0,1\}^n \times \{0,1\}^n \to \{0,1\}$, $D^{CC}(F)$, is upper bounded by a polynomial of the logarithm of the rank the communication matrix $M_F = [F(x,y)]_{x,y}$, where the rank is taken over the reals. Although a lot of effort has been devoted to the conjecture in the past two decades, very little progress has been achieved and the

N. Xie—Research supported in part by NSF grant 1423034.
S. Zhang—Research supported in part by RGC of Hong Kong (Project no. CUHK419413).

A.S. Kulikov and G.J. Woeginger (Eds.): CSR 2016, LNCS 9691, pp. 409–424, 2016.
DOI: 10.1007/978-3-319-34171-2_29

best upper bound known to date is $\mathsf{D}^{\mathsf{CC}}(F) = O\left(\sqrt{\mathrm{rank}(M_F)}\log\left(\mathrm{rank}(M_F)\right)\right)$, due to Lovett [12]. Note that there is still an exponential gap between this and the best known lower bound, which is $\mathsf{D}^{\mathsf{CC}}(F) = \tilde{\Omega}\left(\log^2 \mathrm{rank}(M_F)\right)$ [5]. For an overview of recent developments in this direction, see [13].

An interesting special class of functions computable by two parties is the so-called *XOR functions*. Specifically, F is an XOR function if there exists an $f : \{0,1\}^n \to \{0,1\}$ such that for all x and y, $F(x,y) = f(x \oplus y)$, where $\oplus$ is the bit-wise XOR. Denote such F by $f \circ \oplus$. Besides including important examples such as Equality and Hamming Distance, XOR functions are particularly interesting for studying the Log-rank Conjecture due to its intimate connection with the analysis of Boolean functions. Specifically, if F is an XOR function, then the rank of M_F is just the Fourier sparsity of f (i.e., the number of non-zero Fourier coefficients of f) [2]. Therefore proving the Log-rank conjecture for XOR functions can be achieved by demonstrating short *parity decision tree*[1] computing Fourier sparse Boolean functions, namely showing

$$D_\oplus(f) = \mathrm{polylog}\|\hat{f}\|_0, \tag{1}$$

which attracted a lot of attention in the past several years [8,14–16,18].

Recently, by viewing Boolean functions as GF(2) polynomials, a new communication protocol based on GF(2)-degree reduction was proposed in [16] for XOR functions: suppose $f(x \oplus y)$ is a degree-d polynomial and r_d is the minimum number of variables (up to an invertible linear transformation) restricting of which reduces f's degree to at most $d-1$, then Alice and Bob both apply the optimal linear map to their inputs and send each other r_d bits of their respective inputs. Repeating this process at most $d-1$ times, the restricted function of f becomes a constant function hence they successfully compute $f(x \oplus y)$. Of course, such a protocol is efficient only if the numbers $r_d, r_{d-1}, \ldots, r_1$, of the restricted variables that they need to exchange, are not large. Studying these quantities, namely *linear ranks* of polynomials, is one the central objectives of this paper.

Definition 1 (Linear Rank of a Polynomial). *Let f be a degree-d polynomial, V be a subspace in $\{0,1\}^n$ and $H = a + V$ be any affine shift of V. Denote by $f|_H$ the restriction of f on H. The linear rank of f, denoted $\mathrm{lin\text{-}rank}(f)$, is the minimum co-dimension of any subspace H s.t. the degree of $f|_H$ is strictly less than d; i.e.*

$$\mathrm{lin\text{-}rank}(f) = \min_{\deg_2(f|_H) < \deg_2(f)} \mathrm{co\text{-}dim}(H).$$

In other words, $\mathrm{lin\text{-}rank}(f)$ is the minimum number of linear functions one needs to fix in order to lower the degree of f. Consider, for example, the degree-3

---

[1] Recall that a *parity decision tree* T for a function $f : \{0,1\}^n \to \{0,1\}$ generalizes an ordinary decision tree in the sense that each internal node of T is now associated with a linear function $\ell(x)$, instead of a single bit, of the input, and T branches according to the parity of $\ell(x)$.

polynomial $f(x_1, \ldots, x_{3n}) = (x_1 + \cdots + x_n)(x_{n+1} + \cdots + x_{2n})(x_{2n+1} + \cdots + x_{3n})$. In the original basis, one needs to fix at least n variables to lower the degree of f. However, fixing one linear function $x_1 + \cdots + x_n = 0$ is enough to lower its degree. Therefore lin-rank$(f) = 1$.

For a Boolean function f, let spar(f) denote the Fourier sparsity of f and $D_{\oplus}(f)$ denote the parity decision tree complexity of f. As restrictions do not increase spar(f) and $\deg_2(f) \leq \log \text{spar}(f)$ for every f, the following *linear rank conjecture*—if true—would readily imply the Log-rank Conjecture for XOR functions.

Conjecture 1. *(Linear rank conjecture [16]) For any $f : \{0,1\}^n \to \{0,1\}$, the linear rank of f is upper bounded by polylogarithmic of the Fourier sparsity of f: lin-rank$(f) = O(\log^c(\text{spar}(f)))$ for some $c = O(1)$. Equivalently, if lin-rank$(f) = r$, then spar$(f) = 2^{r^{\Omega(1)}}$.*

Although it is still open whether the linear rank conjecture is equivalent to the Log-rank Conjecture for XOR functions, it is worthwhile to note that these two would be equivalent due to an algorithm in [16], had the well-believed fact that parity decision tree being the most efficient communication protocol for XOR functions been proved to be true.

1.1 Large Fourier Sparsity Guaranteed by Highest Degree Monomials Only

Before further discussing the linear rank conjecture, let us first state a lemma of [16] (Lemma 19) in a slightly stronger form and give an alternative simple proof (another simple proof used polynomial derivatives [3]). The lemma says that, once the linear subspace V in Definition 1 is identified, it does not matter which affine shift is used in the definition of linear rank: all affine subspaces of V are equally good. More specifically, if f restricted to $a + V$ has degree at most $d - 1$ (where $d = \deg_2(f)$), then f restricted to any other $a' + V$ also has degree at most $d - 1$. This can be seen by the following argument. Call a monomial in f a *maxonomial* if it is of the maximal degree (i.e., degree d). Apply a linear map to $\{0,1\}^n$ so that $V = \{x : x_1 = \cdots = x_r = 0\}$, where $r = \text{co-dim}(V)$. Then $f|_{a+V}$ becomes a polynomial of degree at most $d - 1$ if and only if every maxonomial of f (under the new basis) contains at least one variable in the set $\{x_1, \ldots, x_r\}$. Moreover, when this happens it does not matter whether x_i ($i \leq r$) is restricted to 0 or 1, the degree of the maxonomial always decreases, thus $\deg_2(f|_{a'+V}) \leq d - 1$ for all $a' \in \{0,1\}^n$.

The above fact also reveals that the linear rank r of any polynomial $f(x)$ is determined by the maxonomials in $f(x)$ *only*. Fourier sparsity in general, on the other hand, should depend on all GF(2) monomials, not only those with the highest degree. However, the linear rank conjecture claims that if the maxonomials in $f(x)$ make the linear rank large, then no matter how the lower-degree monomials behave, the Fourier sparsity is large. Therefore, for the effect of forcing the Fourier sparsity of GF(2) polynomial to be large, there exists a surprising fact (assuming the linear rank conjecture) that can be summarized by paraphrasing a famous quote from *Animal Farm*: "All monomials are equal, but some monomials are more equal than others".

In retrospect, this phenomenon is known for some extremal cases. When $\deg_2(f) = 2$, the lower degree terms form a linear function χ_α, adding which only shifts Fourier spectrum by α and thus does not affect the Fourier sparsity. When $\deg_2(f) = n$, the Fourier sparsity is at least $2^{\deg_2(f)} - 1 = 2^n - 1$, which is again determined by the (unique) maxonomial. *But for general $2 < d < n$, especially $d = n^{o(1)}$ which is the interesting range of d for Log-rank Conjecture, maxonomials by themselves do not necessarily lead to large Fourier sparsity.* For instance, if there is only one maxonomial $x_1 \ldots x_d$, then the Fourier sparsity can be as small as 2^d (when, say, the lower degree part is $x_1 + \cdots + x_n$), and as large as 2^{n-d} (when, say, the lower degree part is a bent function[2] over $x_{d+1}, \ldots, x_n$). Therefore it is not true for general $2 < d < n$ that maxonomials can control large Fourier sparsity. However, we will show that when the maxonomials form *certain patterns*, the Fourier sparsity is guaranteed to be large, regardless of the lower degree terms (whose number can be much larger than that of maxonomials). One sufficient condition for the pattern is that the linear rank, which depends on maxonomials only, is large. And we will showcase some specific classes of good patterns.

Therefore, apart from leading directly to a proof of the Log-rank Conjecture for XOR functions, studying the linear rank conjecture is interesting in its own right, due to its close connection to the Fourier analysis of Boolean functions in the GF(2) polynomial representation.

1.2 Our Work

We study the linear rank conjecture and in particular investigate how the maxonomials of a GF(2) polynomial could possibly determine by themselves the Fourier sparsity of the polynomial. We develop a new technique which is able to show that, under certain circumstances, the Fourier sparsity is large for all possible settings of lower degree monomials. The precise statement of the general theorem needs some technical definitions; see Theorem 1 in Sect. 3. We next use the result to lower bound Fourier sparsity for arbitrary function with maxonomials forming certain patterns.

Dense Maxonomials. The first pattern we consider is when all $\binom{n}{d}$ maxonomials appear. For this "complete pattern", we are able to determine the exact value of the linear rank. Specifically, for $f = \sum_{|S|=d} \prod_{i \in S} x_i + f'$, where f' is an arbitrary polynomial of degree at most $d - 1$, it holds that

$$\text{lin-rank}(f) = \begin{cases} \lfloor \frac{n}{2} \rfloor - \frac{d}{2} + 1 & \text{if } d \text{ is even,} \\ 1 & \text{if } d \text{ is odd.} \end{cases}$$

---

[2] A Boolean function $f : \{0,1\}^m \to \{-1,1\}$ is *bent* if its Fourier coefficients satisfy that $|\hat{f}(\alpha)| = 2^{-m/2}$ for all $\alpha \in \{0,1\}^m$.

The proof exploits the symmetry of maxonomials and goes through a careful induction on n and d. In particular the "step-function" type behaviour of the linear rank (with respect to n) is proved by showing both upper and lower bounds for the number of linear functions one needs to fix in order to decrease the degree of the polynomial. Due to the space limit, the proof is deferred to Appendix A.

If the linear rank conjecture is true, then for any polynomial with complete d-uniform maxonomials (d is even), the Fourier sparsity must be $2^{n^{\Omega(1)}}$ regardless of the lower degree monomials. We are able to verify this when d is a power of 2: if $f : \{0,1\}^n \to \{0,1\}$ is a degree-d polynomial with complete d-uniform maxonomials, then regardless of what lower degree monomials appear, it holds that

$$\mathsf{spar}(f) \geq 2^{d \cdot \lfloor n/d \rfloor} - 1 = \Omega(2^n).$$

Zhang and Shi [18] proved that any symmetric boolean function has Fourier sparsity $2^{\Omega(n)}$, unless it is constant, the parity function over n bits or its negation. However, as the polynomials considered there are symmetric, their result requires the degree-d' monomials to be either empty or complete d'-uniform, for every $d' \leq d$. On the contrary, our lower bound applies to a broader class of functions as it holds for all possible choices of lower degree monomials, as long as the highest-degree monomials are symmetric.

Sparse Maxonomials. We then go to the other end of the spectrum and consider patterns formed by a few number of maxonomials. In particular, we show lower bounds on the Fourier sparsity of polynomials whose maxonomials are pair-wise disjoint. We also exhibit patterns in which maxonomials have very small pair-wise intersection, and the number of maxonomials is much smaller than that of lower-degree monomials, yet the maxonomials by themselves guarantee a high Fourier sparsity of 2^n.

The techniques developed can be used to show more results. Gopalan et al. [6] studied the *granularity* of a function's Fourier spectrum, which is the smallest integer k such that all Fourier coefficients of the function can be expressed as integer multiples of $1/2^k$. They showed that for any Boolean function $f : \{0,1\}^n \to \{0,1\}$, $\mathsf{gran}(f) \leq \log \mathsf{spar}(f)$. On the other hand, by Parseval's identity, $\log \mathsf{spar}(f) \leq 2\mathsf{gran}(f)$. The granularity of a linear functions is 1 and the maximum granularity of any n-variate quadratic polynomial is $n/2$. It thus natural to conjecture that, for any n-variate low-degree polynomial $f(x)$, although $\mathsf{spar}(f)$ can be as large as 2^n, the granularity of $f(x)$ is always bounded away from n. We are able to apply our technique to show the following *upper bound* on the granularity of low-degree polynomials: for any degree-d polynomial f, $\mathsf{gran}(f) \leq n - \lceil \frac{n}{d} \rceil + 1$. It is easy to see this bound is tight as it is attained by the "generalized inner product function": $f(x) = x_1 x_2 \cdots x_d + \cdots + x_{(k-1)d+1} x_{(k-1)d+2} \cdots x_{kd}$, where $n = kd$.

Techniques. The main challenge in proving sparsity lower bounds based on *only* the maxonomials of a polynomial is how to isolate the effect of *all* lower degree monomials. To the best of our knowledge, there is no prior method or result of this kind. Our method is to first apply the standard procedure to transform a degree-d polynomial f into a Fourier polynomial, and then define a "weight function" $w_f(T)$ on each set $T \subseteq [n]$ such that the Fourier coefficient of f at any set S can be written as $\sum_{T \supseteq S} w_f(T)$. This implies that the weight function at $[n]$ is the most important term as it contributes to all the Fourier coefficients of f. Another nice property of the weight function is that for any T, $2^{|T|} w_f(T)$ can be expressed as a sum of alternating terms in which the kth term is $(-2)^k N_k(T)$, where $N_k(T)$ is the number of ways to cover T with (the supports of) exactly k monomials of $f(x)$. Therefore, the problem of computing the Fourier coefficients of a GF(2) polynomial is now reduced to a combinatorial problem of counting the numbers of covers of all subsets of $[n]$ using various numbers of sets from the set family defined by the monomials of the polynomial. Moreover, the parity of $2^{|T|} w_f(T)$ is likely to be determined by the numbers of smaller covers due to the factor $(-2)^k$ in each term of the sum. Using the notion of "granularity" introduced in [6], our strategy for showing sparsity lower bound is to argue that $w_f([n])$ is the single one with the highest granularity among all weight function values. Note that if $n = kd$ and we can cover $[n]$ with (the supports of) maxonomials of $f(x)$ only, then these covers would be the minimum covers as they require only $k = n/d$ sets while any cover involving lower monomials is of size at least $k + 1$. Hence to prove that $w_f([n])$ has the highest possible granularity, it suffices to show that the number of k-covers of $[n]$ is odd, as we did for the several sparsity lower bounds.

2   Preliminaries

All logarithms in this paper are base 2. For $f : \{0,1\}^n \to \{0,1\}$, we use $f^\pm = 1 - 2f$ to denote the equivalent Boolean function with range converted to $\{+1, -1\}$. For $S \subseteq [n]$, the monomial x_S is $x_S = \prod_{i \in S} x_i$, and S is called the *support* of the monomial. We say a set T *meets* a monomial x_S if $T \cap S \neq \emptyset$. The *degree* of $f : \{0,1\}^n \to \{0,1\}$, denoted $\deg_2(f)$, is the degree of f as a multi-linear polynomial on variables $x_1, \ldots, x_n$.

For any real function $f : \{0,1\}^n \to \mathbb{R}$, the *Fourier coefficients* are defined by $\hat{f}(\alpha) = 2^{-n} \sum_x f(x) \chi_\alpha(x)$, where $\alpha \in \{0,1\}^n$ and $\chi_\alpha(x) = \prod_{i=1}^n (-1)^{\alpha_i x_i}$. Its *Fourier sparsity*, denoted $\|\hat{f}\|_0$, is the number of nonzero Fourier coefficients of f. The Fourier coefficients of $f : \{0,1\}^n \to \{0,1\}$ and $f^\pm := 1 - 2f$ are almost the same:

$$\|\hat{f}\|_0 - 1 \leq \|\widehat{f^\pm}\|_0 \leq \|\hat{f}\|_0 + 1. \tag{2}$$

Sometimes we employ the one-to-one mapping between vectors in $\{0,1\}^n$ and subsets of $[n]$: $x \leftrightarrow \{i \in [n] : x_i = 1\}$, and use the subsets of $[n]$ to index the Fourier coefficients. Parseval's identity asserts that $\mathbf{E}[f^2(x)] = \sum_S \hat{f}^2$.

Definition 2 (Granularity [6]**).** *Let* $r \in \mathbb{Q}$, *the granularity of* r, *denoted* $\text{gran}(r)$, *is the smallest nonnegative integer* k *such that* $2^k r \in \mathbb{Z}$, *and* $\text{gran}(r) = \infty$ *if no such exists. The* Fourier granularity *of a Boolean function* f, *denoted*

gran(f), *is the maximum granularity over all the Fourier coefficients of f; i.e.,*
gran(f) = max$_{\alpha \in \{0,1\}^n}$(gran($\hat{f}(\alpha)$)).

Clearly, gran($-x$) = gran(x) for any $x \in \mathbb{Q}$. An easy but useful fact is that gran($x + y$) $\leq$ max(gran(x), gran(y)) for all $x, y \in \mathbb{Q}$. More generally, gran($\sum_{i=1}^{k} x_i$) $\leq$ max$_{1 \leq i \leq k}$ gran(x_i), where $x_i \in \mathbb{Q}$ for every $1 \leq i \leq k$.

Fact 1. *Let $f^{\pm}, g^{\pm} : \{0,1\}^n \to \{-1,1\}$ be two Boolean functions. Let $h = f \oplus g$. Then $|$gran($f^{\pm}$) $-$ gran($g^{\pm}$)$| \leq$ gran($h^{\pm}$) $\leq$ gran($f^{\pm}$) + gran($g^{\pm}$).*

Gopalan *et al.* [6] showed that, if a Boolean function has only a small number of non-zero Fourier coefficients, then all these non-zero Fourier coefficients have small granularities.

Lemma 1 ([6]). *Suppose $f^{\pm} : \{0,1\}^n \to \{-1,1\}$ is s-sparse with $s > 0$, then all the Fourier coefficients of $f^{\pm}$ have granularity at most $\lfloor \log s \rfloor - 1$.*

This lemma implies that the logarithm of the sparsity and granularity of a Boolean function are in fact equivalent up to a constant factor.

Proposition 1. *Let $f^{\pm} : \{0,1\}^n \to \{-1,1\}$ be a Boolean function, then*

$$\text{gran}(f^{\pm}) + 1 \leq \log \text{spar}(f^{\pm}) \leq 2\text{gran}(f^{\pm}).$$

For a function $f : \{0,1\}^n \to \mathbb{R}$, define two subfunctions f_0 and f_1, both on $\{0,1\}^{n-1}$: $f_b(x_2, \ldots, x_n) = f(b, x_2, \ldots, x_n)$. It is easy to see that for any $\alpha \in \{0,1\}^{n-1}$, $\hat{f}_b(\alpha) = \hat{f}(0\alpha) + (-1)^b \hat{f}(1\alpha)$, thus

$$\|\hat{f}_b\|_0 \leq \|\hat{f}\|_0 \text{ and } \|\hat{f}_b\|_1 \leq \|\hat{f}\|_1. \tag{3}$$

where $\|\hat{f}\|_p = (\sum_{\alpha} |\hat{f}(\alpha)|^p)^{1/p}$ and $\|\hat{f}\|_0 = |\{\alpha : \hat{f}(\alpha) \neq 0\}|$. The notion of subfunctions can be generalized to restrictions to affine subspaces. It is worth noticing that, for any Boolean function, its $\mathbb{F}_2$-degree, Fourier sparsity and granularity are all invariant under invertible linear maps.

3   Fourier Spectra of GF(2) Polynomials

In this section, we present a lower bound for Fourier sparsity. It first needs to compute the Fourier spectrum of a GF(2) polynomial based on its monomials. We suspect that such a formalism was known before but we could not track any previous sources.

For a fixed $S \subseteq [n]$, a collection $\{S_1, \ldots, S_k\}$ of k (distinct) subsets of $[n]$ form a k-*cover* of S if $\cup_{i=1}^{k} S_i = S$. The main result of this section is the following lemma, which shows that the Fourier coefficients of a GF(2) polynomial can be computed by counting the number of k-covers of subsets of $[n]$ — for different values of k — using the supports of monomials in the GF(2) polynomial as subsets. Of particular importance is the number of $k_{\min}$-covers of $[n]$, where $k_{\min}$ is minimum number of subsets that are required to cover $[n]$.

For a family $\mathcal{F} = \{S_i\}_{i \in [m]}$ of subsets S_i of the base set $[n]$ and an index set $M \subseteq [m]$, let $S_M \stackrel{\text{def}}{=} \cup_{k \in M} S_k$, the union of the subsets with indices in M.

Let $f(x_1, \ldots, x_n) = \sum_{i=1}^{m} x_{S_i}$ be the GF(2) polynomial representation of f. Define a *weight* function $w_f : \{0,1\}^n \to \mathbb{Q}$ as

$$w_f(T) = \sum_{M \subseteq [m]: S_M = T} c(M), \quad \text{where } c(M) = \frac{(-2)^{|M|}}{2^{|S_M|}}. \tag{4}$$

Equivalently, if we denote $\mathcal{F} = \{S_i\}_{i \in [m]}$ and let $N_k(T)$ be the number of k-covers of T using sets in $\mathcal{F}$, then

$$w_f(T) = \frac{1}{2^{|T|}} \sum_{k=1}^{m} (-2)^k N_k(T). \tag{5}$$

Lemma 2. *Let* $f(x_1, \ldots, x_n) = \sum_{i=1}^{m} x_{S_i}$ *be a GF(2) polynomial, then the Fourier coefficients of* $f^{\pm}$ *are given by*

$$\widehat{f^{\pm}}(S) = (-1)^{|S|} \sum_{T \supseteq S} w_f(T). \tag{6}$$

Proof. For a Boolean variable $x_i \in \{0,1\}$, let $\tilde{x}_i = (-1)^{x_i} = 1 - 2x_i$ be its $\{+1, -1\}$ representation, with the inverse transformation given by $x_i = (1 - \tilde{x}_i)/2$. Recall that $f^{\pm} = 1 - 2f$. We next express $f^{\pm}$ as a multilinear polynomial over $\mathbb{R}$ from which its Fourier coefficients can be readily read out.

Note that x_S corresponds to $1 - 2 \prod_{i \in S} \frac{1 - \tilde{x}_i}{2}$ and $\prod_{i \in S} \tilde{x}_i$ corresponds to $\tilde{x}_S$, thus

$$f^{\pm}(\tilde{x}_1, \ldots, \tilde{x}_n) = \prod_{i \in [m]} \left(1 - 2 \prod_{j \in S_i} \frac{1 - \tilde{x}_j}{2}\right) \tag{7}$$

Fact 2. *For* $x \in \{-1, 1\}$ *and integer* $k \geq 1$*, we have* $(1 - x)^k = 2^{k-1}(1 - x)$.

By (7), the Fourier polynomial of $f^{\pm}$ in terms of $\tilde{x}$ is

$$
\begin{aligned}
f^{\pm}(\tilde{x}) &= \prod_{i=1}^{m} \left(1 - \frac{\prod_{j \in S_i}(1 - \tilde{x}_j)}{2^{|S_i|-1}}\right) \\
&= \sum_{k=0}^{m} (-1)^k \sum_{1 \leq i_1 < i_2 < \ldots < i_k \leq m} \frac{\prod_{j_1 \in S_{i_1}}(1 - \tilde{x}_{j_1}) \prod_{j_2 \in S_{i_2}}(1 - \tilde{x}_{j_2}) \cdots \prod_{j_k \in S_{i_k}}(1 - \tilde{x}_{j_k})}{2^{|S_{i_1}| + |S_{i_2}| + \cdots + |S_{i_k}| - k}} \\
&= \sum_{k=0}^{m} (-1)^k \sum_{1 \leq i_1 < i_2 < \ldots < i_k \leq m} \frac{\prod_{j \in S_{i_1} \cup \cdots \cup S_{i_k}}(1 - \tilde{x}_j)}{2^{|S_{i_1} \cup \cdots \cup S_{i_k}| - k}} \quad \text{(by Fact 2)} \\
&= \sum_{M \subseteq [m]} (-1)^{|M|} \frac{\prod_{j \in S_M}(1 - \tilde{x}_j)}{2^{|S_M| - |M|}} \\
&= \sum_{S \subseteq [n]} (-1)^{|S|} \left(\sum_{M \subseteq [m]: S_M \supseteq S} (-1)^{|M|} \cdot \frac{2^{|M|}}{2^{|S_M|}}\right) \tilde{x}_S.
\end{aligned}
$$

Since the coefficient of $\tilde{x}_S$ in $f^{\pm}(\tilde{x})$ is just the Fourier coefficient $\widehat{f^{\pm}}(S)$, this completes the proof of the lemma. $\qquad\square$

Theorem 1. *If $f : \{0,1\}^n \to \{0,1\}$ has $d|n$ where $d = \deg_2(f)$, and $N_{n/d}([n])$ is odd, then $\|f^{\pm}\|_0 = 2^n$.*

Proof. Note that for $T = [n]$, one cannot cover T by less than n/d sets S_i, thus $N_k(T) = 0$ for all $k < n/d$. By assumption, $N_{n/d}([n]) = t$ is odd, thus $w_f(T) = (-1)^{n/d}\frac{t}{2^{n-n/d}}+s$, where $\operatorname{gran}(s) < n-n/d$. So $\operatorname{gran}(w_f([n])) = n-n/d$. For general $S \subseteq [n]$, consider $\widehat{f^{\pm}}(S)$ and note that among all $T \supseteq S$ in (6), the largest granularity of $w_f(T)$ is n/d, achieved by $T = [n]$. Indeed, for all other T, $|T| < n$, the largest possible granularity is $|T| - \lceil |T|/d \rceil$, which is strictly less than $n - n/d$. Thus all Fourier coefficients $\widehat{f^{\pm}}(S)$ have granularity $n - n/d > 0$, implying that $\widehat{f^{\pm}}(S) \neq 0$. $\qquad\square$

4 Fourier Sparsity of Polynomials with Complete d-uniform Maxonomials

This section is devoted to the proof of the following Fourier sparsity lower bound for polynomials whose maxonomials are the complete d-uniform monomials.

Theorem 2. *Let d be a power of 2. For any degree-d polynomial $f \in \mathbb{F}_2[x_1,\dots,x_n]$ whose maxonomials include all $\binom{n}{d}$ degree-d monomials, its Fourier sparsity has the following lower bound*

$$\operatorname{spar}(f) \geq 2^{d\cdot\lfloor n/d \rfloor} - 1 = \Omega(2^n),$$

regardless of the lower degree monomials.

Proof. In the rest of this section, we fix $k = \lfloor n/d \rfloor$. First we apply a restriction to set, say the last $n - kd$ variables in f to zero. This leaves us with a function g on $n' = kd$ variables, and by (3), we have $\operatorname{spar}(f) \geq \operatorname{spar}(g)$. Furthermore, the maxonomials of g are still complete d-uniform monomials (now over n' variables).

Let $\mathcal{F}$ be the set of the supports of all monomials in g. In particular, $\mathcal{F}$ contains all d-subsets of $[n']$: $\binom{[n']}{d} \subseteq \mathcal{F}$. By Theorem 1, it suffices to prove the following.

Lemma 3. $N_k([n']) \equiv 1 \pmod 2$.

Proof. Clearly any k-cover of $[n']$ consists of k distinct sets in $\binom{[n']}{d}$, and there are exactly $\frac{\binom{n'}{d,\dots,d}}{k!}$ such k-covers. Hence we have

$$N_k([n']) = \frac{\binom{n'}{d,\dots,d}}{k!} = \frac{1}{k}\binom{kd}{d} \cdot \frac{1}{k-1}\binom{(k-1)d}{d} \cdots 1 \cdot \binom{d}{d}$$

$$= \binom{kd-1}{d-1} \cdot \binom{(k-1)d-1}{d-1} \cdots \binom{d-1}{d-1}.$$

Recall the following Lucas' theorem:

Theorem 3 (Lucas' Theorem, c.f. [4]). *Let s and t be non-negative integers and p be a prime. Let $s = s_0 + s_1 p + \cdots s_i p^i$ and $t = t_0 + t_1 p + \cdots t_i p^i$, $0 \le s_j, t_j < p$, be the base-p expansions of s and t respectively, then*

$$\binom{s}{t} \equiv \prod_{j=0}^{i} \binom{s_j}{t_j} \pmod{p}.$$

In fact, what we need is the a simple corollary of Lucas' theorem (known as Kummer's theorem) for the special case of $p = 2$: the largest integer j such that 2^j divides $\binom{s}{t}$ is equal to the number of carries that occur when s and $s - t$ are added in the binary.

Since d is a power of 2, the binary representation of $d - 1$ is $\underbrace{1 \cdots 1}_{\log d}$ and the binary representation of $jd - 1 - (d-1) = (j-1)d$ is $\cdots \underbrace{0 \cdots 0}_{\log d}$, for every $j \ge 1$. Therefore no carry occurs when adding $(j-1)d$ to $d-1$ and thus, by Kummer's theorem, $\binom{jd-1}{d-1} \equiv 1 \pmod 2$ for all $j \ge 1$. It follows that

$$N_k([n']) \equiv 1 \pmod 2,$$

which also implies Theorem 2.   □

5 Fourier Sparsity for Functions with Sparse Maxonomials

In the previous two sections, we see cases that when all $\binom{n}{d}$ monomials of the highest degree appear, then the function has large Fourier sparsity, no matter what other lower-degree monomials exist or not. In this section, we will consider the other end of the spectrum when there are only a small number of the maxonomials, and show that the same phenomena can occur in this case as well.

The first example is the class of functions with disjoint maxonomials.

Proposition 2. *Suppose that $f : \{0,1\}^n \to \{0,1\}$ has $\deg_2(f) = d$ where $d|n$. If there are exactly n/d monomials of degree d, and their supports are pairwise disjoint, then $\mathsf{spar}(f) \ge 2^n - 1$, regardless of the lower degree monomials.*

Proof. Note that there is only one cover of $[n]$ of size n/d, i.e. $N_{n/d}([n]) = 1$. By Theorem 1, $\mathsf{spar}(f) \ge \mathsf{spar}(f^{\pm}) - 1 = 2^n - 1$.   □

The second example extends the first class by allowing "regular" overlaps between maxonomials. Assume that $\deg_2(f) = d$ is an odd prime power, and $d^2|n$. Divide $[n]$ into n/d^2 piles of equal size, with each pile identified with a $d \times d$ grid. All maxonomials are linear functions in a pile. More precisely, for the first pile $[d] \times [d]$, for each pair $(a, b) \in \mathbb{F}_d^2$, define univariate polynomial $p_{a,b} \in \mathbb{F}[x]$ by $p_{a,b}(x) = ax + b$. Now define sets

$$S_{a,b} = \{(0, p(0)), (1, p(1)), \ldots, (d - 1, p(d - 1))\}.$$

The first pile thus has d^2 sets inside. Similarly define d^2 sets for each other pile. These sets are supports of the maxomonials. Note that there are $d^2 \cdot n/d^2 = n$ maxonomials, a number much smaller than the possible number of lower degree monomials, which is $\sum_{i=0}^{d-1} \binom{n}{i}$. Yet the next theorem says that the this small number of maxonomials determines a large Fourier sparsity, regardless of how the vast majority of other (lower-degree) terms behave.

Theorem 4. *For any function $f : \{0, 1\}^n \rightarrow \{0, 1\}$ with the maxonomials defined as above, $\mathsf{spar}(f) \geq 2^n - 1$, regardless of the lower degree monomials.*

Proof. Clearly the set $[n]$ can be partitioned using supports of n/d maxonomials. We will show that the number of such partitions is d^{n/d^2}, which is an odd number given that d is odd.

Since the piles are disjoint and all maxonomials are defined within each pile, it suffices to show that there are d ways of partitioning each pile into maxonomials. We consider the first pile and the same argument applies to others. Note that for each fixed a, if we vary b over $\mathbb{F}_d$, then we get d maxonomials that are pairwise disjoint. Since there are d different choices of a, there are at least these d ways to partition the pile into d maxonomials. We next show that there are actually no other partition of the pile using d maxonomials. Indeed, assume that a partition uses d maxonomials and not all these maxonomials have the same a, then there are two maxonomials corresponding to $a_1 x + b_1$ and $a_2 x + b_2$ and $a_1 \neq a_2$. But now these two "lines" intersect at exactly one point $x = (a_1 - a_2)^{-1}(b_1 - b_2)$, where the existence of $(a_1 - a_2)^{-1}$ uses the assumption that $a_1 \neq a_2$. Note the trivial fact that the union of d maxonomials of degree d is at most d^2, and it is d^2 only if they are pairwise disjoint. So the existence of intersecting maxonomials in the selected d maxonomials make them impossible to cover the d^2 points in the pile. This shows that the number of partitions of one pile using d maxonomials is exactly d, and thus the number of covers of $[n]$ using n/d^2 maxonomials is d^{n/d^2}, an odd number by the assumption that d is odd. Applying Theorem 1 completes the proof of Theorem 4. □

6 Granularity Upper Bound for Low-Degree Polynomials

Note that there is a gap of factor 2 in characterizing the logarithm of Fourier sparsity of a Boolean function by means of its granularity (cf. Proposition 1). Note also that both lower and upper bounds in Proposition 1 are tight, but one is attained by the AND function (a degree-n polynomial) and the other by any bent function, e.g. the Inner Product function (a degree-2 polynomial). It thus natural to conjecture that, for any low-degree polynomial $f(x)$, although $\mathsf{spar}(f)$ can be as large as 2^n, the granularity of $f(x)$ is always bounded away from n. We now apply our technique developed in Sect. 3 to prove the following upper bound for the granularity of low-degree polynomials.

Theorem 5. *For any Boolean function* $f : \{0,1\}^n \to \{0,1\}$, *if* $d = \deg_2(f)$ *is the* $\mathbb{F}_2$-*degree of* f, *then* $\mathrm{gran}(f^{\pm}) \leq n - \lceil \frac{n}{d} \rceil$, *and consequently,* $\mathrm{gran}(f) \leq n - \lceil \frac{n}{d} \rceil + 1$.

Proof. Suppose $\widehat{f^{\pm}}(T)$ achieves $\mathrm{gran}(f^{\pm})$, i.e., $\widehat{f^{\pm}}(T) = c/2^{\mathrm{gran}(f^{\pm})}$ for some odd integer c. Without loss of generality, we may assume that $T \neq \emptyset$. Actually, if $\widehat{f^{\pm}}(0)$ is the single Fourier coefficient that achieves $\mathrm{gran}(f^{\pm})$, then the sum of the squares of all Fourier coefficients of $f^{\pm}$ would be a rational number with granularity $2\mathrm{gran}(f^{\pm})$ instead of 1, contradicting Parseval's identity.

Now we apply an invertible linear map L such that $(L^T)^{-1}(T) = [n]$. Let $g = f \circ L$ and note that $\deg_2(g) = d$. It is not difficult to show that $\widehat{g^{\pm}}([n]) = \widehat{f^{\pm}}(T)$.

Suppose $g(x) = \sum_{i=1}^m \prod_{j \in S_i} x_j$, where $|S_j| \leq d$ for every $1 \leq j \leq m$. Applying Lemma 2 and notice that, since $|S_j| \leq d$, the minimum number k such that there exists a collection of k subsets from $\{S_j\}_{j \in [m]}$ that cover $[n]$ is $k \geq \lceil \frac{n}{d} \rceil$. Therefore, by (6),

$$\widehat{g^{\pm}}([n]) = (-1)^n w_g([n]) = (-1)^n \sum_{j=k}^m \frac{(-2)^j N_j([n])}{2^n}.$$

Note that the granularity of the jth term in the above summation is at most $n - j$ (we only have inequality here as $N_j([n])$ may be an even number), and the granularity of a sum of rational numbers is at most the maximum granularity in the summands:

$$\mathrm{gran}\left(\sum_{j=1}^\ell y_j\right) \leq \max_{1 \leq j \leq \ell} \mathrm{gran}(y_j),$$

where $y_j \in \mathbb{Q}$ for $1 \leq j \leq \ell$, we therefore have $\mathrm{gran}\left(\widehat{g^{\pm}}([n])\right) \leq n - k = n - \lceil \frac{n}{d} \rceil$. This finally gives

$$\mathrm{gran}(f^{\pm}) = \mathrm{gran}\left(\widehat{f^{\pm}}(T)\right) = \mathrm{gran}\left(\widehat{g^{\pm}}([n])\right) \leq n - \lceil \frac{n}{d} \rceil.$$

The upper bound of the granularity of f follows from the easy fact that $\mathrm{gran}(f) \leq \mathrm{gran}(f^{\pm}) + 1$. □

Acknowledgements. We are indebted to the anonymous reviewers for their detailed helpful comments.

A Linear Rank of Complete d-uniform Maxonomials

In this section we compute the exact value of the linear rank of a degree d polynomial whose set of maxonomials consists of all $\binom{n}{d}$ degree-d monomials.

Define $\mathcal{C}_{d,n}(x) = \sum_{I \subseteq [n], |I| = d} \prod_{i \in I} x_i$, the summation of all degree-d monomials over variables $x_1, \ldots, x_n \in \mathbb{F}_2$. The subscript n is dropped when it is clear from the context. We use the equivalence relation $\equiv_d$ for polynomials with the same maxonomials, *i.e.* $p \equiv_d q$ if both p and q have $\mathbb{F}_2$-degree d and $p+q$ has $\mathbb{F}_2$-degree strictly less than d. It is clear that if $p \equiv_d q$, then lin-rank(p) = lin-rank(q).

Theorem 6. *Let $n \geq d \geq 0$ be integers. Then the following hold:*

1. *If d is odd, then* lin-rank$(\mathcal{C}_{d,n})$ = 1.
2. *If d is even, then* lin-rank$(\mathcal{C}_{d,n}) = \lfloor \frac{n}{2} \rfloor - \frac{d}{2} + 1$, *i.e.*

$$\text{lin-rank}(\mathcal{C}_{d,n}) = \begin{cases} \frac{n-d}{2} + 1 & \text{if } n \text{ is even,} \\ \frac{n-d-1}{2} + 1 & \text{if } n \text{ is odd.} \end{cases}$$

Proof. The first item follows simply by the factorization $\mathcal{C}_{d,n} \equiv_d \mathcal{C}_{1,n}\mathcal{C}_{d-1,n}$. Indeed, when we multiply $\mathcal{C}_{1,n} = \sum_{i \in [n]} x_i$ and $\mathcal{C}_{d-1,n} = \sum_{|I|=d-1} x_I$, for $i \notin I$, $x_i x_I = x_{I \cup \{i\}}$, and each J with $|J| = d$ comes from d many (i, I). For each $i \in I$, $x_i x_I = x_I$, and each resulting x_I with $|I| = d - 1$ comes from $d - 1$ many $i \in I$. Thus

$$\mathcal{C}_{1,n}\mathcal{C}_{d-1,n} = d\left(\sum_{|J|=d} x_J \right) + (d-1)\left(\sum_{|I|=d-1} x_I \right) = d\mathcal{C}_{d,n} + (d-1)\mathcal{C}_{d-1,n}$$

$$= \mathcal{C}_{d,n},$$

for all odd d.

Now we consider the second item in the statement and assume from now on that d is even and $d \leq n$. The second item follows from the following two claims.

Claim 1. *If* lin-rank$(\mathcal{C}_{d,n+1})$ = lin-rank$(\mathcal{C}_{d,n})$, *then* lin-rank$(\mathcal{C}_{d,n+2})$ > lin-rank $(\mathcal{C}_{d,n+1})$.

Claim 2. lin-rank$(\mathcal{C}_{d,n+2}) \leq$ lin-rank$(\mathcal{C}_{d,n}) + 1$.

Let us first show Theorem 6 assuming these two claims. We prove by induction on the number of variables that for all $k \geq d/2$,

$$\text{lin-rank}(\mathcal{C}_{d,2k}) = \text{lin-rank}(\mathcal{C}_{d,2k+1}) = k - \frac{d}{2} + 1. \tag{8}$$

which is just a restatement of the second item of Theorem 6.

Base Case $k = d/2$. We have

$$\mathcal{C}_d(x_1, \ldots, x_{2k}) = \mathcal{C}_d(x_1, \ldots, x_d) = \mathcal{C}_{d-1}(x_1, \ldots, x_{d-1}) \cdot x_d, \tag{9}$$

so lin-rank$(\mathcal{C}_{d,2k})$ = 1. For $n = 2k + 1$, note that

$$\mathcal{C}_d(x_1, \ldots, x_{2k+1}) = \mathcal{C}_d(x_1, \ldots, x_{d+1})$$
$$= \mathcal{C}_{d-1}(x_1, \ldots, x_{d-1})(x_d + x_{d+1}) + \mathcal{C}_{d-2}(x_1, \ldots, x_{d-1})x_d x_{d+1}, \tag{10}$$

Putting restriction $x_d = x_{d+1}$ makes the first summand vanish and decreases the degree of the second summand, hence lin-rank$(\mathcal{C}_{d,2k+1}) = 1$.

General k. Now we assume that (8) holds for k and will prove the case for $k+1$. The following sequence of inequalities hold.

$$k - \frac{d}{2} + 1 < \text{lin-rank}(\mathcal{C}_{d,2(k+1)}) \leq \text{lin-rank}(\mathcal{C}_{d,2(k+1)+1}) \leq k - \frac{d}{2} + 2,$$

where the first inequality follows by Claim 1; the second follows by the facts that $\mathcal{C}_{d,n-1}$ can be obtained from $\mathcal{C}_{d,n}$ by restricting $x_n = 0$ and restriction does not increase lin-rank; and the last inequality follows by Claim 2. Therefore (8) also holds for $k+1$. □

Now it remains to prove the two claims. We start with Claim 2, which is simpler.

Proof (Proof of Claim 2). We first observe the following identity:

$$\begin{aligned}
\mathcal{C}_d(x_1, \ldots, x_{n+2}) &= \mathcal{C}_d(x_1, \ldots, x_n) + \mathcal{C}_{d-1}(x_1, \ldots, x_n)(x_{n+1} + x_{n+2}) \\
&\quad + \mathcal{C}_{d-2}(x_1, \ldots, x_n)x_{n+1}x_{n+2} \\
&\equiv_d \mathcal{C}_d(x_1, \ldots, x_n) + \mathcal{C}_{d-1}(x_1, \ldots, x_n, x_{n+1})(x_{n+1} + x_{n+2}).
\end{aligned}$$
(11)

Therefore the restriction $x_{n+2} = x_{n+1}$ reduces $\mathcal{C}_d(x_1, \ldots, x_{n+2})$ to

$$\mathcal{C}_d(x_1, \ldots, x_{n+2})|_{x_{n+1}=x_{n+2}} \equiv_d \mathcal{C}_d(x_1, \ldots, x_n).$$

Since each restriction can reduce lin-rank by at most 1, we have

$$\text{lin-rank}(\mathcal{C}_{d,n+2}) - 1 \leq \text{lin-rank}(\mathcal{C}_{d,n+2}|_{x_{n+2}=x_{n+1}}) = \text{lin-rank}(\mathcal{C}_{d,n}),$$

as desired. □

Proof (Proof of Claim 1). For the sake of contradiction, assume that

$$\text{lin-rank}(\mathcal{C}_{d,n+2}) = \text{lin-rank}(\mathcal{C}_{d,n+1}) = \text{lin-rank}(\mathcal{C}_{d,n}) = r.$$

Fix an optimal set of linear restrictions for lin-rank$(\mathcal{C}_{d,n+2})$. Without loss of generality, we can assume it contains a restriction of the form $x_{n+2} = \ell(x_1, \ldots, x_{n+1}) = \ell(x)$ for some linear form ℓ. It is clear that such restriction will reduce the lin-rank by exactly 1. So we have

$$\text{lin-rank}(\mathcal{C}_{d,n+2}|_{x_{n+2}=\ell(x)}) \leq \text{lin-rank}(\mathcal{C}_{d,n+2}) - 1 = r - 1.$$
(12)

But by the expansion

$$\mathcal{C}_d(x_1, \ldots, x_{m+1}) = \mathcal{C}_d(x_1, \ldots, x_m) + \mathcal{C}_{d-1}(x_1, \ldots, x_m)x_{m+1},$$

we have

$$\begin{aligned}
\mathcal{C}_d(x_1, \ldots, x_{n+2})|_{x_{n+2}=\ell(x)} &= \mathcal{C}_d(x_1, \ldots, x_{n+1}) + \mathcal{C}_{d-1}(x_1, \ldots, x_{n+1})\ell(x) \\
&= \mathcal{C}_d(x_1, \ldots, x_n) + \mathcal{C}_{d-1}(x_1, \ldots, x_n)x_{n+1} \\
&\quad + \mathcal{C}_{d-1}(x_1, \ldots, x_{n+1})\ell(x).
\end{aligned}$$
(13)

Now, consider to further restrict $x_{n+1} = x_1 + x_2 + \cdots + x_n = \mathcal{C}_1(x_1, \ldots, x_n)$. By the fact that $\mathcal{C}_{d-1}(x_1, \ldots, x_m) \equiv_d \mathcal{C}_{d-2}(x_1, \ldots, x_m)\mathcal{C}_1(x_1, \ldots, x_m)$ for every even $d \geq 4$, the second term on the right of (13) is $\equiv_d$-equivalent to

$$\mathcal{C}_{d-2}(x_1, \ldots, x_n)\mathcal{C}_1(x_1, \ldots, x_n)x_{n+1}|_{x_{n+1}=\mathcal{C}_1(x_1,\ldots,x_n)}$$
$$=\mathcal{C}_{d-2}(x_1, \ldots, x_n)\mathcal{C}_1^2(x_1, \ldots, x_n)$$
$$=\mathcal{C}_{d-2}(x_1, \ldots, x_n)\mathcal{C}_1(x_1, \ldots, x_n) \equiv_d 0,$$

and the last term becomes

$$\mathcal{C}_{d-2}(x_1, \ldots, x_{n+1})\mathcal{C}_1(x_1, \ldots, x_{n+1})\ell(x)|_{x_{n+1}=\mathcal{C}_1(x_1,\ldots,x_n)} = 0.$$

Plugging these two back to (13),

$$\mathcal{C}_{d,n+2}|_{x_{n+2}=\ell(x),x_{n+1}=x_1+\cdots+x_n} \equiv_d \mathcal{C}_{d,n}.$$

As restriction does not increase linear rank, we have from (12) that

$$r = \mathsf{lin\text{-}rank}(\mathcal{C}_{d,n}) = \mathsf{lin\text{-}rank}(\mathcal{C}_{d,n+2}|_{x_{n+2}=\ell(x),x_{n+1}=x_1+\cdots+x_n})$$
$$\leq \mathsf{lin\text{-}rank}(\mathcal{C}_{d,n+2}|_{x_{n+2}=\ell(x)})$$
$$\leq r - 1,$$

which is a contradiction. □

As a simple application of Theorem 6, for any symmetric function f, let r_1, r_0 be the largest and smallest integers such that $f(x)$ is constant or parity on $\{x \in \{0,1\}^n : r_0 \leq |x| \leq n - r_1\}$. The quantity $r := r_0 + r_1$ turns out to be an important complexity measure for symmetric functions. For example, the randomized and quantum communication complexity of symmetric XOR functions is characterized by this r [9,10,18], and $\log \|\hat{f}\|_1 = \Theta(r \log(n/r))$ for all symmetric functions f [1].

Here we relate this measure to the $\mathbb{F}_2$-degree of f. It is clear that we can fix $x_1 = x_2 = \cdots = x_{r_0} = 1$ and $x_n = x_{n-1} = \cdots = x_{n-r_1+1} = 0$ to reduce the degree of f to at most 1. We therefore have the following corollary.

Corollary 1. *Let f be a symmetric function with even $\mathbb{F}_2$-degree d, then*

1. $\lfloor \frac{n}{2} \rfloor - \frac{d}{2} + 1 \leq r_0 + r_1$.
2. $\log \|\hat{f}\|_1 = \Omega(n/\log n)$, if $d = (1 - \Omega(1))n$.

References

1. Ada, A., Fawzi, O., Hatami, H.: Spectral norm of symmetric functions. In: Gupta, A., Jansen, K., Rolim, J., Servedio, R. (eds.) APPROX/RANDOM 2012. LNCS, vol. 7408, pp. 338–349. Springer, Heidelberg (2012)
2. Bernasconi, A., Codenotti, B.: Spectral analysis of boolean functions as a graph eigenvalue problem. IEEE Trans. Comput. 48(3), 345–351 (1999)

3. Cohen, G., Tal, A.: Two structural results for low degree polynomials and applications. ECCC. TR13-145 (2013)
4. Fine, N.: Binomial coefficients modulo a prime. Am. Math. Mon. **54**, 589–592 (1947)
5. Göös, M., Pitassi, T., Watson, T.: Deterministic communication vs. partition number. In: Proceedings of the 56th Annual Symposium on Foundations of Computer Science, pp. 1077–1088 (2015)
6. Gopalan, P., O'Donnell, R., Servedio, R., Shpilka, A., Wimme, K.: Testing Fourier dimensionality and sparsity. SIAM J. Comput. **40**(4), 1075–1100 (2011)
7. Kushilevitz, E., Nisan, N.: Communication Complexity. Cambridge University Press, Cambridge (1997)
8. Lee, T., Zhang, S.: Composition theorems in communication complexity. In: Gavoille, C., Kirchner, C., Meyer auf der Heide, F., Spirakis, P.G., Abramsky, S. (eds.) ICALP 2010. LNCS, vol. 6198, pp. 475–489. Springer, Heidelberg (2010)
9. Leung, M.L., Li, Y., Zhang, S.: Tight bounds on communication complexity of symmetric XOR functions in one-way and SMP models. In: Ogihara, M., Tarui, J. (eds.) TAMC 2011. LNCS, vol. 6648, pp. 403–408. Springer, Heidelberg (2011)
10. Liu, Y., Zhang, S.: Quantum and randomized communication complexity of XOR functions in the SMP model. ECCC **20**, 10 (2013)
11. Lovász, L., Saks, M.E.: Lattices, Möbius functions and communication complexity. In: Proceedings of the 29th Annual Symposium on Foundations of Computer Science, pp. 81–90 (1988)
12. Lovett, S.: Communication is bounded by root of rank. In: Proceedings of the 46th Annual ACM Symposium on Theory of Computing, pp. 842–846 (2014)
13. Lovett, S.: Recent advances on the log rank conjecture. Bull. EATCS **112**, 18–36 (2014)
14. Montanaro, A., Osborne, T.: On the communication complexity of XOR functions (2010). http://arxiv.org/abs/0909.3392v2
15. Shpilka, A., Tal, A., Volk, B.L.: On the structure of boolean functions with small spectral norm. In: Proceedings of the 5th Innovations in Theoretical Computer Science (2014)
16. Tsang, H.Y., Wong, C.H., Xie, N., Zhang, S.: Fourier sparsity, spectral norm, and the log-rank conjecture. In: Proceedings of the 54th Annual IEEE Symposium on Foundations of Computer Science, pp. 658–667 (2013)
17. Yao, A.: Some complexity questions related to distributive computing. In: Proceedings of the 11th Annual ACM Symposium on Theory of Computing, pp. 209–213 (1979)
18. Zhang, Z., Shi, Y.: Communication complexities of symmetric XOR functions. Quant. Inf. Comput. **9**(3), 255–263 (2009)

Author Index

Printed in the United States
By Bookmasters